Springer Proceedings in Earth and Environmental Sciences

The series Springer Proceedings in Earth and Environmental Sciences publishes proceedings from scholarly meetings and workshops on all topics related to Environmental and Earth Sciences and related sciences. This series constitutes a comprehensive up-to-date source of reference on a field or subfield of relevance in Earth and Environmental Sciences. In addition to an overall evaluation of the interest, scientific quality, and timeliness of each proposal at the hands of the publisher, individual contributions are all refereed to the high quality standards of leading journals in the field. Thus, this series provides the research community with well-edited, authoritative reports on developments in the most exciting areas of environmental sciences, earth sciences and related fields.

More information about this series at https://link.springer.com/bookseries/16067

Andrei Kosterov · Nikita Bobrov ·
Evgeniy Gordeev · Evgeniy Kulakov ·
Evgeniya Lyskova · Irina Mironova
Editors

Problems of Geocosmos—2020

Proceedings of the XIII International Conference and School

Springer

Editors
Andrei Kosterov
Earth Physics Department
Saint Petersburg State University
St. Petersburg, Russia

Evgeniy Gordeev
Saint Petersburg State University
St. Petersburg, Russia

Evgeniya Lyskova
Saint Petersburg State University
St. Petersburg, Russia

Nikita Bobrov
Saint Petersburg State University
St. Petersburg, Russia

Evgeniy Kulakov
Centre for Earth Evolution and Dynamics
University of Oslo
Oslo, Norway

Irina Mironova
Saint Petersburg State University
St. Petersburg, Russia

ISSN 2524-342X ISSN 2524-3438 (electronic)
Springer Proceedings in Earth and Environmental Sciences
ISBN 978-3-030-91466-0 ISBN 978-3-030-91467-7 (eBook)
https://doi.org/10.1007/978-3-030-91467-7

This Springer imprint is published by the registered company Springer Nature Switzerland AG
The registered company address is: Gewerbestrasse 11, 6330 Cham, Switzerland

Preface

This volume is a collection of papers presented at the 13th conference and school "Problems of Geocosmos" held online via Zoom on March 24–27, 2021. Organized primarily by faculty of the Department of the Earth Physics, St. Petersburg State University, this conference is one of the few in Russia to cover a broad range of geophysical research. About 150 participants from Russia and abroad registered for the meeting, including over 50 early-career scientists, graduate and undergraduate students.

The conference was organized into four parts: Exploration and Environmental Geophysics (EG), Paleomagnetism and Rock Magnetism (P), Seismology (S), and Solar-Terrestrial Physics (STP).

Program of the EG part covered studies of the Earth's crust using various geophysical methods, processing and interpretation of geophysical data, and results of regional studies. Bataleva and Matyukov summarize the outcome of several decades of geological studies in the Tien Shan area, where magnetic, gravimetric, seismological, and magnetotelluric data were of crucial importance. The boundaries of the largest tectonic terrains of Tien Shan are traced, and the largest fault structures identified. Shevtsov et al. describe results of the FENICS-2019 geoelectric experiment which used grounded industrial power transmission lines to measure the geoelectric signal at the distances up to 1470 km from the source revealing the structure of the lithosphere of the Fennoscandian Shield. Three papers by Petrova et al. present the results of investigation of the Arctic Basin and coastal-shelf cryolithozone based on the interpretation of the Earth's magnetic field anomalies, gravity anomalies, and seismic data. The first paper considers possible influence of thermal flow from deep faults on the localization of ice cover weak zones in the Arctic Ocean. The second paper discusses the distribution and stability of gas hydrates in the Arctic Basin. The third paper deals with the verification of the Arctic magnetic field component model based on satellite observations. Technical issues arising in magnetic field measurements on board of a ferromagnetic vessel are discussed by Ivanov et al. Results obtained during the Russian round the world expedition in 2019–2020 are presented including in particular determinations of the actual position of the South Magnetic

Pole. Isaev et al. introduce a new approach to solving inverse geophysical problems making the use of artificial neural networks. The effectiveness of integration of three geophysical methods—gravimetry, magnetometry, and magnetotellurics—is evaluated in a set of numerical experiments. A new versatile software for geospatial data statistical analysis and correlation (UnCorr) is presented by Krivopaltsev et al. The software is a web service to perform coordinates conversion, data interpolation, spatial correlation and visualization, and can be useful for researchers working in the Earth sciences.

Since topics of contributions to the part on paleomagnetism and rock magnetism proved broader than was initially envisaged, we decided to entitle the respective Proceedings part "Paleo-, Earth-, Rock-, environmental magnetism, and geophysical fluid dynamics" (PERM). Papers in the PERM section cover a variety of topics including magnetostratigraphy, rock- and environmental magnetism, magnetic seaborne survey, modeling of core dynamics, and fluid dynamics. Pavlov et al. report new paleomagnetic data from Ordovician sedimentary rocks of the Moyero river valley (the northern part of the Siberian platform) and discuss the magnetostratigraphic position of the Tremadocian—Arenigian (Floian) boundary within the studied part. Gnibidenko et al. present a new magnetostratigraphic study of Upper Cretaceous sedimentary rocks from drill cores from five wells in Kha-Rampur oil-gas-condensate field (Northern West Siberia). The obtained composite magnetostratigraphic scale was then compared to the regional magnetostratigraphic record for southern West Siberia and the global magnetic polarity scale. Kazansky et al. present detailed paleomagnetic and rock-magnetic data from the Late Eopleistocene Ust-Obor sedimentary section exposed in the Western Transbaikalia (Buryatia). The authors discuss the implications of these data for the age and depositional conditions during the formation of the studied part. Schetnikov et al. report first rock- and paleomagnetic data from lacustrine sediments of the small isolated lake Kotokel (Lake Baikal region) and discuss climatic and environmental fluctuations in the area during deposition of the investigated lake sediments (46 Kyr). Demina et al. determine the position of the South Magnetic Pole based on the historical observations made during 1820 round-the-world expedition of F. Bellingshausen, and on modern observations in 2020 onboard R/V "Admiral Vladimirsky." Starchenko and Yakovleva present a novel view of the statistics of the geomagnetic field records covering a relatively short (~200 years) interval, revealing its possible connections with much longer time scales characteristic for operation of geodynamo. Finally, two papers by Rani et al. discuss various aspects of numerical analysis of fluid flow dynamics.

Papers in Part S cover a broad range of topics in seismology and related disciplines. The results by Rodkin et al. provide new insights on the temporal clustering of seismic events before and after the main quake and can help improve statistical modeling and forecasting of major earthquakes. The approach proposed by Nekrasova and Kossobokov makes use of the unified scaling law for earthquakes as a part of the alternative, scenario-based, NDSHA approach contributing to the improvement of the accuracy and reliability of reproducible seismic hazard maps. Possible effects of industrial explosions on weak earthquake seismicity in Turkey are discussed by Zhuravlev at al. Solovey reports on a new study of the upper mantle structure in Central Asia. The crust structure in South Karelia region is investigated by Isanina et al. combining the active seismic data and the results of the PS converted wave method. Nepeina and Bataleva develop a new approach to study the relationships between the spatial distribution of the electrical conductivity and earthquake hypocenters in the Tien Shan. The series of papers by Gravirov and Kislov are devoted to the development of rotational seismometers. The authors analyze the instrumental noise caused by variations in ambient temperature and consider advantages of adaptive filtering to reduce the seismic noise.

Solar-Terrestrial Physics (STP) part is a traditional part of "Problems of Geocosmos" conferences. Subjects of the section include space climate and space weather, magnetospheric phenomena, magnetosphere—ionosphere—atmosphere coupling and its effects on the Earth's atmosphere. The collection of papers covers a wide range of problems broadly related to solar-terrestrial interaction. Antonyuk et al. analyze variability in the radiation environment at the International Space Station under different levels of geomagnetic activity. Chernogor et al. analyze the peculiarities of August 5–6, 2019 magnetic and ionospheric storms. Chernogor discusses major processes in the Earth—atmosphere—ionosphere—magnetosphere system that accompanied the fall of the Kamchatka meteoroid. In the theoretical investigation by Erkaev and Semenov, a two-dimensional compressible Hall magnetohydrodynamic model is used to study magnetic field reconnection in symmetric, non-symmetric antiparallel, and skewed magnetic fields. Karagodin et al. utilize the chemistry-climate model SOCOLv3 to evaluate the effect of the interplanetary magnetic field B_y-component on the surface meteorological parameters through a change in the stratiform-cloud autoconversion rate. Klimov and Sigaeva present the observational statistics of localized luminosity pulsations in the equatorial part of auroral oval observed from the Lomonosov satellite. Kozyreva et al. use multi-instrument geomagnetic and ionospheric observations in Antarctica to examine the mechanisms of the ultra-low-frequency variations in the polar cap. Kubyshkina et al. test the hypothesis that for asymmetric configurations with a curved current sheet and large curvature of magnetic field lines in the magnetospheric tail, the substorm breakdown threshold decreases. The authors also consider several specific problems, such as a statistical study of substorm activity with respect to changes in solar wind parameters; a study of the influence of the dipole tilt on substorm activity; an estimation of the accumulated magnetic flux in the tail for symmetric and asymmetric configurations of the magnetosphere. Malkin et al. report the observational results for electromagnetic pulses accompanying separate volcanic explosions during the period 2017–2019. In two papers, Sivokon and Cherneva report that scintillation index and amplitude distribution of the radiation scattered on ionospheric magnetically oriented irregularities depend on impact power, and show the dependence of motion velocity and direction of ionospheric magnetically oriented irregularity motion on heating wave power. Zolotova and Vokhmyanin reconstruct the parameters of solar activity from astronomical observations in the seventeenth century, revealing a significant uncertainty in the modern evaluations of sunspot numbers from these early observations.

St. Petersburg, Russia — Andrei Kosterov
St. Petersburg, Russia — Nikita Bobrov
St. Petersburg, Russia — Evgeniy Gordeev
Oslo, Norway — Evgeniy Kulakov
St. Petersburg, Russia — Evgeniya Lyskova
St. Petersburg, Russia — Irina Mironova
September 2021

Contents

Exploration and Environmental Geophysics

Segmentation of the Earth's Crust of the Tien Shan by Geophysical Data

E. A. Bataleva and V. E. Matiukov

Abstract The paper presents the results of a comprehensive analysis of geological studies with data on the deep structure obtained by various geophysical methods for the Tien Shan territory. The areas of junction of different tectonic segments are characterized by anomalies of increased electrical conductivity, are clearly manifested in the magnetic and gravity fields and correspond to the zones of deep faults. Studying of these objects having high permeability and providing heat and mass flow as well as transport of fluids, is the main object of this work. In order to identify the boundaries of the largest tectonic fragments of the Tien Shan the geoelectric and velocity sections along the profiles intersecting the main tectonic elements of the Tien Shan were analyzed. Comparison of velocity anomalies in seismotomographic sections and objects of abnormally low electrical resistivity in geoelectric sections of the Earth's crust indicates the existence of zones of increased fracturing or plasticity in the indicated intervals of the Earth's crust, as well as the coincidence of waveguides and zones of increased conductivity. The distribution of seismic events clearly shows the zones of the Northern, Middle and Southern Tien Shan and the segmentation of the Western and Central Tien Shan.

Keywords Tectonic segments · Fault structures · Anomalies of geophysical fields · Tien Shan

1 Introduction

According to the geological prehistory, the rocks of the modern Tien Shan, which formed continental terranes in the Middle Paleozoic, were separated by the oceanic basins that existed at that time [5, 10, 19]. Later, the oceanic basins were closed, and in their place suture structures were formed, marking the trace of the paleoocean. The

E. A. Bataleva (✉) · V. E. Matiukov
Research Station RAS in Bishkek (RS RAS), Bishkek-49, 720049 Bishkek, Kyrgyzstan
e-mail: bataleva@gdirc.ru

V. E. Matiukov
e-mail: vitaliy@gdirc.ru

A. Kosterov et al. (eds.), *Problems of Geocosmos—2020*, Springer Proceedings in Earth and Environmental Sciences, https://doi.org/10.1007/978-3-030-91467-7_1

collision of continental plates (terranes) led to orogeny, which continues to this day and is reflected in modern geodynamic processes. Geological and geophysical studies carried out in recent years in the Tien Shan have allowed us to significantly expand our knowledge of the deep geological structure of the Tien Shan lithosphere [1, 3, 4, 18]. On the basis of regional geophysical data, geological and geophysical analyzes of interfaces (for example, Konrad and Mohorovichich ones), structural features of the lithosphere [8, 15], tectonic development and geodynamics of processes was carried out.

The geophysical fields reflect the fundamental features of the tectonic structure of the Tien Shan, they are the main objects of study of such deep methods as magnetotelluric and seismic sounding. Inhomogeneities in the structure of the lithosphere are one of the key factors in the study of modern geodynamic processes, for example, the seismic regime, which determines the relevance of comprehensive geophysical studies [1, 3, 4, 8, 15, 17, 18] of the structure of the lithosphere both at the regional and local scales. The obtained geoelectric and velocity sections along the profiles of deep magnetotelluric and seismic sounding (Fig. 1) reflect the inhomogeneities of the Earth's crust both vertically and horizontally. At the same time, quite often, against the background of a general increase in velocity with depth, in some areas we distinguish zones of low-velocity volumes (waveguides), which coincide with areas of increased electrical conductivity [3, 4, 17, 18].

The main goal of this study was to analyze how the largest fault structures being the boundaries of the main tectonic elements of the Tien Shan are reflected in various geophysical fields.

2 Results

As a first approximation, we considered such tectonic segments as the Western and Eastern Tien Shan (division relative to the Talas-Fergana fault), as well as the Northern, Middle and Southern Tien Shan (Fig. 2). At the same time, special attention was paid to the characteristics of geophysical fields within the identified tectonic fragments and boundary regions.

The deep structure of the largest fault, which serves as the boundary of the main tectonic elements of the Tien Shan, is best manifested in geoelectric (Fig. 3) and seismic models, as well as the position of conductive layers and waveguides. We have compared geophysical models obtained from the results of long-term field observations, both within the framework of the implementation of large international projects and during various experiments (for example, in the course of detailed study of the geoelectric structure of the Bishkek geodynamic test site), with the data of geological studies and new tectonic concepts [7]. This comparison made it possible to more reasonably discuss the similarities and differences of the identified geophysical anomalies for various tectonic elements of the Tien Shan, to draw conclusions about their nature and find geophysical anomalies-indicators of geodynamic settings, that

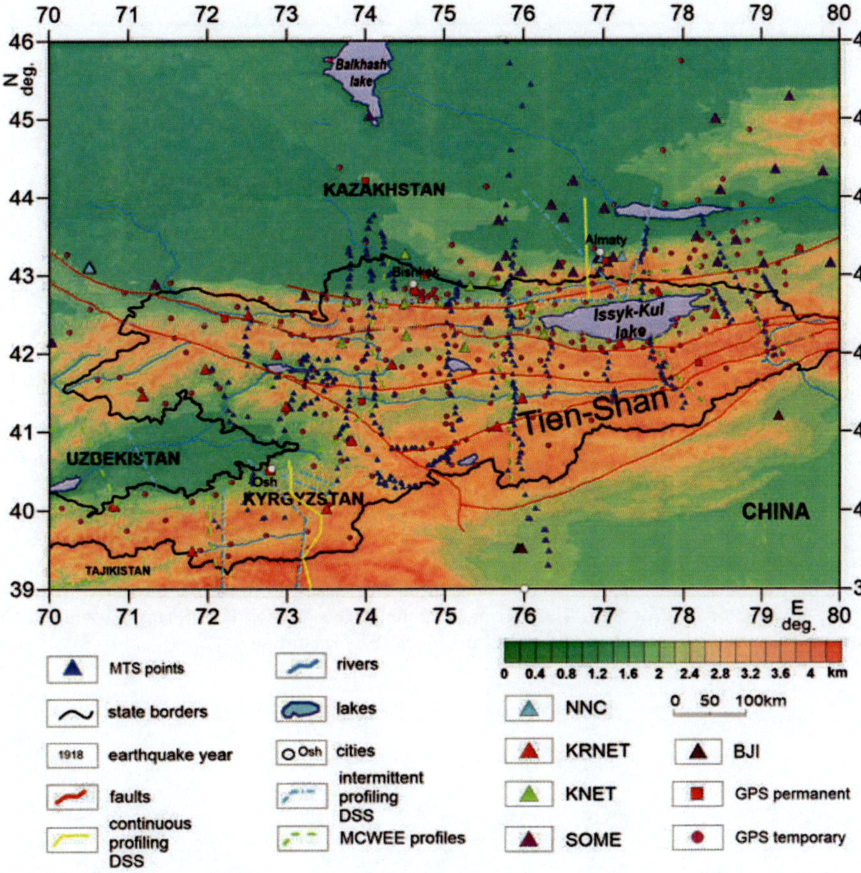

Fig. 1 Scheme of geological and geophysical exploration of the territory of the Tien Shan: 1—MTS points; 2—state borders of the Kyrgyz Republic; 3—major faults; 4—rivers; 5—lakes; 6—cities; deep seismic sounding (DSS): 7—continuous DSS profiling, 8—intermittent DSS profiling; 9—seismic refraction method—exchange waves of earthquakes (MEWE); seismic networks: 10—National Nuclear Centre (NNC), Kazakhstan, 11—National Kyrgyz Republic Network (KRNET), Kyrgyzstan, 12—Kyrgyz Network of the Research station (KNET), Kyrgyzstan, 13—Seismological experimental and methodological expedition of the Science Committee of the Ministry of Education and Science of the Republic of Kazakhstan (SOME), Kazakhstan; 14—Beijing (BJI), China; GPS networks: 15—permanent, 16—temporary

is fundamentally important for the study and understanding of geodynamic processes and in combination with geological data.

As structural features, we consider: the presence of deep fault structures, gradient zones at the boundaries of the Conrad, Moho, Curie, the base of the lithosphere. As indicators we used the depths of the hypocenters of seismic events, the distribution of heat flow at definite depth, outcrops of mineral water sources and the presence

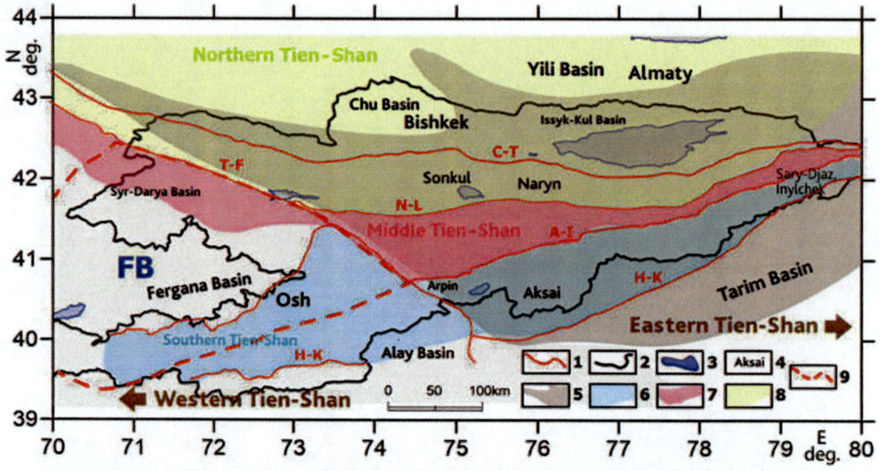

Fig. 2 The main tectonic elements of the Tien Shan and adjacent territories (after 11): 1—the largest faults: T-F—Talas-Fergana; L-N—Nikolaev line, C-T—Central-Terskeysky; A-I—Atbashi-Inylchecksky; H-K—Gissar-Kokshaalsky; S-F—South Fergana; 2—borders of the Kyrgyz Republic; 3—lakes; 4—basins; 5—zone of the Central Tien Shan according to the division according to [5]; 6—zone of the Southern Tien Shan; 7—zone of the Middle Tien Shan; 8—zone of the Northern Tien Shan; 9—borders of the Fergana block (FB) according to [2]

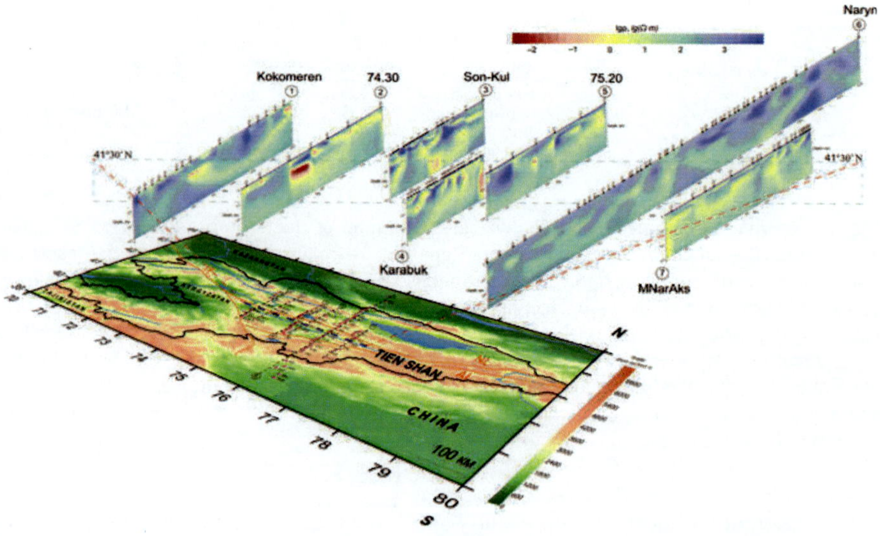

Fig. 3 Distribution of resistivity in geoelectric models along magnetotelluric profiles. Resistivity values are shown in rainbow colors in Ωm. Red arrows indicate borders of the main tectonic elements of the Tien Shan

of mantle helium. During this detailed study it was found that the velocity characteristics and geoelectric properties correlate well with each other: structures with increased electrical conductivity often coincide with low-velocity structures, and vice versa, areas with low electrical conductivity coincide with high-velocity structures in seismic tomography models. The results of geological interpretation of geophysical anomalies showed that, in general, the anomalies correspond to the main geological objects of the Tien Shan and adjacent territories, the depth structure of which can be traced up to ~ 100 km [11]. Fault structures, which serve as boundaries of the main tectonic elements, can be expressed in geophysical sections both in the form of flexure-like structures and in the form of "flower structures" of the Earth's crust. The latter are quite often observed in geoelectric models (Fig. 3) of the Tien Shan crust [1, 3, 13, 14]. In accordance with the results of the analysis of geoelectric models built on three geotraverses (74°, 75°, 76°), intersecting the Central Tien Shan (Fig. 3), it is possible to draw conclusions about the correspondence of the ancient suture zone of the Nikolaev Line, which is represented by a series of subvertical conductive structures of northern vergence, to the border between the Northern and Central Tien Shan. In plan, these structures are represented by listric faults extending from the Earth's surface almost vertically to the depths of the middle crust and deeper, where they flatten out and turn into sub-horizontal conductors [1, 3, 13, 14]. The fault zones are composed of conductive rocks, the resistivity of which varies from 10–15 to 50–70 Ω m. The flattening of the faults is observed at the depths of 10–12 km and 17–20 km, that indicates the existence of weakened zones in the interval of the upper and middle crust.

However, the position of fault structures on the day surface, which we can observe as a result of mapping on geological-tectonic formations, correlates very well with maps of the behavior of the anomalous magnetic field (Fig. 4). The distribution of

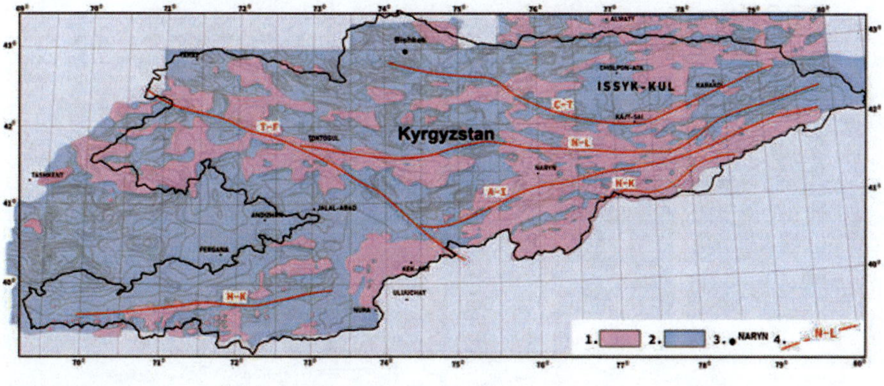

Fig. 4 Map of the anomalous magnetic field (ΔT) and the Kyrgyz Tien Shan. Compiled based on materials from [9, 16], compiled by Pogrebnoy and Grebennikova [12]. Legend: 1—positive field isodynamics (ΔT) a in nT; 2—negative field isodynamics (ΔT) and in nT; 3—settlements of Kyrgyzstan; 4—fault structures: T-F—Talas-Fergana; L-N—Nikolaev line, C-T—Central-Terskey; A-I—Atbashi-Inylcheck; H-K—Gissar-Kokshaal

positive and negative anomalies in Fig. 4 corresponds to the main tectonic fragments of the Northern, Middle and Southern Tien Shan. The Talas-Fergana fault zone separating the Western and Eastern Tien Shan is very well reflected not only in the distribution of the anomalous magnetic field, but also in the behavior of the Curie surface. The 2D model of the Curie surface presented in [12] steadily determines the thickness of the magnetoactive layer. In the zone of dynamic influence of the Talas-Fergana fault, the depth of its occurrence varies from − 26 to − 44 km. The greatest rise of the Curie surface up to − 26 km is noted southwest of the Talas-Fergana fault, within the northern framing of the Fergana depression. In terms of the strike of the iso-depths of the Curie surface, Talas-Fergana is one of the most distinct faults.

The concentration of density inhomogeneities, reflected in the characteristic features of the relief, is due to geodynamic processes occurring in the lithosphere of the Tien Shan, which lead to significant changes in near-surface structures. The distribution of gravity anomalies (Fig. 5), which we observe in Fig. 4, indicates not only the behavior of density inhomogeneities in the upper part of the Earth's crust, but also the configuration of fault structures. The Talas-Fergana fault, the Nikolaev Line and other large tectonic faults of the Tien Shan correlate well with the zones of increased values of the horizontal component of the gradients of gravity anomalies.

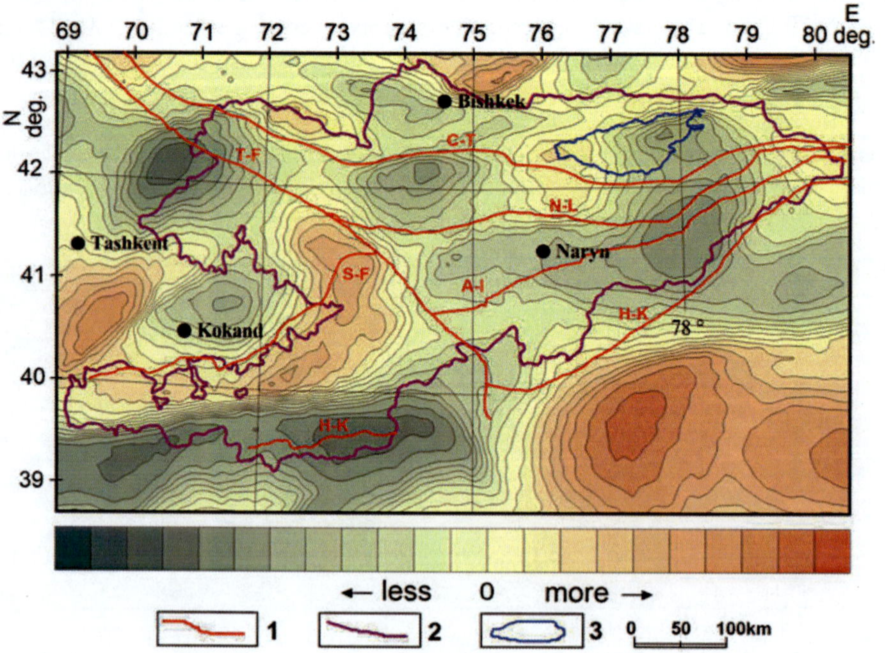

Fig. 5 Map of the distribution of averaged gravity anomalies in Bouguer reduction, taking into account the impact of the relief of the Meso-Cenozoic basins for the territory of the Central Tien Shan [6]

The zone of dynamic influence of the Talas-Fergana fault corresponds to a strip of negative anomalies of north-west strike, which discordantly dissects gravity anomalies of the north-east direction. It is found the clearest interrelation of faults, to which the hypocenters of seismic events are often correspond, with gravity anomalies in isostatic reduction [6]. The Tien Shan gravity field is dominated by significant negative Bougeur anomalies, the values of which increase to the east of the Talas-Fergana fault zone. According to [17], the western part of the Tien Shan is compressed to a much greater extent than the eastern one, which "spreads out" to the north and south.

The largest fault structures of the Tien Shan, which serve as the boundaries of the main tectonic elements, are also reflected in the thermal field of the Tien Shan. At the same time, the zone of secant deformations of the Talas-Fergana fault is expressed inhomogeneously. The northwestern segment of the fault is characterized by abnormally high values of heat flow—up to 108 mW/m^2; whereas in the southeastern part the heat flow is about 70 mW/m^2, and in the central segment—less than 50 mW/m^2. As for the Nikolaev Line, this fault zone is not distinguished in any way in the thermal field of the Tien Shan; the reason may be in the history of the development of the suture zone. For the Atbashi-Inylchek fault, anomalous values of the heat flow exist in the zone of junction with the TFF; in general, the structure is characterized by increased intensity of the heat flow. Measurements of the heat flow values in the Tien Shan indicate that the Northern Tien Shan is characterized by normal values (40–56 mW/m^2), which are observed in the Ili, Talas, Chuya and Kochkor basins. In the basins of the Middle Tien Shan (Naryn and Jumgal), the heat flow data are abnormally high for the depressions—77.2–83.1 mW/m^2. The heat flow in the zones of the newest uplifts of the Tien Shan is generally more intensive (over 100 mW/m^2, and in the Chatkal-Kuramin zone it reaches 120–150 mW/m^2).

Basing on the hypothesis that earthquakes are confined to fault structures, let us consider the nature of the distribution of seismic events for the main tectonic segments of the Tien Shan and their boundaries (Fig. 6). The Gissar-Kokshaal fault (Southern Tien Shan) and the central part of the Talas-Fergana fault are well manifested in the distribution of earthquake epicenters [1]. The Atbashi-Inylchek fault is not marked with earthquake epicenters and does not manifest itself as an active fault structure in the seismic distribution pattern. The Nikolaev Line can be divided into two sections—western and eastern (relative to the 76th meridian), wherein the western section of the fault is more seismically active compared to the eastern one.

3 Conclusions

The correlation of regional anomalies of electrical conductivity, velocity characteristics of the geological environment and anomalies of potential fields with the boundaries of structural-tectonic zones of the Tien Shan has been carried out.

The performed analysis of geological and geophysical materials shows that the boundaries of geoelectric inhomogeneities, as well as magnetic and gravity anomalies, associated with the largest fault structures, correlate well with the boundaries of

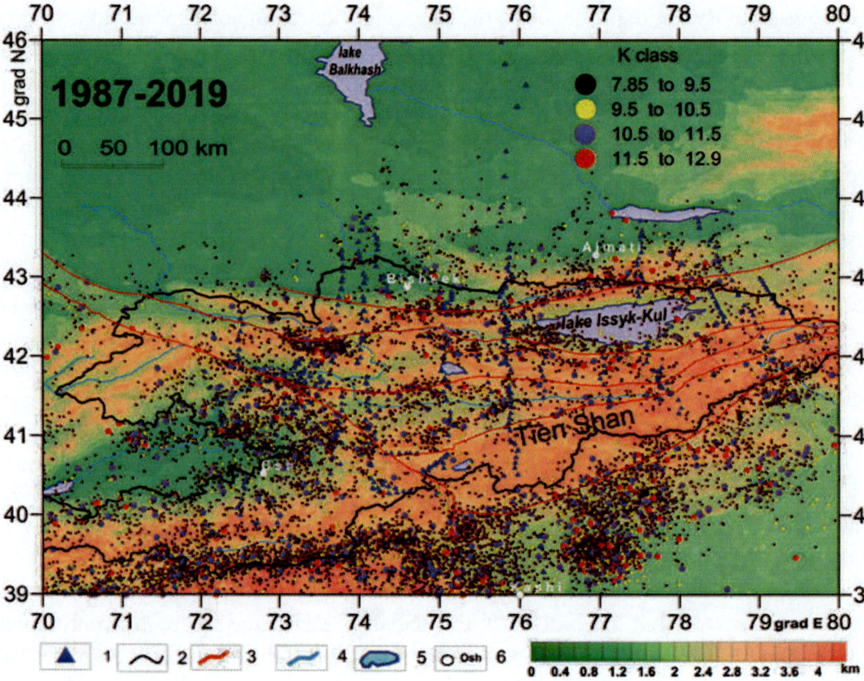

Fig. 6 Distribution of earthquake epicenters for the Tien Shan territory with an energy class of 8–12: 1—MTS points, 2—border of Kyrgyzstan, 3—major faults, 4—rivers, 5—lakes, 6—cities

the main tectonic elements. It is necessary to state the fact that the sectional deformation zone of the Talas-Fergana fault has significant correlation with the anomalies presented in all geophysical models.

The flattening of fault structures in geoelectric models, which is observed at depths of 10–12 km and 17–20 km, is of particular interest and can indicate the existence of weakened zones in this interval of the upper and middle crust.

The aseismicity of the Middle Tien Shan is clearly expressed in the spatial distribution of earthquake epicenters, with the most clearly manifested zones of the Gissar-Kokshaal (South Tien Shan) and Talas-Fergana faults.

Acknowledgements This work was supported by the State Assignment of the Research Station of the Russian Academy of Sciences in Bishkek (AAAA-A19-119020190063-2).

References

1. Bataleva, E.A., Buslov, M.M., Rybin, A.K., Batalev, V.Yu., Safronov, I.V.: Crustal conductor associated with the Talas-Fergana fault and deep structure of the southwestern Tien Shan: geodynamic implications. Russ. Geol. Geophys. **46**(9), 1036–1042 (2006)

2. Bazhenov, M.L.: Cretaceous paleomagnetism of the Fergana Basin and adjacent ranges, Central Asia: tectonic implications. Tectonophysics 221, 251–267 (1993). https://doi.org/10.1016/0040-1951(93)90335-H
3. Batalev, V.Yu., Bataleva, E.A., Matyukov, V.E., Rybin, A.K., Egorova, V.V.: The lithospheric structure of the Central and Southern Tien Shan: MTS data correlated with petrology and laboratory studies of lower-crust and upper-mantle xenoliths. Russ. Geol. Geophys. **52**(12), 1592–1599 (2011). https://doi.org/10.1016/j.rgg.2011.11.005
4. Bataleva, E.A., Batalev, V.Yu., Rybin, A.K.: Interrelation of electrical conductivity anomalies, velocity characteristics and seismicity regime of the lithosphere of the Central Tien Shan. Lithosphere (5), 81–89 (2015)
5. Burtman, V.S.: Tien Shan and High Asia. Geodynamics in the Cenozoic, 188 p. GEOS, Moscow (2012) (in Russian with English abstract)
6. Laverov, N.P.: Recent Geodynamics of Intracontinental Areas of Collision Mountain Building (Central Asia), 400 p. Scientific World, Moscow (2005)
7. Leonov, M.G., Przhiyalgovskiy, E.S., Lavrushina, E.V., Rybin, A.K.: Postmagmatic tectonics of granites of the basement of the Northern Tien Shan. Litosfera **6**, 5–25 (2016)
8. Makarov, V.I., Alekseev, D.V., Leonov, M.G., Batalev, V.Y., Bataleva, E.A., Bragin, V.D., Rybin, A.K., Shchelochkov, G.G., Belyaev, I.V., Dergunov, N.T., Efimova, N.N., Roslov, Y.V., Munirova, L.M., Pavlenkin, A.D., Roecker, S.: Underthrusting of Tarim beneath the Tien Shan and deep structure of their junction zone: main results of seismic experiment along MANAS profile Kashgar-Song-Köl. Geotectonics **44**(2), 102–126 (2010). https://doi.org/10.1134/S0016852110020020
9. Marinchenko, G.G.: High-precision high-altitude aeromagnetic survey on a scale of 1:200,000 of the territory of Kyrgyzstan. Report. Funds KGFE "Gera", Shopokov, Kyrgyz Republic. THF (1993)
10. Morin, J., Jolivet, M., Bourquin, S., Shaw, D., Bataleva, E.: New sedimentological and palynological data from the Yarkand-Fergana basin (Kyrgyz Tian Shan): insights on its mesozoic paleogeographic and tectonic evolution. Geosci. Front. **12**(1), 183–202 (2021). https://doi.org/10.1016/j.gsf.2020.04.010
11. Medved, I., Bataleva, E., Buslov, M.: Studying the depth structure of the Kyrgyz Tien Shan by using the seismic tomography and magnetotelluric sounding methods. Geosciences **11**, 122 (2021). https://doi.org/10.3390/geosciences11030122
12. Pogrebnoy, V.N., Grebennikova, V.V.: Distinctive features of anomalous magnetic (ΔT) a field in individual seismogenic blocks of the territory of the Kyrgyz Tien Shan. In: Proceedings of the Fifth International Symposium "Modern Problems of Geodynamics and Geoecology of Intracontinental Orogens", 19–24 June 2012, Bishkek, Moscow-Bishkek, pp. 58–64 (2012)
13. Rybin, A.K., Leonov, M.G., Przhiyalgovskii, E.S., Batalev, V.Y., Bataleva, E.A., Matyukov, V.E., Lavrushina, E.V., Zabinyakova, O.B., Schelochkov, G.G.: Upper crust structural and morphological ensembles of the Pamir-Tien Shan segment of Central Asia and their reflection in geophysical fields. Vestnik of Saint Petersburg University. Earth Sci. **63**(4), 478–501 (2018). https://doi.org/10.21638/spbu07.2018.405
14. Rybin, A.K., Bataleva, E.A., Matiukov, V.E., Morozov, Y.A., Nepeina, K.S.: Deep structure of the lithosphere in the Central Tien Shan along the Son-Kul magnetotelluric sounding profile. Dokl. Earth Sci. **496**(2), 101–106 (2021). https://doi.org/10.1134/S1028334X21020161
15. Rybin, A.K.: Deep Structure and Modern Geodynamics of the Central Tien Shan Based on the Results of Magnetotelluric Research, 232 p. Scientific World, Moscow (2011)
16. Schematic map of the anomalous magnetic field of the USSR (sheets K-43 and K-44) (ΔTa) scale 1:1,000,000. Materials of the Geology Department of the Kyrgyz SSR (1987)
17. Typical Geological and Geophysical Models of the Environment of Seismic and Aseismic Areas, 230 p. Ilim, Bishkek (1992)
18. Trapeznikov, Yu.A., Andreeva, E.V., Batalev, V.Yu., Berdichevsky, M.N., Vanyan, L.L., Volykhin, A.M., Golubtsova, N.S., Rybin, A.K.: Magnetotelluric soundings in the mountains of the Kyrgyz Tien Shan. Phys. Earth (1), 3–20 (1997)

19. Yudakhin, F.N., Chediya, O.K., Sabitova, T.M.: Modern Geodynamics of the Tien Shan Lithosphere, 192 p. Science, Moscow (1991)

Experiment FENICS-2019: Exploration of Electrical Conductivity of the Eastern Fennoscandinavian Shield with Grounded Sections of Power Transmission Lines (In Memory of Abdulkhay Azimovich Zhamaletdinov)

A. N. Shevtsov, A. A. Skorokhodov, A. E. Hannibal, T. G. Korotkova, V. V. Kolobov, M. B. Barannik, and V. V. Ivonin

Abstract The deep electromagnetic (EM) sounding of the lithosphere carried out in the frame of the FENICS-2019 experiment from 12 to 21 September 2019 in terms of experimental layout was a frequency domain sounding with the source of two grounded industrial power lines operating in the frequency range of 192.2–0.382 Hz. The measurements were mostly carried out by the research team from Geological Institute of the KSC RAS (GI KSC) within 841 km from the source. The additional measurements by stations of the Polar Geophysical Institute, Luleå University, Nizhny Novgorod Radiophysical Research Institute (NIRFI) and the Institute of Geosphere Dynamics of the Russian Academy of Sciences (IDG RAS) were carried out at distances of up to 1470 km. The results obtained in GI KSC RAS were processed and consolidated into a database of apparent resistivity and impedance phase curves, calculated for the absolute values of the total electric and magnetic fields. The principal novelty of the FENICS-2019 experiment, distinguishing it from the AMT-MT sounding, was the quantitative account for static shift at each point. The method of correction for static shift is based on measurements of the apparent resistivity from the total horizontal magnetic field of the controlled source. Magnetometer sensors are not galvanically coupled to the Earth and therefore not affected by static shift. The method for assessing the static shift was applied within the far zone; it accounts for the influence of the ionosphere and displacement currents. Geoelectric sections were constructed for four observation profiles. This work was carried out in the frame of the state assignment of the Ministry of Education and Science of the Russian Federation for GI KSC RAS (topic No. 0226-2019-0052, FMEZ-2022-0025) and supported by the Russian Foundation for Basic Research (grant No 18-05-00528).

Keywords Electromagnetic sounding · Lithosphere · Static shift

---

A. N. Shevtsov (✉) · A. A. Skorokhodov · A. E. Hannibal · T. G. Korotkova
Geological Institute of the Kola Science Center RAS, Fersman Street 14, 184209 Apatity, Russia

V. V. Kolobov · M. B. Barannik · V. V. Ivonin
Institute of Physical and Technical Problems of Power Engineering of the North of the Kola Science Centre RAS, Fersman Street 14, 184209 Apatity, Russia

A. Kosterov et al. (eds.), *Problems of Geocosmos—2020*, Springer Proceedings in Earth and Environmental Sciences, https://doi.org/10.1007/978-3-030-91467-7_2

1 Introduction

The final stage of the deep electromagnetic sounding of the Fennoscandian Shield in the field of natural and controlled sources (Fennoscandian shield Electrical conductivity from Natural Induction and Control Source deep soundings—FENICS-2019) was carried out in September 2019, by the research team from the Geological Institute of the KSC RAS. The FENICS-2019 frequency-domain sounding was carried out in the field of two grounded industrial power transmission lines (PTL) with a length of 125 and 102 km, respectively, directed at an angle of 136° to each other. The current frequency in the power transmission line was within the range of 0.194–194.2 Hz and was supplied by a 200-kW generator developed and manufactured by Center for Physical and Technical Problems of the Power Industry of the North, KSC RAS. An important technological novelty was application of three identical 32-bit certified stations of the 5th generation VMTU-10 (OOO VEGA) with synchronization via GPS satellite communication with an accuracy being within 1 μs. One VMTU-10 station was used to record the generator current in industrial transmission lines and two stations were used to record signals at field points. Soundings were carried out along two profiles, in East Karelia, with a maximum distance of up to 841 km from the source and in the Central Karelia, at a distance of up to 603 km from the source. An important scientific novelty of processing the FENICS experimental data and an advantage over the traditional AMTS method was the original methodology for the quantitative corrections for the static shift for each individual CSAMT point. The amplitude of the static shift was estimated from the apparent resistivity curves calculated from the total horizontal magnetic field within the wave zone, taking into account the influence of the ionosphere and displacement currents. The quality and reliability of CSAMT results were verified by additional processing of FENICS time series using the AMTS scheme in the field of natural variations.

2 Layout and Main Goals of Experiment FENICS-2019

Figure 1 shows the L-154 and L-403 transmitter lines, position of the profiles and measuring GI KSC stations on the profiles of the FENICS-2019 experiment.

Main goals of the experiment:

1. To perform electromagnetic sounding of various blocks of the Earth's crust of the Fennoscandian Shield in order to provide a framework for a quasi-three-dimensional model of the lithosphere and assess the possible relationship between deep electrical conductivity and metallogenic characteristics. (One of the main features of metallogeny of Early Proterozoic ore-bearing structures associated with rifting on the Baltic Shield is pyrite, sulfide copper-nickel and platinum mineralization.)
2. To study the properties of the transitional zone of increased resistivity between the upper and lower crust in the depth interval of 10–30 km using a complex

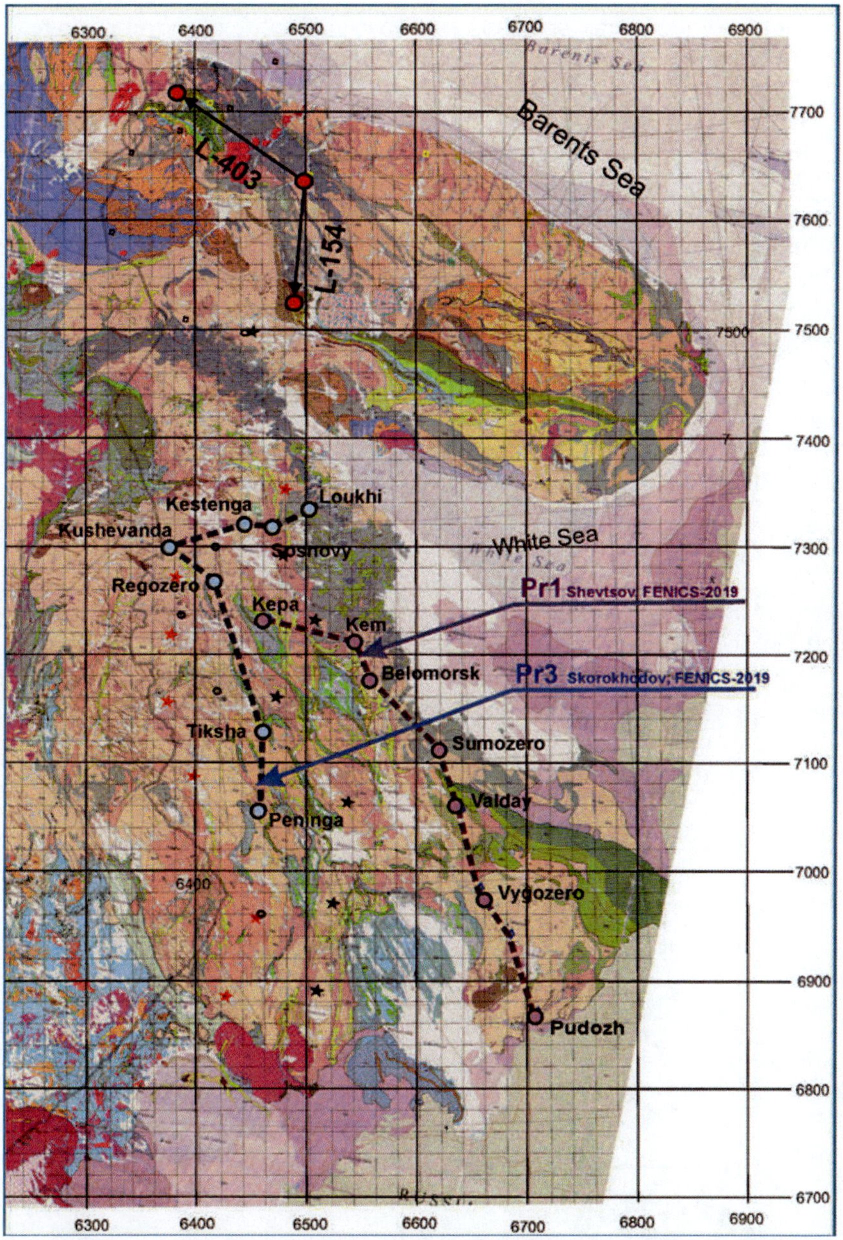

Fig. 1 Transmitter lines and measured profiles of FENICS-2019 pictured on the geological map of the region as the background. The Pr-1 (Kepa-Pudozh) and Pr-3 (Loukhi-Peninga) profiles in 2019 were measured with the L-403 (125 km—Murmansk-Nickel) and L-154 (102 km—Murmansk-Monchegorsk) lines. Asterisks indicate points of the measurements in 2014

solution of the inverse problem based on the data of frequency-domain remote sensing.

3. To investigate the anisotropic properties of the lithosphere of the Fennoscandian Shield by means of measuring the field polarized in two different directions.
4. To carry out soundings in the area of the anomalous seismic Moho boundary in Central Finland at a depth of 50–60 km to explore possible relationship between seismic (elastic) and electromagnetic properties of the lithosphere.
5. To perform sounding of the ground-ionosphere waveguide boundaries, estimate the anisotropic properties of the near-Earth space by measuring the input impedance and field components using different polarizations of the primary field.
6. To account for the effect of static shift by calculating the influence of the ionosphere and displacement currents from the moduli of the total horizontal components of the electric and magnetic fields and the effective impedance.

3 The Array for Measurements

To achieve the above goals, the following technique of measuring the MT-AMT field and the electromagnetic field of the controlled source was used (see Fig. 2).

A scheme of the FENICS experiment shown in Fig. 2 is based on tensor measurements using two mutually orthogonal grounded electrical lines, which we called audio-magnetotelluric sounding with controlled sources (CSAMT). In addition, natural AMT-MT fields were recorded before and after recording the field of the

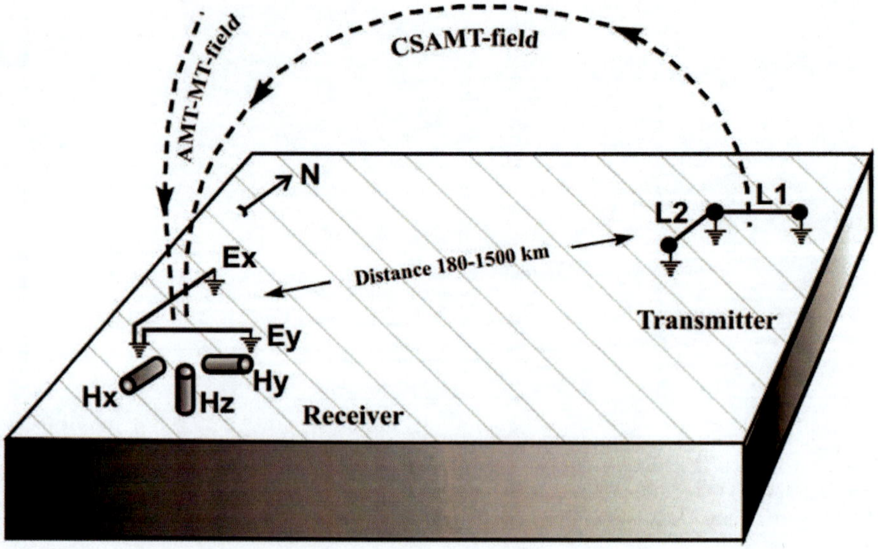

Fig. 2 Scheme of the experiment

controlled source. The method differs from that described in [11] as the CSAMT sounding procedure in the FENICS experiment was designed to work with long (over 100 km) grounded lines in the ELF range (0.1–200 Hz) and was used to study the electrical conductivity of the lithosphere at greater depths, from several hundred meters to 50–70 km. Note that a large depth of field penetration can only be achieved in this frequency range in the areas of the Earth's surface with high electrical resistivity, i.e. so-called "transparency windows", characterized by a relatively high horizontal homogeneity of the electrical conductivity of rocks. The experimental technique was the same as in FENICS-2014 experiment and was presented in detail in [9]. The sub-latitudinal line L-401 (109 km Murmansk-Serebryanka) located within the Murmansk block used during FENICS-2014 experiment, was replaced by a longer line L-403 (125 km Murmansk-Nikel) located within the Kola block. The Kola block is also characterized by a high average resistivity (over 10,000 Ω m), which made it possible to ensure reliable reception of the electric and magnetic components of the source field at distances over 840 km from its center. Moreover, the station Staraya Pustyn located at a distance of over 1600 km from the line managed to record magnetic components of the signal in the frequency band of 0.642–6.422 Hz [1]. To ensure synchronization, the registration of the current in the source and the components of the electromagnetic field at the observation points was carried out using the same type of receiver stations, VMTU-10 (LLC VEGA). To reduce the effect of static distortions caused by an inhomogeneous moraine deposits, the length of receiving electrical lines of GI KSC RAS was extended to 500 m. Additional measurements with Phoenix MTS-50 sensors and 100 m receiving lines were carried out simultaneously with the main measurements with the VMTU-10 equipment and IMS-007 sensors of the Russian production at some points of the Pr-1 profile.

4 Results and Interpretation

4.1 Data Treatment

The most interesting data on the electrical conductivity of the lithosphere were obtained by the GI KSC RAS in central and southern Karelia at a maximum distance from the source. The measured values of the components of the electromagnetic field and the current in the transmitting lines were recalculated into apparent resistances by normalizing to the source field over a homogeneous half-space in the dipole approximation [2–5, 10]

$$\rho_{tot}^{E} = K_{tot} \cdot \frac{E_{tot}}{I \cdot L_{AB}},\ E_{tot} = \sqrt{E_x^2 + E_y^2} \tag{1}$$

$$\rho_{tot}^{H} = \omega\mu_0 \cdot \left(K_{tot} \cdot \frac{H_{tot}}{I \cdot L_{AB}}\right)^2,\ H_{tot} = \sqrt{H_x^2 + H_y^2} \tag{2}$$

$$\rho_{tot}^{Z} = \frac{Z_{tot}^{2}}{\omega \cdot \mu_{0}}, \; Z_{tot} = \frac{E_{tot}}{H_{tot}} \tag{3}$$

where E_{tot} and H_{tot} are the moduli of the electric and magnetic field horizontal components, respectively. E_x and E_y are the horizontal components of the electric field (V/m), H_x and H_y are the horizontal components of the strength of the magnetic field of the source (A/m) at the observation points along the source line (x-component) and perpendicular to it (y-component). Cyclic frequency $\omega = 2\pi f$, where $f = 1/T$ is the frequency (Hz), and T is the period of current oscillations in the source (s), I is the current in the source (A), L_{AB} is the length of the transmitter line (m).

Geometric coefficient K_{tot} in the far zone of a dipole source is defined as

$$K_{tot} = \frac{K_1 \cdot K_2}{\sqrt{(K_1)^2 + (K_2)^2}}, \tag{4}$$

$$K_1 = \frac{2\pi r^3}{3 \cdot \cos\theta \cdot \sin\theta}, \; K_2 = \frac{2\pi r^3}{3 \cdot \cos^2\theta - 2} \tag{5}$$

Here r is the distance (m) from the center of the transmitting dipole to the receiving point, θ is the angle formed by the line from the center of the dipole to the receiving point and the dipole axis (degrees). Figure 3a (upper panels) shows the curves of apparent resistivity at the points of Valdai (left panels) and Pudozh (right panels) on the profile (Pr-1) in the field of the L-403.

Figure 4 shows apparent resistivity curves for the same points in the field of L-154.

Static shift was quantitatively assessed using the technique described in [2, 6–8]. Corrected CSAMT curves supplemented with MT-AMT sounding data or low-frequency asymptotes for the normal model in their absence were used for one-dimensional inversion in the horizontally homogeneous layered media.

Let us compare the results obtained for the point Pudozh obtained with the power lines L-403, L-154 and measurements in the AMT-MT natural field (Fig. 5).

As it follows from Fig. 5, the curves of apparent resistivity and impedance phase coincide with each other and with the data of AMT-MT measurements within the error limits, which validates a one-dimensional approximation in the interpretation of measurement results. The discrepancy of the phase curves in the region of large periods (T > 1 s) appears to be due to the influence of the near-field range of the source.

The results of one-dimensional interpretation in the form of a quasi-three-dimensional layer-by-layer model are shown in Fig. 6.

To build the model of the conductivity of lithosphere, we used the measurement data acquired in 2007, 2009, 2014 and 2019 using industrial power transmission lines. The model is characterized by a relatively small range of changes in conductivity within 10^3–10^4 Ω m on the depth of 1 km. At the depth 4 km we have resistivity from interval 5×10^4–2×10^5 Ω m. The resistivity on the sections has maximum at depth 10 km—about 2×10^6 Ω m in south part of the area of investigation. With increasing of the depth resistivity decreases from 2×10^3 to 2×10^4 Ω m at the 40 km

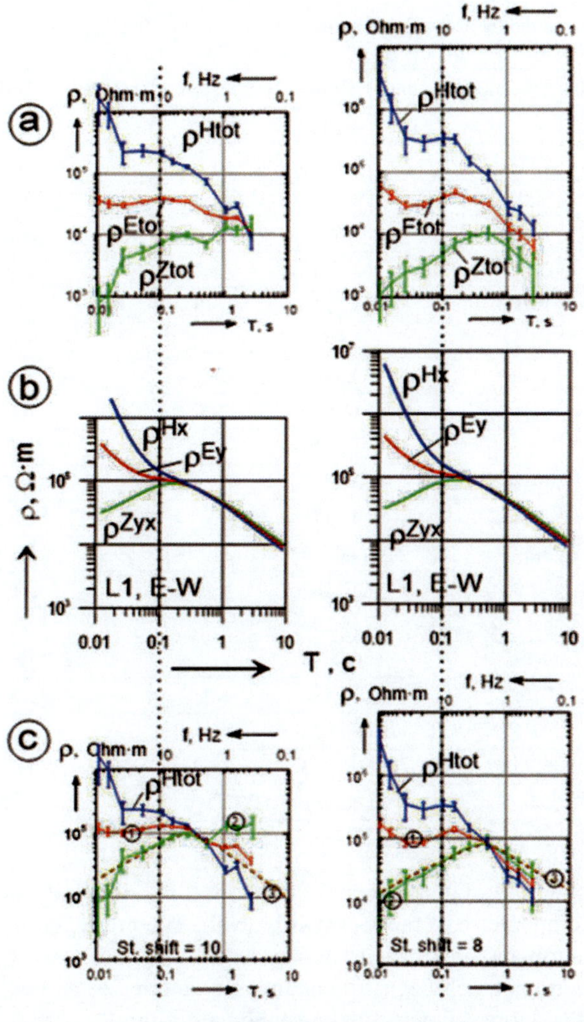

Fig. 3 Apparent resistivity curves calculated from the moduli of the total horizontal components in the field of the power line L403—top panels (**a**). Left panels are for p. Valdai r = 721 km, θ = 44.2°. Right panels are for p. Pudozh r = 844 km, θ = 43.3°. For the results of numerical simulation taking into account the influence of the ionosphere (ionosphere was assumed to be a layer of resistivity $\rho_j = 10^5\ \Omega$, at an altitude of $H_j = -100$ km above the Earth, with a thickness of $h_j = 2$ km, $j = -1$) for a normal resistivity section (Table 1)—central image (**b**). Bottom image (**c**): measured apparent resistivity curves, corrected for static shift. Indexes of corrected curves (1)—ρE_{tot}, (2)—ρZ_{tot}, (3)—smoothed curve for ρZ_{tot} supplemented by low-frequency asymptote for the normal section. The value <St. shift> value indicated at the bottom of the plot is a correction factor for ρE_{tot}; the correction factor for impedance curves equals to the squared <St. shift>^{2}. The dotted line shows the lower boundary of influence of the ionosphere and displacement currents

Table 1 The normal model of the Earth-ionosphere waveguide. In all space electrical permeability is space $\varepsilon = 8.85 \times 10^{-12}$ F/m, and magnetic permeability in H/m is $\mu = 4\pi \times 10^{-7}$

Layer number, j	Thickness, h_j, km	Altitude (<0)/Depth(>0), H_j, km	Resistivity, ρ_j, Ω m
−2	∞	−102	10^{14}
−1	2.000	−100	10^5
0	100.0	0	10^{14}
1	1.03	1.031	20,541.7
2	2.542	3.573	41,061.6
3	1.018	4.591	24,903.8
4	1.364	5.955	34,029.9
5	3.629	9.584	69,713.1
6	2.086	11.670	103,013.8
7	5.174	16.844	167,757.6
8	14.466	31.310	227,112.1
9	11.868	43.178	144,264.1
10	5.397	48.575	93,969.7
11	3.608	52.183	50,358.0
12	5.113	57.296	26,338.5
13	9.122	66.418	12,045.1
14	0.162	66.580	6029.3
15	23.452	90.032	3521.2
16	6.451	96.483	1134.3
17	∞	∞	662.9

to 5×10^1–2×10^2 Ω m at the depth of 100 km. It possible indicates the influence of increasing temperature of the rocks with depth. The presented model is the very first approximation and needs to be supplemented with both new experimental data for more detailed image of electrical conductivity anomalies and with the approach involving two- and three-dimensional inversion algorithms.

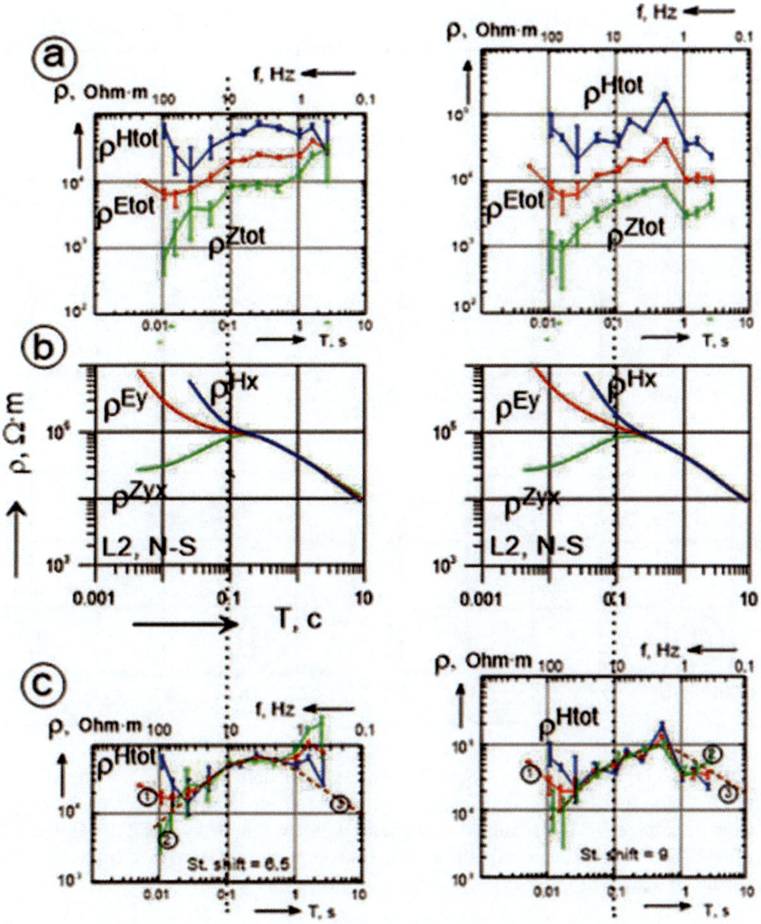

Fig. 4 Apparent resistivity curves calculated from the moduli of the total horizontal components in the field of the power line L154—top panels (**a**). Left panels are for p. Valdai r = 554 km, θ = 20.6°. Right panels are for p. Pudozh r = 751 km, θ = 10°. For the results of numerical simulation taking into account the influence of the ionosphere (ionosphere was assumed to be a layer of resistivity $\rho_j = 10^5\ \Omega$, at an altitude of $H_j = -100$ km above the Earth, with a thickness of $h_j = 2$ km, $j = -1$) for a normal resistivity section (Table 1)—central image (**b**). Bottom image (**c**): measured apparent resistivity curves, corrected for static shift. Indexes of corrected curves (1)—ρE_{tot}, (2)—ρZ_{tot}, (3)—smoothed curve for ρZ_{tot} supplemented by low-frequency asymptote for the normal section. The value <St. shift> value indicated at the bottom of the plot is a correction factor for ρE_{tot}; the correction factor for impedance curves equals to the squared <St. shift>^{2}. The dotted line shows the lower boundary of influence of the ionosphere and displacement currents

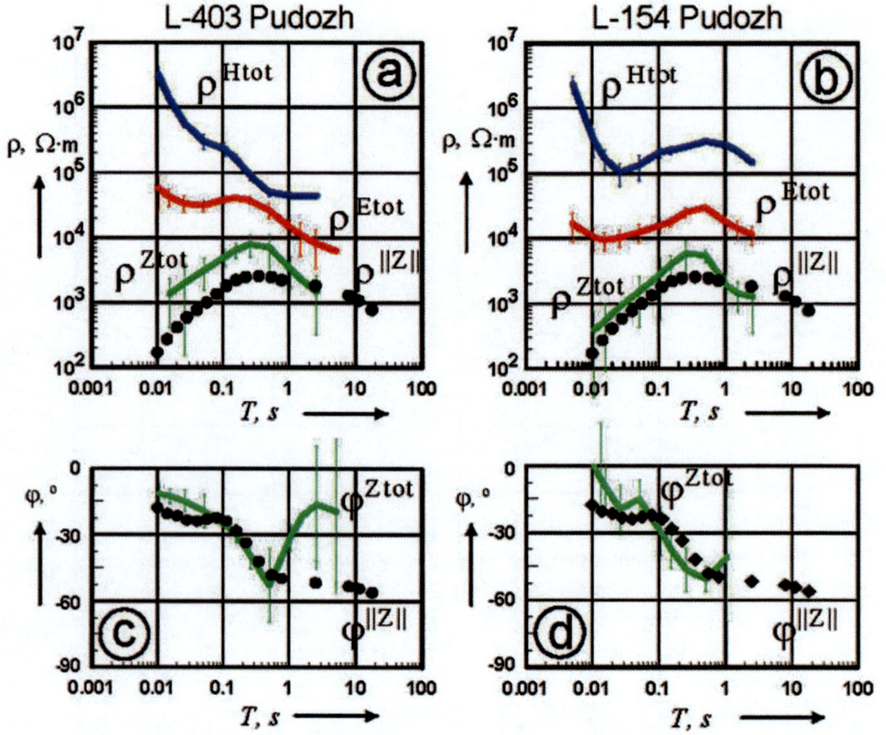

Fig. 5 Comparison of the results of CSAMT and AMT-MT data at the Pudozh point. Left images present the data for line L-403, right images—for line L-154. Upper images (**a**, **b**) are apparent resistivity, lower images (**c**, **d**) are impedance phases $\phi Ztot = 0.5(\phi Zxy + \phi Zyx)$. Filled circles are the data of AMT-MT measurements (in the absence of source current) obtained from the determinant of the impedance matrix $\|Z\| = Zxy \cdot Zyx - Zxx \cdot Zyy$

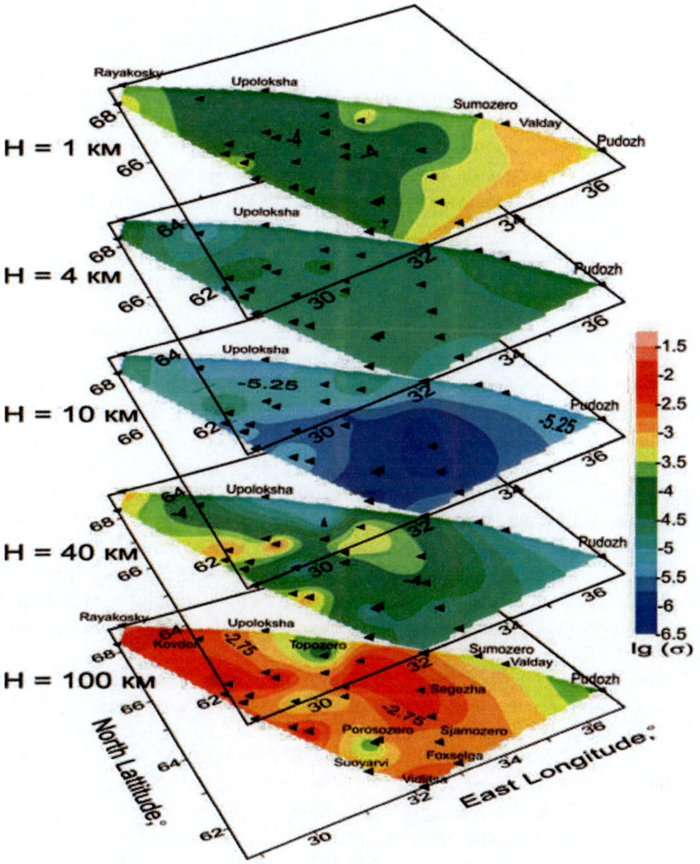

Fig. 6 Quasi-three-dimensional model of the electrical conductivity of the lithosphere for depths of 1, 4, 10, 40 and 100 km, based on one-dimensional inversions obtained using the method of conjugate gradients

Acknowledgements The authors are grateful to the chief engineers of the enterprise of the main electrical networks of the North-West of the Russian Federation and "Kolenergo" A. V. Maslov, O. Yu. Gorokhov and S. A. Zayets, as well as the head of substation "200" A. N. Dudchenko for the opportunity to work with industrial power lines. The authors are grateful to B. M. Samsonov, E. A. Kopytenko and A. S. Scriabin (VEGA LLC) for the provided sets of measuring stations VMTU-10 and participation in measurements.

Funding The study was supported by the Russian Foundation for Basic Research (project no. 018-05-00528) and by the Ministry of Education and Science of the Russian Federation (Geological Institute of the Kola Science Center, Russian Academy of Sciences, research topic no. 0226-2019-0052; Center for Physical-Technical Problems of the Energy Sector in the North, Kola Science Center, Russian Academy of Sciences, research topic no. 0226-2019-0067).

References

1. Ryabov, A.V., Pilipenko, V.A., Ermakova, E.N., Mazur, N.G., Fedorov E.N., Zhamaletdinov, A.A., Shevtsov, A.N.: Recording and modeling of ULF–ELF signals at the Staraya Pustyn station during the Fenics-2019 experiment. Seismic Instrum. **57**, 329–342 (2021)
2. Shevtsov, A.N.: Processing and interpreting the data of deep frequency sounding in a complex with audiomagnetotelluric measurements (Murman-2018 experiment). Seismic Instrum. **56**, 599–608 (2020)
3. Shevtsov, A.N.: On some methods of normalizing and transforming the results of electromagnetic soundings. Deep geoelectric studies using industrial power lines. Apatity 90–95 (1989)
4. Vanyan, L.L.: Fundamentals of Electromagnetic Sounding. Nedra, Moscow (1965)
5. Veshev, A.V.: AC and DC Electrical Profiling. Nedra, Leningrad (1980)
6. Zhamaletdinov, A.A.: Method of quantitative accounting of static distortions in the magnetic field of the controlled source CSAMT. Sci. Technol. Dev. **98**(4), 5–18 (2019)
7. Zhamaletdinov, A.A., Velikhov, E.P., Shevtsov, A.N., Skorokhodov, A.A., Kolobov, V.V., Ivonin, V.V., Kolesnikov, V.V.: The Murman-2018 experiment on remote sensing in order to study the "Impenetrability" boundary at the transition between brittle and plastic states of the crystalline earth's crust. Dokl. Earth Sci. **486**, 575–579 (2019)
8. Zhamaletdinov, A.A., Velikhov, E.P., Shevtsov, A.N., Kolobov, V.V., Kolesnikov, V.E., Skorokhodov, A.A., Korotkova, T.G., Ivonin, V.V., Ryazantsev, P.A.A., Birulya, M.A.: The Kovdor-2015 experiment: study of the parameters of a conductive layer of dilatancy–diffusion nature (DD Layer) in the Archaean crystalline basement of the Baltic Shield. Dok. Earth Sci. **474**, 641–645 (2017)
9. Zhamaletdinov, A.A., Shevtsov, A.N., Velikhov, Ye.P., Skorokhodov, A.A., Kolesnikov, V.E., Korotkova, T.G., Ryazantsev, P.A., Efimov, B.V., Kolobov, V.V., Barannik, M.B., Prokopchuk, P.I., Selivanov, V.N., Kopytenko, Yu.A., Kopytenko, Ye.A., Ismagilov, V.S., Petrishchev, M.S., Sergushin, P.A., Tereshchenko, P.Ye., Samsonov, B.V., Birulya, M.A., Smirnov, M.Yu., Korja, T., Yampolski, Yu.M., Koloskov, A.V., Baru, N.A., Poljakov, S.V., Shchennikov, A.V., Druzhin, G.I., Jozwiak, W., Reda, J., Shchors, Yu.G.: Study of interaction of ELF–ULF range (0.1–200 Hz) electromagnetic waves with the Earth's crust and the ionosphere in the field of industrial power transmission lines (FENICS experiment). Izv. Atmos. Oceanic Phys. **51**, 826–857 (2015)
10. Zhdanov, M.S.: Elektrorazvedka (Electrical Survey). Nedra, Moscow (1986)
11. Zonge, K.L., Hughes, L.J.: Controlled source audio-frequency magnetotellurics. Electromagn. Methods Appl. Geophys. Application, Parts A and B **2** (1991)

Specifics of the Earth's Crust Structure in the Potential Gas Hydrate Accumulation Zones of the Arctic Basin

A. A. Petrova, O. V. Latysheva, and A. I. Petrova

Abstract Due to the favorable thermobaric conditions of gas hydration in the Arctic Basin, mathematical modeling gives a positive forecast for potential zones of methane hydrate accumulation, assuming a very extensive distribution of hydrocarbons. The gas hydrate zones in the Arctic Basin are very extensive and represent a huge resource of hydrocarbons. Layers of hydrates serve as an impervious cover for oil deposits. They create conditions for the formation of near-surface oil and gas accumulation in the form of solid gas hydrates. The unstable state of gas hydrates can cause explosive destruction of these arrays with the sudden release of methane. The fluid systems of the Earth's crust of the main types of the Arctic cryolithozone, the western and eastern oceanic sectors of the Arctic Basin and coastal-shelf cryolithozones are studied on the basis of density and magnetic sections. Thermofluid channels and lenses of fluid layers in the form of zones of reduced density and magnetization are traced on deep sections. Analysis of the lithosphere density sections of confirmed and potential zones of gas hydrate accumulation revealed deep-focus fluid lens systems at depths of 40–50 and 70–90 km. The areas of fluid flow exits along fault zones from fluid-saturated layers to areas with favorable conditions for the accumulation of gas hydrates are outlined on the basis of a complex interpretation of geophysical data. In the presence of a seismic event, vertical thermofluid channels can activate gas fluid flows. Thermofluidic flows can act as a trigger for explosive destruction of hydrates, creating extra risks for navigation and infrastructure of the Northern Sea Route.

Keywords Arctic · Lithosphere · Gas hydrate accumulation zones · Thermofluid channels · Northern Sea Route

A. A. Petrova (✉) · O. V. Latysheva (✉) · A. I. Petrova
Pushkov Institute of Terrestrial Magnetism, Ionosphere and Radio Wave Propagation of the Russian Academy of Sciences, St.-Petersburg Branch (SPbF IZMIRAN), Universitetskaya nab., 5B, St. Petersburg 199034, Russia
e-mail: aa_petrova@inbox.ru

A. Kosterov et al. (eds.), *Problems of Geocosmos—2020*, Springer Proceedings in Earth and Environmental Sciences, https://doi.org/10.1007/978-3-030-91467-7_3

1 Introduction

The purpose of this study is to study the deep structure of the Earth's crust in the zones of the expected distribution of gas hydrates in the Arctic Ocean on the basis of geophysical data. Gas hydrates are solid crystalline compounds formed under certain thermobaric conditions (low temperatures and high pressures) by combining gas (primarily methane) with water. According to the assumption of most experts, large volumes of gas hydrates are concentrated in the cryolithozone of the Russian Arctic shelf [1–20]. The presence of gas hydrates is confirmed in zones where gas exists in situ or there is an influx of gas from the depths through faults, sub-vertical cracks and disjunctive disturbances that serve as channels for the supply of hydrocarbons [4]. The flow of gas fluids from the gas lenses of the Earth's crust is quite intense. Spreading through loose sediments, the gas creates a gas-saturated layer, where gas hydrates can form under favorable conditions.

The stability zone of the gas hydrates depends on such parameters as the water temperature at the bottom, pressure, gas composition, water salinity, geothermal gradient [21–40]. Thermobaric conditions for the formation of gas hydrates exist in most of the Arctic Ocean because the temperature at a depth of 300 m is negative (about − 1 °C) [8, 25]. The critical temperature sufficient for the dissociation of gas hydrates depends on the lithology (particle size) and salinity of the host frozen sediments and ranges from − 3.0 to − 0.3 °C [36, 38].

The gas hydrates stability zone in the Arctic corresponds to the depths of the lower roof of the cryolithozone of more than 200 m with a geothermal heat flow of 60–70 mW/m^2. With an increase in the heat flow to 100 mW/m^2, there is a possibility of through taliks formation. It can lead to a disturbance of the conditions for the existence of gas hydrates and the release of methane [27, 28, 42].

To study the deep structure of the Earth's crust of potential zones of gas hydrates accumulation in the layer, the following tasks were solved:

- the analysis of the conditions of gas hydrate presence and the zones of gas hydrates possible accumulation in the Arctic Ocean region is carried out; the analysis is performed on the basis of mathematical modeling [8, 24, 26–35];
- density and magnetic sections of the Earth's crust are constructed for the predicted and confirmed gas hydrate zones;
- thermofluid channels of fault zones and lenses of fluid-saturated layers of the Earth's crust and mantle are revealed;
- the outputs of channels with thermal fluid flows to the areas of potential formation of gas hydrates in the water area of the Northern Sea Route are determined;
- the most probable explosive areas of an endogenous nature have been identified in the Arctic Ocean.

One of the first works on potential zones of gas hydrate formation in the entire Arctic basin was published by Russian researchers Ginsburg G. D., Gramberg I. S., Solovyov V. A. [2]. Gas hydrate regions of different types for the Russian sector of the Arctic Ocean are presented taking into account the peculiarities of the formation of gas hydrates [9].

Sergienko [42] and Romanovsky [30, 31] distinguish two main types of cryolithozone: the inner part of the shelf (coastal shelf) and the outer part of the shelf (oceanic). As a result of modeling the current state of the underwater permafrost, N. N. Romanovsky comes to the conclusion about the stability of the coastal shelf cryolithozone up to the isobate of 50–70 m.

In [8], a forecast of gas hydrate distribution zones in the Arctic Ocean was published, obtained by mathematical modeling based on the analysis of thermobaric conditions using the geographic information system "Arctic and World Ocean" (GIS AWO), which summarizes various sources of information about the distribution of gas hydrates. Cartographic schemes for the distribution of gas hydrate stability zones in the Arctic Ocean with a bottom water temperature of −2 and 0 °C were created for two different models. The difference between the models was the amount of water mineralization in shallow sediments. At the same temperatures, the gas hydrates stability zone corresponds to lower depths for fresh water than for salt water in shallow sediments. This is explained by the fact that only fresh water participates in the reaction of gas hydrate formation, and salt ions resist this process, reducing the activity of water [8, 26, 34]. In Model 2, the zone of possible formation of methane hydrates is somewhat wider than in Model 1. On the basis of modeling, the minimum depths of the seabed at which gas hydrates can occur and the areas of their possible distribution are predicted [8].

The map of the conditions of gas hydrate content of the Arctic seas [12], created on the basis of [5], allowed us to estimate the types of gas hydrate regions of the layer where gas hydrate accumulations are possible.

The formation of gas hydrates stability zones depends on several factors: the thickness of the water column; the necessary thickness of Neogene-quaternary deposits; a sufficiently low temperature of bottom waters; the presence of permafrost zones, etc. [24–35]. Thermobaric conditions under which gas hydrates are formed are observed in most of the Arctic Ocean and the Russian Arctic shelf. The detection of wave signs helps to detect gas hydrates and gas-saturated thicknesses of possible sites of the location of gas hydrates [4–8, 33].

Studies of the genesis of gas hydrates have shown that their composition is complex. Origin of natural gas inside permafrost can be biochemical as well as thermogenic. Microbial gas is widely spread within the permafrost. Thermogenic gas is locally distributed and its placement depends on the mechanisms for gas migration from deep layers or the presence of ancient hydrocarbon-containing rocks in the permafrost [40–45].

The preservation of the relict (metastable) hydrate is confirmed by various studies [27–40]. It is assumed that the relic hydrate can be located to a depth of 120 m [8, 34, 36–40]. The area of the predicted subaqual distribution of the relict hydrate on the Arctic shelf serves as a metastability zone.

Potential zones of accumulation of gas hydrates in the Arctic Ocean were selected to study the deep structure of the Earth's crust. Existing of these zones confirmed by

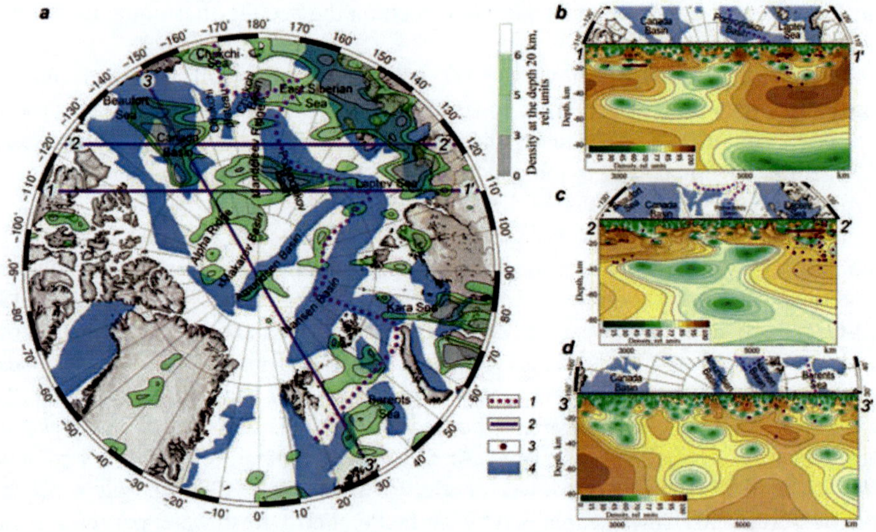

Fig. 1 Scheme of distribution of gas hydrates predictive zones [19] and density minima (depth is 20 km). 1—North border of the Northern Sea Route; 2—sections placement; 3—earthquake foci; 4—hydrate areas

direct studies, including drilling, predicted by logging data, seismic data (BSR) and other indirect evidence.

Hydrocarbon prospects of Russia are concerned with the development of the Arctic seas resources. In the Arctic Basin there are wide potential areas of gas hydrate content [1–3, 7–12, 43, 44]. Gas hydrates of the shelf seas of the Arctic Ocean can contain a huge resource of hydrocarbons (Fig. 1) [4, 11–14]. They create conditions for the formation of a zone of near-surface oil and gas accumulation of hydrocarbon deposits in the form of solid gas hydrates in the Arctic.

Geoecological safety in the development of oil and gas resources is ensured during the seismic situation control. This situation in the Arctic is characterized by an uneven, focal distribution of epicenters of seismic events (earthquakes) associated with areas of tectonic activity [45–50].

The analysis of magnetic and density sections of the lithosphere in the regions of possible zones of gas hydrates accumulation visualizes the features of thermofluid channels, the areas of their exits to the bottom surface and the location of the lenses of fluid layers. It allows to make a forecast of the location of potentially dangerous zones of endogenous origin.

The studies were carried out on the base of interpretation of the Earth's magnetic field anomalies [51–53], gravity anomalies [54] and seismological data [55]. The features of the Earth's crust in the form of fault zones, thermofluid channels and fluid-saturated layers are presented on deep geophysical sections. Density and magnetic sections were obtained by spectral-spatial analysis of magnetic field and gravity anomalies with the conversion of the spectral-spatial representation of the fields into deep sections [56–59]. The density sections are constructed from gravity anomalies, while the magnetic sections are constructed from the anomalies of the modulus (T), vertical (Z), and horizontal (H) components of the Earth's magnetic field calculated using the IZMIRAN SPbF model [53, 60].

Fluid-conducting channels are weakened zones with low magnetization, reduced density properties and may have increased values of heat flow. Channels of thermofluid processing and lenses of liquid layers are visualized on sections as zones of reduced density and magnetization in the lower crust and mantle [59, 61, 62]. The lenses of fluid systems are characterized by an inversion of seismic velocities and a reduced electrical resistance [63]. The maximum effect of the heat flow was found during the upward migration of fluids along vertical faults [9, 14–16]. Fluid channels take part in the formation of hydrocarbon deposits and, under certain conditions, gas hydrate zones due to the upward migration of hydrocarbon fluids.

The locations of the thermofluid channels and the lenses of the feeding fluid layers of the gas hydrate zones in the Arctic Basin are identified from the analysis of the magnetic and density sections of the lithosphere.

2 The Specifics of the Lithosphere Structure of Potential Gas Hydrate Zones in the Arctic Basin

In this paper, the areas of potential zones of gas hydrate formation in the Canadian Basin, in the Amundsen, Nansen and Submariners Basins, in the Beaufort, Barents, Kara, Laptev and East Siberian Seas are studied (Fig. 1, 4) [2, 6, 9, 11].

The gas fluid can flow from deep formations through the fault zones of thermofluid channels. The maximum effect of the heat flow was found during the upward migration of fluids along vertical faults [59, 64–66]. Studies of samples of methane gas hydrates have shown that even small changes in temperature can cause significant emissions of greenhouse gases CO_2 and CH_4 [25, 27, 32–33, 35–36].

An experiment on metastable relic methane hydrates has established that small temperature differences in units of degrees have a great influence on the metastability of hydrates [27, 36].

Our studies have shown that in these areas there are subvertical thermofluid channels of disjunctive disorders, which can serve as channels for the supply of hydrocarbon fluids of endogenous origin.

For each region, an analysis of deep sections (Fig. 1, 4) was carried out, showing the most likely ways of the inflow of gas fluids as a result of vertical migration. In addition, an assumption is made about the probable trajectories of feeding with deep gases in the case of activation of fluid-saturated lenses under the influence of seismic events.

Gas hydrate deposits can suddenly collapse due to its temperature instability. The near-surface complex of solid gas hydrates in the shelf seas of the Arctic Ocean creates extra risk. It complicates navigation, the operation of shelf oil and gas fields and the transportation of hydrocarbon raw materials. This problem declares the increased fire hazard for large hydrocarbon deposits on the Barents and Kara shelves and adjacent land: Urengoy, Yamburgskoe, Bovanenkovskoe, Shtokman gas condensate field etc. [10, 13].

One of the main reasons for the explosive release of methane gas hydrates is the activation of the thermofluid channels coming to the surface because the fluid temperature depends on the depth of the channel. Studies of gas hydrate zones on the shelf of the Eastern Arctic seas have shown that even small temperature changes can cause significant emissions of greenhouse gases CO_2 and CH_4 [29, 30, 33, 35, 36, 40]. An increase in the activity of thermofluidic flows with an increase in seismotectonic processes can cause dangerous phenomenon in the zones of gas hydrate development.

A forecast of the possible distribution of the decompressed areas of fault zones at different depths is made based on a two-dimensional analysis of the Earth's crust density properties (Fig. 1, 3). The depth of the level of 20 km is selected by the characteristic minima of density in the sections of the Earth's crust. Thermofluidic processing of endogenous rocks with the formation of through taliks is possible from this depth. The two-dimensional calculation allows us to determine the location of channels at a depth of about 20 km in the middle part of the Earth's crust from the density minima. As a result of the study the most likely zones of occurrence of natural hazards in the waters of the North Sea Route caused by the release of ancient methane were identified.

Deep sections of the gas hydrate zones were constructed using magnetic anomalies, gravity anomalies, seismic and seismological data to identify the specific structure of the lithosphere of the Arctic Ocean (Fig. 1, 4). At studying the mechanism of occurrence of explosive natural phenomena of a deep nature the greatest attention is focused on the areas adjacent to the water area of the North Sea Route to ensure the safety requirements of all-season navigation.

Deep density and magnetic sections are constructed on the data of anomalies of the Z-and H-components of the Earth's magnetic field across the profile passing through the Amerasian and Eurasian Basins of the Arctic Ocean (Fig. 2). Extended deep sections demonstrate a significant difference in the structure of the lithosphere of these basins.

Analysis of the sections showed that the Alpha ridge of the Amerasian Basin is located above a sub-vertical fault zone extending from a depth of 40–50 km (Fig. 2). Magnetic sections show that the fault is filled with magnetoactive formations in the upper and lower parts of the Earth's crust, whose anomalies near the ocean surface are 500–1500 nT [64]. Layers magnetized vertically create intense anomalies of the Z-component, magnetized horizontally-intense anomalies of the H-component.

Comparison of the density sections with the seismic sections in the Amerasian Basin showed that the velocity characteristics of the rocks are in good agreement with the obtained distribution of the density formations [64, 67–71]. Identification of the marking layers made it possible to clarify the depth and location of the decompressed and weakly magnetic zones of fluid-saturated lenses, and to estimate the depth of the thermofluid channels. Earthquake foci tend to contact rocks of different densities on the top, base and on the lateral boundaries of density inhomogeneities.

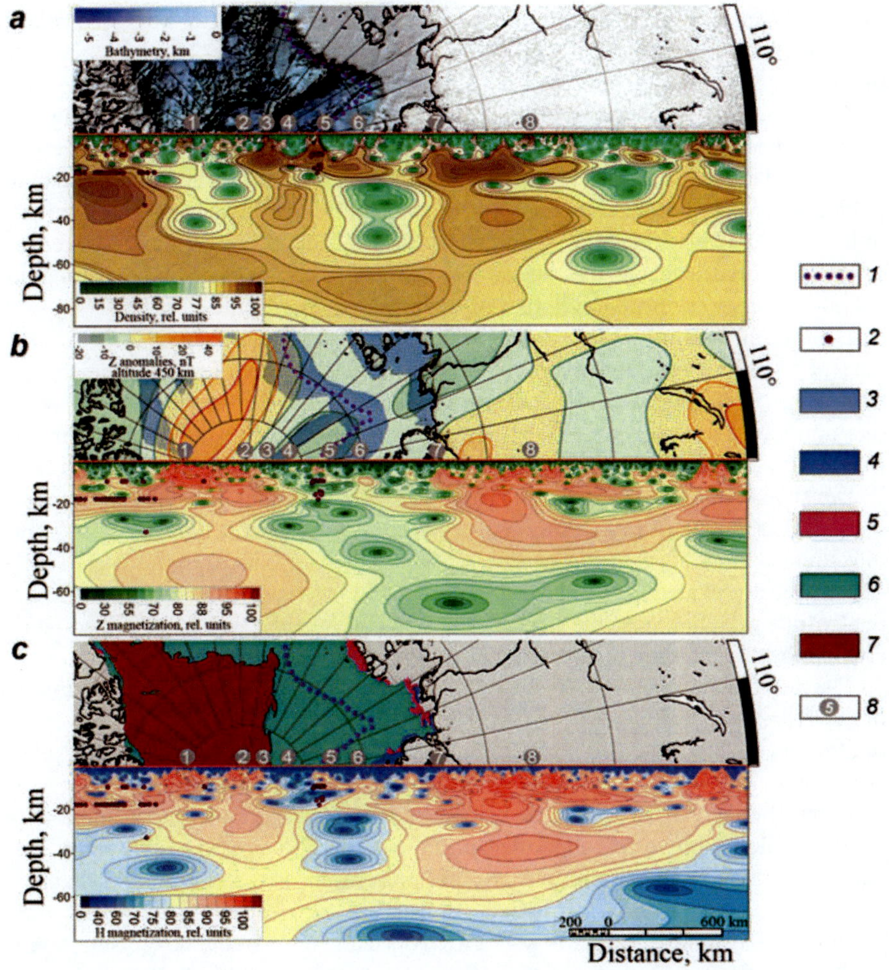

Fig. 2 Sections of the lithosphere of gas hydrates predictive zones in the Arctic Basin: **a** density section (bottom) and bathymetry (top) [77]; **b** magnetic section on the anomalies Z-components (bottom) and anomaly at an altitude of 450 km (top) [53, 60]; **c** magnetic section on the anomalies H-components (bottom) and winter ice cover at 2020, March (top) [73]. 4—nilas; 5—young ice; 6—first-year ice (30–200 cm); 7—old ice; 8—tectonic structures: 1—the Alpha Ridge, 2—the Makarov Basin, 3—the Lomonosov Ridge, 4—the Amundsen Basin, 5—the Gakkel Ridge, 6—the Nansen Basin, 7—the Taimyr, 8—the Siberian Platform. Symbols 1–3 see Fig. 1

Magnetic and density sections demonstrate the specifics of the deep structure of the Earth's crust of the Eurasian basin (Fig. 2). The sections show that the thermal channel enters the zone of possible accumulation of filterogenic gas hydrates and biochemical gas of the deep water of the Amundsen basin from a depth of almost 60 km [12]. The channel is traced as a weakened zone with reduced density and magnetic properties. A large decompressed weakly magnetic lens of the fluid system feeding the channel is located in the mantle at a depth of 70–100 km. It correlates with the global mathematical model of seismic wave speeds [72], which confirms the presence of reduced velocities zone at depths of about 100 km.

The zone of thermofluid processing is most clearly observed on the magnetic sections. The upward vertical migration of fluids is well expressed in the sections constructed on the data of the Z- and H-component anomalies of the Earth's magnetic field. It means the thermal flow can suddenly be withdrawn to the surface in the area of hydrate deposits. It makes conditions for the melting of the ice sheet, the release of ancient methane from the permafrost and the occurrence of explosive natural phenomena near the northern border of the Northern Sea Route.

Long-term satellite observations in the Arctic in the period from 2008 to 2020 made it possible to estimate the ice cover condition [73]. It confirms the transformation of the boundaries as a result of the melting of perennial ice in the zones of the thermofluid channels outlets [64, 74–76]. The shape of the ice edge in the Amundsen Basin (Fig. 2b) verifies the influence of the thermal flow from a depth of ~ 20 km.

Density sections 1–1′ and 2–2′ (Fig. 1) in the Podvodnikov Basin intersect areas of possible accumulations of bottom hydrates of the filterogenic type and deep-water biochemical gas [12]. Significant areas of breaking disturbances of the Earth's crust and mantle are placed in the Podvodnikov Basin. One can see thermofluidic flow to the hydrate region is possible from a depth of 40 km and the fluid-saturated layers feeding the channel are traced at depths of 50 and 80 km.

The 3-3′ density section passes through the Beaufort Sea (Fig. 1). Accumulation of gas hydrates of the filter type in this zone is confirmed by drilling [8, 12]. A vertical fault enters this zone. A thermofluid flow along this fault is possible from a depth of 40 km (Fig. 1).

In the Barents Sea, a fault zone extending to the area of potential accumulation of filtrogenic hydrates was identified in the 3-3′ section (Fig. 1) [12]. Norwegian researchers have studied a compact area of large craters on the bottom of the Norwegian Sector and discovered gas hydrates located in the Western Arctic Sector to the northwest of Svalbard. The study was carried out using high-frequency seismic equipment with subsequent analysis of velocities and hodographs [18, 19]. Norwegian scientists concluded that gas seeped through the fault zone from deep hydrocarbon reservoirs, frozen in the near-surface layer of the bottom in the form of deposits of solid gas hydrates [20]. Samples of gas hydrate were taken during marine expeditions on the northern shelf of Norway, including on the area of the Haakon Mosby mud volcano. The analysis of the samples showed that the genesis of gas hydrates is complex and contains methane of biogenic and endogenous origin.

Our calculations have shown that in the water area of the Norwegian Sector, a fluid flow into the hydrate region is possible from a depth of 30 km, and the fluid-saturated layer is located at a depth of 45–50 km [66]. A similar situation is presented in the section 2–2′ (Fig. 3b), crossing the border of the NSR in the Barents Sea in the zone of possible accumulation of filtrogenic hydrates in the Shtokman gas field (Fig. 4) [12].

The density sections of the most likely zones of occurrence of natural hazards are located in the East Siberian Sea to the north of Wrangel Island. Accumulations of

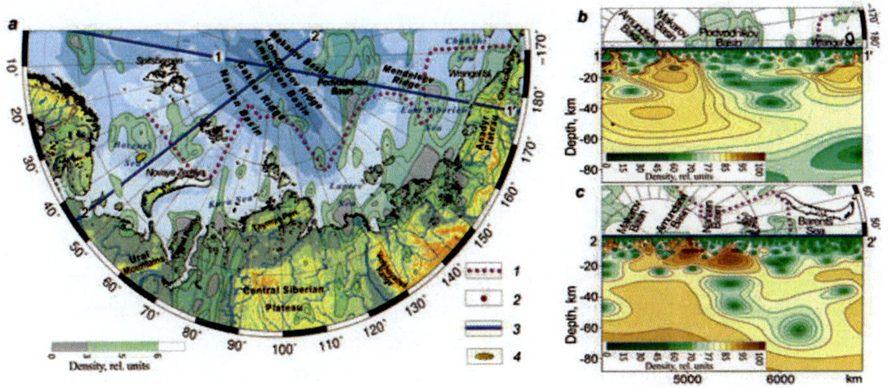

Fig. 3 The position of the fluid system of the shelf seas of the Arctic Ocean. **a** Density minima at a depth of 20 km. **b** Density depth section 1–1′; **c** density depth section 2–2′; 4—Shtokman gas field. Symbols 1–3 see Fig. 1

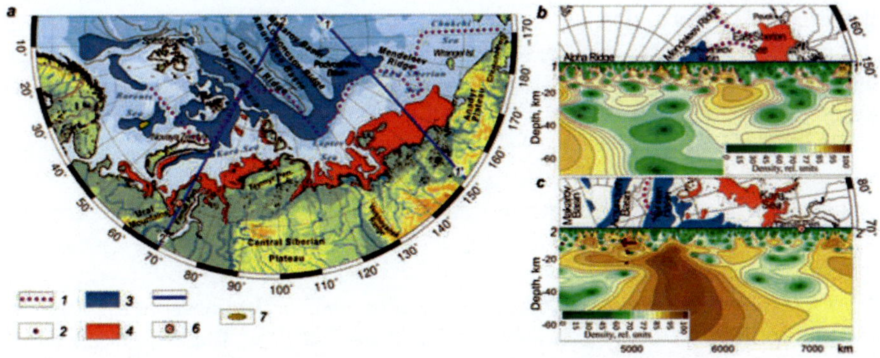

Fig. 4 Potentially dangerous areas of the Arctic zone. **a** Section locations; **b** density deep section 1–1′ (East Arctic); **c** density deep section 2–2′ (Yamal). Predictive gas hydrate zones: 3—near-seabed hydrates; 4—cryogenic hydrates; 5—location of sections; 6—Yamal crater; 7—Shtokman gas field. For symbols 1–2, see Fig. 1

filterogenic and cryogenic gas hydrates are possible in this region [12]. The analysis of the selected samples of bottom water showed the presence of intrusions of deep fluid [11, 25, 29–31]. The zone of disturbance can be traced from a depth of 20 km in the form of a vertical channel in the section 1–1′ (Fig. 3b). Beneath this zone at a depth of ~ 30 and ~ 40 km there are low-density lenses. The activation of these lenses can create an explosive natural phenomenon near the border of the Northern Sea Route due to the presence of a vertical fault.

A joint analysis of the two-dimensional calculation of density properties at a depth of 20 and 40 km and sections of potential gas hydrate zones showed the location of fluid-saturated layer lenses and the trajectories of thermofluid channels in the Arctic Basin (Fig. 1, 4). The lenses of fluid systems are identified by density sections and located in the mantle at depths of ~ 40–50 km and ~ 70–90 km. Seismic events can trigger the activation of fluid-saturated layer lenses. An increase in the fluid flow contributes to the occurrence of dangerous phenomena in gas-hydrate-bearing areas.

One of the most notable examples of explosive natural phenomena of endogenous origin in the gas hydrate zone is the Yamal Peninsula. There are 17 cryogenic geological events occurred at the peninsula with the formation of craters and the release of methane [37, 38, 40, 43, 78]. The analysis of the density meridional section showed that the Yamal Peninsula is placed above the outlet of a powerful vertical fluid channel coming from a depth of 25–45 km. The feeding lens of the fluid layer is observed in the mantle at a depth of $\geq$ 65–70 km (Fig. 4b) [65]. It is possible that the reason for the sudden explosion and the thermokarsts appearance was the pulsation of the lens of the fluid-saturated layer and the activation of the thermofluid channel [59, 65, 66]. In the areas of gas hydrate deposits it can leads to explosive phenomena threatening the gas production and transport infrastructure.

A comprehensive study of the oil and gas provinces of the Russian sector of the Arctic Ocean found the areas the fluid flows of deep faults exit to the areas of a favorable forecast for the accumulation of gas hydrates (Fig. 1, 4). Fault zones of the Earth's crust influence to the process of destruction of solid gas hydrates because thermofluidic flows goes through these zones to the seabed. Temperature and pressure of the thermofluidic flows depend on the depth of the fluid-saturated layers and channels [10, 20–22].

The analysis of the density deep sections of the lithosphere provided the information about the location of the deep-focus systems of the fluid layers of the Arctic zone of the Russia. Seismic event may cause pulsating vibrations in the fluid-saturated layer and activate the vertical migration of thermofluidic flows into the zones of gas hydrate deposits.

As a result of the potential hydrates areas studying near the border of the Northern Sea Route the most likely zones of unwanted natural phenomenon caused by the release of ancient methane are identified. This allows to make a forecast of explosive zones in the waters of the Northern Sea Route and to find the safest routes for navigation.

As part of the Arctic digitalization program the first stage of the project for the construction of the Arctic Connect transarctic underwater communication line was completed in December 2020. Laying cables near hazardous areas requires an risk evaluation from unwanted phenomena of endogenous origin.

3 Conclusions

1. A model of the deep structure of the waters of the Canadian Basin, the Amundsen, Nansen and Podvodnikov Basins, the Beaufort, Barents, Kara, Laptev and East Siberian Seas was obtained from the Earth's magnetic field anomalies and gravity field anomalies for potential zones of hydrate accumulation. A number of researchers have made a forecast of methane gas hydrates

distribution for the entire Arctic basin based on mathematical models. The forecast results for the water areas considered in this paper are confirmed by direct studies, including drilling, seismic data (BSR), logging data and indirect evidence [8, 12, 27].
2. The influence of the deep processes on the melting of the Arctic Ocean ice cover. The maximum effect of thermal action with the formation of through taliks was recorded under the upward mantle migration of fluids along the channels of vertical faults [11, 25, 29, 32, 33].
3. The most possible areas of natural hazards phenomena associated with the release of ancient methane from the permafrost are identified in the water area of the Arctic Ocean. To solve the problems of stable economic development of the Arctic zone of the Russian Federation, it is necessary to continue conducting comprehensive scientific research. It will make possible to clarify the boundaries of different types regions of cryolithozone and to assess the permissible degree of stability of gas hydrates.
4. The localization of the outlets of the deep-lying fluid channels along the routes of the Northern Sea Route was studied to reduce the risks of navigation.
5. The study of the Arctic Basin deep structure has a scientific and applied importance for solving geological and geophysical exploration problems, issues of safe navigation and the development of the transport infrastructure of the shelf seas.

References

1. Ginsburg, G.D., Gramberg, I.S., et al.: Underwater mud-volcanic type of gas hydrate accumulations. Rep. USSR Acad. Sci. **300**(2), 416–418 (1988)
2. Ginsburg, G.D., Gramberg, I.S., Soloviev, V.A.: Geology of submarine gas hydrates. Soviet Geol. **11**, 12–19 (1990)
3. Ginsburg, G.D., Soloviev, V.A.: Submarine Gas Hydrates. VNIIOkeangeologia, St. Petersburg (1994)
4. Gramberg, I.S., Laverov, N.P and Dodin, D.A.: Arctic at the Threshold of Third Millennium (Resource Potential and Ecology). Nauka, St. Petersburg (2000)
5. Soloviev, V.A., Ginsburg, G.D., Telepnev, E.V., Mikhalyuk, Yu.N.: Cryogeothermy and Natural Gas Hydrates in the Bowels of the Arctic Ocean. PGO "Sevmorgeologiya", Leningrad (1987)
6. Soloviev, V.A.: Gas-hydrate-prone areas of the ocean and gas-hydrate accumulations. In: Proceedings of the Sixth International Conference on Gas in Marine Sediments, St. Petersburg, Sept 2000
7. Soloviev, V.A., Ginsburg, G.D.: Conditions of gas-hydrate content and potentially gas-hydrate-bearing water areas. In: Geology and Mineral Resources of the Russian Shelf Areas, Atlas. Scientific World, Moscow (2004)
8. Bogoyavlenskiy, V., Kishankov, A., Yanchevskaya, A., Bogoyavlenskiy, I.: Forecast of gas hydrates distribution zones in the arctic ocean and adjacent offshore areas. Geosciences **8**(12), 453–470 (2018)
9. Pavlenko, V.I.: The Arctic zone of the Russian Federation in the system of ensuring the national interests of the country. Arct. Ecol. Econ. **4**(12), 16–25 (2013)

10. Popkov, V.I., Soloviev, V.A., Solovieva, L.P.: Gas-hydrates—Earth's abyssal degasification product. Geol. Geogr. Global Energy **3**, 56–67 (2012)
11. Shakhova, N., Semiletov, I., Chuvilin, E.: Understanding the permafrost–hydrate system and associated methane releases in the East Siberian arctic shelf. Geosciences **9**, 251 (2019). https://doi.org/10.3390/geosciences9060251
12. Logvina, E.A., Matveeva, T.V., Bochkarev, A.V., Semenova, A.A., Nazarova, O.V.: Analysis of technological and technical advances in the study of subaqueous gas hydrates and the possibility of their application in the Arctic seas of Russia. Arct. Ecol. Econ. **4**(40), 66–76 (2020)
13. Zakharenko, V.S., Kazanin, G.S., Pavlov, S.P.: Prerequisites and conditions of gas hydrates formation on the Shtokman field of the Barents Sea. Vestnik MSTU **17**(2), 394–402 (2014)
14. Matveeva, T.V., Soloviev, V.A.: Gas hydrates of the Sea of Okhotsk: regularities of formation and distribution. Russ. Chem. J. **157**(3), 101–111 (2003)
15. Makogon, Y.F.: Gas hydrates. History of development and prospects of study. Geol. Miner. World Ocean **2**, 5–21 (2010)
16. Makogon, Y.F.: Natural gas hydrates—a promising source of energy. J. Nat. Gas Sci. Eng. **2**, 49–59 (2010)
17. Makogon, Y.F.: Natural gas hydrates: distribution, education models, resources. Russ. Chem. J. **47**(3), 70–79 (2003)
18. Andreassen, K., Hogstad, K., Berteussen, K.A.: Gas hydrate in the southern Barents Sea indicated by a shallow seismic anomaly. First Break **8**, 235–245 (1990)
19. Andreassen, K., et al.: Massive blow-out craters formed by hydrate-controlled methane expulsion from the Arctic seafloor. Science **356**, 948–953 (2017)
20. Waage, M., Serov, P., Andreassen, K., et al.: Geological controls of giant crater development on the Arctic seafloor. Sci. Rep. **10**, 8450 (2020). https://doi.org/10.1038/s41598-020-65018-9
21. Valyaev, B.M.: Tectonic control of oil and gas accumulation and hydrocarbon degassing of the Earth. In: Gavrilov, Yu.O., Kurenkov, S.A. (eds.) Theoretical and Regional Issues of Geodynamics, Moscow, pp. 222–241 (1999)
22. Valyaev, B.M.: Endogenous factors of structural-tectonic and geodynamic control of oil and gas accumulation processes. In: Proceedings of the XVIII International Scientific Conference on Marine Geology, vol. 2, pp. 27–30 (2009)
23. Smith, S.L., Judge, A.S.: Estimates of methane hydrate volumes in the Beaufort-Mackenzie region, Northwest Territories. In: Geological Survey of Canada, Current Research 1995-B, pp. 81–88 (1995)
24. Matveeva, T.V., Semenova, A.A., Shchur, N.A., Logvina, E.A., Nazarova, O.V.: Prospects of gas hydrate presence in the Chukchi sea. J. Min. Inst. **226**, 387 (2017). https://doi.org/10.25515/pmi.2017.4.387
25. Shakhova, N.E., Semiletov, I.P., Sergienko, V.I.: The contribution of the East Siberian Shelf to the modern methane cycle. Her. Russ. Acad. Sci. **79**(3), 237–246 (2009)
26. Sloan, E.D.: Offshore Hydrate Engineering Handbook. Center for Hydrate Research, Colorado School of Mines, Golden, CO, USA (1998)
27. Yakushev, V.S., Semenov, A.P., Bogoyavlensky, V.I., Medvedev, V.I., Bogoyavlensky, I.V.: Experimental modeling of methane release from intrapermafrost relic gas hydrates when sediment temperature change. Cold Reg. Sci. Technol. **149**, 46–50 (2018)
28. Malakhova, V.V.: Mathematical modeling of the submarine permafrost long-term dynamics of the Arctic shelf. In: Proceedings of the 10th International Scientific Congress "Interexpo GEO-Siberia-2014", vol. 4(1), pp. 136–140 (2014)
29. Malakhova, V.V.: Mathematical modeling of the submarine permafrost long-term dynamics and gas hydrate stability zone in the Siberian Arctic shelf. Bull. Nov. Comp. Center. Series: Numerical Modeling in Atmosphere, Ocean, and Environment Studies **14**, 41–54 (2014)
30. Romanovskii, N.N., Eliseeva, A.A., Gavrilov, A.V., Tipenko, G.S., Hubberten, H.-W.: The long-term dynamics of the permafrost and gas hydrate stability zone on rifts of the East Siberian Arctic shelf (report 2). Earth's Cryosphere **10**(1), 29–38 (2006)
31. Romanovskii, N.N., Hubberten, H.-V., Gavrilov, A.V., Eliseeva, A.A., Tipenko, G.S.: Offshore permafrost and gas hydrate stability zone on the shelf of East Siberian Seas. Geo-Mar. Lett. **25**, 167–182 (2005)

32. Eliseev, A.V., Malakhova, V.V., Arzhanov, M.M., et al.: Changes in the boundaries of the permafrost layer and the methane hydrate stability zone on the Eurasian Arctic shelf, 1950–2100. Dokl. Earth Sci. **465**, 1283–1288 (2015)
33. Golubeva, E.N., Malakhova, V.V., Platov, G.A., Kraineva, M.V., Yakshina, D.F.: Dynamics and tendencies of the Laptev Sea hydrology and cryolitozone state in the 20th-21st centuries. Opt. Atmosfery Okeana **30**(6), 529–535 (2017)
34. Malakhova, V.V., Eliseev, A.V.: Salt diffusion effect on the submarine permafrost state and distribution as well as on the stability zone of methane hydrates on the Laptev Sea shelf. Ice Snow **60**(4), 533–546 (2020)
35. Malakhova, V.V., Golubeva, E.N.: Estimation of the permafrost stability on the East Arctic shelf under the extreme climate warming scenario for the xxi century. Ice Snow **56**(1), 61–72 (2016)
36. Chuvilin, E., Davletshina, D., Ekimova, V., et al.: Role of warming in destabilization of intra permafrost gas hydrates in the Arctic shelf: experimental modeling. Geosciences **9**, 407 (2019). https://doi.org/10.3390/geosciences9100407
37. Chuvilin, E.M., Yakushev, V.S., Perlova, E.V.: Gas and possible gas hydrate in the permafrost of Bovanenkovo gas field, Yamal peninsula West Siberia. Polarforschung **68**, 215–219 (1998)
38. Chuvilin, E., Bukhanov, B., Davletshina, D., Grebenkin, S., Istomin, V.: Dissociation and self-preservation of gas hydrates in permafrost. Geosciences **8**(12), 431 (2018). https://doi.org/10.3390/geosciences8120431
39. Chuvilin, E., Davletshina, D.: Formation and accumulation of pore methane hydrates in permafrost: experimental modeling. Geosciences **8**(12), 467 (2018). https://doi.org/10.3390/geosciences8120467
40. Yakushev, V.S., Chuvilin, E.M.: Natural gas and gas hydrate accumulations within permafrost in Russia. Cold Reg. Sci. Technol. **31**, 189–197 (2000)
41. Malakhova, V.V.: Methane hydrates as a possible source of methane in a glacial-interglacial cycle. Opt. Atmosfery Okeana **24**(1), 84–87 (2011)
42. Sergienko, V.I., Lobkovskiy, L.I., Semiletov, I.P., et al.: The degradation of submarine permafrost and the destruction of hydrates shelf of East Arctic seas as a potential cause of the «methane catastrophe»: some results of integrated studies in 2011. Repo. Acad. Sci. **446**(3), 330–335 (2012)
43. Vorob'yev, A.E.: Prospects of nanotechnologies of developing gaseous-hydrate resources of the Russian Arctic shelf. Vestnik MSTU **19**(1/1), 70–81 (2016)
44. Dmitrievsky, A.N., Valyaev, B.M.: Gas hydrate distribution and resources. Sci. Technol. Gas Ind. **1–2**, 5–13 (2004)
45. Dmitrievsky, A.N., Valyaev, B.M.: Hydrocarbon degassing through the ocean floor: localized manifestations, scale, significance. In: Dmitrievsky, A.N., Valyaev, B.M. (eds.) Degassing of the Earth and the Genesis of Hydrocarbon Fluids and Deposits, pp. 7–36. GEOS, Moscow (2002)
46. Bogoyavlensky, V.I.: Emergency situations in developing oil and gas resources in the Arctic and the ocean. Arct. Ecol. Econ. **4**, 48–59 (2014)
47. Bogoyavlensky, V.I., Bogoyavlensky, I.V.: Accidental and catastrophic gas blowouts during hydrocarbon resources development in the Arctic zones of the USA and Canada. Drill. Oil **12**, Researches (2019)
48. Bogoyavlensky, V.I.: Threat of catastrophic gas blowouts from the Arctic permafrost zone. Cones at Yamal and Taimyr. Drill. Oil **9**, Researches (2014)
49. Obzhirov, A.N.: Gas component increase during seismo-tectonics and the role of gas in earthquake origination (Okhotsk Sea). Tikhookeanskaya Geol. **32**(2), 86–89 (2013)
50. Zakharenko, V.S.: Potential ecological risk related to gas-hydrates deposits within West Arctic continental margin. Environ. Prot. Oil Gas Complex **4**, 21–26 (2011)
51. Petrova, A.A.: Digital maps of the components of the magnetic field induction vector. In: Kuznetsov, V.D. (ed.) Proceedings of IZMIRAN Electromagnetic and Plasma Processes from the Depths of the Sun to the Depths of the Earth, pp. 412–423. IZMIRAN Pub, Moscow (2015)

52. Kopytenko, Yu.A., Petrova, A.A.: Results of the development and application of a component model of the Earth's magnetic field in the interests of magnetic cartography and geophysics. Fundam. Appl. Hydrophys. **9**(2), 88–106 (2016)
53. Kopytenko, Yu.A., Petrova, A.A.: World maps of the Earth's magnetic field components of the epoch 2020.In: Proceedings of the XV All-Russian Conference Applied Technologies of Hydroacoustics and Hydrophysics, pp. 288–291. St. Petersburg (2020)
54. Bonvalot, S., Balmino, G., Briais, A., et al.: Gravity Map of the World. Commission on the Geological Map of the World (eds.). BGI-CGMW-CNES-IRD, Paris (2012)
55. International Seismological Centre. On-line Bulletin. http://www.isc.ac.uk/iscbulletin. Last accessed 15 Mar 2021
56. Petrova, A.A.: Methods of spectral-spatial analysis of the geomagnetic field. Geofisicheskii Sbornic AN UkSSR **76**, 55–66 (1977)
57. Petrova, A.A., Kolesova, V.I.: Method of geophysical exploration. A.S.1289232 USSR (1986)
58. Petrova, A.A., Kolesova, V.I., Domaratsky, S.N.: Method of spatial-spectral analysis in applied geophysics. In: Russian Air Geophysics and Remote Sensing, pp. 525–534. Golden, Colorado (1992)
59. Petrova, A.A., Kopytenko, Yu.A.: Fluid systems of the Mamsko-Bodaibin mineragenic zone in north Transbaikal. Vestnik KRAUNTs Nauki Zemle **41**(1), 37–53 (2019)
60. Petrova, A.A., Latysheva, O.V.: Verification of the model of anomalies of the Arctic magnetic field components. In: Gliko, A.O., Baryakh, A.A., Lobanov, K.V., Bolotov, I.N. (eds.) Proceedings of the All-Russian Conference Global Problems of the Arctic and Antarctic, pp. 279–284. FECIAR UrB RAS, Arkhangelsk (2020)
61. Petrishchev, M.S., Petrova, A.A., Kopytenko, Yu.A., Latysheva, O.V.: Magnetic anomalies of the Precambrian in near-Earth space. In: Proceedings of XVII Conference on Modern Problems of Remote Sensing of the Earth from Space, vol. 395. IKI RAS, Moscow (2019)
62. Petrova, A.A., Kopytenko, Yu.A., Petrishchev, M.S.: Deep fluid systems of the greenstone belts of Fennoscandia. In: Proceedings of the 45th Uspensky International Geophysical Seminar: Practical and Theoretical Aspects of the Geological Interpretation of the Gravitational, Magnetic and Electric Fields of Russia, pp. 239–247, Kazan (2019)
63. Pavlenkova, N.I.: The nature of regional seismic boundaries in the Earth's crust and upper mantle. In: Proceedings of the XIX International Conference Physicochemical and Petrophysical Research in Earth Sciences, V.1, pp. 250–253. IGEM RAS, Moscow-Borok (2018)
64. Petrova, A.A., Latysheva, O.V., Kopytenko, Yu.A.: Natural phenomena of endogenous origin in the Arctic basin. Vestnik KRAUNTs Nauki Zemle **4**(48), 37–53 (2020)
65. Petrova, A.A., Latysheva, O.V., Kopytenko, Yu.A.: Dangerous natural phenomena of endogenous nature in the Arctic zone of the Russian Federation. In: Gliko, A.O., Baryakh, A.A., Lobanov, K.V., Bolotov, I.N. (eds.) Proceedings of the All-Russian Conference Global Problems of the Arctic and Antarctic, pp. 815–820. FECIAR UrB RAS, Arkhangelsk (2020)
66. Petrova, A.A., Petrishchev, M.S., Kopytenko, Yu.A., Latysheva, O.V.: Identification of fluid-carrying channels in the Arctic seas by anomalies of magnetic and gravitational fields In: Gliko, A.O., Baryakh, A.A., Lobanov, K.V., Bolotov, I.N. (eds.) Proceedings of the All-Russian Conference Global Problems of the Arctic and Antarctic, pp. 810–815. FECIAR UrB RAS, Arkhangelsk (2020)
67. Pavlenkova, N.I.: The role of fluids in the formation of inhomogeneity of the Earth's crust and upper mantle. In: Proceedings of the Conference Modern Tectonophysics. Methods and Results, pp.56–68, Moscow (2013)
68. Koulakov, I.Yu., Gaina, C., Dobretsov, N.L., et al.: Plate reconstructions in the Arctic region based on joint analysis of gravity, magnetic, and seismic anomalies. Russ. Geol. Geophys. **54**(8), 859–873 (2013)
69. Kaban'kov, V.Ya., Andreeva, I.A., Krupskaya, V.V., et al.: New data on the composition and origin of bottom sediments of the Southern Mendeleev Ridge of the Arctic Ocean. Dok. Earth Sci. **419A**(3), 403–405 (2008)

70. Lebedeva-Ivanova, N.N., Zamansky, Yu.Ya., Langinen, A.E., Sorokin, M.Yu.: Seismic profiling through the Mendeleev Ridge at 82°N: evidence of the continental crust. Geophys. J. Int. **165**, 527–544 (2006)
71. Gusev, E., Rekant, P., Kaminsky, V., et al.: Morphology of seamounts on the Mendeleev Rise, Arctic Ocean. Polar Res. **36**(1), 1–10 (2017)
72. Simmons, N.A., Forte, A.M., Boschi, L., Grand, S.P.: GyPSuM: a joint tomographic model of mantle density and seismic wave speeds. J. Geophys. Res. **115**, B12310 (2010). https://doi.org/10.17611/DP/9991624
73. Center "Sever" FGBU AANI. http://www.aari.ru. Last accessed 31 Mar 2020
74. Petrova, A.A., Petrishchev, M.S., Kopytenko, Yu.A., Latysheva, O.V.: Identification of fluid-carrying channels in the Arctic seas by anomalies of magnetic and gravitational fields In: Gliko, A.O., Baryakh, A.A., Lobanov, K.V., Bolotov, I.N. (eds.) Proceedings of the All-Russian Conference Global Problems of the Arctic and Antarctic, pp. 810–815. FECIAR UrB RAS, Arkhangelsk (2020)
75. Petrova, A.A., Latysheva, O.V., Petrova, A.I.: Deep factors of ice destruction of the Arctic Ocean. In: Materials of the 13th International School-Conference "Problems of Geocosmos", pp. 67–76. SPb (2021) (in Russian)
76. Kopytenko, Yu.A., Latysheva, O.V., Petrova, A.A.: Influence of fault zones of the Earth's crust on the evolution of the thickness and edge of the Arctic ice cover. In: Proceedings of the Military Space Academy named after A. F. Mozhaisky, vol. 674, pp. 207–212 (2020)
77. Amante, C., Eakins, B.W.: ETOPO1 1 Arc-minute global relief model: procedures, data sources, and analysis. NOAA NESDIS NGDC-24 Technical Memorandum. National Geophysical Data Center, NOAA, pp. 1–19 (2009)
78. Bogoyavlensky, V.I., Sizov, O.S., Bogoyavlensky, I.V., et al.: Earth degassing in the arctic: comprehensive studies of the distribution of frost mounds and thermokarst lakes with gas blowout craters on the Yamal Peninsula. Arct. Ecol. Econ. **4**(36), 52–68 (2019)

Deep Factors of Ice Destruction of the Arctic Ocean

A. A. Petrova, O. V. Latysheva, and A. I. Petrova

Abstract The analysis of satellite observations of annual and seasonal changes in the thickness of perennial Arctic ice in the period from 2007 to 2020 showed that the vertical fluid channels of deep-lying fault zones play an important role in the process of transformation and degradation of the ice cover. They contribute to the local destruction of ice with the formation of melt ponds, affect the variability of the thickness of perennial ice and the formation of the edge of ice of different ages. The results of the complex interpretation of magnetic anomalies, gravity anomalies, and seismic and seismological data allowed us to improve the model of the deep structure of the main tectonic structures of the Arctic Ocean. The directed influence of thermal flows of deep fault zones on the localization of ice cover destruction in the zones of fluid-carrying channels exits to the ocean bottom surface in the water area of the Northern Sea Route is revealed.

Keywords Arctic Ocean · Satellite observations of ice cover · Fluid flow · Northern Sea Route

1 Introduction

In recent decades, one of the significant changes in the Arctic Ocean is the reduction of the ice cover and the predominance of thin annual ice. As a result of these processes, there was a decrease in the volume of ice in the Arctic, which led to an increase in the drift of the ice cover of the Arctic Ocean.

The paper presents the results of a study of the influence of deep-lying faults on the destruction of the perennial ice cover of the Arctic Ocean. A joint analysis of overview

A. A. Petrova · O. V. Latysheva (✉) · A. I. Petrova
Pushkov Institute of Terrestrial Magnetism, Ionosphere and Radio Wave Propagation of the Russian Academy of Sciences, St.-Petersburg Branch (SPbF IZMIRAN), Universitetskaya nab., 5B St. Petersburg 199034, Russia

A. A. Petrova
e-mail: aa_petrova@inbox.ru

A. Kosterov et al. (eds.), *Problems of Geocosmos—2020*, Springer Proceedings in Earth and Environmental Sciences, https://doi.org/10.1007/978-3-030-91467-7_4

maps of the ice cover state with a complex interpretation of magnetic anomalies [1–3], gravity anomalies [4], taking into account seismic [5] and seismological data [6] was carried out. The overview maps were produced by State Research Center AARI and Mercator Ocean International on the basis of satellite data [7, 8]. The performed study shows the thermofluid channels going to the ocean bottom surface make a significant contribution to the process of melting of perennial ice and affect the configuration of the boundaries of old and annual ice. A similar assumption about the correlation between the faults of the Earth's crust and the destruction of the ice cover was supposed at the end of the twentieth century by Kupetsky [9, 10]. The influence of thermofluid channels predicted by our method is confirmed by the calculations and results of the experiment of V. I. Bogoyavlensky in the oil and gas basin of the Yamalo-Nenets Autonomous District [11, 12].

Regular satellite observations of the ice cover state in the Arctic allowed to estimate the dynamics of the boundaries displacement of different age ice, the interannual and seasonal transformation of the foci of perennial ice destruction and the configuration of the annual ice edge [7, 8, 13–15]. This made it possible to study the nature of local destruction of perennial ice, depending on the depth of the faults of the thermofluid channels and lenses of the mantle fluid system.

2 The Deep Factor of the Arctic Ice Cover Destruction

In order to study phenomena of an endogenous origin in the water area of the Arctic Ocean, complex studies of the deep structure of the lithosphere were carried out on the basis of the analysis of geophysical fields. Based on the results of the interpretation of the Earth's magnetic field anomalies and gravity anomalies, taking into account seismic and seismological data, the model of the deep structure of the underwater uplifts of the Arctic Ocean bottom and large fault faults of the Earth's crust is refined. Satellite observations of annual and seasonal changes in ice thickness [7, 8] made possible to estimate the influence of thermofluid channels of fault zones of the Earth's crust on the condition of the Arctic Ocean ice cover.

The inhomogeneities of the Earth's crust and faults are most clearly observed on deep sections. Geodynamic processes of tectonothermal processing are reflected in the magnetic and density properties of the Earth's crust. These properties are studied by the spectral-spatial analysis method that converts the spectral-spatial representation of the anomalous field into a deep section (Fig. 1) [16].

Density sections of the Arctic Ocean tectonic structures are constructed taking into account seismic and seismological data. The identification of the Earth's crust layers for the sections was carried out taking into account the seismological data on earthquakes in the water area of the Arctic Ocean. Comparison of density sections with seismic sections [5] showed that the velocity characteristics are in good agreement with the distribution of density formations with depth and the boundaries of the density layers of the Earth's crust correlate with the velocity ones.

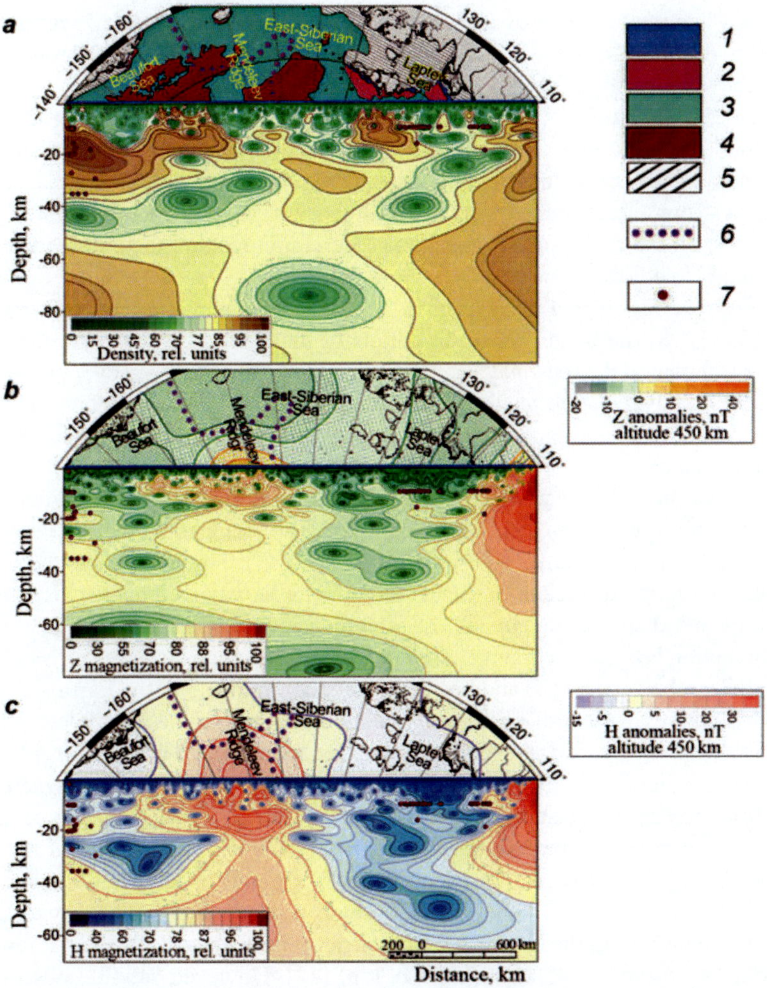

Fig. 1 Sections of the Arctic Ocean lithosphere at the boundary of ice of different ages: **a** density section, top part—ice cover (02.2015) [7]; **b** magnetic section for anomalies of the Z-component, top part—Z-anomalies at an altitude of 450 km [1–3, 19]; **c** magnetic section for anomalies of the H-component, top part—H-anomalies at an altitude of 450 km [20]. 1—nilas; 2—young ice; 3—annual ice (30–200 cm); 4—old ice; 5—fast ice; 6—border of the Northern Sea Route; 7—earthquake foci

Earthquake hypocenters showed at the deep sections emphasize the position of rock contacts of different densities and control fault zones. This information is important to estimate the depth of channels and lenses of fluid-saturated layers (Figs. 1, 2, 3 and 4).

Fluid channels and lenses of the fluid system of the Earth's crust and mantle are clearly distinguished in deep sections due to reduced density and magnetic properties. The migration paths of fluid flows that destroy the ice cover are well observed as weakly magnetic zones of low density on magnetic and density sections. The trajectory of the channels and tracks of the thermofluidic processing of rocks, along which they approach to the surface (Fig. 1) are visible at the sections.

Satellite observations of annual and seasonal changes in ice thickness in the period from 2007 to 2020 emphasized the crucial role of vertical thermofluid channels in the process of ice cover degradation. The analysis of density and magnetic sections showed a strong effect of the directional influence of fluid flows of the Earth's crust and mantle on the melt ponds formation of perennial ice and the boundaries of old and annual ice edges.

To estimate the impact of deep faults on the destruction nature of perennial and annual ice [7, 8] the locations of the outlets of thermofluid channels to the seabed were identified and the seasonal and annual degradation of the ice cover of the Arctic Ocean was analyzed. The results of the study made possible to predict the further localization of the melting zones of perennial ice and the nature of the formation of the annual ice edge.

Deep section of the Beaufort, Laptev and the East Siberian Seas show density and magnetic inhomogeneities of the lithosphere structure, fault zones of deep deposition, thermofluid channels and fluid-saturated layers (Fig. 1). The profile intersects a large structure of the Arctic Ocean Earth's crust—the underwater Mendeleev Ridge. This ridge is clearly distinguished by intense anomalies of the magnetic field [17, 18]. The density section is based on gravity anomalies and the magnetic section is based on the anomalies of the vertical (Z) and horizontal (H) components of the Earth's magnetic field [1–3] are constructed. The density section shows that the Mendeleev Ridge is located above a vertical fault zone that can be traced from a depth of 40–50 km (Fig. 1a). Magnetic sections demonstrated that the fault is filled with magnetoactive formations in the upper and middle parts of the Earth's crust (Fig. 1b, c). Layers magnetized vertically create intense anomalies of the Z-component near the ocean surface up to 700–1000 nT. These anomalies detected in near-Earth space at an altitude of 450 km (Fig. 1b, top) [19, 20]. Layers magnetized horizontally can create magnetic anomalies of the H-component with amplitude up to 250–500 nT, remained visible at an altitude of 450 km (Fig. 1b, top) [20]. However, satellite observations show that the Mendeleev Ridge fault zone is a channel that brings the thermal fluid flow to the perennial ice cover. As a result of the destruction of perennial ice, the boundary of old and annual ice is formed near the Northern Sea Route (Fig. 1a, top).

The boundary of the ice cover of different ages (old perennial and annual ice of the East Siberian Sea) the Beaufort and Laptev Seas are located in the zone of fluid flows ascending through vertical channels from a depth of ~ 20 km (Fig. 1a). The low density weakly magnetic lenses of the fluid system are distinguished at a depth of 30–50 km. The fluid-saturated layer is observed as a regional minimum in the magnetic field at altitudes of 400 and 450 km. It is clearly detected in the module and components of the Earth's magnetic field [19–21]. The regional minimum is controlled by fluid-saturated layer extending from the Gakkel Ridge across the Laptev Sea to the Verkhoyansk upper mantle asthenolens of the North Asian Craton in Eastern Siberia. The asthenolens fluid system feeds a powerful vertical thermofluid channel that goes into the Laptev Sea from a depth of 50 km. In the region of the Verkhoyansk upper mantle lens, it is an ore-controlling channel that forms the Verkhoyansk-Cherskaya gold-bearing province [22, 23]. The thermofluidic processing is well expressed in the magnetic sections of the Northern Sea Route water area, where it is represented by a vertical fault zone in the Laptev Sea and a lateral one in the East Siberian Sea (Fig. 1).

Large thermofluid channels are identified in the Canadian Basin (Fig. 2b, d). Sections 1–1′ and 2–2′ intersect the ice degradation zones of 2019 and 2020 (Fig. 2). The thermal flow can be traced from a depth of ~ 20–40 km. The lenses of the fluid system are identified at a depth of 50–60 km and 70–100 km. It is an accordance with the location of the reduced velocities zone of the global velocity model GyPSuM for the layer in the depth range of 100–175 km [24].

In the East Barents Sea oil and gas province the fault zone near the Northern Sea Route border can be traced from 30 km (Fig. 2d) [25]. It is possible this fault is under the influence of the fluid system observed in the form of low density lenses at a depth of 50–70 km. Satellite observations of seasonal changes in ice thickness confirmed that during the formation of the September ice minimum in 2019 the fluid channel controlled the edge of the annual ice cover [7, 13].

An increase in the navigation period is expected in the Barents Sea near the northern border of the Northern Sea Route due to the directional effect of the mantle

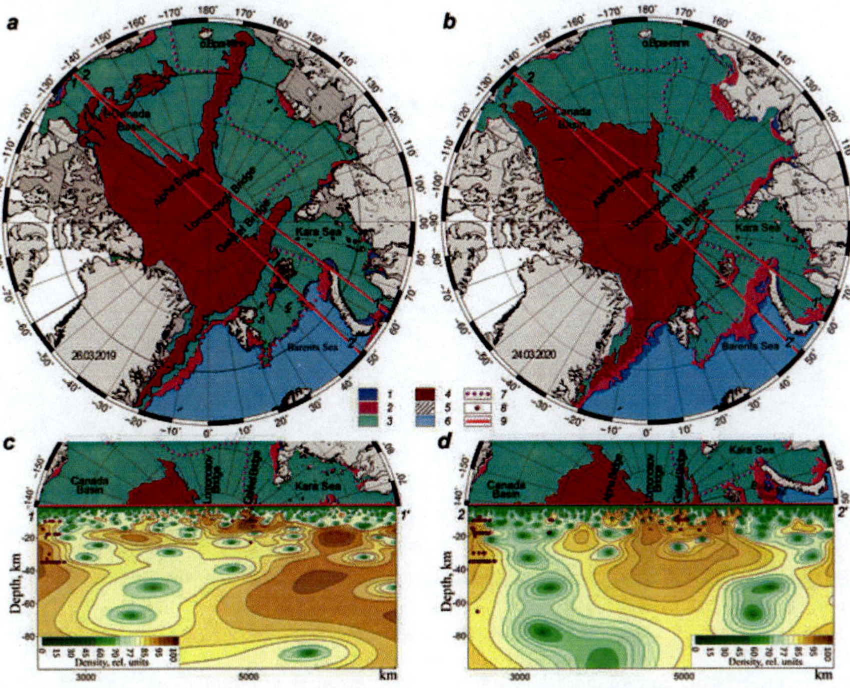

Fig. 2 The boundaries and edges of the winter ice cover (March): **a** 2019; **b** 2020 [7]. Depth density sections: **c** 1–1′; **d** 2–2′. 6—no ice; 9—position of sections. Symbols 1–5, 7–8 see Fig. 1

heat flow along the sub-vertical fault system. This flow increases degradation of the ice cover. However, in this region there are widely distributed deposits of gas hydrates [26]. These deposits can suddenly destroy due to temperature instability provide the phenomenon of natural hazards in the waters of the Northern Sea Route.

An even stronger effect of explosive zones can be expected in the Canadian Basin where an extensive gas hydrate zone is located [26]. It arises from the depth of the fault zones mantle formation and the high power of the vertical fluid channels (Fig. 2d). At the sections deep faults are distinguished as zones of reduced density. The faults output fall on the areas of annual changes in the boundaries of old and annual ice. Sections 1–1′ and 2–2′ show the destruction of the edge of the perennial ice in the Canadian Basin is associated with a powerful weakened zone of breaking disturbances of the Earth's crust and mantle. These disturbances could be traced from a depth of more than 80 km.

A large fluid channel is identified in the Chukchi Sea (Fig. 3b). The density section 1–1′ shows a powerful thermal flow goes out from a depth of 30–40 km and come to the border of the Northern Sea Route along the listric fault between the Chukchi Plateau and the Wrangel Island. In the Chukchi Sea on the south side of the Wrangel Island a vertical fault approaches the seabed that could be traced along the

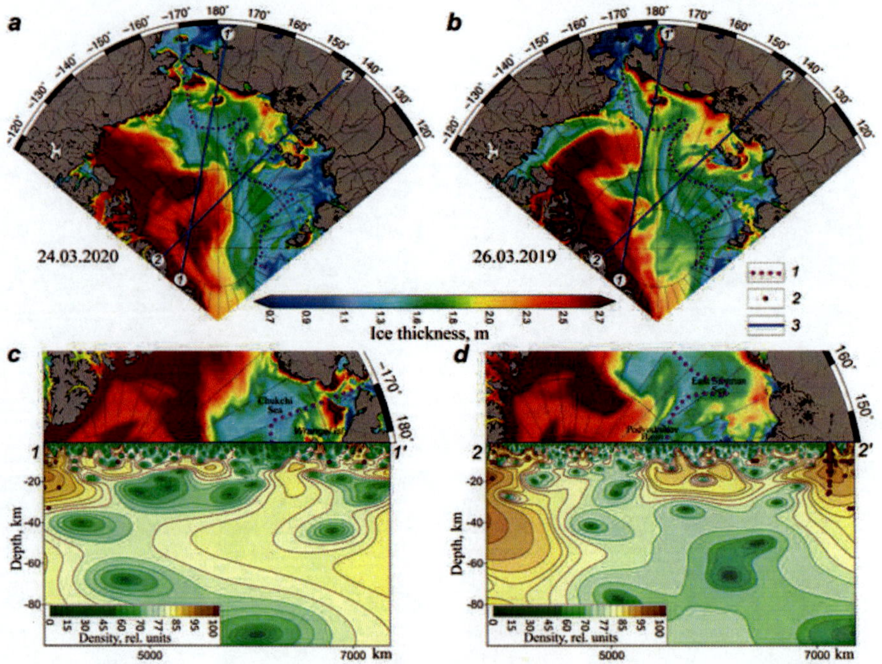

Fig. 3 Channels of thermofluid processing in the eastern part of the Northern Sea Route water area. Winter ice cover: **a** 2020; **b** 2019 [8]. Density sections: **c** 1–1′; **d** 2–2′. 3—The position of the sections. Symbols 1–2 see Fig. 1

density section from 50 km. It affects the thickness of the ice, destroys the edge of the annual ice and makes Northern Sea Route free for shipping. The lenses of fluid-saturated layers and the fluid system of the reduced velocity zone of the GyPSuM model [24] feeding the thermofluid channels are located at depths of 60–100 km.

The increased heat flow of the Chukchi Sea is clearly observed in the magnetic sections obtained from the anomalies of the module and the H-component of the Earth's magnetic field [27, 28]. Exactly this thermofluidic flow destroys the edge of the annual ice in the autumn period and makes the Northern Sea Route in the Chukchi Sea completely freeing for navigation (Fig. 3a, b).

Visual representation of sections based on a complex interpretation of gravity and magnetic anomalies has shown the efficiency of fluid-carrying channels influence on the formation of perennial ice thickness and the boundaries of old and annual ice. The maximum effect of the ice cover destruction is observed under the influence of the heat flow during the upward migration of fluids along vertical faults.

A scheme of the most likely location of dangerous zones has been drawn up to ensure the safety requirements of all-season navigation of the Northern Sea Route. This scheme is made on the basis of the mechanism of occurrence of explosive phenomena of an endogenous nature taking into account the placement of fluid layers allocated at the sections.

In order to identify the location of the lenses of the fluid-saturated layers feeding the thermofluid channels a two-dimensional calculation of the density of the Earth's crust in the water area of the Arctic Ocean was performed (Fig. 4b). The density properties are calculated for two levels of depth – 20 and 40 km. The depths of the levels are determined on the basis of sections taking into account the thickness of the Earth's crust layers.

The performed studies of the influence of thermofluid channels of the Earth's crust fault zones on the process of ice melting made possible to estimate the distribution of local accelerated destruction zones of the ice edge in the water area of the Northern Sea Route.

The profiles that intersect the minima of the horizontal density sections show these minima are associated with fluid channels that form local zones of seasonal ice melting (Figs. 3 and 4). In the density sections that intersect the density minima in the Canadian Basin, the Podvodnikov Basin and the Amundsen Basin, in the Barents Sea fluid channels could be traced from a depth of 50–60 km. Thermal flows affect the thickness of the ice. Producing ice erosion and melting areas these flows form the boundaries of ice coverings of different ages (Fig. 4b–d). In addition, there are extensive areas of gas hydrate deposits in the Canadian Basin, the Podvodnikov and the Amundsen Basins, the Laptev Sea, and the Kara and Barents Seas [26]. Explosive natural phenomena are very likely in these regions.

Along the Northern Sea Route, the areas of the largest exits of deep-lying fluid-carrying faults are determined based on the results of the shelf seas sections and two-dimentional density calculations (Fig. 4b). Potentially dangerous zones of endogenous origin are identified in the East Siberian Sea, the Laptev Sea, and the Kara and Barents Seas based on the density minima at a depth of 20 km and deep magnetic and density sections [5]. The study of the deep structure features of the lithosphere

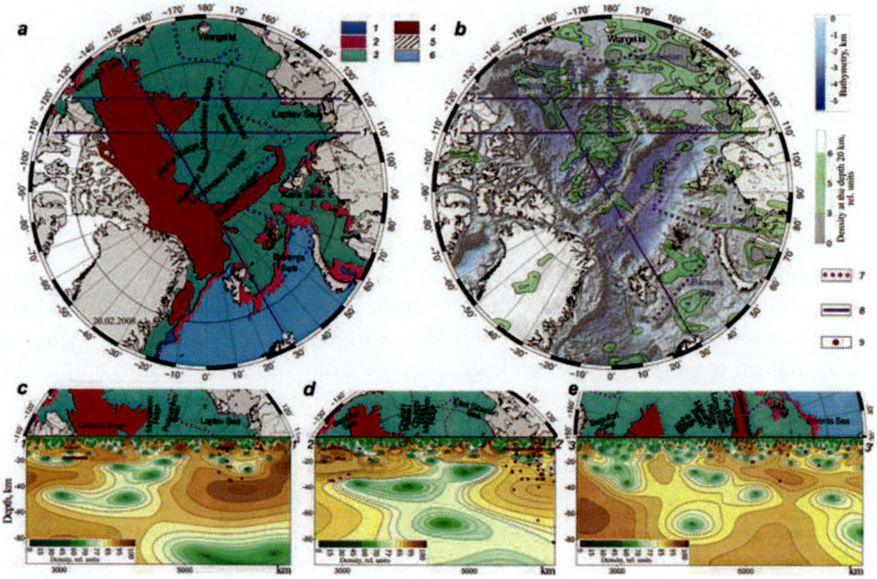

Fig. 4 Formation of the boundaries of the old ice in the zones of the thermal channels output. **a** Ice cover of small development (February 2008) [7]; **b** bathymetry [29] and density at the depth of 20 km; **c**–**e** density sections. 6—No ice, 8—position of the sections. For symbols 1–5, 7, 9, see Fig. 1

clarified the specificity of the localization of seasonal melting of the ice cover in the waters of the Northern Sea Route.

A joint analysis of horizontal sections and density sections of the Earth's crust demonstrated the location of deep-focus systems of fluid-saturated layers in the water area of the Northern Sea Route. The deep zones of faults leaving the mantle are identified from the areas of reduced density in the sections. In the presence of a seismic event these areas can activate the fluid flows of fault zones and make the process of destruction of the ice cover more intensive.

The destruction of perennial ice edge under the influence of fluid channels of fault zones was most distinctly observed on the ice cover of small development in the winter of 2007–2008 (Fig. 4).

Satellite observations of the ice conditions together with the interpretation of gravitational and magnetic data made possible to determine the localization of dangerous zones of endogenous origin caused by the influence of thermofluid flows of the Earth's crust. Our studies of the ice cover for 2007–2021 showed that during this period a significant role in the ice destruction was played by thermofluid channels coming out of the mantle from depths of 50–100 km. In the Arctic basin more than 1800 combinations of magnetic and density sections with the state of the ice cover during the autumn–winter period were analyzed. The coincidence of the areas of ice melting with the fault zones of deep laying has a 95% probability. In 5% of cases, there are no coincidences. This may be due to the situation when the profiles do not intersect the areas of ice degradation transversely. The performed analysis allows to assess the nature of variability in the thickness of perennial ice, the rate of the annual ice edge destruction and the occurrence of melting ponds.

The reduction of perennial ice and the increasing predominance of thin annual ice is currently one of the most significant changes in the Arctic Ocean has led to increased drift of the Arctic Ocean ice cover [13–15, 30].

Fluid-carrying channels permeate the Earth's crust and being the cause of natural hazards creates risks for navigation and operation of oil and gas production platforms. Such hazards can to occur in the waters of the Northern Sea Route as a result of the sudden release of methane from the permafrost.

As part of the Arctic digitalization program the first stage of the project for the construction of the Arctic Connect transarctic underwater communication line was completed in December 2020. It is necessary to take into account the explosive zones of deep origin during laying the cable.

3 Conclusions

1. The model of the deep structure of the Arctic Ocean is improved on the results of the interpretation of magnetic field and gravity field anomalies and taking into account seismic and seismological data.
2. On the basis of a joint analysis of satellite observations of annual and seasonal changes in the thickness of perennial ice, magnetic and density sections it is shown that the thermofluid channels of vertical faults of deep deposition have a significant impact on the formation of the configuration of the perennial ice cover, the boundaries of old and annual ice.
3. The directional influence of fluid-carrying channels contributes to the local destruction of ice with the formation of melt ponds, reduces the thickness of perennial ice and forms the edges of ice of different ages.
4. The complex interpretation of the geophysical data made it possible to identify the outputs of deep-lying faults along the Northern Sea Route.
5. The results obtained made it possible to predict the occurrence of zones of erosion and degradation of perennial ice and dangerous zones of an endogenous nature. This degradation will change operation mode of shipping routes and avoiding dangerous regions will reduce the destruction of marine infrastructure in the waters of the Northern Sea Route.

The main results of the work were presented at a meeting of the Scientific Council for the Study of the Arctic and Antarctic under the Presidium of the Russian Academy of Sciences.

References

1. Petrova, A.A.: Digital maps component of the magnetic field induction vector. In: Kuznetsov, V.D. (ed.) Proceedings of IZMIRAN Electromagnetic and Plasma Processes from the Depths of the Sun to the Depths of the Earth, pp. 412–423. IZMIRAN Pub, Moscow (2015)
2. Kopytenko, Yu.A., Petrova, A.A.: Results of the development and application of a component model of the Earth's magnetic field in the interests of magnetic cartography and geophysics. Fundam. Appl. Hydrophys. **9**(2), 88–106 (2016)
3. Kopytenko, Yu.A., Petrova, A.A.: World maps of the Earth's magnetic field components of the epoch 2020. In: Proceedings of the XV All-Russian Conference Applied Technologies of Hydroacoustics and Hydrophysics, pp. 288–291, St. Petersburg (2020)
4. Bonvalot, S., Balmino, G., Briais, A., et al.: Gravity map of the world. Commission on the Geological Map of the World (eds.). BGI-CGMW-CNES-IRD, Paris (2012)
5. Petrova, A.A., Latysheva, O.V., Kopytenko, Yu.A.: Natural phenomena of endogenous origin in the Arctic Basin. Vestnik KRAUNTs Nauki Zemle **48**(4), 37–53 (2020)
6. International Seismological Centre. On-line Bulletin. http://www.isc.ac.uk/iscbulletin. Last accessed 15 Mar 2021
7. Center "Sever" FGBU AANI. http://www.aari.ru. Last accessed 30 Nov 2020
8. Mercator Ocean International. http://bulletin.mercator-ocean.fr. Last accessed 09 Oct 2020
9. Kupetsky, V.N.: The Ice of the Chukchi Seas, 58 p. MAOBTI, Magadan (1997)
10. Kupetsky, V.N.: We Will Return to the Arctic ..., 299 p. New Polygraphy, Magadan (2005)
11. Bogoyavlensky, V.I., Sizov, O.S., Bogoyavlensky, I.V., Nikonov, R.A., Kargina, T.N.: Earth degassing in the Arctic: comprehensive studies of the distribution of frost mounds and thermokarst lakes with gas blowout craters on the Yamal Peninsula. Arct. Ecol. Econ. **4**(36), 52–68 (2019)
12. Bogoyavlenskiy, V.I., Bogoyavlenskiy, I.V., Nikonov, R.A.: Results of aerial, space and field investigations of large gas blowouts near Bovanenkovo field on Yamal peninsula. Arct. Ecol. Econ. **3**(27), 4–17 (2017)
13. Repina, I.A., Tikhonov, V.V.: Melt pond on the sea ice surface during summer and its connection with Arctic climate change. Russ. Arct. **2**, 15–30 (2018)
14. Shalina, E.V., Bobylev, L.P.: Changes in ice conditions in the Arctic according to satellite observations. Mod. Prob. Remote Sens. Earth Space **14**(6), 28–41 (2017)
15. Yulin, A.V., Vyazigina, N.A., Egorova, E.S.: Interannual and seasonal variability of the ice area in the Arctic Ocean according to satellite observations. Russ. Arct. **7**, 28–40 (2019)
16. Petrova, A.A., Kopytenko, Yu.A.: Fluid systems of the Mamsko-Bodayba mineragenic zone of Northern Transbaikalia. Vestnik KRAUNTs Nauki Zemle **41**(1), 37–53 (2019)
17. Kopytenko, Yu.A., Petrova, A.A.: Components of magnetic anomalies of the Amerasian basin. In: Proceedings of the XV All-Russian Conference Applied technologies of Hydroacoustics and Hydrophysics, pp. 292–295, St. Petersburg (2020)
18. Ouki, G.N., Saltus, R.V.: Geophysical analysis of the Alpha–Mendeleev ridge complex: characteristics of the high-altitude Arctic large magmatic province. Tectonophysics Part A, 65–84 (2016)
19. Thebault, E., Vigneron, P., Langlais, B., Hulot, G.: A Swarm lithospheric magnetic field model to SH degree 80. Earth Planets Space **68**, 126 (2016)
20. Petrova, A.A., Latysheva, O.V.: Verification of the model of anomalies of the Arctic magnetic field components. In: Gliko, A.O., Baryakh, A.A., Lobanov, K.V., Bolotov, I.N. (eds.) Proceedings of the All-Russian Conference Global Problems of the Arctic and Antarctic, pp. 279–284. FECIAR UrB RAS, Arkhangelsk (2020)
21. Kopytenko, Y.A., Petrova, A.A., Guryev, I.S., Labetsky, P.V., Latysheva, O.V.: Analysis of the informativity of the Earth's magnetic field in near-Earth space. Cosm. Res. **59**(3), 143–156 (2021)
22. Starostin, V.I., Isakov, E.D., Razin, L.V., Sakiya, D.R.: Prospects for the discovery of large and unique deposits of precious metals in the North-East of the Siberian platform. Vestnik Mosk. un-ta. Ser. 4. Geologiya **2**, 34–43 (2016)

23. Stogniy, G.A., Stogniy, V.V.: The structure of the south-eastern framing of the North-Asian craton. In: Proceedings of the XXXVIII Tectonic Conference Tectonics of the Crust and Mantle. Tectonic Patterns of Mineral Placement, vol. 2, pp. 238–241. GEOS, Moscow (2005)
24. Simmons, N.A., Forte, A.M., Boschi, L., Grand, S.P.: Gypsum: a joint tomographic model of mantle density and seismic wave velocities. J. Geophys. Res. **115**, B12310 (2010). https://doi.org/10.17611/DP/9991624
25. Pavlenko, V.I.: Arctic zone of the Russian Federation in the system of ensuring national interests of the country. Arct. Ecol. Econ. **4**(12), 16–25 (2013)
26. Shakhova, N., Semiletov, I., Chuvilin, E.: Understanding the permafrost–hydrate system and associated methane releases in the East Siberian arctic shelf. Geosciences **9**, 251 (2019)
27. Kopytenko, Yu.A., Latysheva, O.V., Petrova, A.A.: Influence of fault zones of the Earth's crust on the evolution of the thickness and edge of the Arctic ice cover. In: Proceedings of the Military Space Academy named after A. F. Mozhaisky, vol. 674, pp. 207–212 (2020)
28. NOAA/ESRL Physical Sciences Laboratory. http://psl.noaa.gov. Last accessed 15 Nov 2020
29. Amante, C., Eakins, B.W.: ETOPO1 1 Arc-minute global relief model: procedures, data sources, and analysis. NOAA NESDIS NGDC-24 Technical Memorandum. National Geophysical Data Center, NOAA, pp. 1–19 (2009). https://doi.org/10.7289/V5C8276M
30. Laxon, S.W., et al.: CryoSat-2 estimates of the thickness and volume of Arctic sea ice. Geophys. Res. Lett. **40**(4), 732–737 (2013)

Verification of the Arctic Magnetic Field Component Model Based on Observations on the CHAMP and Swarm Satellites

A. A. Petrova, O. V. Latysheva, and A. I. Petrova

Abstract The article presents the results of comparing anomalies of the Arctic's magnetic field in near-earth space with empirical data measured by the CHAMP and Swarm spacecraft at altitudes of 400 and 450 km. Magnetic anomalies were calculated using a three-dimensional component model of the Earth's magnetic field of the SPbF IZMIRAN. To identify the features of the deep structure of the lithosphere of Arctic's magnetoactive zones observed in near-Earth space, sections were constructed based on the surface values of magnetic anomalies. The paper studies the density properties of the magnetic zones of the lithosphere, determines the location, magnetization, thickness and density characteristics of the layers of the lithosphere that create magnetoactive zones in outer space. Density sections are constructed based on gravity anomalies, taking into account seismological data. The study of the information content of magnetic anomalies in near-Earth space in the altitude range from 300 to 600 km is of great scientific and practical importance for solving geological and geophysical problems and questions of spacecraft navigation.

Keywords Component model of the Arctic's magnetic field · Magnetic anomalies · Lithosphere of magnetoactive zones

1 Introduction

In order to verify the three-dimensional component model of the Earth's geomagnetic field [1–3], the anomalies of the Arctic's model were compared with the observations of the CHAMP and Swarm satellites. The models of the vertical (Z) and horizontal (H) components, magnetic declination and inclination of the Earth's magnetic field are constructed on the basis of the measured and calculated values of the magnetic

A. A. Petrova · O. V. Latysheva (✉) · A. I. Petrova
Pushkov Institute of Terrestrial Magnetism, Ionosphere and Radio Wave Propagation of the Russian Academy of Sciences, St. Petersburg Branch (SPbF IZMIRAN), Universitetskaya Nab., 5B, St. Petersburg 199034, Russia

A. A. Petrova
e-mail: aa_petrova@inbox.ru

A. Kosterov et al. (eds.), *Problems of Geocosmos—2020*, Springer Proceedings in Earth and Environmental Sciences, https://doi.org/10.1007/978-3-030-91467-7_5

field elements according to the data of airborne and hydro-magnetic surveys carried out at the ground level. Estimates of the error of the Earth's field component model near the ground surface are obtained by comparing the model values with vector data from airborne and hydro-magnetic surveys and geomagnetic observatories of the World Network [2].

In this paper, to estimate the error of the three-dimensional component model Earth's magnetic field in the Arctic comparisons are made with the data of spacecraft measurements in near-Earth space.

To refine the model of the deep structure of the Arctic lithosphere, a complex interpretation of the anomalies of the modulus and components of the geomagnetic field [3], gravity anomalies [4], and seismological data [5] was carried out. The features of the lithosphere of the magnetically active zones of the Arctic Ocean and the ancient blocks of the Earth's crust of Eurasia and North America are studied.

2 Verification of the Component Model Based on the Data of the CHAMP and Swarm Spacecraft Measurements

The SPbF IZMIRAN component model of the Earth's magnetic field [1–3] was tested on the basis of magnetic anomalies measurements by the CHAMP and Swarm spacecraft at flight altitudes of 400 and 450 km [6]. To verify the Arctic model, the anomalies of the modulus and the vertical component at an altitude of 400 km were calculated from the measurements on the CHAMP spacecraft. For comparison with the measurements of the Swarm spacecraft, the vertical component of the Earth's magnetic field was calculated at an altitude of 450 km. The paper presents the anomalies of the modulus and the components of the Earth's magnetic field in the Arctic for near-Earth space according to the IZMIRAN SPbF model in a stereographic equiangular projection north of 50°N (Fig. 1, *1* to *3*).

The results of the Swarm spacecraft measurements of the Z-component anomalies showed that intense magnetic anomalies are observed in the near-Earth space over the Alpha and Mendeleev ridges in the Amerasian Basin of the Arctic Ocean and over the ancient platforms of the Arctic zone of Eurasia and North America (Fig. 1, *1–3*).

Near the Arctic Ocean surface, the anomalies of the Earth's magnetic field modulus reach 1000–1500 nT, the anomalies of the vertical Z-component on the Alpha and Mendeleev ridges –800 nT, and the horizontal H-component –300–600 nT. In the near-Earth space at an altitude of 450 km, intense magnetic anomalies of the modulus and component over the Alpha and Mendeleev ridges are clearly distinguished. At the altitude of the Swarm spacecraft, the anomalies of the Z-component of the magnetoactive zone reach ± 15 nT, according to the Earth's magnetic field model, the anomalies of the Z-component are 10–14 nT, and the modulus and H-component are 10–15 nT.

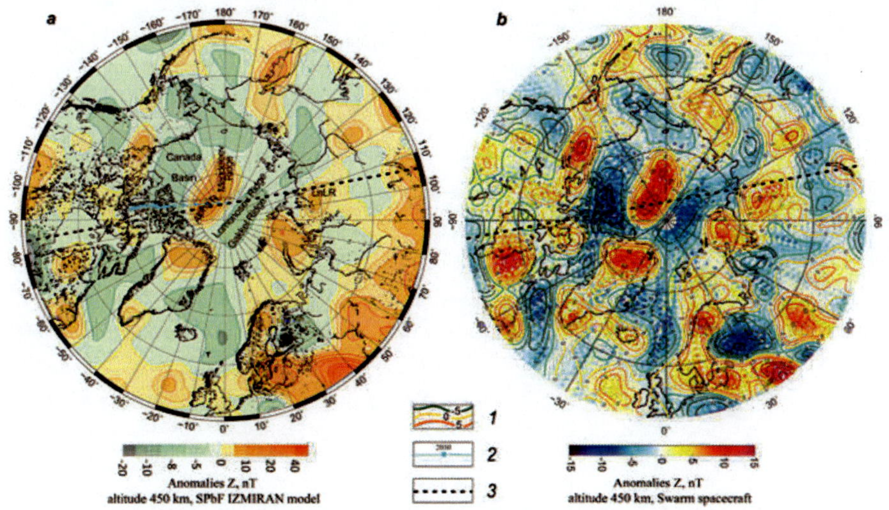

Fig. 1 Anomalies of the Z-component of the Earth's magnetic field in the Arctic at an altitude of 450 km: **a** the IZMIRAN SPbF model [3], **b** measurements of the Swarm spacecraft [6]. *1* isolines of the Z anomalies, nT according to measurements of the CHAMP spacecraft [6], *2* the magnetic pole (MP) (1900–2020); *3* the position of the deep section

Comparisons of the values observed by the CHAMP and Swarm spacecraft with the component model of the Arctic at an altitude of 400–450 km revealed a good agreement between the model and the measurements of the spacecraft: when the intensity of the modulus magnetic anomalies and the anomaly of the Z-component is up to ± 15 nT, the difference of the anomalies in the Arctic Ocean are 1–2 nT (Fig. 2).

The analysis of the information content of the anomalies of the modulus and the components of the Earth's magnetic field in the altitude range from 300 to 600 km showed that the intensity of the anomalies in the Arctic Ocean in near-Earth space exceeds the level of the instrumental error of the Earth's magnetic field measurements by more than 5–10 times. This makes it possible to use the Earth's magnetic field anomalies to solve geological, geophysical and navigational problems—marine, air and space magnetic navigation [2, 3, 6–10].

Component maps clearly reflect information about the nature of the magnetization of rocks that produce magnetic anomalies in the earth's crust on continents and in the oceans. This information is widely used for scientific and practical purposes.

The anomalies of the Earth's magnetic field components are of particular interest for the study of hard-to-reach regions of the Arctic. The optimal set of the Earth's magnetic field components is determined in accordance to the specifics of the structure of anomalies associated with mineral resources.

Regional search criteria for long-lived hydrocarbon deposits, geothermal zones, ferruginous quartzites, and thermal zones of the earth's crust have been developed

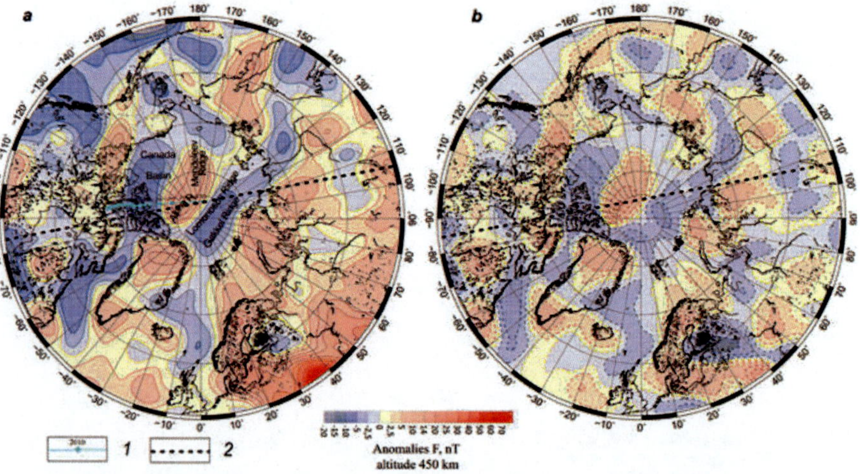

Fig. 2 Modulus anomalies of the Earth's magnetic field in the Arctic at an altitude of 400 km: **a** the SPbF IZMIRAN model [3], **b** the isolines of the modulus anomalies, in nT according to the measurements of the CHAMP spacecraft [6]. For symbols *1* and *2*, see Fig. 1

on the basis of the component model of the Earth's magnetic field. Thermal zones are promising for the search for indigenous gold and diamond deposits [7, 8].

The three-dimensional component model allows to make forecasts of anomalies of Earth's magnetic field components for different altitudes from the ocean level to the ionosphere layers, which is of great applied importance for solving problems of sea and air navigation in difficult conditions of the Arctic Ocean (Fig. 3).

3 Deep Structure of Magnetoactive Zones of the Arctic

In order to study the features of the deep structure of the Arctic and to identify the nature of the magnetoactive zones whose anomalies are observed in near-Earth space, density and magnetic sections were constructed based on the surface magnetic anomalies of the component model and gravity anomalies (Figs. 1, 2, 3 4). The map of magnetic anomalies of the Z-component of the Earth's magnetic field in the Arctic was calculated using the component model for the Swarm spacecraft flight altitude (Fig. 4b).

The study of the deep structure of the Arctic lithosphere is based on the interpretation of the modulus anomalies, the vertical and horizontal components of the Earth's magnetic field [3], taking into account gravity anomalies and seismological data. Visualization of the inhomogeneities of the Earth's crust is performed in the form of geophysical sections using the method of spectral–spatial analysis of anomalous fields converted into deep sections [13].

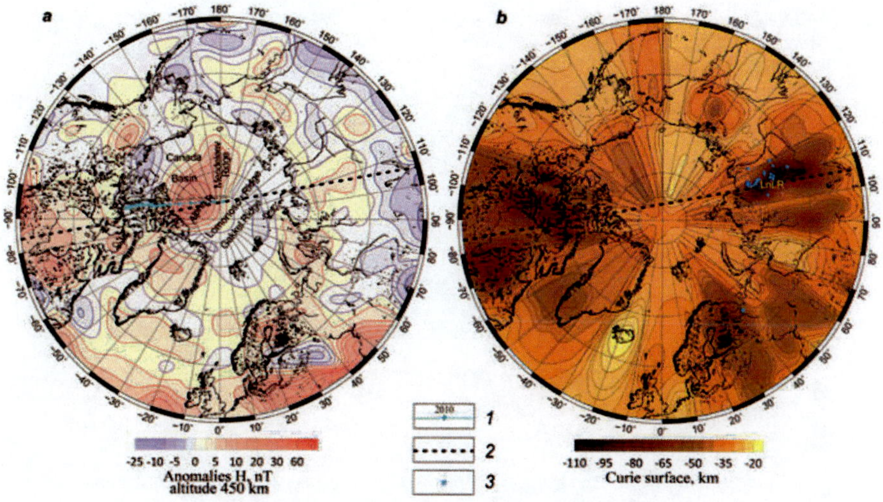

Fig. 3 The Arctic. **a** Anomalies of the H-component of the Earth's magnetic field according to the IZMIRAN SPbF model [3] (altitude 450 km; **b** Curie surface [11]. *3* kimberlites [12]. For symbols *1*, *2*, see Fig. 1

The physical properties of the Earth's crust were evaluated by comparing density sections with seismic sections [14]. Studies have confirmed that the velocity characteristics are in good agreement with the distribution of density inhomogeneities in the sections.

The lengthy sections through the Arctic Ocean make it possible to conduct a comparative analysis of the anomalous fields of the underwater uplifts of the Amerasian Basin and the ancient blocks of North America and Eurasia [15].

The deep sections (Fig. 4) intersect the magnetoactive zones of the North American Platform in Arctic Canada, the Alpha Ridge in the Arctic Ocean, and the Lena Lithospheric Root in Eastern Siberia. Near the Earth's surface, the magnetic anomalies of the modulus and the Z-component in the Arctic Canada have values of ~1000 nT, the anomalies of the Alpha Ridge ~500–1500 nT, the Lena Root -2000–4000 nT [3, 16, 17], at an altitude of 450 km, respectively, 2–5, 8–15 and 5–10 nT (Fig. 1, *1* to *3*) [7, 9]. At the same regions anomalies of the H-component near the surface are 500–700 nT for first and second regions and 1000–3500 nT for third one, at an altitude of 450 km, 7–13 nT [3, 7].

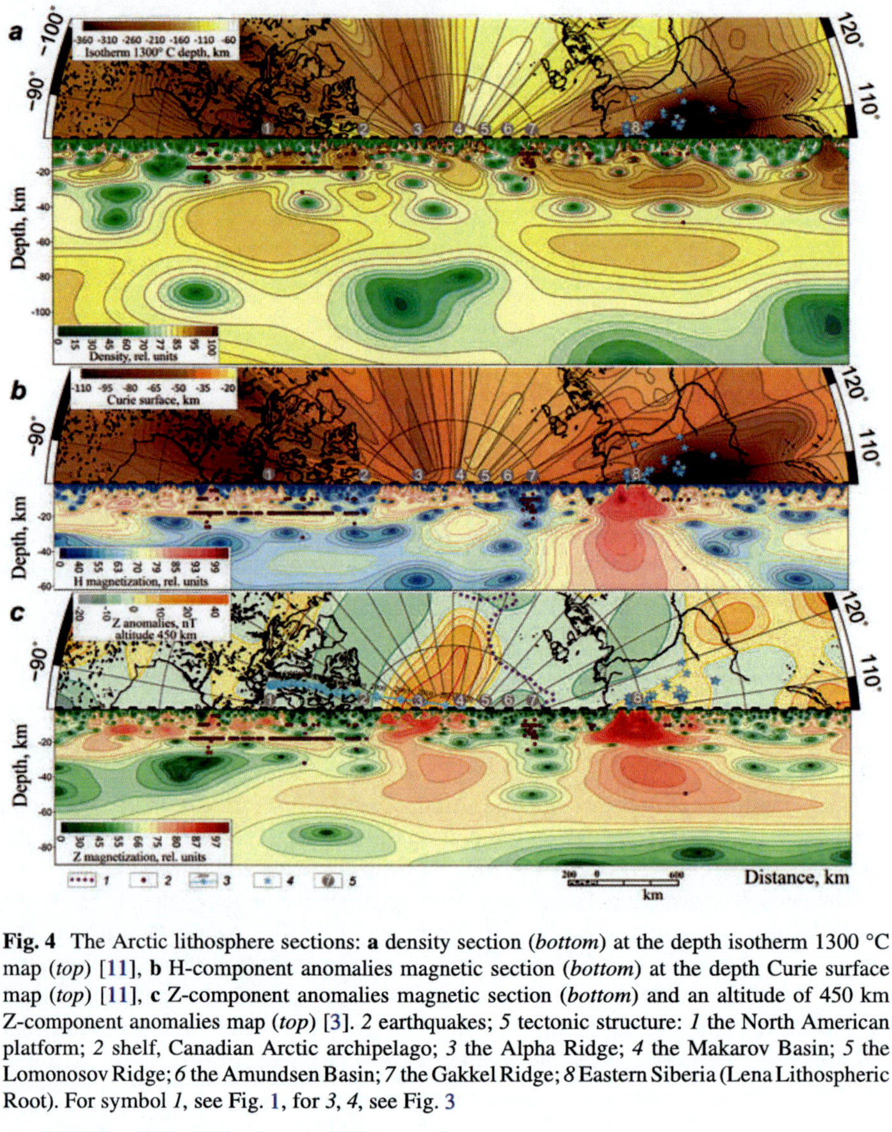

Fig. 4 The Arctic lithosphere sections: **a** density section (*bottom*) at the depth isotherm 1300 °C map (*top*) [11], **b** H-component anomalies magnetic section (*bottom*) at the depth Curie surface map (*top*) [11], **c** Z-component anomalies magnetic section (*bottom*) and an altitude of 450 km Z-component anomalies map (*top*) [3]. *2* earthquakes; *5* tectonic structure: *1* the North American platform; *2* shelf, Canadian Arctic archipelago; *3* the Alpha Ridge; *4* the Makarov Basin; *5* the Lomonosov Ridge; *6* the Amundsen Basin; *7* the Gakkel Ridge; *8* Eastern Siberia (Lena Lithospheric Root). For symbol *1*, see Fig. 1, for *3*, *4*, see Fig. 3

4 The Deep Structure of the Central Magnetic Zone of the Arctic Ocean

The central magnetic zone of the Arctic Ocean covers the Alpha and Mendeleev Ridges, the Chukchi Plateau and Northwind Ridge, the Canada Basin, Makarov and Podvodnikov Basins.

Magnetoactive and dense horizons of the oldest (AR2-PR1) North American platform formed by the magnetite zone are developed at depths from 5–10 to 20 km (Fig. 4).

The main sources of anomalies of the Alpha Ridge magnetoactive zone are represented by two horizons: the upper magnetic and dense horizon at a depth of 6–12 km and the layer at a depth of 15–21 km. This is a vertically magnetized horizon related to a fault zone extending from a depth of 40–50 km (Fig. 4a). Magnetic sections show that the fault zone is filled with magnetic formations that are magnetized vertically and horizontally (Fig. 4b, c).

The Curie surface map shows that the physical conditions of the Alpha Ridge lithosphere allow for the existence of magnetized rocks up to a depth of 40 km (Fig. 4b, top) [11, 14]. Magnetite zones create intense magnetic anomalies of the modulus and components near the Earth's surface and are traced in near-Earth space (Figs. 1, 2, 3, 4). The intensity of the anomalies depends on the concentration of magnetite in rocks located above the Curie surface [18, 19]. The anomalies of the modulus and components reflect the nature of the inductive and thermoremanent magnetization of the ancient crust horizons enriched with iron minerals.

In the Arctic, the main values of the modulus and the Z-component near the Magnetic Pole (MP) are ~ 57,000 nT. The increased values of the modulus and Z-component anomalies are due to the total inductive and thermoremanent magnetization of the rocks.

Intense magnetic anomalies are typical for the magnetite zones of the Precambrian crust. Magnetite zones arise near the base of the upper crust at the boundary of the granulite–basitic and granite layers as a result of tectonothermal processes of regional metamorphism with the release of magnetite [18]. The analysis of the deep sections of the Earth's magnetic field components allowed us to suggest the possible nature of the regional anomalies sources of the Arctic Ocean Central Magnetic Zone [19, 20].

Intense anomalies of the modulus and the Z-component in near-Earth space were detected not only in the area of the Alpha Ridge (Fig. 1, *1* to *3*). In the water area of the Arctic Ocean, they are typical for the entire Central Magnetic Zone, including the Alpha and Mendeleev Ridges, the Chukchi Plateau and Northwind Ridge, as well as the Canada Basin, Makarov and Podvodnikov Basins. In addition, the Central magnetic zone creates significant anomalies of the H-component at an altitude of 450 km (Fig. 3a) [2, 3, 19]. However, in the Central magnetic zone there is an MP, which creates a minimum of the main values of the H-component of the Earth's magnetic field along the line of displacement (Figs. 1, 2, 3 4) [21]. The main value of the H-component at the MP is zero, and in the ±300 km band it is less than 500–1000 nT. Therefore, the anomalies of the H-component in the area of the Alpha Ridge, reaching values of ~500–700 nT, are probably created by thermoremanent magnetization. This magnetization is occurred in an earlier period, when the magmatic formations of the underwater ridges of the Amerasian Basin were cooling at a large value of the main field of the H-component. It is possible that the magnetoactive block of the Earth's crust of the Central Magnetic Zone is formed near the equator, where the main values of the H-component of the Earth's magnetic field have values of ~40,000 nT [2, 3, 14, 19, 22]. The analysis of magnetic sections showed that the main sources of anomalies of the H-component of the Alpha Ridge are the marker horizon at a depth of 6–15 km and the magnetic layer at 10–20 km.

Based on the system of deep density sections taking into account the surface of the 1300 °C isotherm, seismological data of earthquake hypocenters, the global model of mantle densities and velocity inversion zones, an idea of the distribution of density inhomogeneities in the Earth's crust and the Arctic mantle was suggested [5, 11, 23]. Earthquake foci control the contact surfaces of layers of different densities and magnetizations. The complex interpretation allowed us to obtain information on the relative positions and ratios of the deep layers of the lithosphere, as well as on the magnetization and density of the regional anomalies sources. Calculations have shown that the main source of Arctic Ocean regional anomalies, which can be traced in near-Earth space at altitudes of 400–450 km, is the magnetized layer of the Central Magnetic Zone at a depth of 15–20 km.

Intense anomalies are typical for the Precambrian crust [18]. Precambrian magma is significantly saturated with iron minerals, so the magnetite zone is a source of intense anomalies caused by inductive, thermoremanent and viscous magnetization. It is possible that the Central magnetic zone of the Arctic Ocean is originated in the Precambrian. The magnetoactive blocks of the basement of this zone are highly enriched in iron minerals. Visualization of the sections allowed us to determine the power and location of the regional magnetic anomalies sources observed by satellites at near-Earth altitudes.

5 The Deep Structure of the Lena Lithospheric Root

Analysis of deep magnetic and density sections, the surface of the 1300 °C isotherm and Curie, earthquake hypocenters and the global model of mantle densities revealed that the roots of the lithosphere are powerful sub-vertical blocks of increased density and magnetization [5, 11, 23]. The thickness of the root density block can be more than 100 km (Fig. 4) [8].

One of the largest known roots of the lithosphere—the Lena Lithospheric Root (LnLR)—is located in Eastern Siberia [8, 24, 25]. Calculations have shown that the magnetized LnLR block, traced to a depth of more than 50 km, creates anomalies of the modulus, vertical and horizontal components of the Earth's magnetic field.

The physical conditions at the bottom of the lithosphere root in Eastern Siberia are favorable for the formation of thermoremanent and inductive magnetization [11]. However, the anomalous field of the H-component of the Earth's magnetic field is most likely caused mainly by the residual magnetization of the rocks, since the LnLR is located on the line of advance of the MP moving in the direction of the World Magnetic Siberian anomaly, in the zone of the minimum of the main field of the H-component of the Earth's magnetic field induction vector (Figs. 3, 4) [8, 20]. Therefore, large values of anomalies of the H-component with amplitudes of ~1000 nT most likely reflect the thermoremanent magnetization of rocks that occurred earlier. It is possible that the thermoremanent magnetization could have been preserved since ancient times, when there were no MP in the Arctic Ocean, and the rocks of the LnLR were formed and magnetized near the equatorial zone, where the main field of the H-component has large values [8, 10, 22].

A joint analysis of magnetic sections with maps of the surface Curie and heat flux [11] showed that the position of magnetized base roots of the Eastern Siberia lithosphere is consistent with data on the temperature of the bottom of the crust and upper mantle (Fig. 4).

At the section LnLR (Fig. 4) his depth is about 100 km and corresponds to the depth of the isotherm 1300 °C [11]. In accordance to the global model of mantle densities and velocity inversion zones a depth of LnLR of more than 250 km is possible [23].

The Yakut diamondiferous province is located within the borders of the LnLR (Figs. 3, 4) [12, 25]. The joint analysis of the magnetic and density sections through the diamondiferous province deposits, taking into account the Curie isotherm and the data of the global mantle density model, confirmed that the indigenous diamond deposits tend to the root of the lithosphere (Fig. 4).

As a result of the complex interpretation of surface anomalies of the Earth's magnetic field components, gravity anomalies and seismological data, an idea of the structure of the lithosphere of the kimberlite provinces is obtained. The analysis showed that the indigenous diamond deposits tend to the root of the lithosphere. The performed research allowed us to estimate the magnetization, thickness and density properties of the lithosphere root.

6 Visualization of Thermofluid Channels and Lenses of the Fluid System

The geodynamic regimes of the formation of the Earth's crust and the tectonothermal processes are reflected in the density and magnetic sections. The density section emphasizes the horizontal stratification and fragmentation of the Earth's crust. The thermofluid channels on the shelf of the Canadian Arctic Archipelago in the Amundsen Basin are visualized on the presented section (Fig. 4a). Thermofluid channels are connected to fluid-saturated lenses located at depths of 60 and 80 km. The fluidic channels are most clearly observed in magnetic sections (Fig. 4b, c).

Density deep sections of the lithosphere make it possible to identify the location of the lenses of the fluid systems deep-focus layers of the Arctic zone. In the presence of a seismic event, pulsating vibrations occur in the lenses of fluid systems, causing vertical migration of fluid through thermofluid channels. This can create explosive situations in the areas of the thermal flow outlet to the surface in the areas of gas hydrate deposits [12].

7 Conclusions

1. A three-dimensional component model of the Earth's magnetic field in the Arctic was verified using the Swarm and CHAMP spacecraft measurements.
2. On the basis of a complex interpretation of surface anomalies of the Earth's magnetic field components, gravity anomalies and seismological data, the idea of the deep structure of the magnetoactive zones of the Arctic and the lithosphere of the kimberlite provinces is refined.
3. The research is of scientific, practical and applied importance for solving geological and geophysical problems and issues of marine and air magnetic navigation.

References

1. Petrova, A.A.: Digital maps component of the magnetic field induction vector. In: Kuznetsov, V.D. (ed.) Proceedings of IZMIRAN Electromagnetic and Plasma Processes from the Depths of the Sun to the Depths of the Earth, pp. 412–423. IZMIRAN Pub, Moscow (2015).
2. Kopytenko, Yu.A., Petrova, A.A.: Results of development and application of the Earth's magnetic field component model in the interests of magnetic cartography and geophysics. Fundam. Appl. Hydrophys. **9**(2), 88–106 (2016)
3. Kopytenko, Yu.A., Petrova, A.A.: World maps of the Earth's magnetic field components of the epoch 2020. In: Proceedings of the XV All-Russian Conference Applied Technologies of Hydroacoustics and Hydrophysics, pp. 288–291. St. Petersburg (2020)
4. Bonvalot, S., Balmino, G., Briais, A. et al.: World Gravity Map, 1:50000000 map, Eds.: BGI-CGMW-CNES-IRD, Paris (2012)
5. International Seismological Centre. On-line Bulletin, http://www.isc.ac.uk/iscbulletin. Accessed 15 Mar 2021
6. Thebault, E., Vigneron, P., Langlais, B., Hulot, G.: A Swarm lithospheric magnetic field model to SH degree 80. Earth, Planets Space **68**, 126 (2016)
7. Petrova, A.A., Latysheva, O.V.: Verification of the model of anomalies of the Arctic magnetic field components. In: Gliko, A.O., Baryakh, A.A., Lobanov, K.V., Bolotov, I.N. (eds.) Proceedings of the All-Russian Conference Global Problems of the Arctic and Antarctic, pp. 279–284. FECIAR UrB RAS, Arkhangelsk (2020)
8. Petrova, A.A., Latysheva, O.V., Kopytenko, Yu.A.: Specificity of the lithosphere of the permafrost zone of Eastern Siberia. Vestnik KRAUNTs Nauki o Zemle **49**(1), 36–52 (2021)
9. Kopytenko, Yu.A., et al.: The use of high-rise models of Earth's magnetic field for the solution of geophysical problems. Cosm. Res. **57**(3), 185–191 (2019)
10. Kopytenko, Y.A., Petrova, A.A., Guryev, I.S., Labetsky, P.V., Latysheva, O.V.: Analysis of the information content of Earth's magnetic field in space. Cosm. Res. **59**(3), 177–190 (2021)
11. Artemieva, I.M.: Global $1° \times 1°$ thermal model TC1 for the continental lithosphere: implications for lithosphere secular evolution. Tectonophysics **416**, 245–277 (2006)
12. Volkov, A.V., Galyamov, A.L., Lobanov, K.V.: The mineral wealth of the circum-arctic belt. Arctic: Ecol. Econ. **1**(33), 106–117 (2019).
13. Petrova, A.A., Kopytenko, Yu.A.: Fluid systems of the Mamsko-Bodayba mineragenic zone of Northern Transbaikalia. Vestnik KRAUNTs Nauki o Zemle **1**(41), 37–53 (2019)
14. Petrova, A.A., Latysheva, O.V., Kopytenko, Yu A.: Natural phenomena of endogenous origin in the Arctic Basin. Vestnik KRAUNTs Nauki o Zemle 4 (48), 37–53 (2020).

15. Verba, V.V., Petrova, A.A.: Comparative characteristics of the anomalous magnetic field of the Amerasian Basin and ancient shields of Eurasia and North America. In: Proceedings of the Structure and History of Development of the Arctic Ocean, pp. 80–86. Sevmorgeologiya, Leningrad (1986)
16. Litvinova, T., Petrova, A.: Features of the structure of the lithosphere of the Arctic Ocean near the Gakkel Ridge, the Alpha and Lomonosov. In: Eriksen, S., Haflidason, H., Olesen, O., Schiellerup, H., Husas, A.M. (eds.) Proceedings of the Geological Society of Norway 2014, vol. 2, pp. 31–34. Tromsø (2014)
17. Oakey, G.N., Saltus, R.W.: Geophysical analysis of the Alpha-Mendeleev ridge complex: characterization of the high Arctic large igneous province. Tectonophysics **691A**, 65–84 (2016)
18. Nalivkina, E.B., Petrova, A.A.: Magnetite zone of the Earth's crust of continents. VSEGEI, SPb (2018)
19. Kopytenko, Yu.A., Petrova, A.A.: Components of magnetic anomalies of the Amerasian Basin. In: Proceedings of the XV All-Russian Conference Applied Technologies of Hydroacoustics and Hydrophysics, pp. 292–295. Politech-Press, St. Petersburg (2020)
20. Petrishchev, M.S., Petrova, A.A., Kopytenko Yu.A., Latysheva, O.V.: Precambrian magnetic anomalies in near-Earth space. In: Proceedings of XVII Conference on Modern Problems of Remote Sensing of the Earth from Space, 395. IKI RAS, Moscow (2019)
21. Kopytenko, Yu.A., Chernouss S., Petrova, A.A. et al.: The Study of Auroral Oval Position Changes in Terms of Moving of the Earth Magnetic Pole. In: Yanovskaya, T.B., Kosterov, A., Bobrov, N., Divin, A.V., Saraev, A.K., Zolotova, N.V. (eds.) Proceedings of XII International Conference and School Problems of Geocosmos, pp. 289–297. Springer, Heidelberg (2020)
22. Iosifidi, A.G., Khramov, A.N.: Paleomagnetism of paleozoic sediments from the Kozhim River section: on the problem of palinspastic reconstructions of the Subpolar Urals and Pai-Khoi. Izvestiya, Phys. Solid Earth **49**, 63–76 (2013)
23. Simmons, N.A., Forte, A.M., Boschi, L., Grand, S.P.: GyPSuM: A joint tomographic model of mantle density and seismic wave speeds. J. Geophys. Res. **115**, B12310 (2010)
24. Stogniy, G.A., Stogniy, V.V.: The structure of the south-eastern framing of the North-Asian craton. In: Proceedings of the XXXVIII Tectonic Conference Tectonics of the Crust and Mantle. Tectonic Patterns of Mineral Placement, vol. 2, pp. 238–241. GEOS, Moscow (2005)
25. Starostin, V.I., Izbekov, E.D., Razin, L.V., Sakiya, D.R.: Prospects for the discovery of large and unique deposits of precious metals in the North-East of the Siberian platform. Vestnic Mosk. un-ta. Ser. 4. Geologiya 2, pp. 34–43 (2016)

Interpretation of Component Geomagnetic Field Measurements Carried Out on Board a Ferromagnetic Vessel from the Round-the-World Expedition of the R/V "Admiral Vladimirsky" in 2019–2020

S. A. Ivanov, S. A. Merkuryev, I. M. Demina, V. A. Soldatov, and D. B. Zaitsev

Abstract During the anniversary round-the-world expedition of the R/V "Admiral Vladimirsky", the magnetic field components were measured using two 3-component magnetometers with torsion and magnetoresistive sensors developed at SPbF IZMIRAN. In total, about 60,000 linear kilometers were covered with the measurements. Testing of the obtained results was carried out by comparison with the data obtained on the non-magnetic schooner "Zarya" in the Gulf of Finland and the Bay of Biscay. For the first time, the hodograph method was used to determine the horizontal *H* component. In particular, this method was used to estimate the position of the South Magnetic Pole. The obtained results showed that despite the complexity of magnetic measurements on a ferromagnetic vessel, such a survey is promising, but further development of the data processing method is required.

Keywords Geomagnetic field components · Hodograph · Ship's magnetic field

1 Introduction

In connection with the 200th anniversary of the first Russian Antarctic expedition of F.F. Bellingshausen and M.P. Lazarev in 1819–1821 the Russian Geographical Society organized a round-the-world expedition of the R/V "Admiral Vladimirsky". The route of this expedition was planned to follow closely the Bellingshausen expedition. One of the tasks of the expedition was to determine the position of the South Magnetic Pole (SMP) (Fig. 1). Along the entire route, the magnetic field components were recorded, and in some sections, the field magnitude was surveyed. Data processing of the geomagnetic field components on a ferromagnetic vessel is fraught

S. A. Ivanov (✉) · S. A. Merkuryev · I. M. Demina · V. A. Soldatov · D. B. Zaitsev
Pushkov Institute of Terrestrial Magnetism of the Russian Academy of Sciences, St. Petersburg Filial, Universitetskaya emb. 5, St. Petersburg, Russian Federation

S. A. Merkuryev
St. Petersburg State University, Universitetskaya emb. 7-9, St. Petersburg, Russian Federation

A. Kosterov et al. (eds.), *Problems of Geocosmos—2020*, Springer Proceedings in Earth and Environmental Sciences, https://doi.org/10.1007/978-3-030-91467-7_6

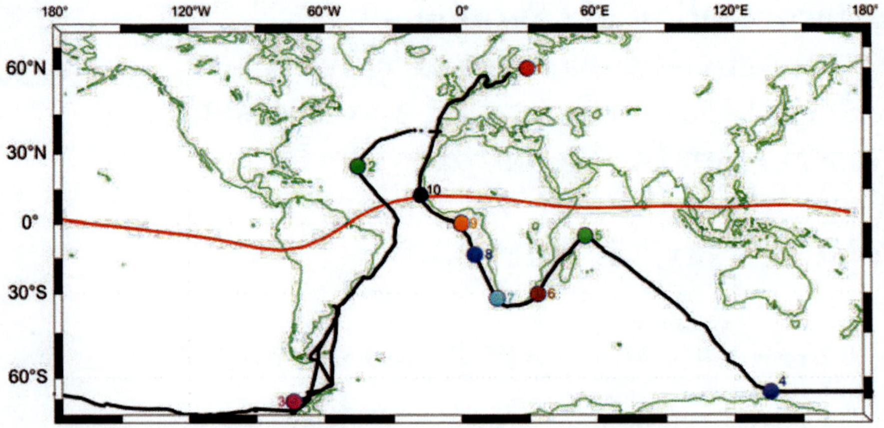

Fig. 1 Route map of the expedition of the R/V "Admiral Vladimirsky". The *black line* shows the route. The position of the magnetic equator is shown in *red*. Places of field estimates are shown and numbered with *colored dots*

with great difficulties. The influence of the ship's magnetic field, which depends on the presence of soft and hard iron, is difficult to determine. In addition, in the absence of systems for stabilization and orientation of the axes of magnetometers along the axes of the ship, when processing data, it is necessary to take into account the pitching, rolling and yaw of the ship. Despite these difficulties, it was possible to show that the use of the component magnetometers is promising for solving a number of geophysical problems, in particular, it was possible to obtain estimates of the horizontal component of the magnetic field near Antarctica and an estimate of the SMP position.

As far as we know, such measurements on such a long-distance route have not been carried out before. A vector magnetometer has (i) been used in marine geophysics to identify "short polarity intervals" with a three-axial flux gate sensor clamped to the aircraft by means of a boom [1] (ii) been used as an onboard three-component magnetometer to study marine magnetic anomalies [2], and (iii) been used by several authors to study vector anomalies by near-bottom magnetic measurements on board a submersible, e.g. [3]. The measurement errors were tens or hundreds nT and usually filtering was applied (see, e.g., [4]) to improve the signal to noise ratio. From the works describing the use of component measurements to find the field components, we point also to [5, 6]. A method for determining the angular components of the induction vector on a moving ship is proposed in [5], which requires solving a system of nonlinear equations. The field components were determined in [6] under simplifications reasonable for high latitudes, in particular, under the assumption that the vertical component significantly exceeds the horizontal one.

2 Magnetometric Laboratory and Verification of the Agreement of Measurements with the Data of the Schooner "Zarya"

The magnetometric laboratory located in the aft part of the R/V "Admiral Vladimirsky" was equipped with the MVC-2 magnetovariational complex, as well as a component magnetometer with three magnetoresistive sensors. The GPS sensor synchronized the operation of the entire system. Comparison of measurements made by component magnetometers with torsion and magnetoresistive sensors showed their high degree of consistency. The sufficient reliability of the instruments is also indicated by the fact that their parameters have changed little after a six-month voyage, as shown by the verification at the Voeikovo Observatory. We compared the data of measuring the vertical component of the field made by the R/V "Admiral Vladimirsky" in the Gulf of Finland and the Bay of Biscay with the measurements made by the schooner "Zarya".

2.1 *Gulf of Finland*

The route map of the schooner "Zarya" and the R/V "Admiral Vladimirsky" is shown in Fig. 2a. Figure 2b shows the Z component measured on each of the tacks. Since the measurements were carried out in calm conditions, the Z component does not contain the ocean-specific short-period disturbances associated with pitching (see below), so data filtering was not performed in this case. As can be seen from Fig. 2b, there is a qualitative agreement between the obtained profiles. Calculations show that the correlation coefficient varies from 0.65 to 0.85, despite the fact that there was no exact match of the routes

2.2 *Bay of Biscay*

Tack diagram of "Zarya" and the R/V "Admiral Vladimirsky" is shown in Fig. 3a. Figure 3b shows the filtered Z component of the anomalous field obtained by the R/V "Admiral Vladimirsky", and the Z component of the anomalous field obtained from the data of the schooner "Zarya". As you can see from this figure, there is a qualitative match of the vertical component on the profiles. Thus, the efficiency of the devices has been proven throughout the entire route.

Let us emphasize the necessity of filtering component measurements in the ocean. When the ship moves along a given azimuth, three short-period oscillatory processes occur: boardside rolling, keel pitching, and yaw. Therefore, a high-pass filter that cuts off periodic components with a period of less than two minutes helps to isolate

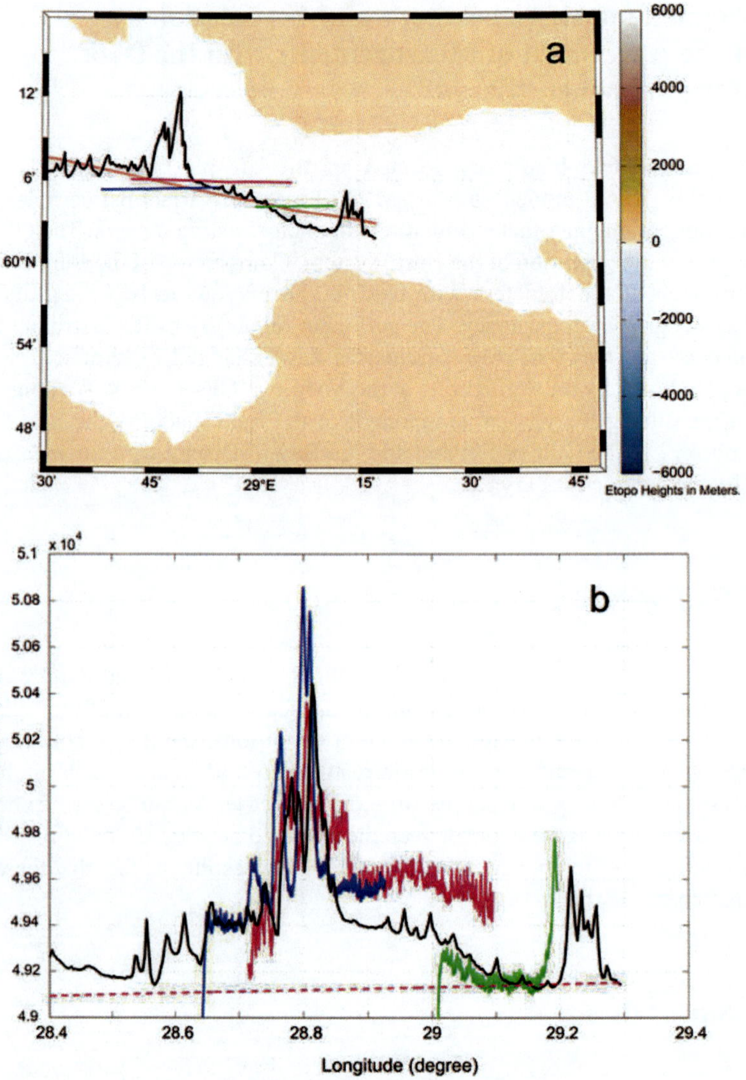

Fig. 2 Comparison of the Z component measurements made by the schooner "Zarya", and the R/V "Admiral Vladimirsky" in the Gulf of Finland. **a** Tack diagram; *brown line* shows the tack of "Zarya", *black line* is the Z component measured by the schooner; *purple* (1), *blue* (2), and *green* (3) lines show the tacks of "Admiral Vladimirsky" made during the trial run on 28 November 2019. **b** *Purple* (1), *blue* (2), and *green* (3) lines indicate the measured values of the Z component for the corresponding tacks

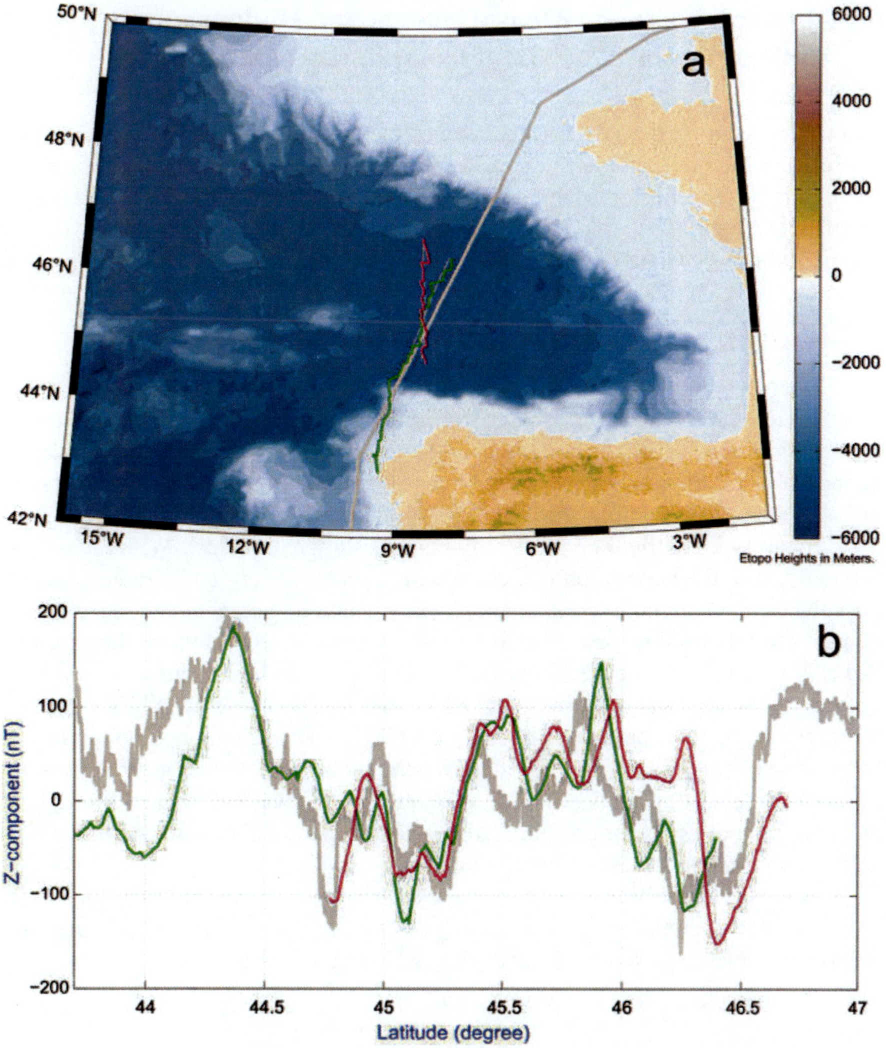

Fig. 3 Comparison of the *Z* component of the anomalous field on the profiles made by the schooner "Zarya" and the R/V "Admiral Vladimirsky" in the Bay of Biscay. **a** Tack diagram, *green* and *pink* lines show the tacks of "Zarya", *gray* line shows the tack of the R/V "Admiral Vladimirsky". **b** *Green* and *pink* lines show the *Z* component obtained from the data of the schooner "Zarya"; *gray* line indicates the smoothed values of the *Z* component measured by the R/V "Admiral Vladimirsky"

the signal from noise. More accurate processing would require the use of data from accelerometers and GPS.

3 Onboard Magnetic Measurements and Hodograph Application for Horizontal Component Estimates

Further, we assume that the vertical axis of the device coincides with the vertical (corrections for pitching are not considered here). The magnetic field in the ship's coordinate system consists of three components: the external $\boldsymbol{T}_{\mathrm{GMF}}$ field (GMF is the geomagnetic field), the part of the ship's magnetic field corresponding to the residual magnetization, $\boldsymbol{T}_{\mathrm{h}}$ (solid iron), and the induced magnetization $\boldsymbol{T}_{\mathrm{s}}$, depending on the course.

$$T_0 = T_{\mathrm{GMF}} + T_s + T_{\mathrm{h}}.$$

In this case, $\boldsymbol{T}_{\mathrm{s}}$ is linearly related to the Earth's magnetic field through the Poisson matrix. To obtain data on the horizontal GMF component, the main tool for us was the hodograph, i.e., the curve (X,Y) of the measured horizontal components, obtained or calculated when turning the ship 360° (theoretically, to build a hodograph, it is sufficient to know the horizontal components on five courses). Without a ship's magnetic field, the hodograph is a circle with a center at zero and a radius equal to the value of the horizontal component H. On a magnetic vessel the hodograph is an ellipse of a general form, see Fig. 5. Usually it was used to eliminate deviation [7] and calibrate magnetometers (on aircraft, it is possible to build a three-dimensional hodograph, which is an ellipsoid). By our knowledge, there are no published papers where hodograph has been used to study the GMF itself, usually hodograph is used for sensor calibrations only [2]. We describe the process of obtaining and processing the hodograph. Let the ship go through the circulation (Fig. 4a), or pass through the polygon (Fig. 4b), or cross one point with several tacks, as in Fig. 4b (for brevity, we will call any such maneuver a circulation).

As a result of performing any of these routes, we get an array of values (X,Y) of the horizontal components that define points on the plane. To approximate these points, a second-order figure is selected, described by the equation

$$x^2 + a_1y^2 + a_2xy + a_3x + a_4y + a_5 = 0$$

The coefficients of the equation are found by least squares and define a general ellipse (obviously, the arc of a parabola cannot approximate the hodograph well). In [8], the least squares method was also used, but for a different functional, and the deviation corrections were expressed in terms of the coefficients a_j of the equation. See also [9, 10].

It is easy to show that if the vertical component of the external field does not induce a horizontal component, the center of this ellipse gives the value of $\boldsymbol{T}_{\mathrm{h}}$. Next, the hodograph is shifted to the origin, rotated and transformed into a circle, see Fig. 5. Note that the axes of all the constructed ellipses were almost parallel to the coordinate axes, which usually happens with a diagonal Poisson matrix. The angles between the axes of the ellipse and the axes of the ship coordinate system were from

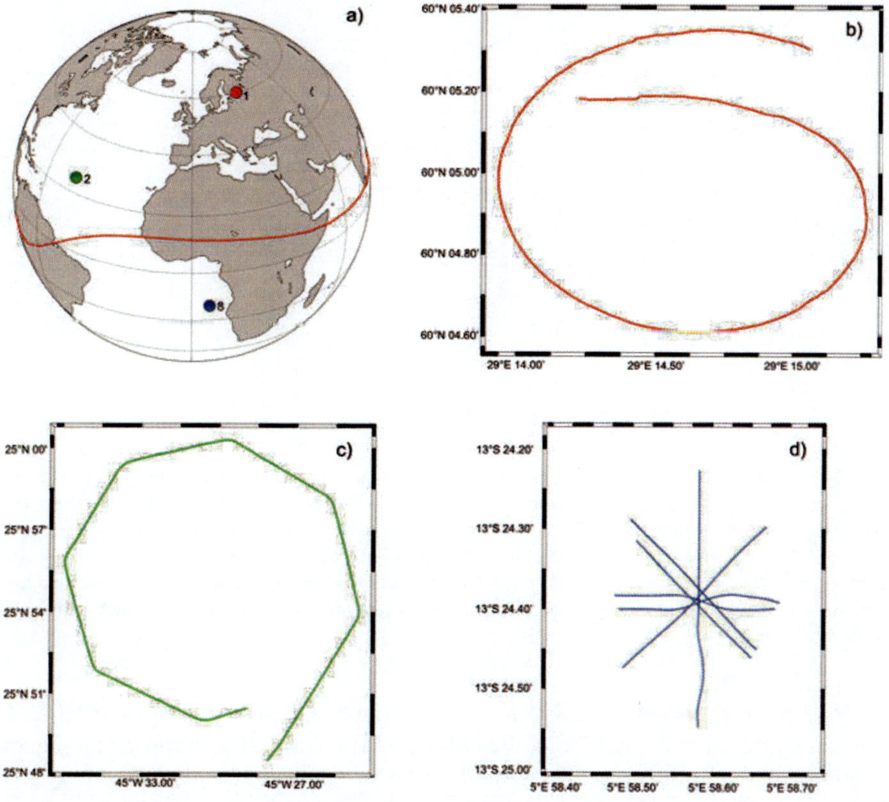

Fig. 4 Three types of ship maneuvers during deviation works and their locations **a** possible circulations, **b** corresponding to the hodograph in Fig. 7, **c** polygon, **d** star

0 to 3°. Knowing the ratio of the axes and the center of the hodograph, we can (in some vicinity of the circulation) find either the value H itself or determine it up to a constant (if there is a suspicion that the calibration has changed under the influence of extraneous magnetic fields). This theory was confirmed during the expedition where several circulations were performed along the route (Fig. 6).

As a result, ten hodograph radii were obtained. For comparison, at the points of circulation, the horizontal component H was calculated using the IGRF model, and it was found that the "radius" of the hodograph (the square root of the product of the semiaxes) is proportional to H. An almost linear relationship is clearly visible. Figure 6 shows the values of the horizontal component at the points of circulation calculated using the IGRF model versus hodograph radii obtained from the data of the R/V "Admiral Vladimirsky". We see from Fig. 6 that the "radius" of a hodograph near the pole is not that small. This can be seen also on the right part of Fig. 7, where it is not easy to find an ellipse. The main reason is the deviation of the upward magnetometer axis from the vertical while rolling or circulating. Near the magnetic poles the influence of such deviation is essential. Indeed, if the vertical axis in the

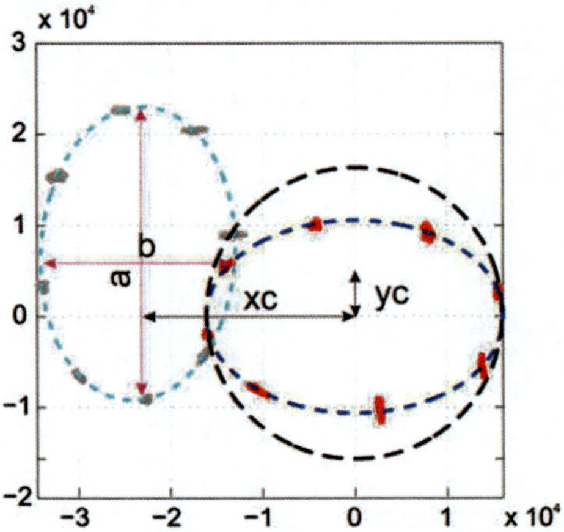

Fig. 5 A hodograph constructed from the results of measuring the *X* and *Y* components during deviation works in the Atlantic Ocean by the star method (Fig. 4b) and its transformation. The hodograph with the center *xc, yc* is shown in *gray*. The major and minor semiaxes are designated "*a*" and "*b*". *Blue color* shows the result of fitting the ellipse to the observed values. *Red color* shows the shift of the ellipse and bringing it to the canonical form using rotation. The *dashed line* shows the ellipse found along this curve. The *black dashed line* shows the circle obtained by scaling with the factor *a/b*

ship coordinate system turns by 1°, the *Z* component gives a contribution to *H* component equal to *Z* sin(1°). On the polygon this contribution of the *Z* component is about 1000 nT while the *H* component does not exceed 100 nT. In the next step of study, we will take the ship rotations into account.

4 Determination of SMP by the Value of the Horizontal Component

This part is described in detail in our article of the same issue [11]. With the help of hodographs, the value of the horizontal component was calculated from the polygon survey made by the R/V "Admiral Vladimirsky" in the D'Urville Sea. Its local minima gave virtual magnetic poles (VMP). The assessment of the SMP position was obtained by averaging the VMP locations. The polygon itself, the VMP positions, and the averaging result are shown in Fig. 7. The SMP position calculated by IGRF-13 is also given here. The corresponding numerical values are: the latitude of the SMP position from IGRF-13 is 64.07° S, the longitude is 135.82° E; the latitude of the SMP position from the VMPs 64.08° S, the longitude is 135.80° E.

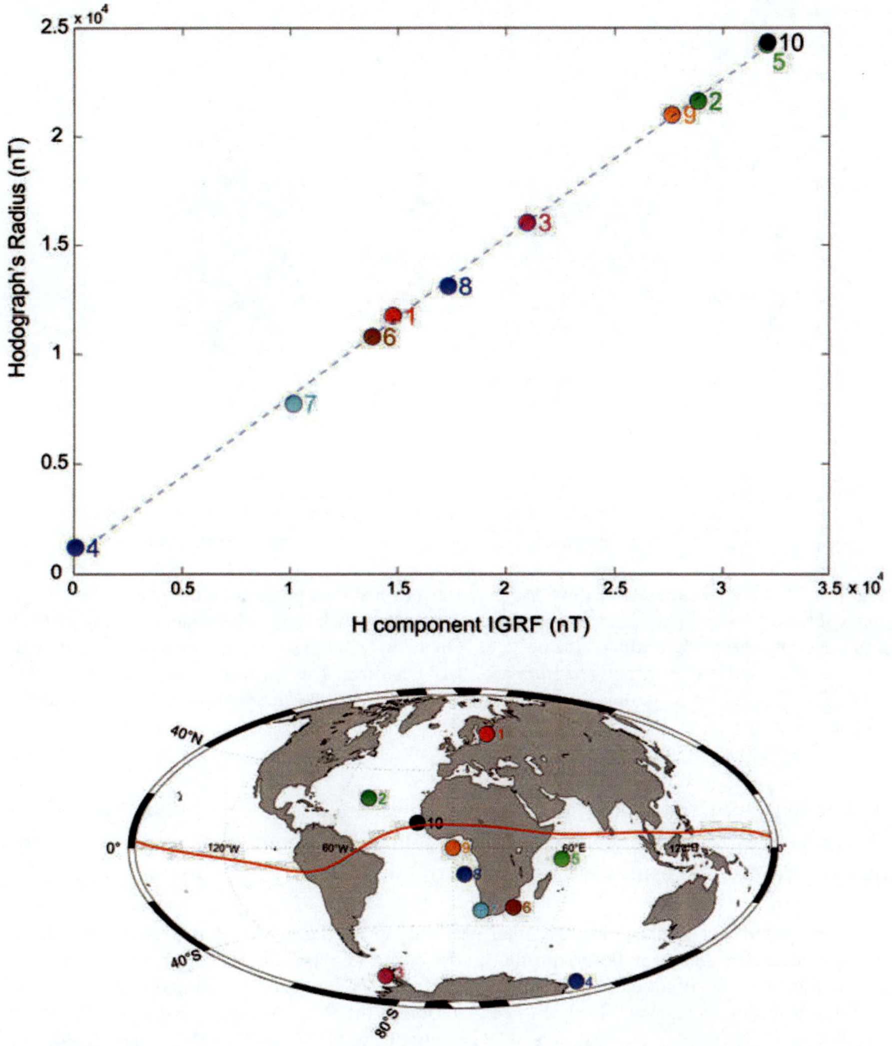

Fig. 6 Dependence of the radius of the hodograph on the value of the horizontal component (the color and numbers of the points correspond to the locations of the circulations shown in the low part of the Fig. 6)

5 Conclusions

The results of the study suggest that the use of component magnetometers on board a ferromagnetic carrier is a promising direction, although it requires solving several methodological problems related to both measurement and data processing.

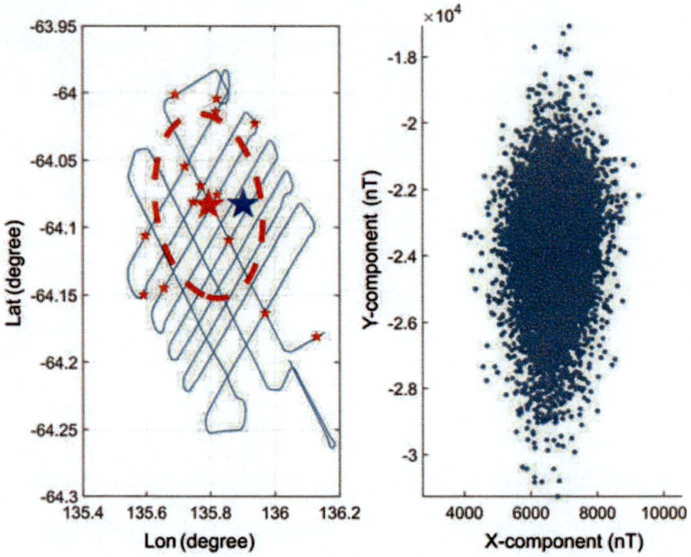

Fig. 7 Obtaining an estimate of the SMP position as a result of processing the horizontal components of the magnetic field. The *right part* of the figure shows a hodograph based on the values of the *X* and *Y* components measured on the polygon. On the *left part* of the figure, the *blue lines* show the tacks of a detailed survey. Small *red stars* are VMP positions. The *red star* is the result of the VMP averaging. The *blue star* is the SMP calculated by IGRF-13. *Red dashed line* is 50%-confidence ellipse

The analysis of the data of the component measurements by the hodograph method allows us to estimate the value of the horizontal GMF component. In particular, this method was used to estimate the position of the SMP.

Acknowledgements Authors express their gratitude to all those who took part in the delivery of equipment for the expedition, promptly organized the transfer of information, provided communication and coordination along the route of the R/V “Admiral Vladimirsky”:

Oleg Dmitrievich Osipov, Head of the expedition of the R/V "Admiral Vladimirsky", Deputy Head of the Department of Navigation and Oceanography of the Ministry of Defense of the Russian Federation.

Research staff of the Department of Geomagnetic Research (SPbF IZMIRAN), Prof. Yu. A. Kopytenko, Ph.D. P.A. Sergushin, V.A. Levanenko, A.V. Petlenko.

Evgeny Konstantinovich Grigoriev, Head of the Department of Gravimetric and Magnetometric Marine Research.

Employee of the 6th Atlantic Expedition, Ph.D. Sergey Vladimirovich Protsaenko.

The GMT program package was used to perform the drawings [12].

References

1. Blakely, R.J., Cox, A., Ernest, J., Iufer, E.I.: Vector magnetic data for detecting short polarity intervals in marine magnetic profiles. JGR **78**(29), 6977–6983 (1973)
2. Isezaki, N.A.: New shipboard three-component magnetometer. Geophysics **51**(10), 1992–1998 (1986)
3. Honsho, C., Dyment, J., Tamaki, K., Ravilly, M., Horen, H. et al.: Magnetic structure of a slow spreading ridge segment: insights from near-bottom magnetic measurements on board a submersible. J. Geophys. Res. Solid Earth, American Geophysical Union, 2009, 114, pp.B05101. ff10.1029/2008JB005915ff. ffinsu-00446462ff
4. Nogi, Y., Seama, N., Isezaki, N.: Preliminary report of three components of geomagnetic field measured on board the icebreaker SHIRASE during JARE-30, 1988–1989. In: Proceedings of the NIPR Symposium on Antarctic Geosciences, 4, 191–200 (1990)
5. Korenaga, J.: Comprehensive analysis of marine magnetic vector anomalies. J. Geophys. Res. **100**(B1), 365–378 (1995)
6. Blednov, V.A.: Geomagnetic component measurements on board of a moving ferromagnetic carrier. Phys. Usp. **37**, 921–925 (1994)
7. Kopytenko, Yu. A., Petlenko, A.V., Petrova, A.A., Kopytenko, E.A., Voronov, P.M., Ismagilov, V.S., Zaitsev, D.B., Timoshenkov, Yu. P.: Peculiarities of interpretation of magnetic field components' data obtained at high-latitudes on the board of moving carrier. In: Proceedings of the International Conference on Marine Electromagnetic: Marelec 97: 23–26 June 1997. London UK, pp. 1–6 (1997)
8. Vasconcelos, J.F., Elkaim, G., Silvestre, C., Oliveira, P., Cardeira, P.: Geometric approach to strapdown magnetometer calibration in sensor frame. IFAC Proceedings Volumes , 41(1), 172–177 (2008)
9. Churikov, D.S., Andrianov, S., Belchenko, K.G.: The use of the least-squares method to determine an ellipsis-locus vector of the magnetic field. Gyroscopy Navig **1**(12), 60–65 (1996)
10. Yiğitler, H., Leblebicioğlu, V.: Online calibration of strapdown magnetometers. In: 9th IFAC Symposium on Robot Control, Nagaragawa Convention Center, Gifu, Japan, September 9–12, (2009)
11. Demina, I.M., Boyarskikh, V.G., Merkuryev, S.A., Ivanov, S.A., Soldatov, V.A.: Determination of the position of the South Magnetic Pole based on experimental data obtained during Russian round-the-world expeditions: 1820 (F. Bellingshausen), 2020 (R/V "Admiral Vladimirsky"). This issue
12. Wessel, P., Luis, J.F., Uieda, L., Scharroo, R., Wobbe, F., Smith, W.H.F., Tian, D.: The generic mapping tools version 6. Geochem. Geophys. Geosyst. **20**, 5556–5564 (2019)

Integration of Geophysical Methods for Solving Inverse Problems of Exploration Geophysics Using Artificial Neural Networks

Igor Isaev, Ivan Obornev, Eugeny Obornev, Eugeny Rodionov, Mikhail Shimelevich, and Sergey Dolenko

Abstract The inverse problem (IP) of exploration geophysics consists in reconstructing the spatial distributionf of the properties of the medium in the Earth's interior from measurements on its surface. This IP is a non-linear ill-posed ill-conditioned problem with high dimensionality both by input and by output. One of the approaches free of many shortcomings inherent for traditional methods of IP solving, is the use of artificial neural networks (NN). In this study, it has been suggested to use an integration of geophysical methods to improve the quality of the solution obtained by NN. The considered model combines three geophysical methods: gravimetry, magnetometry, and magnetotellurics. The problem considered is that of determining the structural boundaries separating the geological layers with constant values of the parameters: density in gravimetry, magnetization in magnetometry, electrical resistivity in magnetotellurics. In this study, a four-layer 2D model was considered. It is demonstrated that integration of geophysical methods provides significantly better results that use of each of the methods separately. It is also shown that in some cases it is also possible to improve the quality of the IP solution using multitask learning—simultaneous determination of the positions of two or all three layer boundaries.

Keywords Inverse problems · Exploration geophysics · Gravimetry · Magnetometry · Magnetotelluric sounding · Integration of geophysical methods · Neural network · Multitask learning

I. Isaev (✉) · I. Obornev · S. Dolenko (✉)
D.V. Skobeltsyn Institute of Nuclear Physics, M.V. Lomonosov Moscow State University, Moscow, Russia
e-mail: dolenko@srd.sinp.msu.ru

I. Isaev
Kotelnikov Institute of Radio Engineering and Electronics, Russian Academy of Sciences, Moscow, Russia

E. Obornev · E. Rodionov · M. Shimelevich
S.Ordjonikidze Russian State Geological Prospecting University, Moscow, Russia

A. Kosterov et al. (eds.), *Problems of Geocosmos—2020*, Springer Proceedings in Earth and Environmental Sciences, https://doi.org/10.1007/978-3-030-91467-7_7

1 Introduction

The solution of the inverse problems (IP) of exploration geophysics (EG) is aimed at studying the structure of the near-surface layers of the earth and searching for mineral resources. The general setting of the EG IP consists in constructing the distribution of the physical parameters of the medium in the earth's crust from the physical fields measured on the earth's surface. The traditional methods for solving the EG IP include: (i) optimization methods based on the multiple solution of the direct problem with the minimization of residuals in the space of the observed fields, and (ii) approximation methods, which also include matrix methods using Tikhonov regularization. Optimization methods have a number of disadvantages: they are characterized by high computational costs and the need for a good first approximation. At the same time, the main drawback of optimization methods is the need to have a correct model for solving the direct problem, in the absence of which this method is not applicable. In addition, due to the inherent incorrectness of many IP, a small residual in the space of the observed fields does not guarantee a small residual in the space of the determined parameters. For matrix methods based on regularization, the main difficulty is the need to choose the regularization parameter. In addition, matrix methods are linear methods, so when using them to solve nonlinear problems, it is necessary to perform nonlinear data preprocessing. To solve such problems, machine learning (ML) methods that are free from the above disadvantages can be considered as an alternative. High computational costs when using ML methods are shifted from the stage of application of the computing system to the stage of its development, which increases the convenience of practical use of such a system.

At the same time, when solving the RG problem, the ML methods can be used at various stages of its solution: in data preprocessing, e.g., for noise removal [1, 2]; as an independent optimization method [3–5] or as a component of optimization methods used to solve the EG problem [6]; as an independent inversion method [7–16]; in solving the classification problem to select a class of geological media [17].

To implement the solution of the IP by means of the ML methods, the initial distribution of the geophysical parameters of the geological section must be described by a finite number of determined parameters, i.e., the so-called parameterization scheme must be introduced. Here, a straightforward approach is to use "universal" parameterization schemes, which can be used to describe any geological structure. For example, in the parameterization scheme used in [9–11], the distribution of electrical conductivity is set by its values in the nodes of the macro grid. The values of electrical conductivity in each node for each pattern of the training sample are set randomly. The disadvantage of this approach is the high degree of incorrectness of the resulting IP described by these parameterization schemes, which leads to a decrease in the quality of the solution and to the need of using special approaches to improve it.

One of the ways to improve the quality of the solution is to use additional information about the studied object. Such information can be a priori assumptions about the structure of the geological section. The implementation of this approach may consist in the use of parameterization schemes with a rigidly defined spatial structure (the so-called "class-generating models" [12–14]), based on the results brought by alternative measurement methods or on assumptions about the structure of the specific area. The disadvantage of this approach is the need to develop individual solution for each problem, as well as the need to have a priori information about the spatial structure of the defined parameterization.

As an intermediate approach, one can use more narrow models of the media that describe some certain class of geological sections. At the same time, such models must have geological validity and prevalence in nature. In particular, in this study we used the horizontal-layered model of the medium [15].

Another way to introduce additional information is using integration of geophysical methods [4, 5, 16]. This approach was considered by the authors in [16], where both gravimetry and magnetometry data were used, and an improvement in the quality of the solution was demonstrated in comparison with the solutions obtained from the individual use of each method.

Another way, which may help to improve the quality of the solution when using ML methods, is the "multitask learning" method [9, 10, 18–22]. It may increase the generalization ability of the algorithms. The efficiency of this approach is confirmed for such ML methods as decision trees [18], the k-nearest neighbor method [18], the support vector machine (SVM) method [19], convolutional neural networks [20], and deep neural networks [21, 22]. At the same time, the results were obtained for completely different types of tasks, such as medical data processing [18], computer vision [20], natural language processing [21], and traffic flow forecasting [22]. For the magnetotelluric sounding (MTS) IP, this approach showed not only an improvement in the quality of the solution [9], but also an increase in the resilience of the solution to noise in the data [10].

The purpose of this work is to test the applicability of the approach associated with the integration of methods of gravimetry, magnetometry, and MTS in the neural network solution of the EG IP for a horizontally layered model of the medium, as well as to test the applicability of the Multitask learning approach for this type of a problem.

2 Physical Statement of the Problem

2.1 Parameterization Scheme

In order to implement the integration of various geophysical methods, it is necessary that the determined parameters for each of the methods were the same. This approach corresponds to the geometric formulation of the problem, which consists in determination of the boundaries of geophysical objects. In particular, in this study we considered the parameterization scheme, which consists in determining the boundaries of geological layers of a layered medium.

The parameterization scheme was a four-layer two-dimensional model (Fig. 1)—a section modeling that for the Norilsk region. The first layer corresponds to the basalt layer, the second and fourth ones—to terrigenous carbonate deposits of the Tunguska series, the third one—to the gabbro-dolerites massive copper–nickel–platinum ores.

The size of the section was 15 km wide and 3 km deep. The physical fields measurements step was 0.5 km—a total of 31 measurement points along the profile. The discreteness of the boundaries of geological layers was 1 km—a total of 15 depth levels for each layer. In the problem, the depths of the lower boundaries of the three upper layers were determined. Each layer was characterized by fixed values of density, magnetization, and resistivity, which did not change within the layer, and which were the same across the entire data set. The physical characteristics of the second and fourth layers were the same. The values of the physical and spatial characteristics of the layers are shown in Table 1.

2.2 Data

For each pattern of the original data set, the layer depth values were set randomly in the ranges shown in Table 1. Further, the direct problem was solved by finite-difference methods for each of the selected geophysical methods. An example of a geological section and its corresponding geophysical fields are shown in Fig. 1.

The input dimension of the problem was:

- *Gravimetry: 1 field component* $\times$ *31 measurement point (picket)* $=$ *31 features*
- *Magnetometry: 1 field component* $\times$ *31 picket* $=$ *31 features*
- *MTS: 6 field components* $\times$ *13 frequencies* $\times$ *31 picket 2418 features.*

The output dimension of the problem was:

- *3 layers* $\times$ *15 values of layer boundary depth* $=$ *45 parameters.*

A total of 30,000 patterns were calculated.

3 Methodical Statement of the Problem

3.1 Datasets

The original data set was divided into training, validation, and test sets in a ratio of 70:20:10. Their dimensions were 21,000, 6,000, and 3,000 patterns, respectively.

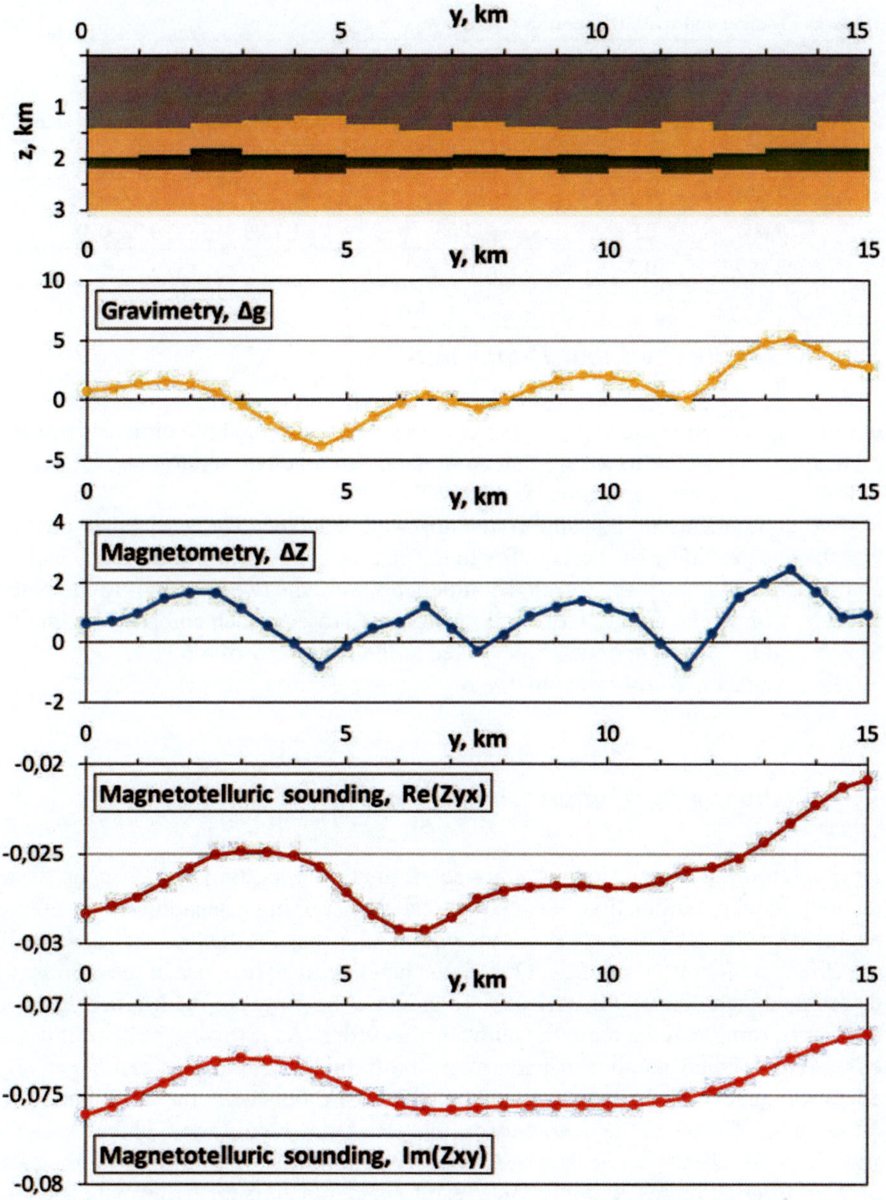

Fig. 1 An example of the geological section within the considered parameterization scheme (*above*), and the corresponding components of the fields used in this study (*below*)

Table 1 Physical and spatial properties of the layers

Layer	Physical properties			Spatial properties	
	Density, σ (kg/m^3)	Magnetization μ (A/m)	Resistivity ρ (Ω m)	Lower bound, min–max (km)	Thickness, min–max (km)
1	2,800	3.0	2,000	1.00–1.48	1.00–1.48
2	2,550	0.5	100	1.80–1.98	0.32–0.98
3	3,000	0.9	1,000	2.20–2.28	0.22–0.48
4	2,550	0.5	100	–	–

3.2 Reducing the Input Dimension

In order to overcome the negative factors associated with the high dimension of the input space, as well as to ensure that all methods are in close conditions, a selection of input features was carried out for the MTS IP.

Two components at the same frequency were selected. The selected frequency was the one providing the best quality in the case of using a single frequency and all components. Similarly, the selected components were the two best ones found when a single component and all frequencies were used [15]. As such compromise inputs, ReZyx and ImZxy components were taken at the frequency of 4.6 Hz.

For gravimetry and magnetometry, no feature selection was performed.

3.3 Reducing the Output Dimension

As a reference solution, autonomous determination was used, which consists in the individual determination of each parameter by training a separate single-output neural network. This approach allows one to tune out the factors that negatively affect the quality of the solution. On the one hand, with an increase in the number of determined parameters, it is necessary to increase the power of the ML architecture used (for example, to increase the number of neurons and layers in the case of using NN). With a small number of patterns in the training sample, this can negatively affect the quality of the resulting solution. On the other hand, the contribution of different parameters to the overall error functional may be different, in the case of a significant difference between simultaneously determined parameters or in the case of their different representativity. As a result, the parameters that contribute less can be ignored during the training process.

3.4 Multitask Learning

In this study, we also considered the Multitask learning approach, which in some cases can improve the quality of the solution [9, 10, 18–22]. In paper [9] for the MTS IP, which consisted in determining the conductivities in the nodes of a rigidly defined spatial grid, it was shown that an improvement in the quality of the solution was observed when simultaneously determining the parameters corresponding to nodes lying in the same vertical.

In this study, we considered the central vertical, for which we considered autonomous determination and multitask learning (group determination of parameters).

3.5 Use of Neural Networks

The type of NN used was the multi-layer perceptron, which is a universal approximator. The architecture used had a single hidden layer with 32 neurons in it.

To reduce the factor associated with the influence of the initialization of weights on the training of NN, five networks were used for each case under consideration, and the statistical indicators of their application were averaged.

To prevent overtraining, early stopping by validation dataset was used.

For individual use of data from the gravimetry and magnetometry methods, the NN input was fed by 31 features, for individual use of MTS data—62 features, for simultaneous use of data from two geophysical methods—62 or 93 features, for simultaneous use of data from all the three methods—124 features.

For autonomous determination of the parameters, each network had 1 output, and when using the Multitask learning approach, it had 2 or 3 outputs.

4 Results

In this study, as an indicator of the quality of the solution, we used the relative error calculated as the mean absolute error (MAE) normalized by the range of change of the determined parameter.

4.1 Autonomous Determination

The results of autonomous determination for various sets of input data are presented in Fig. 2.

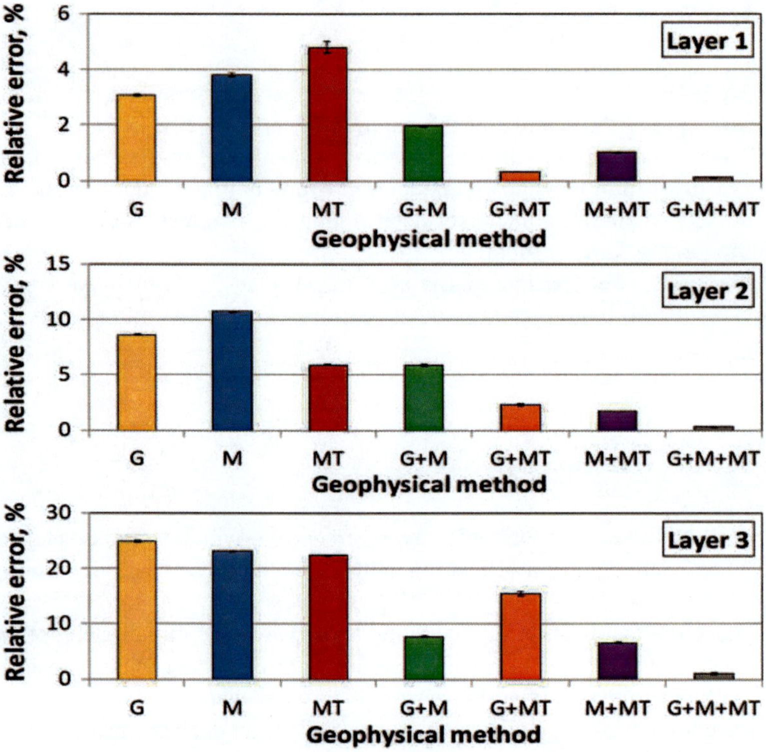

Fig. 2 Results of autonomous determination for different input data. G, M, MT—individual use of gravimetry, magnetometry and MTS data; G + M, G + MT, M + MT, G + M + MT—simultaneous use of data from several geophysical methods (integration of methods)

For all the three layers, simultaneous use of data from any two geophysical methods reduces the error compared to the individual use of data from any of them. The best result was demonstrated in the simultaneous use of the data from three geophysical methods.

Thus, for the considered problem, a positive effect is observed from the integration of geophysical methods.

4.2 Multitask Learning

The results of applying the Multitask learning approach together with the integration of geophysical methods are shown in Fig. 3. The displayed values are the difference between the results of using multitask learning (group determination) and the results of autonomous determination. Therefore, negative values correspond to the situations when use of multitask learning improves the results.

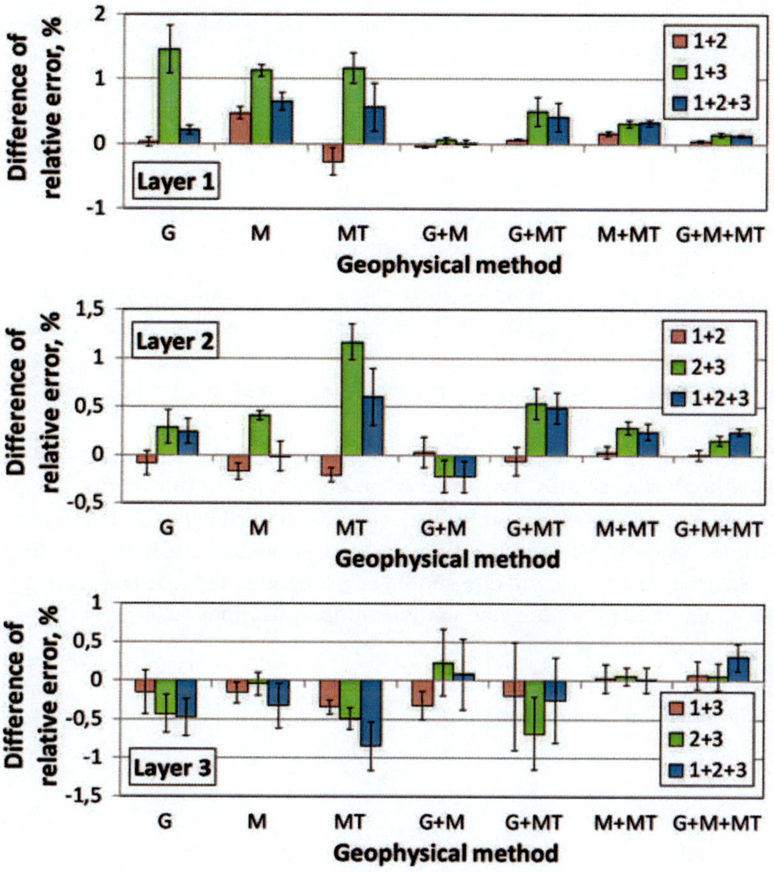

Fig. 3 Results of applying the Multitask learning approach to various input data. G, M, MT—individual use of gravimetry, magnetometry and MTS data, G + M, G + MT, M + MT, G + M + MT—simultaneous use of data from several geophysical methods. The legend specifies the numbers of layers whose parameters were determined simultaneously

There is a decrease in the error when using this approach to determine the boundary of the third layer with the individual use of geophysical methods and with the integration of gravimetry and magnetotellurics data. In the other cases, no positive result was observed; however, in all these cases, the error in the autonomous determination mode was lower.

For the second layer, the improvement in the quality of the solution was observed only when it was determined simultaneously with the first layer. For the first layer, in almost all cases, this approach did not lead to a positive result.

Therefore, the grouping approach is mostly applicable to improve the worse results. If the quality of autonomous determination is relatively high, this approach leads to the increase of the error. As it was observed in our previous studies, and also in this study, the determined parameter should be preferably grouped with those that are determined better than it in the autonomous determination mode.

5 Conclusions

Based on the results of this study, the following conclusions can be made:

- Integration of the data of different physical methods allows one to improve the quality of the neural network solution in comparison with the individual use of each method separately.
- The best result was provided by simultaneous use of the data from all three geophysical methods.
- The Multitask Learning (group determination) approach has demonstrated its partial applicability also for the problem considered in this study. The positive effect of its use was observed for the second and third layers of the model.
- The conclusion made by the authors in their previous studies was confirmed: to improve the results, a parameter should be grouped with those parameters that are determined better than this one in the autonomous mode.

Acknowledgements This study has been performed at the expense of the grant of the Russian Science Foundation (project no. 19-11-00333).

References

1. Manoj, C., Nagarajan, N.: The application of artificial neural networks to magnetotelluric time-series analysis. Geophys. J. Int. **153**(2), 409–423 (2003)
2. Wu, X., Xue, G., Xiao, P., Li, J., Liu, L., Fang, G.: The removal of the high-frequency motion-induced noise in helicopter-borne transient electromagnetic data based on wavelet neural network. Geophysics **84**(1), K1–K9 (2019)
3. Yuan, S., Wang, S., Tian, N.: Swarm intelligence optimization and its application in geophysical data inversion. Appl. Geophys. **6**(2), 166–174 (2009)
4. Roux, E., Moorkamp, M., Jones, A.G., Bischoff, M., Endrun, B., Lebedev, S., Meier, T.: Joint inversion of long-period magnetotelluric data and surface-wave dispersion curves for anisotropic structure: application to data from Central Germany. Geophys. Res. Lett. **38**(5), L05304 (2011)
5. Akca, İ., Günther, T., Müller-Petke, M., Başokur, A.T., Yaramanci, U.: Joint parameter estimation from magnetic resonance and vertical electric soundings using a multi-objective genetic algorithm. Geophys. Prospect. **62**(2), 364–376 (2014)
6. Conway, D., Alexander, B., King, M., Heinson, G., Kee, Y.: Inverting magnetotelluric responses in a three-dimensional earth using fast forward approximations based on artificial neural networks. Comput. Geosci. **127**, 44–52 (2019)
7. Al-Garni, M.A.: Inversion of residual gravity anomalies using neural network. Arab. J. Geosci. **6**(5), 1509–1516 (2013)

8. Al-Garni, M.A.: Interpretation of some magnetic bodies using neural networks inversion. Arab. J. Geosci. **2**(2), 175–184 (2009)
9. Dolenko, S., Isaev, I., Obornev, E., Persiantsev, I., Shimelevich, M.: Study of influence of parameter grouping on the error of neural network solution of the inverse problem of electrical prospecting. In: Iliadis, L., Papadopoulos, H., Jayne, C. (eds.) Engineering Applications of Neural Networks. EANN 2013. Communications in Computer and Information Science, vol. 383, pp 81–90. Springer, Berlin, Heidelberg (2013)
10. Isaev, I., Obornev, E., Obornev, I., Shimelevich, M., Dolenko, S.: Increase of the resistance to noise in data for neural network solution of the inverse problem of magnetotellurics with group determination of parameters. In: Villa, A., Masulli, P., Pons Rivero, A. (eds.) Artificial Neural Networks and Machine Learning—ICANN 2016. ICANN 2016. Lecture Notes in Computer Science, vol. 9886, pp. 502–509. Springer, Cham (2016)
11. Isaev, I., Dolenko, S.: Adding noise during training as a method to increase resilience of neural network solution of inverse problems: test on the data of magnetotelluric sounding problem. In: Kryzhanovsky, B., Dunin-Barkowski, W., Redko, V. (eds.) Advances in Neural Computation, Machine Learning, and Cognitive Research. NEUROINFORMATICS 2017. Studies in Computational Intelligence, vol. 736, pp. 9–16. Springer, Cham (2018)
12. Spichak, V., Popova, I.: Artificial neural network inversion of magnetotelluric data in terms of three-dimensional earth macroparameters. Geophys. J. Int. **142**(1), 15–26 (2000)
13. Spichak, V., Fukuoka, K., Kobayashi, T., Mogi, T., Popova, I., Shima, H.: ANN reconstruction of geoelectrical parameters of the Minou fault zone by scalar CSAMT data. J. Appl. Geophys. **49**(1–2), 75–90 (2002)
14. Montahaei, M., Oskooi, B.: Magnetotelluric inversion for azimuthally anisotropic resistivities employing artificial neural networks. Acta Geophys. **62**(1), 12–43 (2014)
15. Isaev, I., Obornev, E., Obornev, I., Rodionov, E., Shimelevich, M., Shirokiy, V., Dolenko, S.: Using domain knowledge for feature selection in neural network solution of the inverse problem of magnetotelluric sounding. In: Samsonovich, A.V., Gudwin, R.R., Simões, A.S. (eds.) Brain-Inspired Cognitive Architectures for Artificial Intelligence: BICA*AI 2020. BICA 2020. Advances in Intelligent Systems and Computing, vol. 1310, pp. 115–126. Springer, Cham (2021)
16. Obornev, E., Obornev, I., Rodionov, E., Shimelevich, M.: Application of neural networks in nonlinear inverse problems of geophysics. Comput. Math. Math. Phys. **60**(6), 1025–1036 (2020)
17. Isaev, I., Obornev, E., Obornev, I., Shimelevich, M., Dolenko, S.: Neural network recognition of the type of parameterization scheme for magnetotelluric data. In: Kryzhanovsky, B., Dunin-Barkowski, W., Redko, V., Tiumentsev, Y. (eds.) Advances in Neural Computation, Machine Learning, and Cognitive Research II. NEUROINFORMATICS 2018. Studies in Computational Intelligence, vol. 799, pp. 176–183. Springer, Cham (2019)
18. Caruana, R.: Multitask learning. Mach. Learn. **28**(1), 41–75 (1997)
19. Evgeniou, T., Pontil, M.: Regularized multi-task learning. In Proceedings of the Tenth ACM SIGKDD International Conference on Knowledge Discovery and Data Mining, pp. 109–117 (2004)
20. Girshick, R.: Fast r-cnn. In: Proceedings of the IEEE International Conference on Computer Vision, pp. 1440–1448 (2015)
21. Collobert, R., Weston, J.: A unified architecture for natural language processing: deep neural networks with multitask learning. In: Proceedings of the 25th International Conference on Machine learning, pp. 160–167 (2008)
22. Huang, W., Song, G., Hong, H., Xie, K.: Deep architecture for traffic flow prediction: deep belief networks with multitask learning. IEEE Trans. Intell. Transp. Syst. **15**(5), 2191–2201 (2014)

A Versatile Software for Statistical Data Analysis and Spatial Correlation

Dmitry Krivopaltsev, Leonid Surovitskii, and Mikhail Lukin

Abstract Development of new tools for interpreting of and maximizing the amount of information extracted from georeferenced experimental and observational data remains an important task of computational geoscience. One of powerful tools for data analysis is the examination of correlation dependence between different, sometimes seemingly unrelated parameters. However, the correlation analysis of geospatial data can often be hampered by inadequate sampling or lack of coincidence between coordinates in the compared datasets. The number and capabilities of affordable software that can be used for these purposes are limited. Here we present a versatile software package for correlation analysis of spatial geo-referenced data with an additional set of tools. The software is implemented as a web service and contains a toolset for performing a geographic coordinate conversion, data interpolation, mapping, spatial correlation, data visualization and other auxiliary functions for spatial data analysis. The performance of the software is checked and confirmed on the real data. We call this service "UNCORR"—the UNiversal CORRelation tool.

Keywords Correlation · Interpolation · Data analysis · Comparison · Software · UNCORR

1 Introduction

Analyses of geodata, even simple tasks such as correlation calculation, require additional steps and are more time-consuming than analyses of non-georeferenced data. For instance, if data are presented in different coordinate systems, additional difficulties arise from the need to convert the coordinates for one or more of the analyzed

D. Krivopaltsev (✉) · M. Lukin
ITMO University, St. Petersburg, Russia

L. Surovitskii
Michigan Technological University, Houghton, MI, USA

St. Petersburg State University, St. Petersburg, Russia

M. Lukin
Sudo, LLC, St. Petersburg, Russia

A. Kosterov et al. (eds.), *Problems of Geocosmos—2020*, Springer Proceedings in Earth and Environmental Sciences, https://doi.org/10.1007/978-3-030-91467-7_8

datasets. If the coordinates of measurement/observation points for the datasets do not coincide, then one has to create an additional grid of intermediate values and make sure that this grid fits both datasets. In addition, there are many other tasks related to processing, verification, and visualizing data of geospatial data.

The program is implemented as a web service; user does not need to install additional packages or have powerful computing resources. The service simplifies all steps in the correlation analysis process, providing an ability to flexibly configure parameters at each stage. In addition to calculating the correlation, this service can be used to perform the intermediate operations: converting coordinates, data visualization, calculating interpolation, and creating a map of interpolated data.

The current version of the service allows for upload of multiple datasets in two coordinate systems: the Universal Transverse Mercator (UTM) system [1] and the spherical coordinates (longitude and latitude) provided either in decimal degrees or in degrees, minutes, and seconds. The following interpolation methods have already been implemented: Inverse Distance Weighted (IDW) [2]; k-nearest neighbors, and several variants of kriging [2, 3]: Simple Kriging (SK), Ordinary Kriging (OK) and Universal Kriging (UK). For each of kriging methods, one can set a type (exponential, spherical, Gaussian) and parameters of the variogram.

To implement the service, Java and JavaScript programming languages, the Spring Boot Framework for creating web applications, PostgreSQL database and Nginx web server have been used.

2 Architecture

The service consists of several independent modules (Fig. 1). The coordinate conversion module contains the logic for converting data between these coordinate systems supported by the application. This part of the service is used to convert the data to a

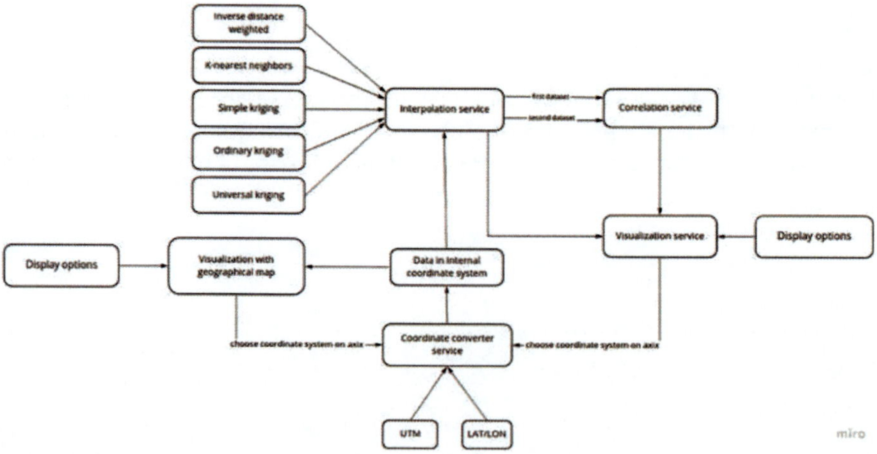

Fig. 1 The architecture of the presented service

common format when it is loaded into the system and then to display the resulting maps and data in a desired coordinate system. The data interpolation module is a tool which implements various interpolation methods with the ability to configure the corresponding parameters. The data visualization module is divided into two parts: one for creating images with a substrate map as a background, and one for images without a background. This part of the program also contains an independently implemented algorithm for drawing isolines, based on the marching squares algorithm [4]. Finally, the data analysis and validation module, monitors if the input data are correct, ensures that data are properly transferred between modules, and informs a user about errors detected in the data.

The implemented architecture makes it easy to increase the functionality of the service, since all components depend only on the internal data representation. If one needs to add a new coordinate system, only the implementation of a converter to the internal coordinate system is needed. Afterwards, a user will be able to use the new coordinate system without changing other components of the service. The same principle is employed for the interpolation methods. It is sufficient to define a new method with appropriate parameters and implement it for unified coordinate system. This implementation method that divides the program into independent parts, allows for minimization of the probability of errors at each stage, providing a simple and safe way to improve/replace any module as needed.

The service presented here has no dependencies on external libraries for working with geodata and their visualization (other than those related to background maps), allowing us to optimize each stage of the calculations. The lack of these dependences allowed us to implement certain algorithms and solutions which led to an improvement to the total performance of the service. For instance, it has been found that using UTM as the internal coordinate system speeds up the interpolation computation by several times compared to using longitude/latitude system, and thus one of the implemented solutions was UTM as the internal coordinate system. Otherwise, we would have to convert the data to the format required by a third-party application programming interface (API), degrading the performance.

3 Operating the Service

3.1 Uploading and Preparing Data

The first step in the service use is loading a dataset. The dataset must be a csv-file containing georeferenced data. After the data are uploaded to the website, part of it appears on the working page (Fig. 2). This allows the user to verify that the uploaded file is correct and to check that the data are correctly identified by the system and assigned to appropriate fields (data is in the correct format and valid). This function allows the user to catch some formatting errors, such as using a comma instead of a dot in fractional numbers, etc.

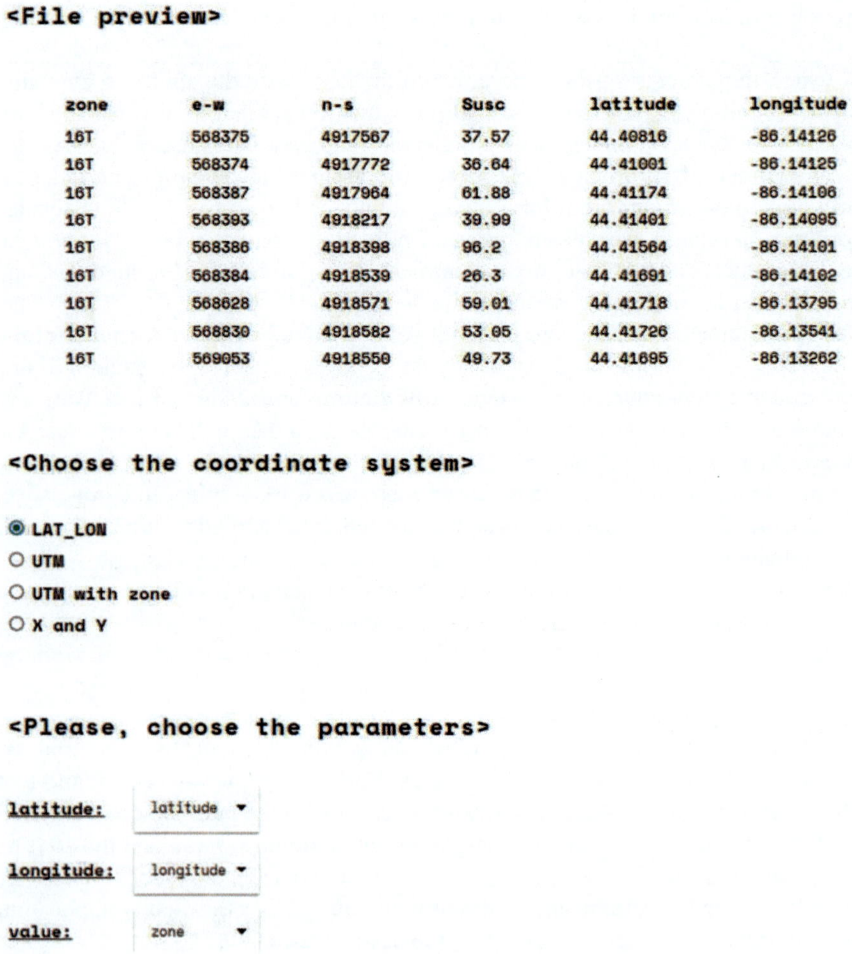

<File preview>

zone	e-w	n-s	Susc	latitude	longitude
16T	568375	4917567	37.57	44.40816	-86.14126
16T	568374	4917772	36.64	44.41001	-86.14125
16T	568387	4917964	61.88	44.41174	-86.14106
16T	568393	4918217	39.99	44.41401	-86.14095
16T	568386	4918398	96.2	44.41564	-86.14101
16T	568384	4918539	26.3	44.41691	-86.14102
16T	568628	4918571	59.01	44.41718	-86.13795
16T	568830	4918582	53.05	44.41726	-86.13541
16T	569053	4918550	49.73	44.41695	-86.13262

Fig. 2 The example of the preview of the data uploaded, displayed on working page

Next, the user is prompted to select a coordinate system which will be used in further mapping and confirm that the fields determined by the system are correct. The columns with coordinates are named in the common (standard) way will be detected automatically. If the service detects two coordinate systems in the dataset uploaded, the UTM-corresponding data will be automatically chosen as a primary one. Otherwise, the user must indicate correct fields that contain coordinates and for the parameter studied manually.

For the current version of the service, the standard method for the data recognition is as follows below.

3.1.1 Degrees, Minutes, Seconds

XX° YY′ ZZ″, where XX is degrees—an integer in the range [–90, 90] for latitude and [–360, 360] for longitude; YY is minutes—a positive integer in the range [0, 59]; ZZ is seconds—a positive integer or fractional number in the range [0, 59.99]. By default, the system uses positive and negative values for north and south latitudes, respectively. The same conversion is used for the longitude, positive values correspond to East, negative values to West. However, the latitude and longitude will also be recognized automatically if the field contains "N" or "S" letters in front of or after the of latitude, and "E" or "W" in the case of longitude. In this case for latitude N-marked values will be recognized as a positive and marked with letter "S" as negative. For longitude, "E" is positive, "W" is negative. For the automatic determination of the field corresponding to latitude and longitude, its header has to contain following sequences of letters: "latitude" or "lat" and "longitude", "long" or "lon" for latitude and longitude, respectively.

3.1.2 Decimal Degrees

XX.XXXXX—a fractional number in the range [–90, 90] for latitude and [–360, 360] for longitude. The service allows five decimal places, because six digits give a redundant value of accuracy of $\pm$ 0.1 m. In this case the system for determining the positive and negative directions and the system for the automatic determination of the fields in the dataset work in the same way as described above.

3.1.3 UTM

XXD YYYYYY ZZZZZZZ, where XX is a zone number (two-digit integer from 01 to 60) and D is zone letter (can be from "A" to "Z"). YYYYYY—the value of easting in meters, ZZZZZZZ—the value of northing in meters. The headers in UTM dataset have to contain following sequences of letters to be recognized automatically be the system: "z" or "zone" for zones, "e", "e-w", "east" or "easting" for east direction and "n", "n-s", "north" or "northing" for the northing, respectively.

This procedure of data uploading and verification can be done for two datasets. Since these two datasets can have different coordinate systems, the system will detect this mismatch and ask the user to choose the system most convenient for further display. Moreover, if the data in one of the fields does not fit the expected format, the service will report it (for example, if the longitude is out of the acceptable range as indicated above).

After the data loaded and verified, the system displays the brief statistics on the dataset (Fig. 3). The statistics contain:

- the number of points in the dataset;
- the maximum and minimum limits for the coordinate components;

Fig. 3 The example of the brief statistics view

- the maximum, minimum, median, and average values for the data.

Additionally, this module allows the user to convert uploaded data to any available coordinate system and download the result as a csv-file.

Users who only need to change the coordinate system in their dataset, can use "Convert and Download" button (Fig. 3), which will offer to choose the path to save resulting csv file. The file will contain additional columns with the coordinates presented in the new coordinate system. This function allows the users to use the service as the coordinate converter, but unlike the majority of similar tools, this service allows to convert the entire dataset at once.

3.2 Coordinate System Conversion

To convert to UTM, the transverse Mercator flattening series calculated from WGS84 data is used [5]. Coordinate conversion is performed by computing Gauss–Kruger series [6, 7]. The transformation from geographic to Cartesian coordinates for isometric latitude ψ is defined as

$$\psi = gd^{-1}\varphi - \frac{2\sqrt{n}}{1+n}\tanh^{-1}\left(\frac{2\sqrt{n}}{1+n}\sin\varphi\right) \tag{1}$$

where φ is geographic latitude, n is the third flattening and $gd(x) = \int_0^x \frac{dt}{\cosh t}$ is the Guderman function.

Holomorphic transformations are then performed to normalize range of variables. As a result, one gets the following expressions for complex planes ζ' and ζ:

$$\zeta' = gd(\psi + i(\lambda - \lambda_0)) \tag{2}$$

$$\zeta = \zeta' + \sum_{j=1}^{k} \alpha_j \sin\left(2j\zeta'\right) \tag{3}$$

where λ is longitude, λ_0 is longitude of the central meridian, and α is flattening series.

From formula (1), it is easy to derive:

$$\psi = \sinh\left(\tanh^{-1}(\sin\varphi) - \frac{2\sqrt{n}}{1+n}\tanh^{-1}\left(\frac{2\sqrt{n}}{1+n}\sin\varphi\right)\right) \tag{4}$$

From formulas (2), (3) and (4), using complex transformations and hyperbolic functions, one can derive following formulas:

$$\xi' = \tan^{-1}\left(\frac{\psi}{cos(\lambda - \lambda_0)}\right) \tag{5}$$

$$\eta' = \tanh^{-1}\left(\frac{\sin(\lambda - \lambda_0)}{\sqrt{1+\psi^2}}\right) \tag{6}$$

Then, using (3), (5) and (6) one gets:

$$E = E_0 + k_0A\left(\eta' + \sum\nolimits_{j=1}^{k}\alpha_j\cos\left(2j\xi'\right)\sinh\left(2j\eta'\right)\right) \tag{7}$$

$$N = N_0 + k_0A\left(\xi' + \sum\nolimits_{j=1}^{k}\alpha_j\sin\left(2j\xi'\right)\cosh\left(2j\eta'\right)\right) \tag{8}$$

where k_0—is the scale factor of central meridian, E is easting, N is northing. Similar equations are used to convert data back from UTM to longitude/latitude.

3.3 Visualization

At the beginning of the mapping stage, user selects the coordinate system that will be used for all following graphs. When the coordinate system is selected, the dataset will be visualized automatically. This stage allows the user to see where the data points are located on a geographic map and correlate the data points for the first and second datasets, depicting them on the same map.

The service provides several visualization options (Fig. 4). One option is a possibility to group points by their values, where a larger value corresponds to a larger point size on the map. If parameter set is "0"—the sizes will be assigned automatically (value by default). However, the user can select the number of such groups in the interface or pick "1" so that all data points have the same fixed size.

Another useful visualization option is the selection of the type of map on a background (substrate map): light, dark, streets, outdoors, satellite, and satellite streets. The substrate maps are taken as static images via the Mapbox API, which internally uses the open-source resource OpenStreetMaps (OSM) [8]. When the map

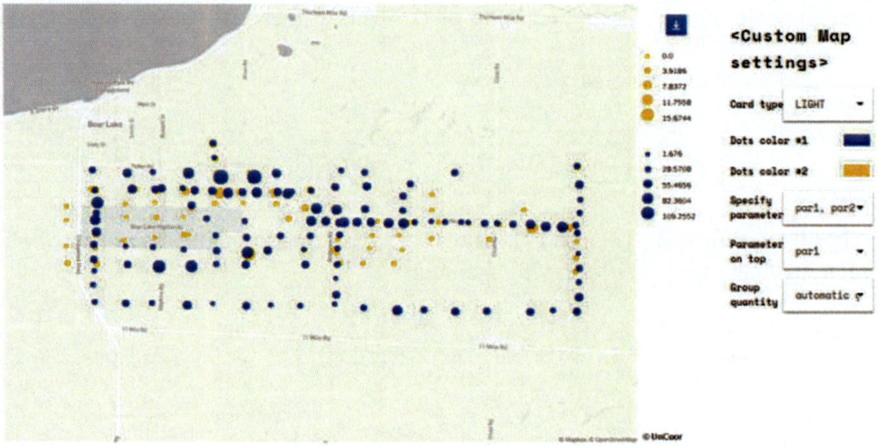

Fig. 4 The example of the data visualization

required for the substrate is received from the OSM server, data points are plotted on it according to the settings selected in the interface. The Mapbox API supports only the longitude/latitude coordinate format, but for higher accuracy in specifying points on the map, the service converts the data to the UTM system. The resulting image can be downloaded immediately in the *.png format.

3.4 Brief Statistical Overview

This tool allows user to examine the brief statistical overview for the initial data analysis (Fig. 5). The overview contains a histogram of the studied data distribution with a cumulative sum chart. The number of groups (columns in the histogram) to merge data is determined automatically. The service provides two overviews for each dataset.

Furthermore, the service provides the combined statistics for two datasets, which contains the total number of points, the area of intersection for the datasets as well as the minimum, maximum, average, and median values of the studied parameters. The intersection area is defined by the common zone covered by both datasets.

Under the combined statistic data, a geographical map is displayed that shows the area where the datasets intersect (using the map type selected in the previous step).

The preliminary statistical overview reflects the following parameters:

- how the data are distributed;
- what values prevail;
- whether there are intersects for the datasets studied;
- whether there are enough data points for the further analysis.

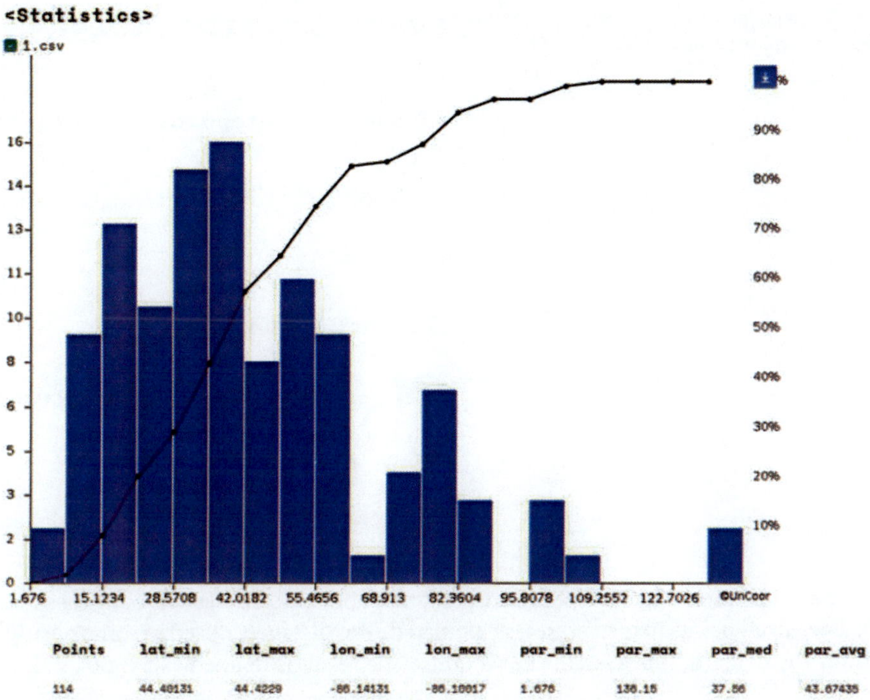

Fig. 5 The example of the brief statistical overview

3.5 Interpolation Methods

The datasets are verified and ready for interpolation. At this stage, the user can select one of the interpolation methods offered (Fig. 6). The list of methods implemented in this service to date, was formed by selecting the most valuable and widely used methods [9, 10] from a wide variety of available interpolation algorithms. The following methods are currently supported by the service:

- inverse weighted distance;
- k-nearest neighbor;
- simple kriging
- ordinary kriging
- universal kriging

For each method, the user can set additional parameters. For IWD, this is the power parameters to which the distance between the points should be raised; for the nearest neighbor method—the number of neighbors; for kriging—the type of the variogram and the values of it is parameters; for universal kriging—the maximum degree of approximating polynomials.

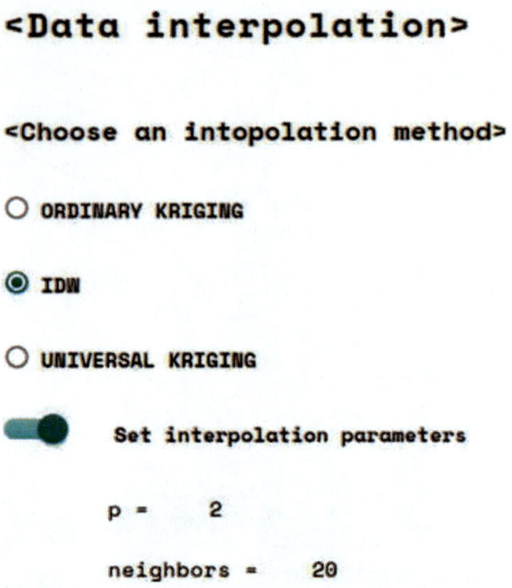

Fig. 6 The interpolation methods available

The calculations for IDW and k-nearest neighbors are similar. In both cases the distance to all points in the dataset is calculated. The distance, raised to some positive power (by default, 2) is used as the weights. The result is defined as a ratio:

$$\frac{\sum w_i * u_i}{\sum w_i(x)} \tag{9}$$

where w is the weight.

The interpolation algorithms differ in how they select the meaningful points from an entire dataset to calculate a new value. One of the three major approaches is commonly used for this purpose:

1. take all points;
2. to take into account only those points, the distance to which is less than the specified value;
3. to take into account only the k-nearest points.

The software implements approach 1 for IDW interpolation method and approach 3 in case of k-nearest neighbors method.

All selected kriging interpolation types (SK, OK and UK) are calculated in the same way. The kriging methods can be written in linear form:

$$V1(x) - m(x) = \sum w_i(x)(V(x_i) - m(x_i)) \tag{10}$$

where $w_i(x)$ are weights assigned to the data, V is the spatial variable, and $m(x)$ is the expected value of the spatial variable V.

Often V is split into two components: $V(x) = R(x) + m(x)$, where m is a deterministic trend and R is a random residual.

The types of kriging implemented in the program differ by the $m(x)$ model:

- in simple kriging, $m(x)$ is a known and constant value;
- in ordinary kriging, $m(x)$ is a constant but unknown value;
- in universal kriging, $m(x)$ changes smoothly throughout the entire field of research. The component $m(x)$ is modeled as a linear combination of the known functions $f(x)$: $m(x) = \sum a_i(x) * f_i(x)$.

A brief description of mathematics behind these three selected kriging interpolation methods is as follows:

Simple kriging (SK). Using of this type of kriging interpolation method requires solving a system of n equations:

$$\sum_{j=1}^{n} w_j^i * C_j^i = C_0^i \tag{11}$$

where C is the covariance of two random variables.

Limitations: This method should be used with caution if real average value in area is unknown or it does not match with the average value for the dataset.

Ordinary kriging (OK). Unlike simple kriging, the interpolation by the ordinary kriging method requires to solve a system of $n + 1$ equations:

$$\sum_{j=1}^{n} w_j * C_j^i + \mu = C_0^i, \tag{12}$$

$$\sum (w_i) = 1$$

where C is the covariance of two random variables.

In this implementation, the variogram is used as the covariance. μ is a Lagrange multiplier. The system of equations is solved with respect to the variables w and μ. If the determinant is not equal to zero, then the LU (lower–upper) decomposition of the matrix is used to optimize the computations; otherwise, the singular value decomposition is used.

Universal kriging (UK). To calculate universal kriging, it is necessary to solve equations of the form:

$$\sum_{j=1}^{n} w_j * C_j^i + \sum a_k(x) * f_k(x) = C_0^i, \tag{13}$$

$$\sum (w_i) = 1$$

Note that ordinary kriging is a special case of universal kriging with $k = 1$ and $f1(x) = 1$. As function f, an approximation of polynomials with increasing degree is often used. In the implementation offered, one can set the maximum degree of a polynomial. For example, for degree 2, the following polynomials will be used: $1, x, y, xy, x^2, y^2$. The result of the transformations is a system of $n+k+1$ equations, where k is the number of functions used for the approximation. This system must be solved for each point at which the interpolation is calculated. Detailed evaluation and derivation of formulas for universal kriging are given in [11].

3.6 Interpolation Map

Often the visualization of data is an especially useful tool in terms of preliminary analysis, which sometimes can be use as a self-contained informative tool. Therefore, the next tool available in the service is the creation of interpolation maps based on the data obtained in the previous step (Fig. 7). The service will calculate the data interpolation for the intersection area of two datasets studied and display interpolation maps with additional settings. The resolution of resulting maps is 800 × 600 pixels. Since the service calculates the interpolation on the intersection area of the datasets uploaded, the result does not depend on the order of data loading. The service allows the user to configure the following display parameters:

- whether the parameters need to be normalized;
- choosing colors for mapping;
- number of isolines;
- whether to show original points on top of the resulting map.

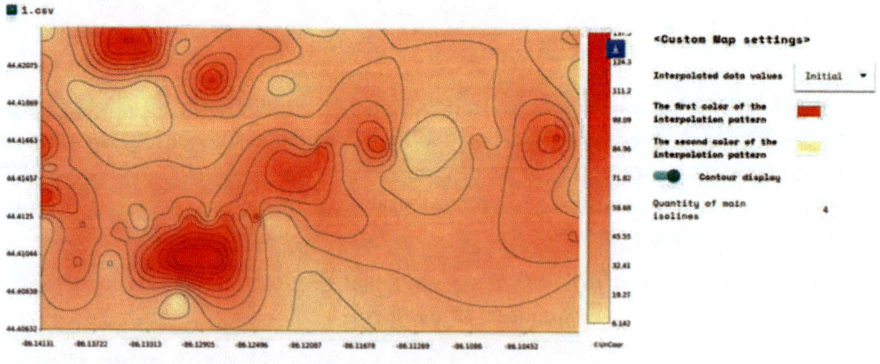

Fig. 7 The example of the interpolation map

The "Download" button located next to the image allows the user to download the table of data interpolated as well as resulting map in a usable format. The resulting table will be downloaded as a csv file and contain the coordinates and the corresponding value for each point displayed on the interpolation map. When creating the interpolation map for the second dataset, the user can specify different interpolation method and different displays parameters (by default, the same method and parameters are used).

In the case of loading and processing the single dataset, the map boundaries will be based on the minimum and maximum values of the coordinates of this dataset.

3.7 Correlation

The final data-analyzing tool created during the current stage of work on the service is calculating and displaying the spatial correlation map. The map shows the correlation between two studied datasets and provides an opportunity to scrutinize the data relationship.

The calculation of correlation between two datasets is performed as follows:

- Two equal-size datasets $\{A_i\}$ and $\{B_i\}$ $(i = 1 \ldots N)$ interpolated over the same spatial grid during the previous step (Sect. 3.6) are taken.
- The service calculate: $\left\{A_i^{'}\right\} = \{A_i\}$—median $\{A_i\}$ and $\left\{B_i^{'}\right\} = \{B_i\}$—median $\{B_i\}$ for $i = 1 \ldots N$.
- Resulting data are normalized: if $\left\{A_i^{'}\right\} \geq 0$, $\left\{A_i^{*}\right\} = \left\{A_i^{'}\right\}/\max\left\{A_i^{'}\right\}$, if $\left\{A_i^{'}\right\} < 0$, $\left\{A_i^{*}\right\} = \left\{A_i^{'}\right\}/\max\left\{A_i^{'}\right\}$ for $i = 1 \ldots N$.
- Calculating the interpolation value in each point: $\left\{C_i^{*}\right\} = \left|\left\{A_i^{*}\right\} - \left\{B_i^{*}\right\}\right|$ for $i = 1 \ldots N$.
- The values of $\left\{C_i^{*}\right\}$ may range from 0 (perfect correlation) to 2 (perfect anticorrelation) (Fig. 8).

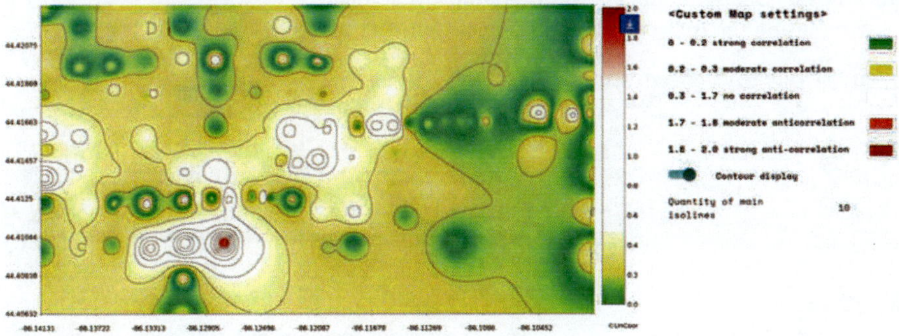

Fig. 8 The example of the correlation map

- The correlation degree is calculating: $D_{corr} = N_c/N$ where N_c is the number of points $\left\{C_i^*\right\} \leq 0.2$ (this is 10% of all range, and this value set by default).

After creating the interpolation maps, it becomes possible to configure display parameters. The user can configure the colors and confidence intervals for displaying data. The isolines and their number can be set in the same way as in interpolation display settings (Fig. 8).

According to math described above, the results of the correlation calculations are a list of points with values ranging from 0 to 2, which covers studied area with resolution 800 $\times$ 600 pixels. If necessary, this range can be shifted to values from $-$ 1 to 1.

The data obtained during the entire sequence, from uploading raw datasets to calculating correlation map with the settings selected, can be downloaded as a single archive with all the tables and processed images. The archive in zip format will contain:

- geographical maps with points from datasets;
- histograms of data distribution;
- two interpolation maps;
- two csv files with the interpolated data;
- correlation map;
- csv file with information on the correlation values at each point.

All images will be identical to those that were shown during the use of the service.

4 Conclusions

The work on this service has already been tested on real data in one of the studies related to the spatial data analysis. The results obtained from this study confirmed the efficiency of the proposed service and the possibility of its further application for the analysis of spatial data in various fields of science.

We outline the main advantages of using the service:

Web service implementation —the service is implemented as a web-application and the user is not required to install additional software. Moreover, this approach will allow use of the service from different devices regardless of OS, since the resource-intensive computing is performed on a server and the end user need only a web browser.

Multitool —The service provides a set of useful tools for effective analysis preparation from coordinate system converter to powerful interpolation and correlation tools.

Flexibility —the user can use different methods, change their parameters, and set the data visualization during the using the service.

Informativeness —Besides the maps and other visualized data the user is able to download all intermediate results to perform additional investigation.

Although the service is a complete set of tools for the spatial correlation analysis, we believe that it should be further improved and evolved to expand the available toolset. At the next stage we intend to consider the following ideas related to the development of the service from the point of view of their implementation:

- improving maps resolution;
- using machine learning to determine the best data interpolation methods for datasets studied;
- creation of graphical tools for more effective and flexible map visualization.

We believe that our service can be useful in various fields of science and can become a highly resourceful tool for spatial data analysis. We intend to work on support and improving the service as well as continuously expanding its functionality.

The service presented is available at the following link: http://uncorr.com/

References

1. Michael, N.: DeMers: Cartesian coordinate systems. Int. Encyclop. Geogr. (2017). https://doi.org/10.1002/9781118786352.wbieg0221
2. Lam, N.S.N.: Spatial interpolation methods: a review. Am. Cartograph. **10**(2), 129–150 (1983). https://doi.org/10.1559/152304083783914958
3. Oliver, M.A., Webster, R.: Kriging: a method of interpolation for geographical information systems. Int. J. Geogr. Inf. Sys. **4**, 313–322 (1990). https://doi.org/10.1080/02693799008941549
4. Maple, C.: Geometric design and space planning using the marching squares and marching cube algorithms. In: Proceedings of 2003 International Conference Geometric Modeling and Graphics, pp. 90–95 (2003). doi:https://doi.org/10.1109/GMAG.2003.1219671
5. Kumar, M.: World geodetic system 1984: a modern and accurate global reference frame. Mar. Geod. **12**, 117–126 (1988). https://doi.org/10.1080/15210608809379580
6. Charles, F., Karne, F.: Transverse Mercator with an accuracy of a few nanometers. J. Geod. **85**, 475–485 (2010). https://doi.org/10.1007/s00190-011-0445-3
7. Kawase, K.: Concise derivation of extensive coordinate conversion formulaein the Gauss-Krüger projection. Bull. Geospatial Inform. Authority Japan 60, 1–6 (2013)
8. Mooney, P., Minghini, M.: A review of OpenStreetMap data. In: Foody, G. et al (eds.) Mapping and the Citizen Sensor, pp. 37–60 (2017). doi:https://doi.org/10.5334/bbf.c
9. Li, J., Heap, A.D. Spatial interpolation methods applied in the environmental sciences: a review. Environment. Modell. Soft. **53**, 173–189 (2013). https://doi.org/10.1016/j.envsoft.2013.12.008
10. Wu, Y.H., Hung, M.C.: Comparison of spatial interpolation techniques using visualization and quantitative assessment. In: Hung, M.C. (ed.) Applications of Spatial Statistics (2015). doi:https://doi.org/10.5772/65996
11. Menafoglio, A., Secchi, P.: A universal kriging predictor for spatially dependent functional data of a Hilbert space. Elect. J. Stat **7**, 2209–2240 (2013). https://doi.org/10.1214/13-EJS843

Paleo-, Earth-, Rock-, Environmental Magnetism, and Geophysical Fluid Dynamics

Magnetostratigraphic Constraints on the Position of the Tremadocian–Floian Boundary at the Key Section of the Moyero River Valley (Siberian Platform)

Vladimir Pavlov, Andrei Dronov, Alexander Larionov, and Tatiana Tolmacheva

Abstract The magnetostratigraphic study of the geomagnetic reversal frequency just before the Ordovician Moyero superchron can be very useful for testing various geodynamic conceptions and different geodynamo models describing the geomagnetic field evolution. In terms of completeness of the geological record, the Ordovician carbonate-terrigenous key section of the Moyero River valley, located in the northern part of the Siberian Platform, is considered to be one of the best Ordovician sections of Northern Eurasia and seems to be very promising to carry out such a study. However, the Tremadocian part of the section, which could record the pre-superchron reversal history, until now remained poorly studied and exact positions of its boundaries with underlying Upper Cambrian and overlying Arenigian (Floian) strata are still not well determined. In this study we present magnetostratigraphic data, which taken together with obtained biostratigraphic data, allow constraining the position of the Tremadocian–Arenigian (Floian) boundary within the key Moyero River section.

Keywords Superchron · Magnetic stratigraphy · Paleomagnetism · Ordovician · Tremadocian–Floian boundary

1 Introduction

The study of the geomagnetic reversal frequency just before the onset of superchrons can be very useful for testing various conceptions and different geodynamo models describing the geomagnetic field evolution. Numerous studies of changes of geomagnetic reversal frequency prior to the youngest Cretaceous superchron result

V. Pavlov (✉)
Institute of Physics of the Earth of Russian Academy of Science, Moscow, Russia

A. Dronov
Geological Institute of Russian Academy of Science, Moscow, Russia

A. Larionov · T. Tolmacheva
Russian Geological Research Institute, St. Petersburg, Russia

A. Kosterov et al. (eds.), *Problems of Geocosmos—2020*, Springer Proceedings in Earth and Environmental Sciences, https://doi.org/10.1007/978-3-030-91467-7_9

in rather ambiguous conclusions on the pattern of its evolution (for discussion see Hulot et al. [1]) and require additional data from other superchrons. Second by age—the Kiaman superchron—is rather difficult to study mainly due to the limited number of suitable sections. On the contrary, the third Phanerozoic superchron–superchron Moyero—seems to be very interesting for such studies as in Siberia there are a number of sections which have been formed over the time preceding the Moyero superchron and covering its beginning. One of them—the carbonate-terrigenous section of Moyero River is perhaps the most promising one to obtain a detailed magnetostratigraphic record of the geomagnetic polarity changes just before the third Phanerozoic superchron. This section is very well exposed, composed mainly by red lithologies favorable for paleomagnetism and its Tremadocian part (recall that Tremadoc is the Early Ordovician subdivision immediately preceding the Moyero superchron) has a thickness which is unusually high for platforms.

However, until now the Tremadocian part of the section remained poorly studied and the positions of its boundaries with underlying Cambrian and overlying Floian (Arenigian) strata were not determined. This situation significantly complicates the use of the Tremadoc deposits of the Moyero section for the study of the geomagnetic field evolution. To determine the position of the lower and upper boundaries of the Tremadoc, a detailed study of distribution of conodonts in the section is necessary. However, the thickness of the Tremadoc in the Moyero valley can reach 400 m and even more, making such detailed studies very laborious and time consuming.

In this paper we present our new magnetostratigraphic data which allow to limit the stratigraphic interval where the Tremadocian–Floian boundary in the Ordovician key section of the Moyero River valley is located.

2 Geology

In terms of completeness of the geological record and quality of exposure, the Ordovician key section of the Moyero River valley, located in the northern part of the Siberian Platform, is considered to be one of the best Ordovician sections of Northern Eurasia. Our new data (paper in preparation) indicate the occurrence in this section of the Cambrian–Ordovician boundary and the existence of the several hundred meters thick Tremadocian sequence. Combination of the excellent section with the exceptional esthetic value of the region (deep canyons, waterfalls, rapids), makes the Moyero River Section an outstanding geological monument of planetary significance.

Despite the fact that the presence of the Tremadocian strata has been known in the region for a long time, they have never been studied in detail and, moreover, even their thickness has never been established correctly. In previous studies [2] the sequence of the Ordovician rocks has been described in a series of outcrops along the Moyero River valley (Fig. 1). Only the interval from the Arenigian (Floian) to Ashgillian (Katian) part of the section, exposed between outcrops nos. 101a and 75, have been thoroughly studied. The presence of the Tremadocian strata in the section

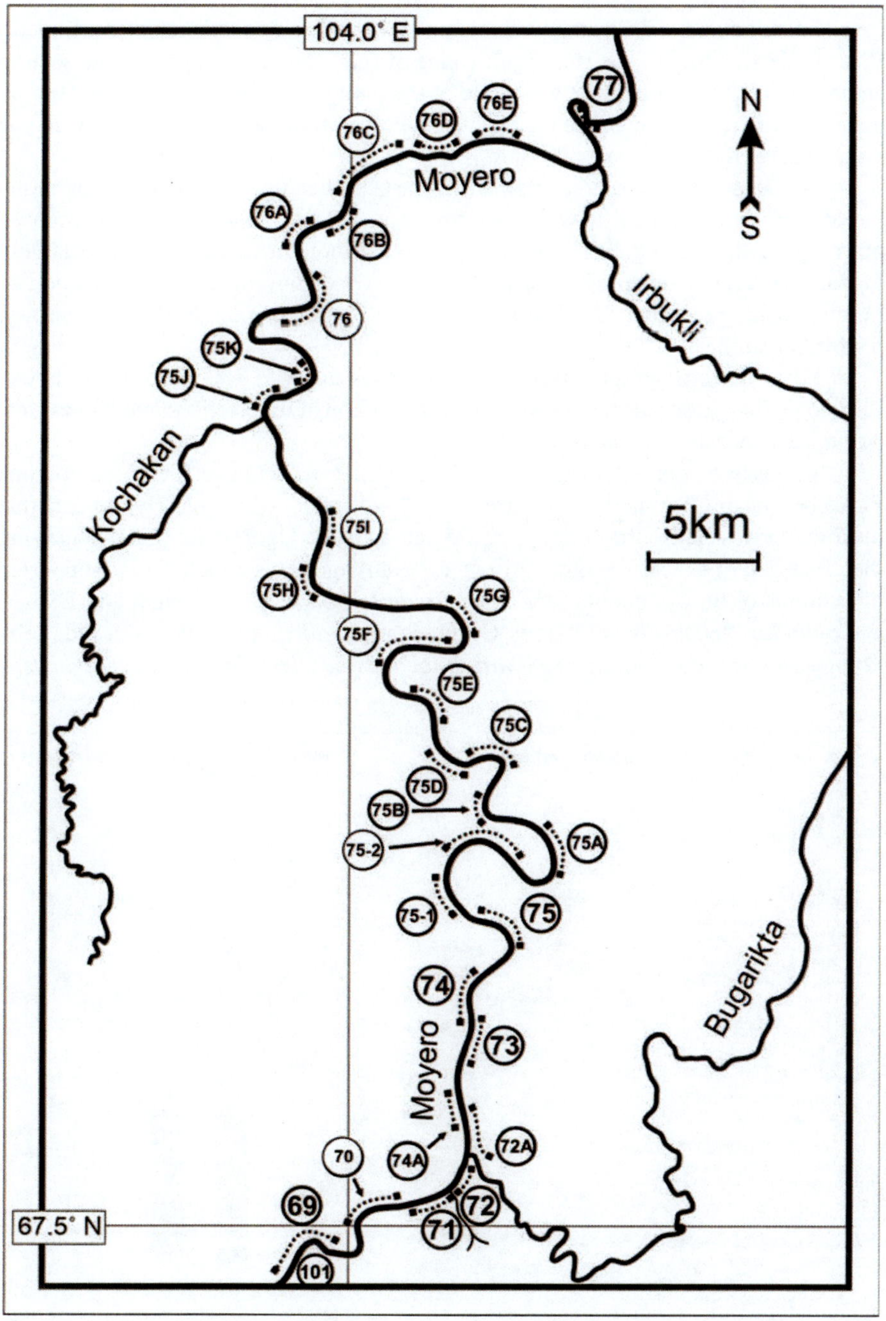

Fig. 1 Location of Ordovician outcrops along the Moyero river valley

(in outcrops nos. 76, 77, 78 and 79) has been just briefly noted as being rather limited. Clear understanding of the true significance of the Tremadocian part of the section came only much later in 2013 when one of the authors of this article—A.V. Dronov discovered that this part of the Ordovician succession in the Moyero River valley is much thicker than previously thought [2].

It has been established that numerous vast, earlier unexplored outcrops along the Moyero River valley between outcrops nos. 75 and 76 (Fig. 1) are composed not only by the Arenigian rocks (as it has been thought earlier) but, also largely of rocks of Tremadocian age. At the same time determining the exact position of Tremadocian–Arenigian (Tremadocian–Floian) boundary remained to be the task for further studies.

In Fig. 2 the stratigraphic position of outcrops nos. 75, 75C and 75F is shown. Lithologically, these outcrops represent an alternation of limestone, clayey limestone, dolomite, and, rarely, siltstone and claystone.

At the base of the outcrop no. 75, Myagkova et al. [2] have found macrofauna Finkelburgia sp., Hormatoma aff. artemosia, Opileta sp., which clearly indicate that the lowermost strata of the outcrops no. 75 belong to the Ugorian Regional Stage and, therefore, are of Lower Arenigian (Lower Floian) age. This conclusion is supported by findings of the conodonts (our data) *Drepanoistodus cf. nowlani* Ji and Barnes, *Scalpellodus? felicitii* Ji and Barnes, Glyptoconus *quadriplicatus* Barnson and Mehl, *Paroistodus sp., Polycostatus sp., Variabiloconus bassleri* Furnish, *Scandodus sp.,*

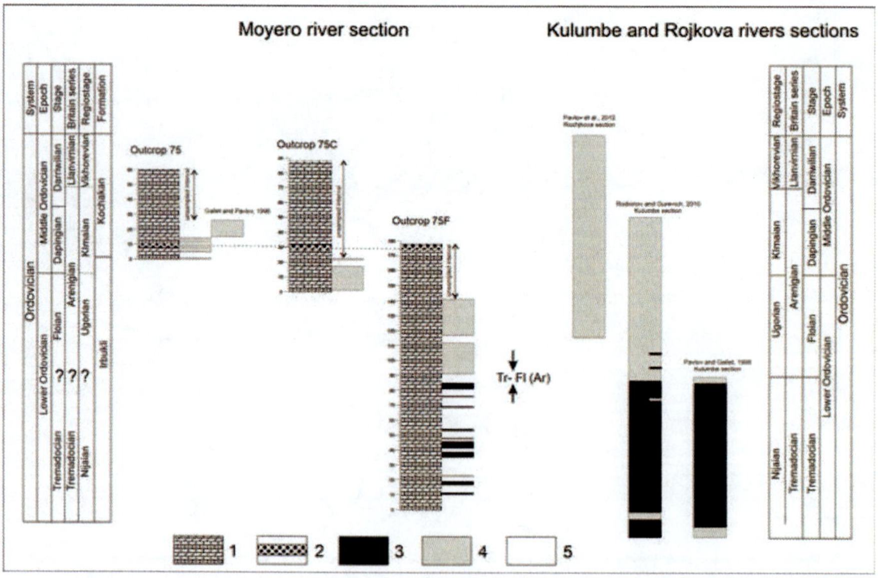

Fig. 2 Magnetic stratigraphy of the studied sections of the Moyero River valley. 1—Alternation of limestone, clayey limestone, dolomite, and, rarely, siltstone and claystone; 2—correlative bed of yellow dolomite; 3—normal polarity; 4—reverse polarity; 5—no primary record. Thick arrows indicate the Tremadocian–Arenigian (Floian) position

Acanthodus sp., Glyptoconus quadriplicatus Barnson and Mehl, *Paroistodus sp., Polycostatus sp.* at approximately the same stratigraphic level.

On the other hand, in the sample taken in the lowermost part of the outcrop no. 75F we have found conodonts *Glyptoconus quadriplicatus* Barnson and Mehl, *Variabiloconus bassleri* Furnish, *Rossodus manitouensis* Repetski and Ethington, *Scalpellodus? felicitii* Ji and Barnes, Acodus sp., which clearly indicate a Late Tremadocian age of the base of the outcrop no. 75F.

Thus, the boundary between Tremadocian and Arenigian (Floian) in the section of the Moyero River is situated in a ~170 m thick stratigraphic interval between the levels corresponding to the lowermost beds of outcrops nos. 75 and 75F.

3 Paleomagnetism and Magnetostratigraphy

For this study, we have sampled the lowermost parts of sections 75 and 75C and most of section 75F with the exception of its uppermost part, which is not accessible without special climbing equipment. Depending on accessibility and lithology these sections were sampled with intervals from 0.2 to 2.5 m. A total of 245 hand samples were collected. Samples were picked by hand with the use of a magnetic compass; the magnetic declinations in the study region were determined from the 13th generation IGRF model.

All samples were subjected to laboratory processing within a standard paleomagnetic procedure [3–5]. The processing was carried out in the Center for Rock physics, Geomechanics and Paleomagnetism of the Institute of Physics of the Earth of the Russian Academy of Sciences (Moscow, Russia). Demagnetization of the samples was conducted in a TD80 thermal demagnetizer (Magnetic Measurements, Great Britain) in a special room shielded from the ambient magnetic field. The measurements of natural remanent magnetization (NRM) were performed in the same room on a cryogenic magnetometer manufactured by 2G Enterprise, USA. Depending on the character of the paleomagnetic record, thermal demagnetization included between 12 to 18 heating steps up to the temperatures corresponding to the Curie point of hematite.

The results of thermal demagnetization show that a considerable part of the samples contains only one component with step positive inclination. The direction of this component perfectly coincides with the direction of the Permo-Triassic intrusions, widespread along the Moyero river valley and studied by us earlier [6]. Therefore, this component is interpreted to be a secondary overprint.

However, at least 40% of the samples, mostly red and reddish in color, also contain primary characteristic magnetization. The presence of this component is suggested by either from straight segments directed to the origin of the Zijderveld's diagrams, or from remagnetization circles (Fig. 3) passing through the area of the expected Lower Ordovician direction [7].

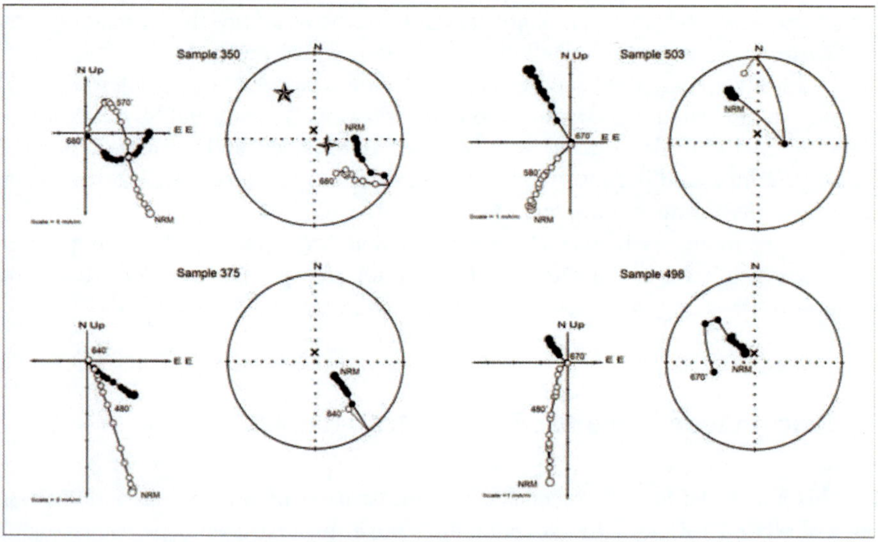

Fig. 3 The character of the paleomagnetic record in samples containing primary component of magnetization. Zijderveld's diagrams: filled (open) circles designate projections of vectors on the vertical (horizontal) plane. Stereograms: filled (open) circles designate projections of vectors on the lower (upper) hemisphere. Oblique cross denotes local present-day field, four-pointed star denotes mean trap intrusions direction, five-pointed star denotes mean Lower Ordovician paleomagnetic direction [7]

The primary origin of the isolated characteristic component follows from the comparison with the primary Lower Ordovician directions of the Moyero River section described and studied in Gallet and Pavlov [7]. Since these samples are distributed more or less evenly across the section, they can be used to construct its magnetostratigraphic profile (Fig. 2). While the lowermost parts of outcrops nos. 75 and 75C are characterized exclusively by a reverse polarity component, outcrop no. 75F contains both polarities. Mean directions of normal and reverse polarity do not pass the reversal test due to sensible Permo-Triassic overprint. Nevertheless, averaging the directions of normal and reverse polarity yields a mean direction (D = 326.8°, I = 35.4°, K = 13.4, alfa95 = 5.4°, N = 54) which is very close to the ones earlier obtained for the Tremadoc and the Arenig of the Moyero river section (Table 1) and differs significantly from any directions expected from younger poles [7].

The layers corresponding to ~85 m of the lower part of this outcrop are magnetized mainly with normal polarity, and only two samples at a level of ~22 m from the base of the outcrop are magnetized reversely. Magnetic polarity changes at the interval between 84 and 92 m. The beds from this interval do not contain a primary paleomagnetic record. Beginning from level ~92 m we observe exclusively a reverse polarity. The studied beds of the outcrops nos. 75 and 75C, which lie stratigraphically higher than considered beds of the outcrop 75F, demonstrate only reverse polarity.

Table 1 Paleomagnetic directions

	N	D (°)	I (°)	k	α95 (°)
Normal polarity	18	140.3	−22.8	15.3	9.1
Reverse polarity	36	329.5	41.6	17.2	5.9
Mean direction	54	326.8	35.4	13.4	5.4
Tremadoc mean direction [7]	15	333.3	37.3	18.3	9.2
Arenig mean direction [7]	144	319.5	31.7	29.7	2.2

N—number of samples, D—declination, I—inclination, k, α95—Fisher's [11] precision parameter, radius of the 95% cone of confidence about the mean direction

4 Discussion and Conclusion

The magnetostratigraphic profile which we have obtained for outcrops nos. 75, 75C and 75F can be compared with the coeval ones described for the Siberian Lower Ordovician key sections Kulumbe and Rojkova [8–10]. The main feature of the Lower Ordovician magnetostratigraphic records of these sections is the absolute predominance of reverse polarity above the Tremadocian–Arenigian (Floian) boundary and the existence of a thick interval of normal polarity below this boundary. The same feature we can observe also in the magnetostratigraphic record of studied outcrops of Late Tremadocian–Lower Arenigian (Floian) age. This similarity is a clear indication that the thick normal polarity interval from the outcrop no. 75F should be correlated with the Tremadocian normal polarity interval from the Kulumbe section and the reverse polarity interval is the same as the reverse polarity interval which begins at the proximity of the Tremadocian–Arenigian boundary in the Kulumbe River section and which indicate the onset of the Lower-Middle Ordovician reverse polarity superchron.

Thus, the Tremadocian–Arenigian (Floian) boundary in the Moyero River section is likely located in the outcrop no. 75F in the interval between 84 and 92 m from the base of the section.

Acknowledgements The field work and the laboratory paleomagnetic studies were supported by RSF grant 20–17–00198. Data interpretation was supported by the Ministry of Science and Higher Education of the Russian Federation under the Agreement no. 14.Y26.31.0029 as part of the implementation of the Russian Federation Government Resolution no. 220.

References

1. Hulot, G., Finlay, C.C., Constable, C.G., et al.: The magnetic field of planet earth. Space Sci. Rev. **152**, 159–222 (2010). https://doi.org/10.1007/s11214-010-9644-0
2. Myagkova, E., Nestor, H., Einasto, R.: Stratigraphic Sequence of Ordovician and Silurian Deposits from the Moyero River. Nauka, Novosibirsk (1977) (in Russian)

3. Butler, R.F.: Paleomagnetism: Magnetic Domains to Geologic Terranes. Blackwell Sci. Publ., Boston (1992)
4. Khramov, A.N. (ed.): Paleomagnitologiya. Nedra, Leningrad (1982) (in Russian)
5. Tauxe, L.: Essentials of Paleomagnetism. University of California Press, Orlando (2010)
6. Pavlov, V.E., Courtillot, V., Bazhenov, M., et al.: Paleomagnetism of the Siberian traps: new data and a new overall 250 Ma pole for Siberia. Tectonophysics **443**, 72–92 (2007)
7. Gallet, Y., Pavlov, V.: Magnetostratigraphy of the Moyero river section (northwestern Siberia): constraint on the geomagnetic reversal frequency during the early Paleozoic. Geophys. J. Int. **125**(1), 95–105 (1996)
8. Pavlov, V.E., Gallet, Y.: Upper Cambrian to Middle Ordovician magnetostratigraphy from the Kulumbe river section (northwestern Siberia). Phys. Earth Planet. Inter. **108**(1), 49–59 (1998)
9. Pavlov, V.E., Veselovskiy, R.V., Shatsillo, A.V., et al.: Magnetostratigraphy of the Ordovician Angara/Rozhkova River section: further evidence for the Moyero reversed superchron. Izv. Phys. Solid Earth **48**(4), 297–305 (2012)
10. Rodionov, V.P., Gurevich, E.L.: Key magneto-stratigraphic sequence of the Lower Ordovician deposits, north-western Siberian platform. Neftegazovaya Geologiya. Teoriya I Praktika. **5**(3), 1–17 (2010) (in Russian with English abstract)
11. Fisher, R.A.: Dispersion on a sphere. Proc. R. Soc. Lond. Ser. A **217**, 295–305 (1953)

Upper Cretaceous Paleomagnetism and Magnetostratigraphy of the Pur-Taz Interfluve (Northern West Siberia)

Z. N. Gnibidenko, A. V. Levicheva, L. G. Smolyaninova, V. A. Marinov, and N. N. Semakov

Abstract Upper Cretaceous magnetostratigraphy of northern West Siberia has been reconstructed using new paleomagnetic data from five wells in the Kharampur oil-gas-condensate field (1049, 109N, 106P-Yu, 105N, and 2073N). The scientific motivation of this investigation was to construct a composite Upper Cretaceous magnetostratigraphic section of the north of Western Siberia based on integrated paleomagnetic and biostratigraphic data, which represent a part of the Upper Cretaceous magnetic polarity scale of all Western Siberia. The study area is located within the Middle Pur trench in the Pur-Taz interfluve. Natural remanent magnetization in the samples of the Pokur, Dorozhkovo, Okhteurievka, Kuznetsovo, Lower Berezovo, and Upper Berezovo Formations was analyzed by stepwise alternated field (AF) and thermal demagnetization and used to compile the magnetostratigraphic sections of the five wells. Correlations among the well sections form a basis for a composite magnetostratigraphic record of Upper Cretaceous (Cenomanian through Campanian) strata in the area. The composite section comprises two normal polarity zones $N\mathrm{K}_2$(sn-st) and $N\mathrm{K}_2$cp and correlates with the regional Upper Cretaceous magnetostratigraphic section of southern West Siberia, as well as with the global polarity time scale.

Keywords Paleomagnetism · Magnetostratigraphy · Geomagnetic polarity · Magnetozone · Upper Cretaceous · Pur-Taz interfluve · Northern West Siberia

Z. N. Gnibidenko (✉) · A. V. Levicheva · N. N. Semakov
Institute of Petroleum Geology and Geophysics SB RAS, Koptug Ave. 3, 630090 Novosibirsk, Russia
e-mail: gnibidenkozn@ipgg.sbras.ru

A. V. Levicheva
e-mail: levichevaav@ipgg.sbras.ru

N. N. Semakov
e-mail: semakovnn@ipgg.sbras.ru

L. G. Smolyaninova
Institute of Geology and Mineralogy SB RAS, Koptug Ave. 3, 630090 Novosibirsk, Russia

V. A. Marinov
Tyumen Petroleum Science Center, 79/1, Osipenko Str., 625000 Tyumen, Russia

A. Kosterov et al. (eds.), *Problems of Geocosmos—2020*, Springer Proceedings in Earth and Environmental Sciences, https://doi.org/10.1007/978-3-030-91467-7_10

1 Introduction

In the recent decade, the Upper Cretaceous and the Cretaceous/Paleogene boundary of southern West Siberia have been investigated systematically by high-resolution paleomagnetic methods. The results include magnetostratigraphic records of the respective time span from three wells (8, 10, 2) in the Omsk basin [1], two wells (S-114 and S-124) in the Bakchar iron province [2], and two wells (23 and 19) in the southern Kulunda basin [3]. Data from those wells are synthesized in a correlation chart of the wells and in a composite regional Upper Cretaceous and Cretaceous/Paleogene geomagnetic polarity scale as a part of the respective Cretaceous scale of West Siberia (Fig. 1) [4].

The studies of the Upper Cretaceous paleomagnetism continue in northern West Siberia, where the Cretaceous section is among most complete sedimentary records of Northern Asia. Data from the northern Krasnoyarsk region cover the lower Upper Cretaceous sedimentary fill of the Yenisei-Khatanga basin (Ust-Yenisei facies zone), which still remains not studied with paleomagnetism. The basin sediments belong to the Dolgan, Dorozhkovo, and Nasonov Formations [5] stripped by eight wells (Khikiglinskaya 1, Suzunskaya 34, East Lodochnaya 1, West Tagulskaya 1, Tagulskaya 21 and 25, and Vankorskaya 10 and 13). The paleomagnetic sections of the three formations sampled in the wells are mainly of normal polarity interrupted by few reverse polarity excursions: a zone of normal polarity corresponding to the Turonian, with thin subzones of reverse polarity. According to geological, stratigraphic, paleontological, and paleomagnetic constraints, the sediments were deposited in Late Cretaceous-Turonian time, which corresponds to the fragment of normal polarity

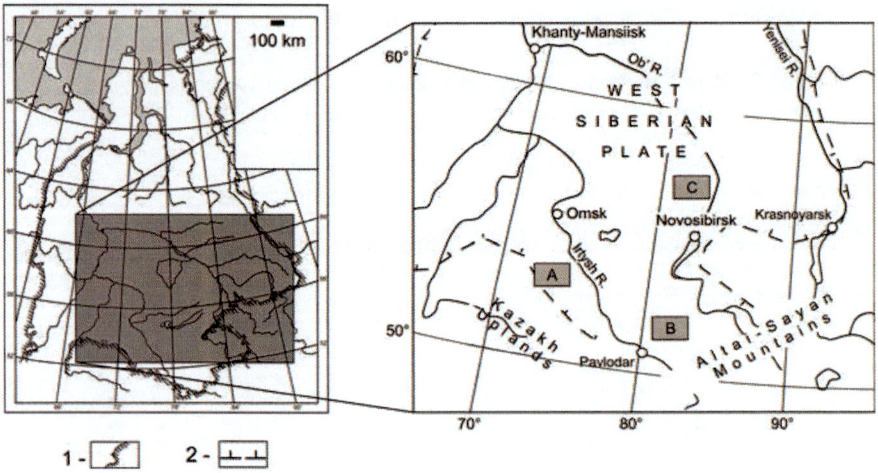

Fig. 1 Location map of wells in Upper Cretaceous and Cretaceous–Paleogene boundary strata in southern West Siberia. A, Omsk Basin (3 wells); B, southern Kulunda Basin (2 wells); C, Bakchar Iron Basin (2 wells). 1, 2, limits of the West Siberian Plate

Chron C34 (~94–89.8 Ma) of the global geomagnetic scale [6]. Recently data on Upper Cretaceous magnetostratigraphy of northern West Siberia were obtained from three wells (Kharampur 106P, Zapadno-Chaselskaya 1P and Novo-Chaselskaya 5P) in the Pur-Taz interfluve (Taz facies zone). The results were partly published in conference proceedings of the 12th International Conference "Problems of Geocosmos" in St. Petersburg (Petergof) in 2018 [7] and in the proceedings of the 25th anniversary Seminar on Rock Magnetism and Paleomagnetism in Borok in 2019 [8].

This paper presents new paleomagnetic data from five Upper Cretaceous drill sections in the Kharampur oil-gas-condensate field (1049, 109N, 106P-Yu, 105N, and 2073N) in the Pur-Taz interfluve (Fig. 2). The scientific motivation of this investigation was to construct a composite Upper Cretaceous magnetostratigraphic section of the north of Western Siberia based on integrated paleomagnetic and biostratigraphic data, which constitute a portion of the Upper Cretaceous magnetic polarity scale of all Western Siberia. The paleomagnetic data, combined with geological, stratigraphic, and paleontological evidence, were used to reconstruct a composite Upper Cretaceous magnetostratigraphic record of northern West Siberia. Regional magnetostratigraphic section of the Upper Cretaceous of southern Western Siberia already has been developed [4]. The specific objectives include rock magnetism and paleomagnetism of the core samples, and correlation of the obtained north West Siberian Upper Cretaceous magnetostratigraphy with that for southern West Siberia and with the global geomagnetic polarity scale [9].

2 Materials and Methods

The study area is located in the northern West Siberian Basin, in the Middle Pur tilted trench, where Upper Cretaceous reservoirs store rich oil and gas resources [10]. The Cretaceous portion of the Mesozoic section in northeastern West Siberia is a thick clastic sequence, more voluminous than Triassic and Jurassic sediments. The Upper Cretaceous stratigraphy in the area has been well documented [11–13]. The sediments are mainly marine facies with representative foraminifers' assemblages, which provide biostratigraphic reference for the basin interior. The foraminifera-based stratigraphic division is limited to West Siberia, and the regional stage division of the Upper Cretaceous is guided also by rare index fauna species of ammonites and inoceramidae bivalves (Fig. 3).

Paleomagnetic parameters were measured and analyzed in 617 "top–bottom" oriented 8 cc cubic samples cut from 225 large core samples that represent individual stratigraphic levels in five oil wells (2–3 specimens per stratigraphic level). The oriented samples were picked manually or extracted using a Bishaev-type corer, from the center of the core, with the Z-axis directed downward, and random X and Y directions. The coring of the oil wells was fragmentary. Samples were taken with approximately 1 m spacing.

Measured parameters included magnetic susceptibility (χ) and its temperature dependence, as well as natural remanent and induced magnetization (NRM, IM,

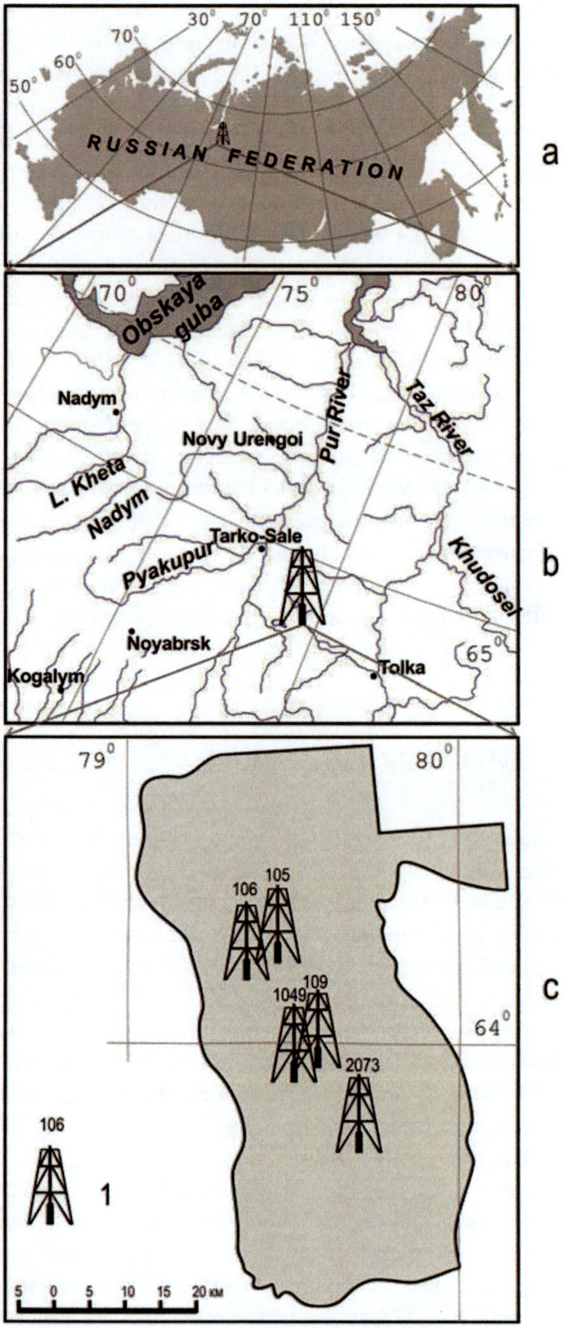

Fig. 2 Location map of wells **a** in Russian Federation; **b** of the Pur-Taz interfluve, northern West Siberia; **c** five wells in the Kharampur oil-gas-condensate field. 1—wells

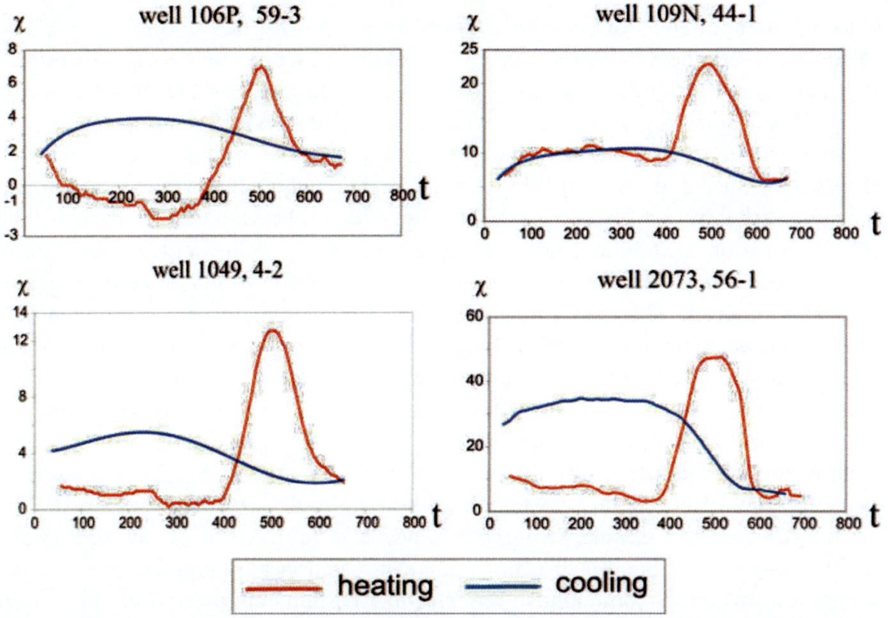

Fig. 3 Temperature dependence of magnetic susceptibility in core samples of Upper Cretaceous sediments

respectively). The Koenigsberger ratio was determined as (Qn = NRM/IM). The samples were subjected to stepwise thermal and alternated field (AF) demagnetizations. Demagnetization data were analyzed using principle component analysis, in order to isolate characteristic NRM component (ChRM).

The used instruments were: a Bartington MS-2 meter with a dual-frequency sensor for magnetic susceptibility and its temperature dependence; 2G Enterprises cryogenic and AGICO JR6A spinner magnetometers for the magnitude and direction of NRM; ASC TD48 nonmagnetic oven for stepwise thermal demagnetization and an AF demagnetizer built into the 2G Enterprises cryogenic magnetometer for stepwise AF demagnetization. The NRM components were analyzed by AGICO Remasoft 3.0 paleomagnetic processing software.

3 Results

The cored rocks of the Pokur, Kuznetsovo, Dorozhkovo, Okhteurievka, and Lower and Upper Berezovo Formations (silt, siltstone, mudstone, sandstone, clay, and silty clay) are weakly magnetic and vary widely in their rock magnetic parameters. Magnetic susceptibility and natural remanent magnetization in cores from all five wells vary, respectively, from 2.3 to 118 $\times$ 10^{-5} SI units (6.2–28 $\times$ 10^{-5} SI units on average) and from 0.04 to 19.9 mA/m (0.15–11.0 mA/m on average). The values are the highest in samples from the Pokur and Upper Berezovo Formations. The Koenigsberger ratio is markedly below unity ($Q_n = 0.05–0.33$) in all samples, which is an implicit evidence of the orientation (detrital) NRM origin [14, 15].

Magnetic minerals were identified proceeding from the temperature dependence of magnetic susceptibility (χ). This dependence is diagnostic of magnetic mineral species and their possible responses to heating, with implications for the choice of the demagnetization technique. Early during heating, the χ values cluster about zero, i.e., the rocks have very low susceptibility and, possibly, bear low percentages of magnetic minerals. The thermomagnetic curves of all experiments were identical, with broad peaks in the 430–520 °C range (Fig. 4). During the cooling process, there is an increase in the magnetic susceptibility. The cooling curves in the temperature range 420–100 °C are located above the heating curves. This thermal behavior records the presence of iron sulfides as magnetization carriers [16, 17], which is confirmed by identification of siderite and pyrite by mineralogical methods. Siltstone, mudstone, and silty clay often contain up to 3% of siderite and 1–2% of cryptocrystalline pyrite. The presence of siderite and pyrite was also revealed by X-ray microprobe analysis (EPMA) of Santonian-Campanian samples. Pyrite is nonmagnetic but can supply ferromagnetic compounds to rocks under certain conditions and can convert to magnetite upon heating (sharp peaks at 570–580 °C) [16, p. 111].

The remanence components were determined by stepwise AF and thermal demagnetization: each technique was applied to one sample from each stratigraphic level. The AF demagnetization was the most efficient for the Pokur, Kuznetsovo, Dorozhkovo, Okhteurievka, and Lower and Upper Berezovo Formations samples and revealed unstable (low coercivity) and stable (high coercivity) NRM components. In Fig. 4 these two components are emphasized with red lines for clarity. The magnetic cleaning procedure comprised most often six-eight demagnetization steps depending on the stability of magnetization the coercivity spectrum of the mineral magnetization carriers. The fields on the 2G-Enterprises cryogenic magnetometer were up to 60–100 mT. The low-coercivity component could be removed by a low field of 20–30 mT, while that of high coercivity, which we consider as the characteristic magnetization (ChRM), held till 60–100 mT (Fig. 4). Thermal demagnetization was possible only within 300–400 °C interval. Upon heating to these temperatures, the magnetic susceptibility and remanence of samples increased substantially indicating phase change in magnetic minerals with formation of a new ferromagnetic phase, likely, magnetite. Accordingly, thermal demagnetization was not applied to the Pur-Taz Upper Cretaceous sediments.

The primary origin of the revealed ChRM component is supported by three main lines of evidence: (i) the presence of samples of normal and reverse polarity in the sections; (ii) the independence of polarity from the lithological composition of sediments; (iii) the correlation of the paleomagnetic pattern in composite the Pur-Taz Upper Cretaceous section with other scales: the respective records of regional magnetostratigraphy of the Upper Cretaceous and the Cretaceous/Paleogene boundary for southern West Siberia [4] and for other regions, such as Tuarkyr, Caucasus, Kopetdag,

Volga valley [18], as well as with the global magnetic polarity scale [9]. Unfortunately, due to the absence of consistent orientation of the samples, relative to the core axis, the formal application of the reversal test is impossible. However, the presence of NRM of both polarities with nearly antipodal paleomagnetic inclinations indicates the primary nature of natural remanence.

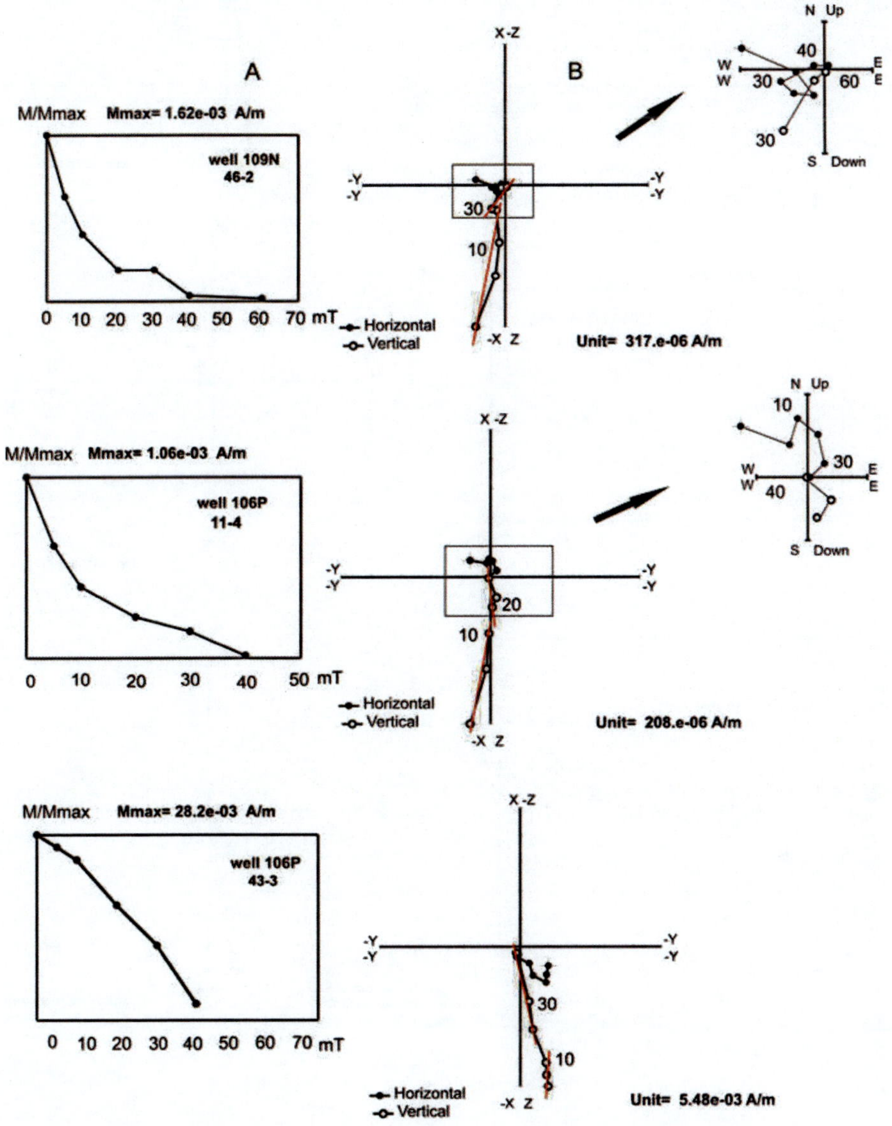

Fig. 4 AF demagnetization of NRM and Zijderveld diagrams. A: normalized demagnetization curve; B: Zijderveld diagrams, horizontal (1) and vertical (2) NRM projection

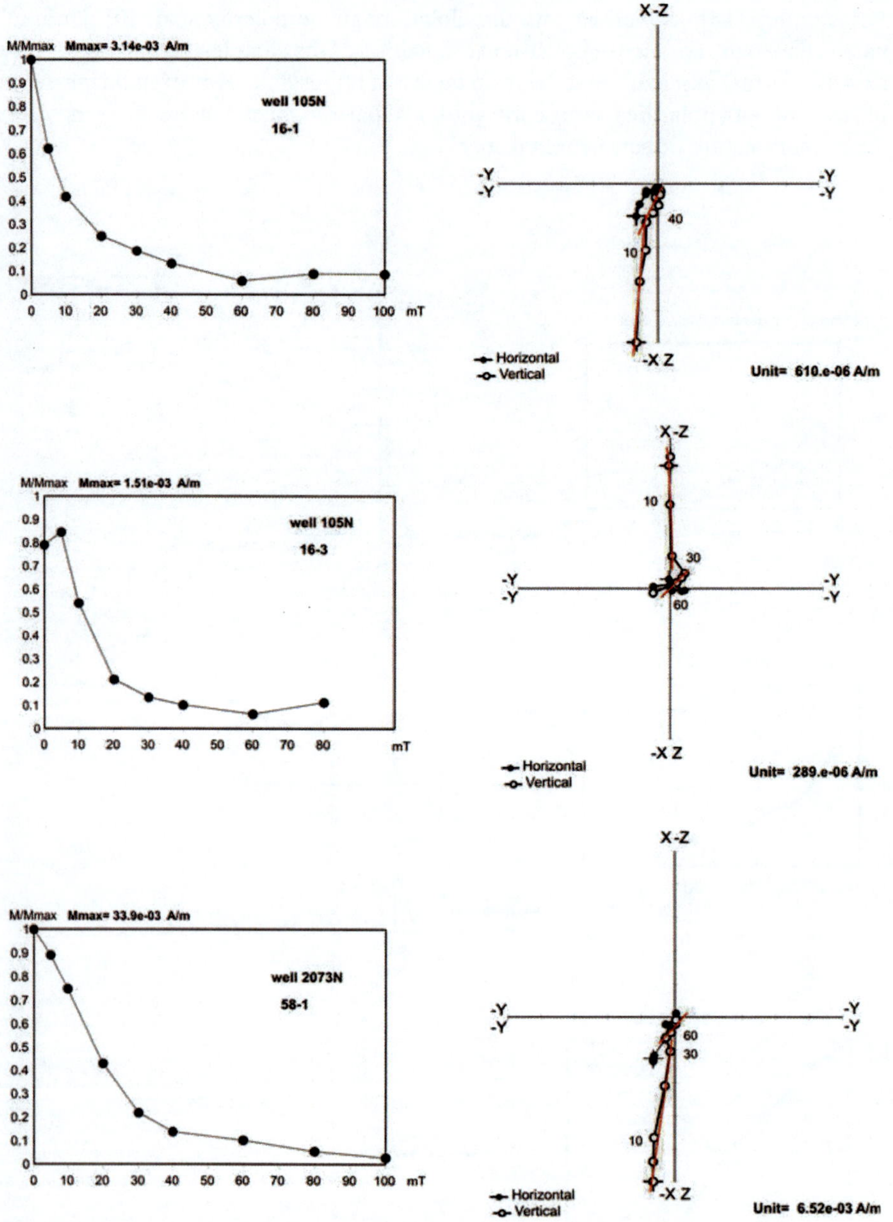

Fig. 4 (continued)

The ChRM component revealed in the Upper Cretaceous samples from the five wells in the Kharampur field (1049, 109N, 106P-Yu, 105N, and 2073N) was used to compile paleomagnetic sections of these wells. They were dominated by a normal polarity interval interrupted by four reversals in the Okhteurievka, and Lower and Upper Berezovo Formations (Figs. 5, 6, 7, 8 and 9).

The correlated magnetostratigraphic sections of five wells from the Kharampur oil-gas-condensate field (Pur-Taz interfluve, northern Siberia) were combined into a composite Upper Cretaceous record spanning Cenomanian through Campanian strata (Figs. 10 and 11), with two zones of normal polarity: $N\mathrm{K}_2$cp and $N\mathrm{K}_2$(sn-st$_2$). The magnetostratigraphy of the Pur-Taz interfluve was correlated further with the respective regional section of southern West Siberia [4] and with the global magnetic polarity time scale [9] on the basis of paleomagnetic, paleontological, and stratigraphic data (Fig. 11). The large zone of normal polarity $N\mathrm{K}_2$(sn-st$_2$) revealed in the regional magnetostratigraphy of northern West Siberia, which encompasses Cenomanian through Campanian formations (Pokur, Kuznetsovo, Dorozhkovo, Okhteurievka, Lower Berezovo, and lower Upper Berezovo Formations), can be correlated, fully or partly, with the large zone $N\mathrm{K}_{1\text{-}2}$(al-st) in the respective record for southern West Siberia [4] and to long C34 Chron of the global scale [9]. The overlying zone $N\mathrm{K}_2$cp has no equivalents in the southern West Siberian record but correlates with Chrons C33n or C32 in the global scale of [9].

4 Conclusions

The reported paleomagnetic, stratigraphic and paleontological data from five cored wells of the Kharampur oil-gas-condensate field in the Pur-Taz interfluve allowed compiling the first regional magnetostratigraphic scale for northern West Siberia. The five Kharampur wells (1049, 109N, 106P-Yu, 105N, 2073N) stripped the Pokur, Kuznetsovo, Dorozhkovo, Okhteurievka, and Lower and Upper Berezovo Formations that span the Cenomanian, Turonian, Coniacian, Santonian, and Campanian strata. The obtained magnetostratigraphic record includes two magnetozones of normal polarity: a short zone of $N\mathrm{K}_2$cp and a long $N\mathrm{K}_2$(sn-st) zone. The magnetostratigraphy of northern West Siberia correlates with the respective regional record for southern West Siberia [4] and with the global polarity scale [9].

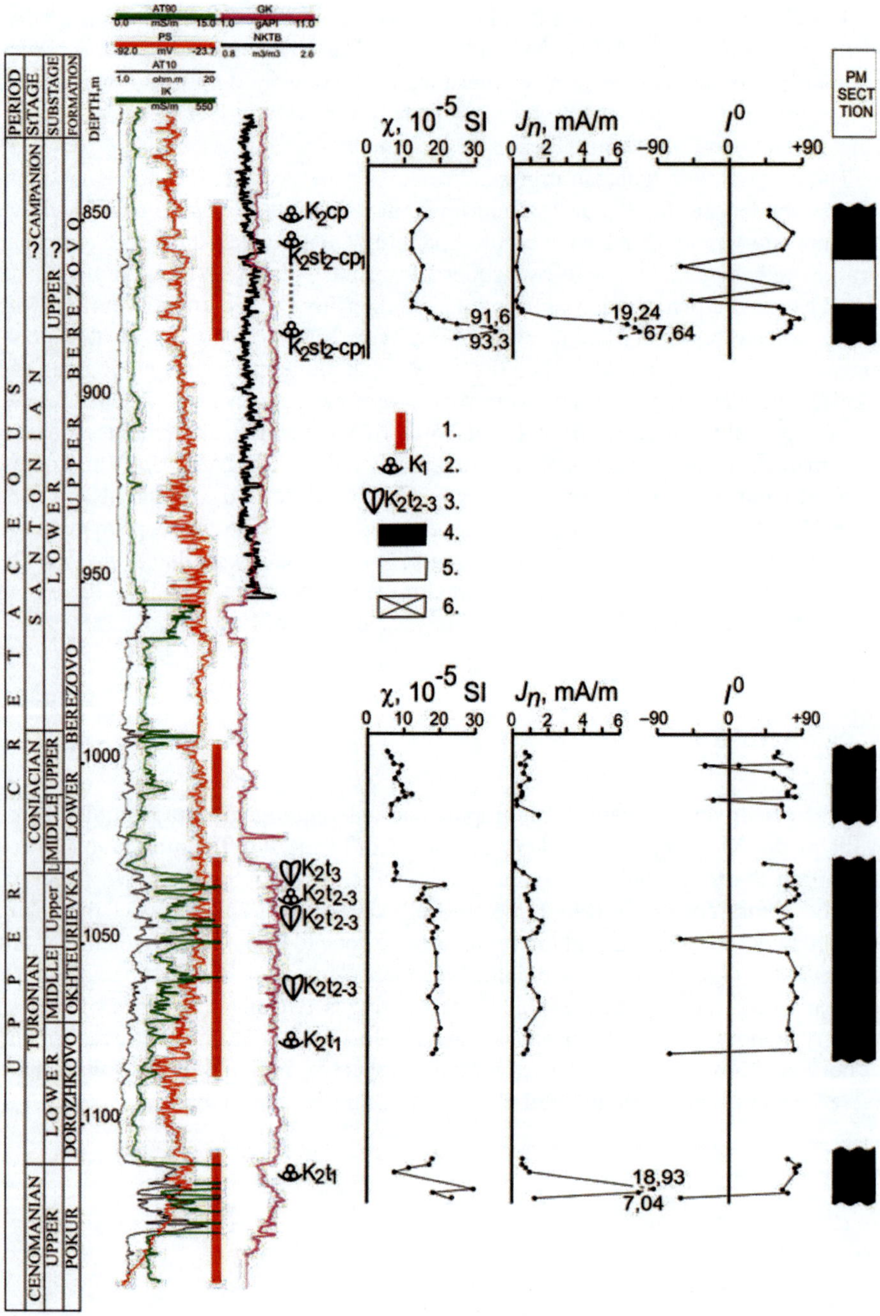

Fig. 5 Upper Cretaceous magnetostratigraphy from core data of Kharampur well 2073. 1 = cored intervals; 2, 3 = fauna constraints from foraminifera (2) and inoceramus (3); 4, 5 = normal (4) and reverse (5) geomagnetic polarity; 6 = no core; χ = magnetic susceptibility; J_n = characteristic remanent magnetization (ChRM); I° = magnetic inclination

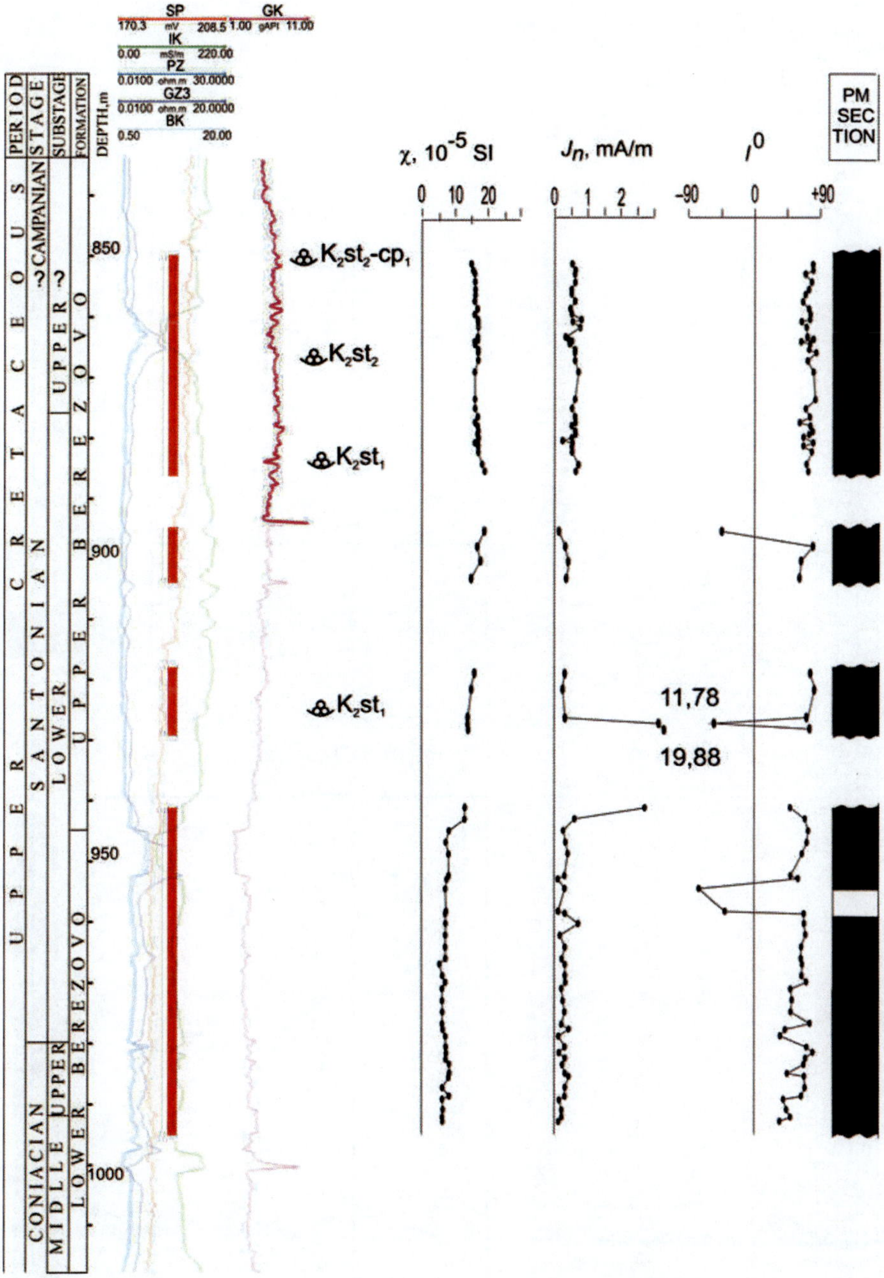

Fig. 6 Upper Cretaceous magnetostratigraphy from core data of Kharampur well 106P-Yu. Legend as in Fig. 5

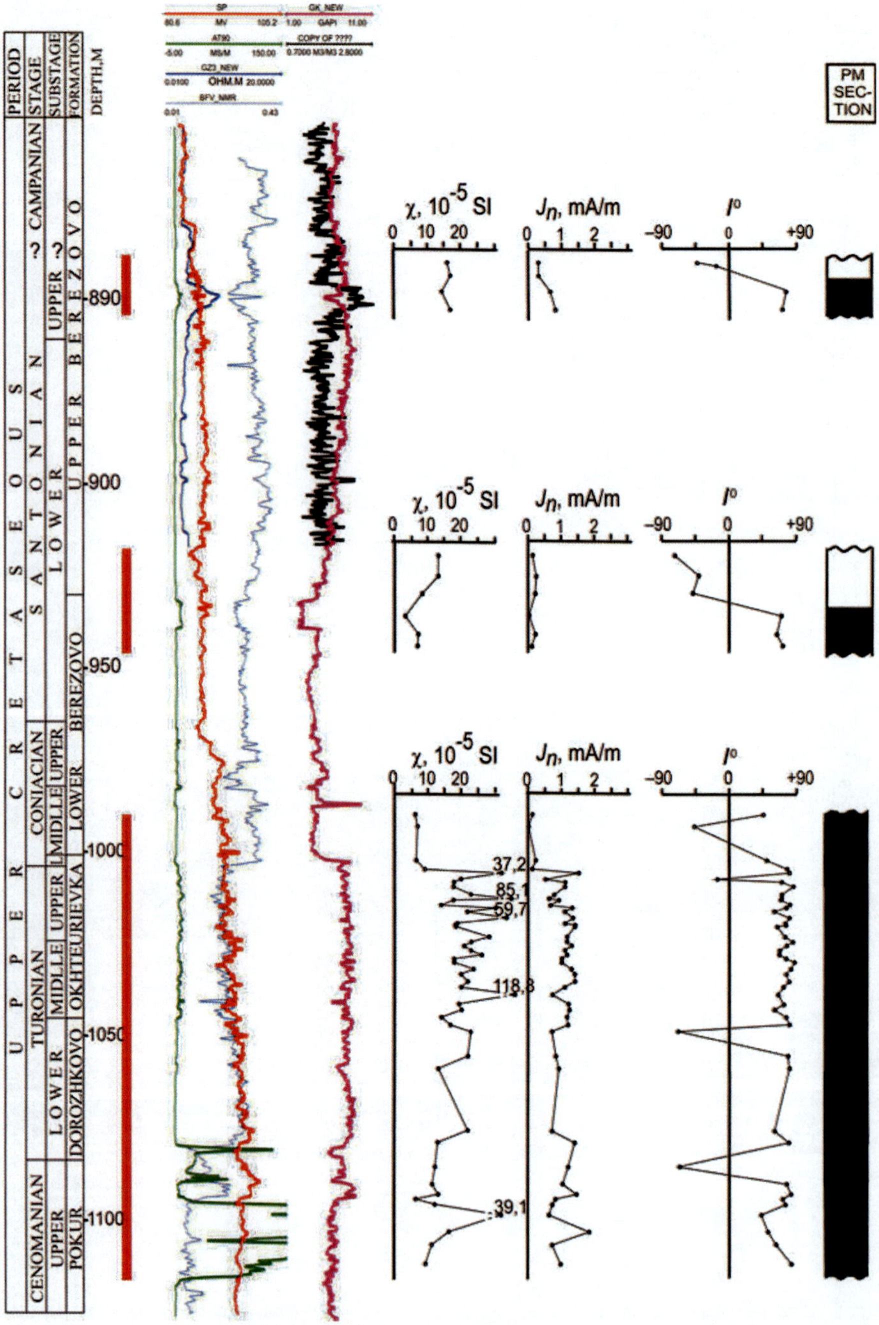

Fig. 7 Upper Cretaceous magnetostratigraphy from core data of Kharampur well 109N. Legend as in Fig. 5

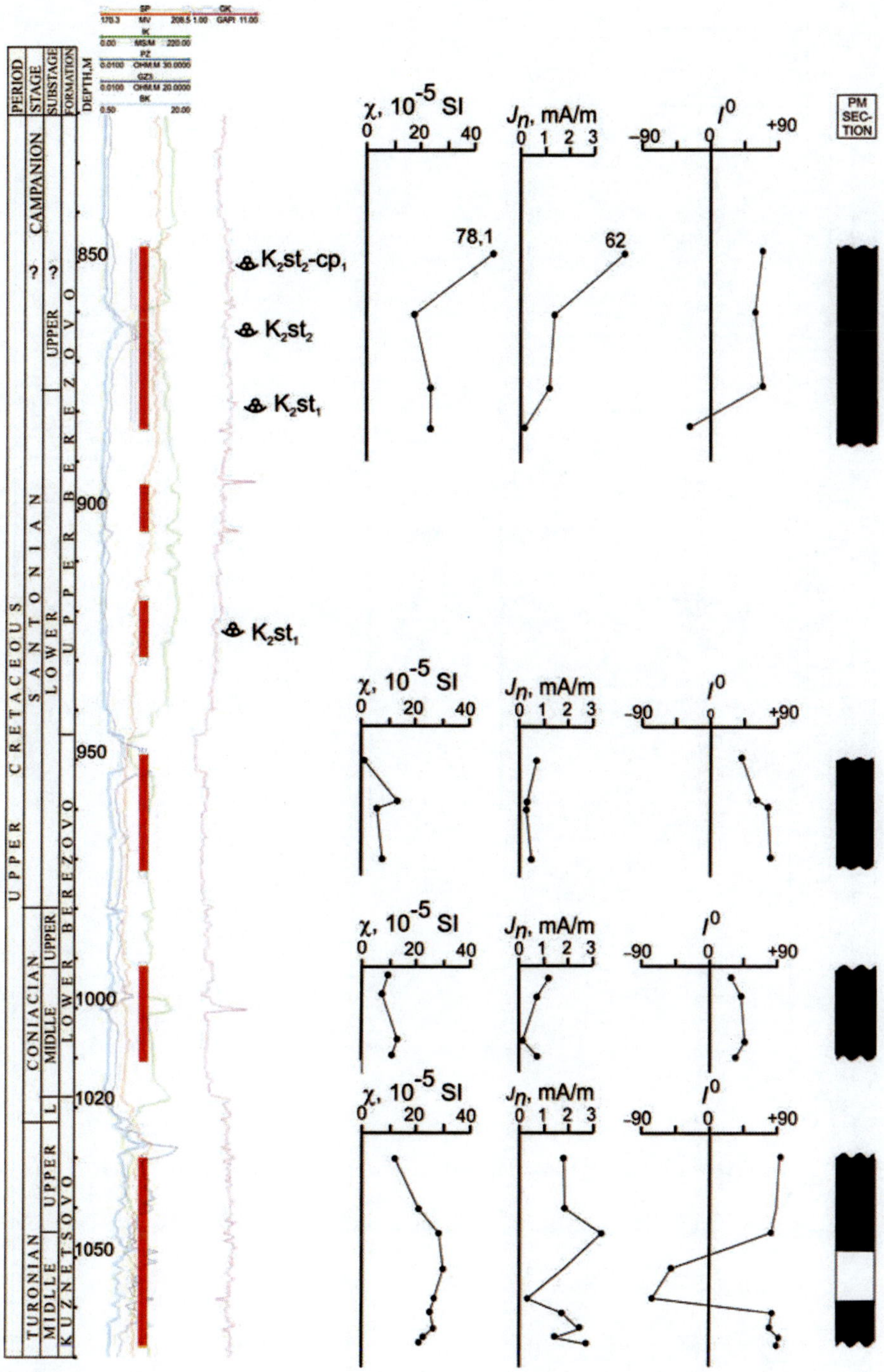

Fig. 8 Upper Cretaceous magnetostratigraphy from core data of Kharampur well 105N. Legend as in Fig. 5

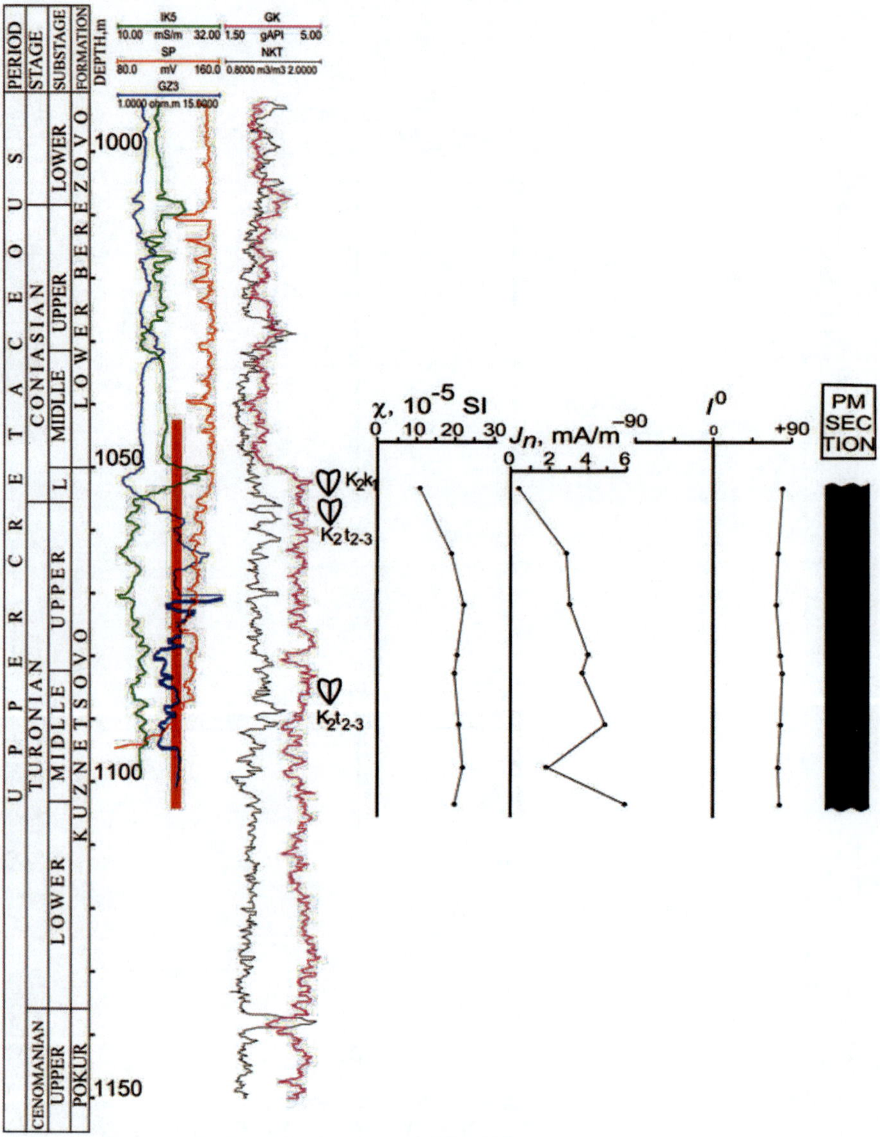

Fig. 9 Upper Cretaceous magnetostratigraphy from core data of Kharampur well 1049. Legend as in Fig. 5

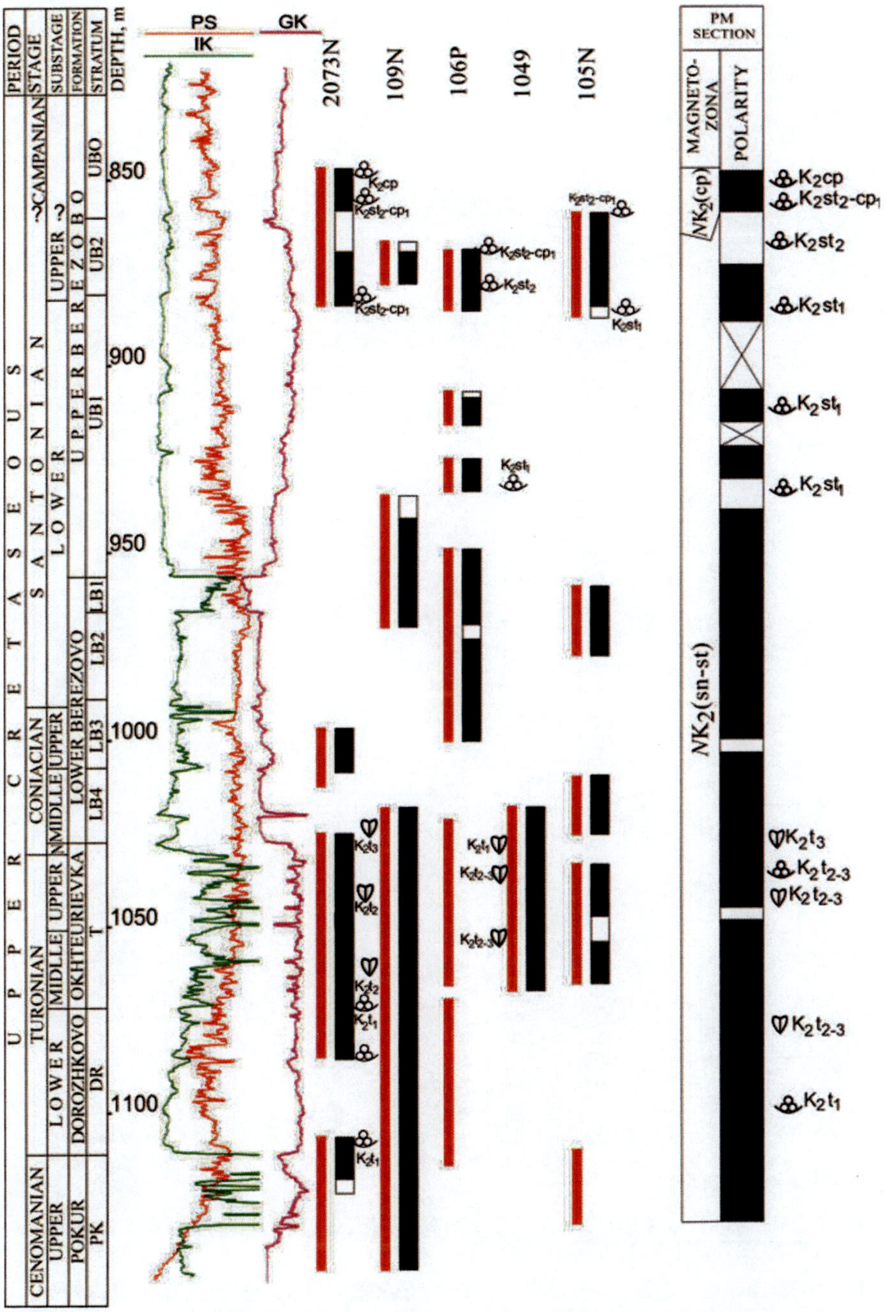

Fig. 10 Magnetostratigraphic record in Kharampur wells 2073, 109N, 106P-Yu, 1049, and 105N, and composite Upper Cretaceous section of the Pur-Taz interfluve (northern West Siberia). Legend as in Fig. 5

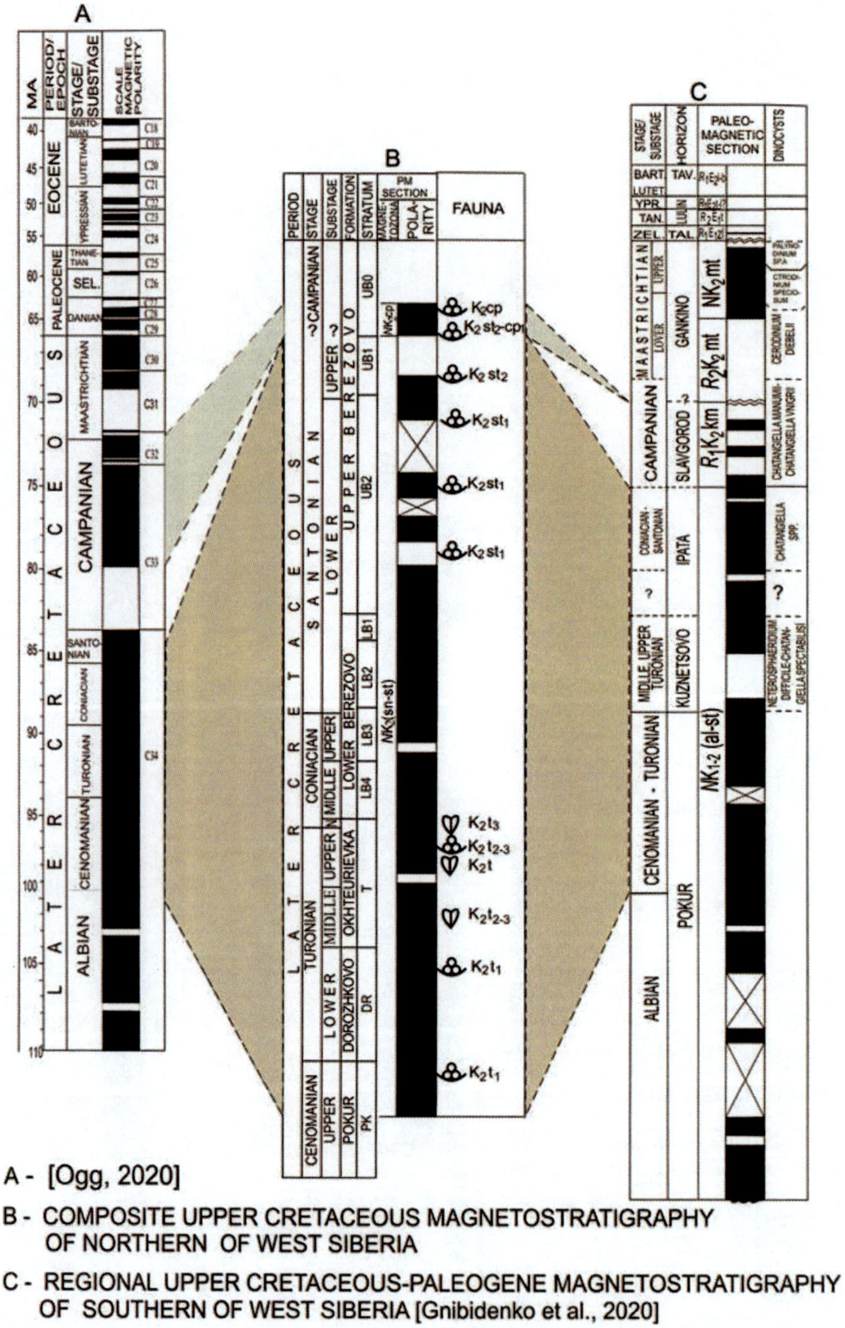

Fig. 11 Upper Cretaceous magnetostratigraphy: composite regional record of the Pur-Taz interfluve and its correlation with the respective record for southern West Siberia [4] and with the global magnetic polarity time scale [9]. Legend as in Fig. 5

Acknowledgements We thank Florian Lhuillier (Department of Earth and Environmental Sciences, Munich University), Roman Veselovsky (Institute of Earth Physics, Russian Academy of Science) and Evgeniy Kulakov (Centre for Earth Evolution and Dynamics university of Oslo) for constructive comments and suggestions, which helped to improve the manuscript.

References

1. Gnibidenko, Z.N., Lebedeva, N.K., Shurygin, B.N.: A regional Upper Cretaceous magnetostratigraphic section of southern West Siberia (Omsk depression). Dokl. Earth Sci. **458**(2), 1107–1111 (2014)
2. Gnibidenko, Z.N., Lebedeva, N.K., Levicheva, A.V.: Magnetostratigraphy of the Campanian-Maastrichtian Bakchar basin (southeastern West Siberia). Russ. Geol. Geophys. **56**(11), 1652–1666 (2015)
3. Gnibidenko, Z.N., Levicheva, A.V., Semakov, N.N., Rusanov, G.G.: Paleomagnetism and magnetostratigraphy of the Upper Cretaceous and Cretaceous-Paleogene boundary deposits in the south of the Kulunda basin (West Siberia). Russ. Geol. Geophys. **58**(1), 87–98 (2017)
4. Gnibidenko, Z.N., Kuzmina, O.B., Levicheva, A.V.: Regional magnetostratigraphy of the Upper Cretaceous and the Cretaceous/Paleogene boundary in southern West Siberia as applied to compilation of the Cretaceous magnetic-polarity scale. Russ. Geol. Geophys. **61**(9), 1028–1035 (2020)
5. Gnibidenko, Z.N., Levicheva, A.V., Marinov, V.A.: Upper Cretaceous paleomagnetism of northeastern West Siberia (from core data). In: Subsoil Use. Mining. Perspectives and Technologies of Mineral Exploration and Development, vol. 4. Economy and Geoenvironment. Proceedings of XIII Interexpo GEO-Sibir-2017. International Conference, Novosibirsk, 17–21 Apr 2017, pp. 196–201 (in Russian with English abstract)
6. Gradstein, F.M., Ogg, J.G., Schmitz, M.D., Ogg, G.M.: The Geological Time Scale 2012 (2012), pp. 793–853
7. Gnibidenko, Z.N., Levicheva, A.V., Marinov, V.A., Smolyaninova, L.G., Semakov, N.N.: Upper Cretaceous paleomagnetism of northern West Siberia. In: Problems of Geocosmos, Proceedings of 12th International Conference, St. Petersburg, Petergof, 8–12 Oct 2018, pp. 58–63 (in Russian)
8. Gnibidenko, Z.N., Levicheva, A.V., Smolyaninova, L.G., Marinov, V.A.: Upper Cretaceous magnetostratigraphy of the Pur-Taz interfluve, northeastern West Siberia. In: Paleomagnetism and Magnetism of Rocks. Proceedings of XXVth, Seminar on Rock Magnetism and Paleomagnetism, Borok, 25–29 Sept 2019, pp. 81–86 (in Russian)
9. Ogg, J.G.: Geomagnetic Polarity Time Scale, Chap. 5, pp. 159–192. Elsevier (2020)
10. Kontorovich, V.A., Belyaev, S.Yu., Kontorovich, A.E., Krasavchikov, V.O., Kontorovich, A.A., Suprunenko, O.I.: Tectonic structure and evolution of the West Siberian geosyneclise in the Mesozoic and Cenozoic. Russ. Geol. Geophys. **42**(11–12), 1740–1753 (2001)
11. Bulynnikova, A.A., Trandafilova, E.F.: The Cretaceous system. Late Cretaceous epoch. In: Correlation of Sedimentary Sections in the West Siberian Petroleum Basin: Stratigraphic and Paleontological Background, pp. 84–96. ZapSibNIGNI, Tyumen (1972) (in Russian)
12. Galerkina, S.G., Alekseichik-Mitskevich, L.S., Kozlova, G.E., Strelnikova, N.I.: Upper Cretaceous stratigraphy of northern West Siberia. Soviet Geol. **12**, 77–95 (1982)
13. Kharitonov, V.M., Marinov, V.A., Ivanov, A.V., Fomin, V.A.: Upper Cretaceous Inoceramus in core samples from the West Siberian Plain and some problems of the Turonian stratigraphy. Izv. Saratov Univ. Nov. Ser. Ser. Nauki Zemle **7**(2), 61–70 (2007) (in Russian with English abstract)
14. Guzhikov, A.Yu., Arkad'ev, V.V., Baraboshkin, E.Yu., Bagaeva, M.I., Piskunov, V.K., Rud'ko, S.V., Perminov, V.A., Manikin, A.G.: New sedimentological, bio- and magnetostratigraphic

data on the boundary Jurassic-Cretaceous interval of eastern Crimea (Feodosiya). Stratigr. Geol. Correl. **20**(3), 261–294 (2012)
15. Guzhikov, A.Yu., Pronin, A.P.: Possible sources of magnetic anomalies in the south of the Caspian Basin. Izv. Saratov Univ. Nov. Ser. Ser. Nauki Zemle **15**(3), 38–45 (2015) (in Russian with English abstract)
16. Burov, B.V., Yasonov, P.G.: Introduction into Differential Thermomagnetic Analysis of Rocks, 160 pp. Kazan University, Kazan (1979) (in Russian)
17. Minyuk, P.S., Tyakova, E.E., Subbotneykova, T.V., Kazansky, A.Yu., Fedotov, A.P.: Thermal magnetic susceptibility data on natural iron sulfides of northeastern Russia. Russ. Geol. Geophys. **54**(4), 464–474 (2013)
18. Guzhikov, A.Yu., Baraboshkin, E.Yu., Fomin, V.A.: Cretaceous magnetostratigraphy: state of the art, problems, and prospects. In: Pervushov, E.M. (ed.) The Cretaceous System of Russian and Former Soviet Countries: Problems of Stratigraphy and Paleogeography, Collection of Papers, pp. 69–86. Saratov University, Saratov (2007) (in Russian)

The First Paleomagnetic Data on the Ust-Obor Section (Western Transbaikalia, Buryatia)

A. Yu. Kazansky, G. G. Matasova, A. A. Shchetnikov, and I. A. Filinov

Abstract We present detailed rock magnetic and paleomagnetic data of the key Ust-Obor section in Transbaikalia (Buryatia, Russia). The section consists of two parts—the lower loamy sediment unit of Eopleistocene age and the upper sandy sediment unit of Neopleistocene age. The lower unit contains a stratotype fauna complex of Late Eopleistocene age. The paleomagnetic study presented herein reveals only normal polarities of natural remanent magnetization throughout the section. Paleomagnetic results appear at odds with paleontological data, which constrain the lower unit to Eopleistocene age and consequently to the reversed-polarity Matuyama chron. Rock magnetic and paleomagnetic data do not support a remagnetization hypothesis to explain this apparent contradiction. Conversely, the statistically similar rock magnetic and paleomagnetic data of the upper (Neopleistocene) and lower (Eopleistocene) units suggest a common depositional environment and thus a similar geological age. The preferred hypothesis explaining the observations calls for a re-deposition of the Late Eopleistocene Ust-Oborian fauna within younger Neopleistocene sediment.

Keywords Quaternary sediments · Paleomagnetism · Rock magnetism · Transbaikalia

---

A. Yu. Kazansky (✉)
Geological Department of Lomonosov, Moscow State University, Moscow, Russia

A. Yu. Kazansky · G. G. Matasova · A. A. Shchetnikov · I. A. Filinov
Geological Institute of Russian Academy of Sciences, Moscow, Russia

G. G. Matasova
Trofimuk Institute of Petroleum Geology and Geophysics, Siberian Branch, Russian Academy of Sciences, Novosibirsk, Russia

A. A. Shchetnikov · I. A. Filinov
Irkutsk State University, Irkutsk, Russia

Institute of the Earth's Crust, Siberian Branch, Russian Academy of Sciences, Irkutsk, Russia

A. A. Shchetnikov
Vinogradov Institute of Geochemistry, Siberian Branch, Russian Academy of Sciences, Irkutsk, Russia

A. Kosterov et al. (eds.), *Problems of Geocosmos—2020*, Springer Proceedings in Earth and Environmental Sciences, https://doi.org/10.1007/978-3-030-91467-7_11

1 Introduction

The stratigraphic scheme of Transbaikalia's Quaternary is insufficiently documented with respect to paleomagnetic data. Paleomagnetic studies in this region, carried out in the 70–90s of the last century, are few, often contradictory [1–3], and require revision and verification using modern experimental techniques. Paleomagnetic data obtained using modern methodologies and instruments [4, 5] established that the sediments containing the Tologoi faunal complex correspond to a time interval spanning from the upper part of the Jaramillo subchron to the lower part of the Bruhnes chron, including the mainly reversed-polarity Matuyama chron [4, 5]. The absence of reliable paleomagnetic data from natural outcrops that contain fauna assemblages older than this chronological level, motivates the acquisition of new paleomagnetic data on reference sites of Quaternary faunal complexes. The situation is complicated by the fact that not all typical fauna locations are currently available. In addition, some sections are insignificant in stratigraphic thickness and thus not informative from a paleomagnetic perspective.

One of the easily accessible and relatively stratigraphically representative localities of key faunal complexes is the Ust-Obor section, which is a stratotype of the Ust-Oborian fauna of the upper part of the Lower Eopleistocene [6].

The aim of this work was to acquire the first paleomagnetic data on the Ust-Obor section in order to better define the stratigraphic scheme of the Quaternary period in Transbaikalia through observed changes in magnetic polarity and positioning of the Ust-Oborian fauna on the Geological time scale [7].

2 Geology and Sampling

The Ust-Obor section (50° 55′ 04″ N; 108° 33′ 24″ E) is located 120 km to the SE from the city of Ulan-Ude (Fig. 1a). The outcrop, located within a low butte on the right bank of the Khilok River near the confluence of the Obor River, is exposed on a vertical face of a ravine that cuts an erosion scarp at the base of a gentle valley pediment (Fig. 1b). The scarp, formed due to erosion by the Khilok River is 15–20 m-high. The depth of the ravine exceeds 10–12 m. At its base, a thick sequence of loose sediments (Fig. 1c) is exposed and overlies the bedrock granitic massif.

According to geological descriptions [3, 6, 8], the section is composed of approximately 10 m of interlayered reddish sandy and silty loams, which overlie weathered Paleozoic granites. Two distinct layers are recognized in the section based on previous geological descriptions and fauna findings. The Neopleistocene predominantly sandy upper part contains the Khilok fauna: *Ochotona daitrica, Spermophilus (Urocitellus)* cf. *unduiatus, Marmota sibirica, Cricetnlus* sp., *Microtus gregaiis, Microlinae* gen. [8]. The mostly silty lower part of upper Eopleistocene age [6] contains the Ust-Oborian fauna complex consisted of the following species: *Ochotona ustoborica, Ochotona* sp., *Allophaiomys* cf. *pliocaenicus, Lagurodon*

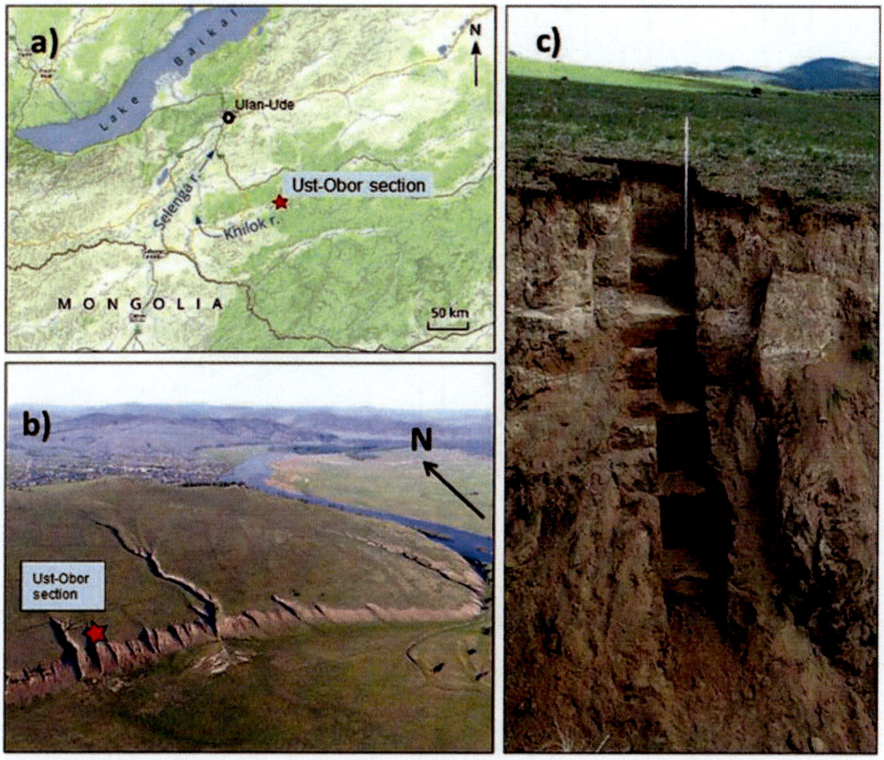

Fig. 1 Location and general view of the Ust-Obor section. **a** Overview map; **b** panoramic photo taken from the drone; **c** section overview, the height of the geodetic rod is 2 m

arankae, Prolagurus temopolitanns, Prosipheneus ex gr. *youngi, Clethrionomys* sp., *Spermophilus* sp., *Leporidae* gen. indet., *Lagurini* sp. [8]. Based on available fauna, the Ust-Oborian complex may be placed between the Eopleistocene Dodogolian and Kudunian complexes [9] (Fig. 2).

For sampling, a man-made trench was dug over the exposure to remove the top cover of weathered sediments (Fig. 1c). A total of 95 samples for rock magnetic and paleomagnetic studies were taken at a 10 cm vertical interval over the entire section. Each sample represents one sampling level. Standard 7 cm^3 internal volume plastic boxes with rounded edges were pressed into the subvertical wall of the trench. The depth from the slope surface was not less than 0.8–1 m. The sample orientation arrow was directed downward and dip direction and dip of the arrow was measured using magnetic compass, checked with solar readings. Loose sedimentary material for rock magnetic analysis was taken from the same levels as paleomagnetic samples.

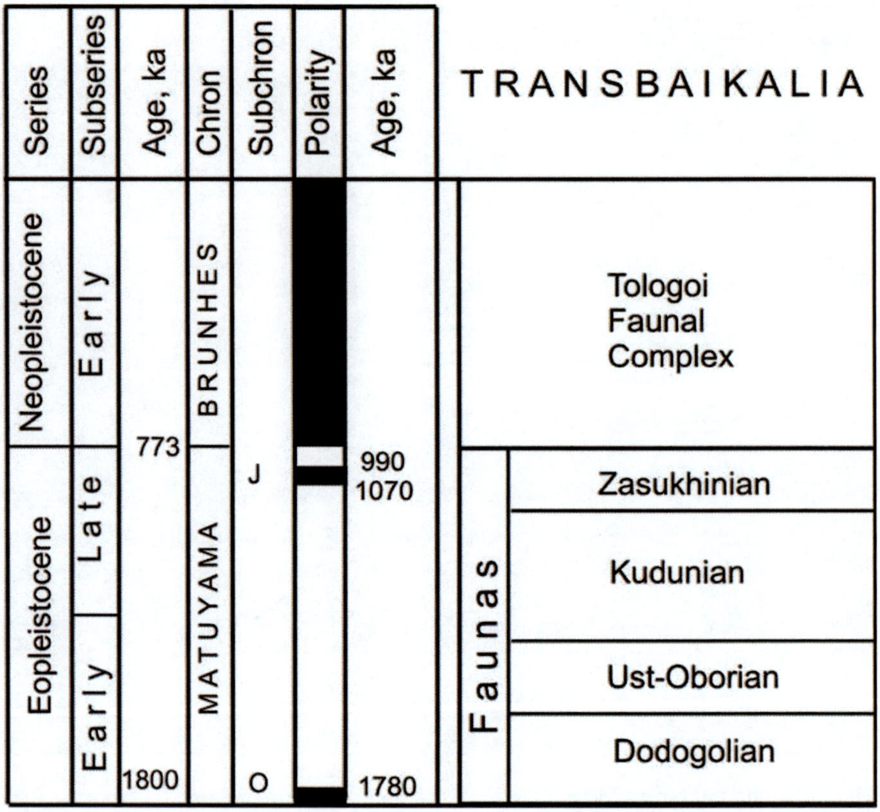

Fig. 2 Chronostratigraphy, geomagnetic polarity time scale (GPTS) and faunal succession of the Eopleistocene and Lower Neopleistocene of Transbaikalia (modified form [9]). Legend: J—Jaramillo; O—Olduvai

3 Methods

Laboratory studies were carried out at the Main Geomagnetic Field and Petromagnetism laboratory of the Schmidt Institute of Physics of the Earth of the Russian Academy of Sciences (Moscow) and the Geodynamics and Paleomagnetism laboratory of Trofimuk Institute of Petroleum Geology and Geophysics, Siberian Branch of the Russian Academy of Sciences (Novosibirsk) following standard procedures [10]. Volume-normalized low field magnetic susceptibility (κ) and its anisotropy (AMS) were measured using an AGICO MFK-1 Kappabridge. Measurements of Natural Remanent Magnetization (NRM) were made using 2G Enterprises 755R cryogenic magnetometer. Koenigsberger ratio was calculated as $Qn = NRM_0/(\kappa \times H_{amb})$, where H_{amb} is the strength of the Earth magnetic field at the sampling site (60 μT) and NRM_0—the NRM intensity before demagnetization. Specimens from each sampling level were subjected to stepwise AF demagnetization, using a built-in Automatic Degaussing System 600 in 3–10 mT steps up to 70 mT (for pilot specimens up to 130 mT). Demagnetization results were analyzed using vector end-point diagrams [11] and principal component analysis [12] implemented in AGICO Remasoft 3.0 paleomagnetic analytical software package.

All magnetic hysteresis parameters were measured using a J_meter coercivity spectrometer manufactured by Kazan Federal University, Russia [13].

Starting from a demagnetized state, first magnetization curves up to a saturating field of 1500 mT followed by measurements of the descending and ascending branches of the major hysteresis loop were acquired. The following hysteresis parameters were determined: saturation magnetization (Ms), remanent saturation magnetization in the field of 1500 mT (Mrs), coercive force (Bc), remanent coercive force (Bcr), and initial magnetization (χint). Initial susceptibility χint was calculated from the linear slope of the first magnetization curve at low fields between 0 and 10 mT. The paramagnetic contribution (χp) to the hysteresis loops was determined by a linear fit through the descending branch of the hysteresis loops in the interval (+1300; + 1500) mT. This correction is also applied before determining Mf—saturation magnetization and Bcf—coercive force of ferrimagnetics. Following [14] superparamagnetic susceptibility (χsp) is calculated from the slope of the Mrs-loss curve at low fields between 10 and 0 mT. The ferromagnetic susceptibility part was calculated as $\chi\text{fer} = \chi\text{int} - \chi\text{p} - \chi\text{sp}$. The relative proportion of soft remanence bearing minerals to hard remanence bearing minerals (S) was calculated as the ratio of the remanent magnetization in back field of 300 mT to the saturation isothermal remanent magnetization in the field of 1500 mT: S = −Mr(−300 mT)/Mrs(1500 mT). The “hard” part of saturation remanent magnetization (HIRM) was calculated as HIRM = (Mrs(1500 mT) + Mr(−300 mT))/2. Domain state was estimated plotting the ratios Mrs/Mf versus Bcr/Bcf [15].

The magnetic mineralogy was determined using temperature dependence of induced magnetization (Ji) in 400 mT. Two induced magnetization runs of successive heating cycles were measured using a Curie balance manufactured Kazan Federal University. Each heating was conducted up to 700 °C in air. Rock magnetic data were analyzed and interpreted according to standard procedures [16].

4 Results

4.1 Lithostratigraphy of the Section

Field study of the Ust-Obor section has shown that the sequence is represented mainly by sandy loams and loams with numerous inclusions of grit-gravel material, sometimes forming hollow wavy lenses and interlayers. Light brown buried soils in the upper part of the section are underlain by intensely carbonized very light-colored horizons. Below there are reddish-brown loams with grit, sand and gravel, resting directly on the mature weathered granite bedrock (Fig. 3b). Ten different layers have been recognized in the section:

Layer 1. Modern soil. Gray-brown sandy loam with scree. The lower contact is diffused.

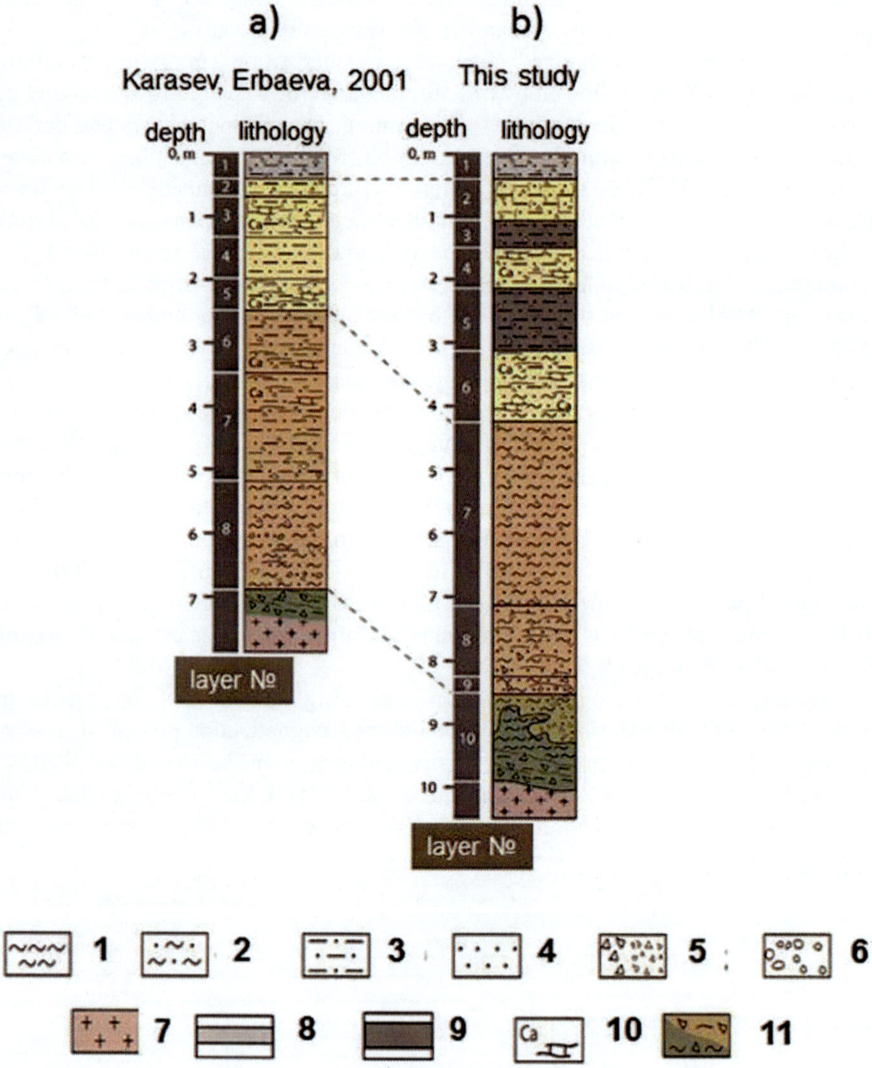

Fig. 3 Different versions of the Ust-Obor section structure and correlation between those versions. **a** Generally accepted [3, 8]; **b** This study and correlation between those versions. Legend: 1—clay; 2—loam; 3—sandy loam; 4—sand; 5—scree (debris); 6—gravel; 7—granite; 8—modern soil; 9—buried soil; 10—carbonatization; 11—weathering crust

Layer 2. Pale yellow sandy loam with scree, strongly carbonated, loessified, massive, whitish, dense. The lower contact is diffused.

Layer 3. Light brown sandy loam with scree, indistinctly layered, the lower contact is very irregular, with penetration from the underlying layer. Buried soil.

Layer 4. Whitish sandy loam, with scree, strongly carbonated, wavy layered, strongly cryoturbated. The lower contact is very irregular.

Layer 5. Light brown sandy loam with scree and gravel, non-layered (massive). The lower contact is sharp and irregular. Buried soil.

Layer 6. Loams of brown and beige color, carbonated, gleyed, platy structure, thin-layered, interbedded with brownish sandy loam with scree. The bedding is thin, wavy. Second carbonate horizon. The lower contact is diffused.

Layer 7. Reddish brown loam, with gravel, unstratified, massive. In the lower part of the layer, the gravel content increases. The bedding is indistinct, lenticular. The lower contact is poorly expressed, wavy.

Layer 8. Reddish brown loam with lenses of well-sorted sand with gravel. Scree inclusions are present. The lower contact is wavy, unclear.

Layer 9. Reddish brown sandy loam gravel with granite debris. The lower contact is irregular, pocket-like.

Layer 10. Clays, scree, crushed stone. Varicolored (green, cherry, brown) weathering crust over granites.

The upper Neopleistocene part of the Ust-Obor section includes layers from 2 to 6, Eopleistocene part—from 7 to 9. Layer 1 represents modern soil, layer 10—weathering crust.

The description above shows that the sedimentary structures proposed in the present study of the Ust-Obor section (Fig. 3b) are quite similar to those (Fig. 3a) proposed in [8] (and other later publications). However, the buried soils (layers 5 and 3) were not reported in previous works. In addition, authors [8] reported the Eopleistocene to Neopleistocene boundary at 2.4 m, whereas our study puts the boundary at 4.2 m. Thickness of the lower part is practically similar in both versions, which suggests that the studied section reflects the principal geological structures of the Ust-Obor area proposed by previous studies.

At present, there is no consensus on the genesis of the studied sediments [3, 8, 9]. We suggest a complex genesis—the combination of aeolian, slope wash and colluvial processes and partly reworked by solifluction.

4.2 Scalar Rock Magnetic Parameters (κ, NRM, Qn)

Sediments of the Ust-Obor section are characterized by relatively high values of magnetic parameters. Most of the samples yielded κ values in a range between 200 and 600 10^{-5} SI units. NRM values vary between 10 and 100 mA/m (Fig. 4). The exception is the weathering crust, which is mostly characterized by low κ and NRM values. Qn ratio for all samples vary between 0.1 and 1. Bilogarithmic plot of scalar magnetic parameters in Fig. 4 shows that the distribution of scalar parameters is

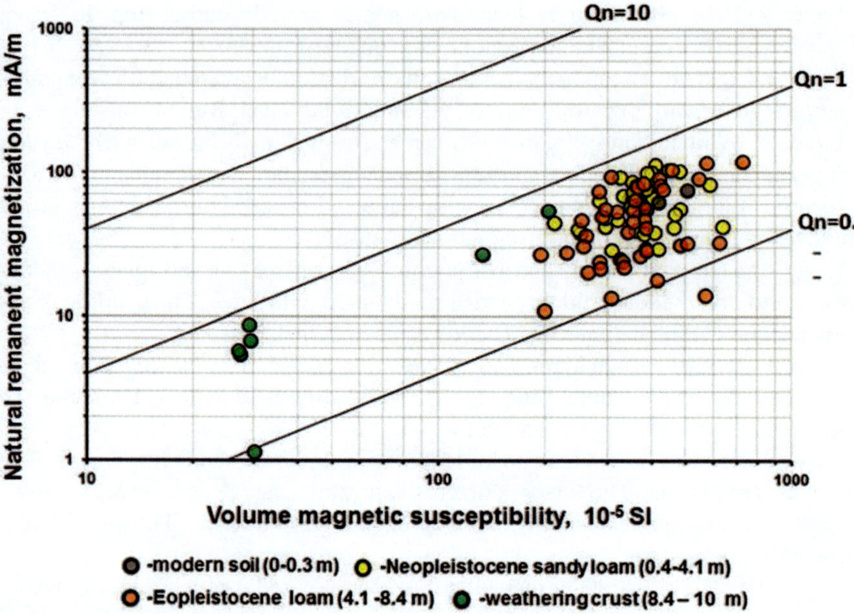

Fig. 4 Bilogarithmic plot of scalar magnetic parameters of different sediments from the Ust-Obor section

very compact (except weathering crust) with data points for the modern soil, the Neopleistocene sandy loam, and the Eopleistocene loam overlapping substantially (Fig. 4).

4.3 Magnetic Hysteresis Parameters

Magnetic hysteresis data allow subdividing the sequence of the Ust-Obor section into four different units, characterized by specific sets of parameters (Figs. 5 and 6). Layer 1 represents the modern soil with enhanced values of concentration-dependent parameters (χfer, χsp, Mf and Mrs) (Fig. 5) and low values of coercive parameter Bcr (Fig. 6). Neopleistocene part of the section (layers 2–6) is characterized by a distinct difference in hysteresis parameters of sandy loam and buried soil (Table 1). The buried soil horizons have reduced concentration-dependent magnetic values (except superparamagnetic content that increases in buried soils) and lower values of magnetic grain size (χfer/Mrs) and Bcr/Bcf ratio. Such rock magnetic behavior corresponds to the "Siberian mechanism" of pedogenic alteration [17], which is widespread across Siberia.

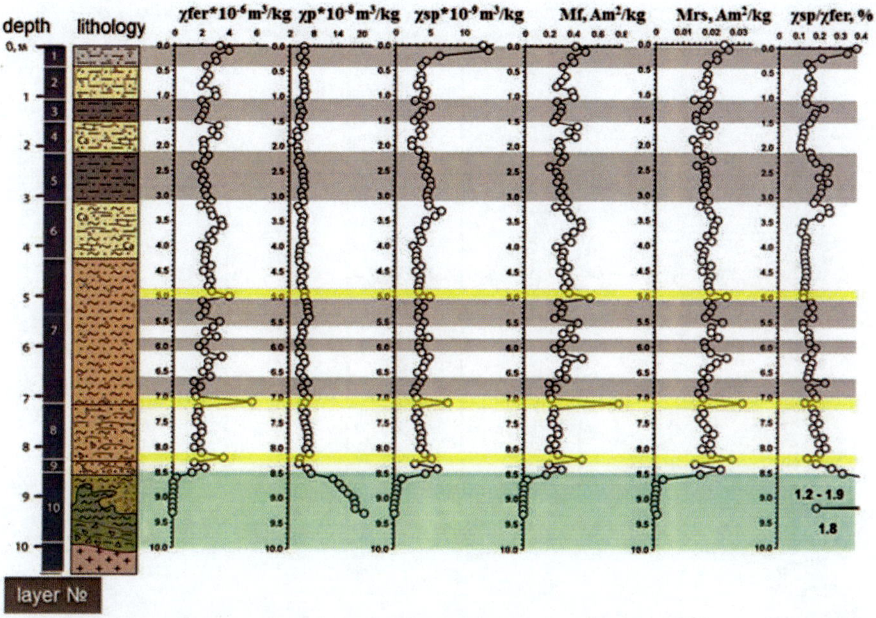

Fig. 5 Variation of concentration-dependent parameters in the Ust-Obor section. Grey bars—modern and buried soils (visible and proposed), yellow bars—sands, green bar—weathering crust. For legend see Fig. 3

Three paleosol horizons within layer 7 of the Eopleistocene section show similar with the Neopleistocene buried soils rock magnetic behavior, characteristic of the "Siberian mechanism".

In addition to these 3 paleosols, three thin horizons with coarse-grained magnetic material were recognized within layers 7 and 8. These horizons are characterized by increased concentration-dependent parameters χfer, χsp, Mf, Mrs, domain state (Bcr/Bcf) and magnetic grain size χfer/Mrs, along with decreases of Bcf, Bcr and HIRM/Mrs percentage (Figs. 5 and 6). These sandy intervals are interpreted as slope-wash basal sands.

Weathering crust yielded strikingly different magnetic properties. A dramatic decrease of all concentration-dependent parameters (except χsp) and magnetic grain size (χfer/Mrs) as well as a sharp increase of magnetic hardness (Bcf, Bcr, Bcr/Bcf, HIRM, S) is observed (Figs. 5 and 6). These data indicate the presence of high-coercivity minerals in clays of the weathering crust. High Bcf and Bcr values suggest the presence of hematite while negative S values, typical for goethite [18] suggest the presence of the latter in addition to hematite.

Domain state of magnetic grains in modern soil, Neopleistocene and Eopleistocene sediments is practically identical (Fig. 7). All magnetic grains are multidomain (MD), Single-domain (SD) particles are practically absent. According to SD-MD mixing curve [19] the percentage of SD particles in this part of the section does

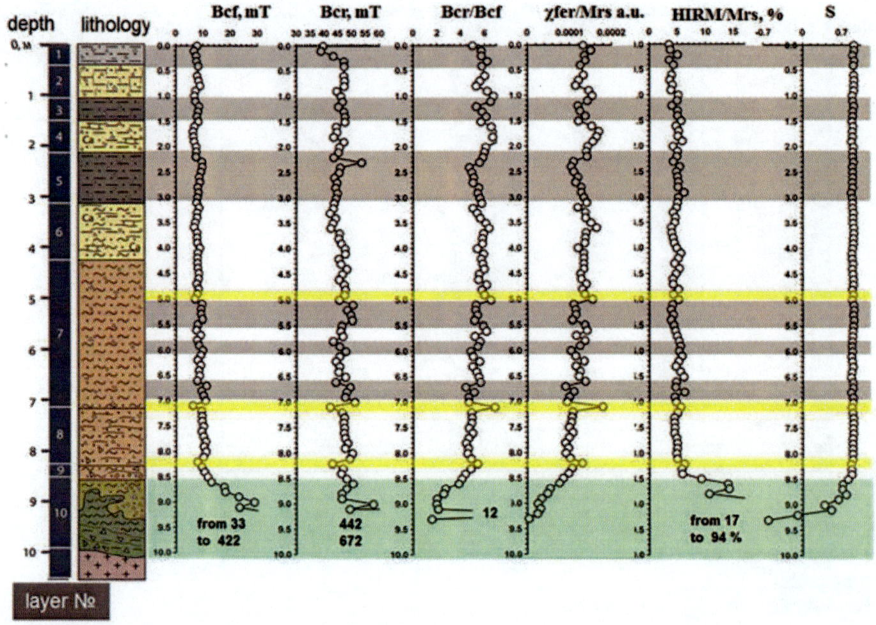

Fig. 6 Variation of hysteresis and structurally sensitive parameters in the Ust-Obor section. Grey bars—modern and buried soils (visible and proposed), yellow bars—sands, green bar—weathering crust. For legend see Fig. 3

not exceed 5% (Fig. 7). In contrast to overlaying sediments, magnetic state of particles in the weathering crust are characterized by quite different—single domain and pseudo-single domain. Half of the samples contain from 30 to 100% of SD grains (Fig. 5). Mrs/Mf > 0.5 highly likely is determined by high coercivity of goethite and hematite.

Differences between mean values of Neopleistocene and Eopleistocene parts of the Ust-Obor section are statistically insignificant according to the student's T test for all concentration-dependent parameters (except Mrs) and magnetic grain size χfer/Mrs (Table 2). On the other hand, all mean values of hysteresis parameters (except S) are statistically different. We attribute this difference to the enrichment of Eopleistocene sediments with magnetically hard minerals due to erosion of the weathering crust during Eopleistocene deposition.

4.4 Magnetic Anisotropy

Parameters of magnetic anisotropy (low field magnetic susceptibility, corrected anisotropy degree, foliation and lineation) in Neopleistocene and Eopleistocene sediments are not statistically different (see the results of the Student's T-test in

Table 1 Mean values, standard deviation of rock magnetic parameters of modern soil, paleosols and sandy loam from Neopleistocene part of the Ust-Obor section

Rock magnetic parameters	Modern soil (layer 1), N = 5		Paleosols (layers 3 and 5 combined), N = 16		Sandy loam (layers 2, 4, 6 combined), N = 27		T value
	Mean	SD	Mean	SD	Mean	SD	Between paleosols and sandy loam
χfer (m^3/kg)	3.16×10^{-6}	$6.06\ 10^{-7}$	2.22×10^{-6}	3.09×10^{-7}	2.63×10^{-6}	4.53×10^{-7}	**3.455**
χsp (m^3/kg)	8.17×10^{-9}	4.59×10^{-9}	4.58×10^{-9}	8.75×10^{-10}	3.60×10^{-9}	7.09×10^{-10}	**3.809**
Mrs (Am^2/kg)	2.30×10^{-2}	3.06×10^{-3}	1.78×10^{-2}	1.70×10^{-3}	1.86×10^{-2}	2.34×10^{-3}	1.320
Mf (Am^2/kg)	4.04×10^{-1}	6.70×10^{-2}	2.92×10^{-1}	3.96×10^{-2}	3.49×10^{-1}	6.05×10^{-2}	**3.711**
χfer/Mrs (arb. unit)	1.37×10^{-4}	9.20×10^{-6}	1.25×10^{-4}	1.20×10^{-5}	1.41×10^{-4}	1.39×10^{-5}	**3.986**

Note T critical value for 41 degrees of freedom is 2020 (difference in means with T > 2.020 are statistically significant). N—number of samples, SD—standard deviation

T values (student's test) are given for paleosols and sandy loam

Bold values mean statistical significance of difference between compared parameters

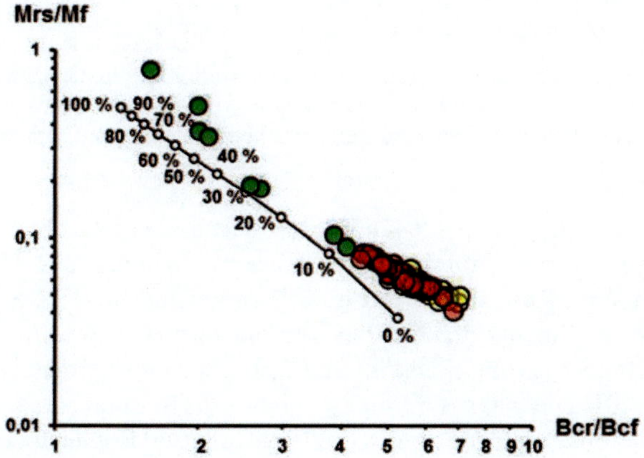

Fig. 7 "Day plot" [15] with SD-MD mixing curve [19] showing domain state of magnetic particles from the sediments of the Ust-Obor section. Legend: gray symbols—modern soil; yellow symbols—Neopleistocene; pink symbols—Eopleistocene; green symbols—weathering crust. Black line—SD-MD mixing curve. Numbers—percentage of the single domain particles in magnetic grain assemblage

Table 2 Mean values, standard deviation and T values (student's test) of rock magnetic parameters of Neopleistocene and Eopleistocene sediments of the Ust-Obor section

Rock magnetic parameter	Neopleistocene (N = 38)		Eopleistocene (N = 43)		T value
	Mean	SD	Mean	SD	
χfer (m^3/kg)	2.48×10^{-6}	4.80×10^{-7}	2.43×10^{-6}	7.83×10^{-7}	0.380
χsp (m^3/kg)	4.08×10^{-9}	9.21×10^{-10}	3.95×10^{-9}	9.24×10^{-10}	0.646
χp (m^3/kg)	5.03×10^{-8}	7.22×10^{-9}	5.91×10^{-8}	8.15×10^{-9}	**5.127**
Mrs (Am^2/kg)	1.83×10^{-2}	2.30×10^{-3}	2.02×10^{-2}	3.36×10^{-3}	**2.898**
Mf (Am^2/kg)	3.28×10^{-1}	6.43×10^{-2}	3.24×10^{-1}	1.06×10^{-1}	0.204
Bcf (mT)	7.9	0.8	9.0	1.2	**4.945**
Bcr (mT)	45.8	2.1	47.2	2.0	**3.041**
Bcr/Bcf	5.9	0.6	5.3	0.6	**4.340**
S	0.90	0.01	0.90	0.01	1.444
HIRM	8.68×10^{-4}	1.06×10^{-4}	1.01×10^{-3}	2.46×10^{-4}	**3.329**
χfer/Mrs (arb. unit)	1.35×10^{-4}	1.62×10^{-5}	1.18×10^{-4}	1.18×10^{-4}	0.876
NRM_0 (mA/m)	63.14	22.88	51.02	29.84	**2.064**
Qn	0.404	0.140	0.421	0.142	0.527
κ (SI)	3.73×10^{-5}	1.13×10^{-5}	3.8×10^{-5}	0.84×10^{-5}	1.54
Magnetic lineation	1.006	0.01	1.003	0.01	0.84
Magnetic foliation	1.028	0.016	1.031	0.02	0.23
Corrected anisotropy degree	1.037	0.021	1.038	0.02	0.31

Note T critical value for 79 degrees of freedom is 1.990 (difference in means with T < 1.990 are statistically insignificant). SD—standard deviation
Bold values mean statistical significance of difference between compared parameters

Table 2), however, ellipsoid shapes in Eopleistocene are more oblate. Ellipsoids of individual samples with prolate shape (T < 0) are rare in both cases. The distributions of AMS principle axes are shown in Fig. 8. Neopleistocene (Fig. 8c) and Eopleistocene (Fig. 8d) sediments demonstrate sedimentary magnetic fabric—minimum axes are subvertical, maximum and intermediate axes approximately correspond to bedding plane [20]. The magnetic foliation plane dips in Eopleistocene and Eopleistocene magnetic foliation plane dips in Eopleistocene and Eopleistocene populations may reflect the secondary reorientation of magnetic gains in percolating water due to slope-wash process [21]. Another explanation of foliation dips suggest that sediments were deposited on the gentle slope, which dip to north–north east at 7–12° angle. AMS axes mean directions of Eopleistocene are different but their confidence ellipses overlap (Table 3).

In the modern soil (Fig. 8a), the axes are rather scattered. Three samples from the modern soil show oblate ellipsoids with only one sample has prolate ellipsoid. The small number of samples does not allow determining a magnetic fabric.

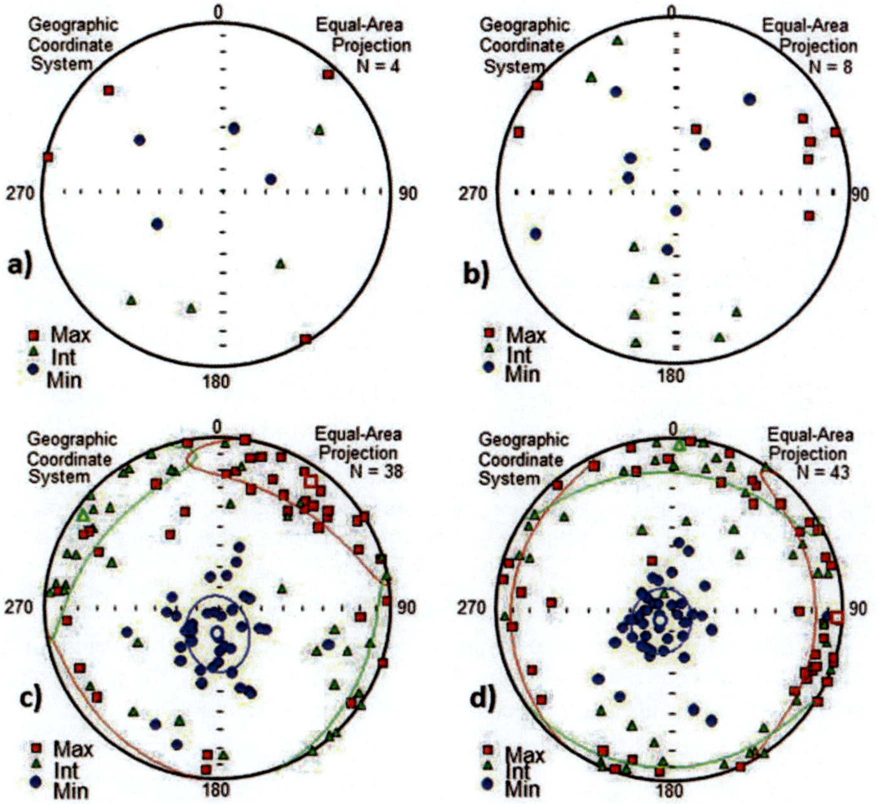

Fig. 8 Distribution of AMS principle axis in Neopleistocene and Eopleistocene sediments of the Ust-Obor section. **a** Modern soil; **b** weathering crust; **c** Neopleistocene; **d** Eopleistocene. Open symbols—mean tensor directions, lines—confidence ellipses

Table 3 Mean tensor direction of principal AMS axes and their confidence angles in Neopleistocene and Eopleistocene sediments of the Ust-Obor section

Strata	Dmax	Imax	Dint	Iint	Dmin	Imin
Neopleistocene (N = 38)	39.5	11.4	305.3	4	189.7	77.9
	Confidence angles					
	48.4	17.7	48	16	19.7	13.1
Eopleistocene (N = 43)	92.8	4.8	2.3	5.6	222.9	82.6
	Confidence angles					
	60.2	13.3	60.2	15.3	15.6	13.8

Note D—declination; I—inclination (all data in degrees); max, int and min indexes correspond to maximal, intermediate and minimal axes of AMS. N—number of samples

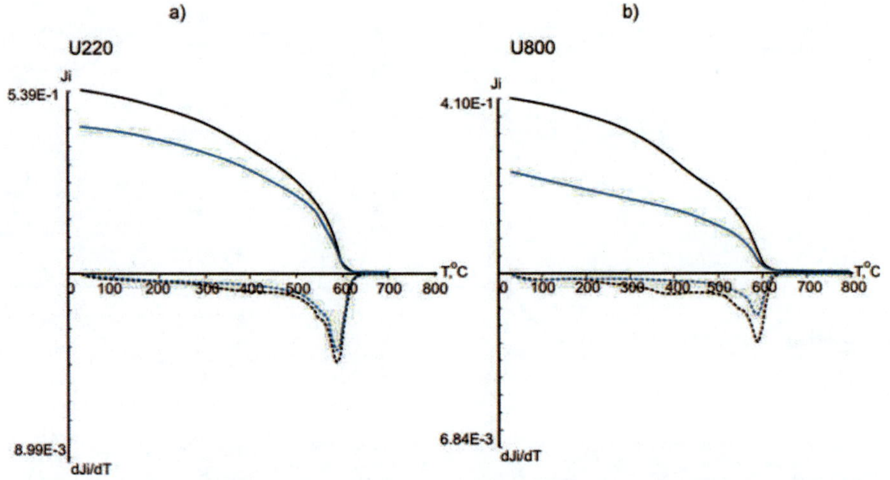

Fig. 9 Representative thermomagnetic results for Neopleistocene (**a**) and Eopleistocene (**b**) sediments of the Ust-Obor section. Black line—first heating; blue line—second heating. Solid lines Ji (T), dashed lines dJi/dT

The AMS axes in the weathering crust are also scattered (Fig. 8b) because of the mechanical breakdown weathering of the basement rock. Most of the samples have prolate AMS ellipsoids due to level because of higher SD and PSD content.

4.5 Magnetic Mineralogy

Temperature dependent experiments were carried out only for Neopleistocene and Eopleistocene sediments. No sufficient difference in thermomagnetic behavior of Neopleistocene and Eopleistocene samples was recognized. Measured Curie temperatures suggest that magnetite with some admixture of hematite is the main magnetic mineral in both parts of the section (Fig. 9). The inflection in the magnetization observed around the 300–350 °C temperature range and the decrease in magnetization after the first heating suggest that the magnetite grains are partially maghemitized. Moreover, judging by the magnitude of this decrease, the degree of maghemitization for the Eopleistocene sediments (Fig. 9b) is somewhat higher than for the Neopleistocene sediments (Fig. 9a).

4.6 Magnetic Polarity Stratigraphy

Stepwise AF demagnetization revealed the presence of only two distinguishable components in most samples. Low-coercivity component was typically isolated in fields between 6 and 24 mT. A characteristic NRM component (ChRM) with straight linear decay to the origin of vector end-point diagrams was demagnetized by 24–60 mT fields. However, the presence of high-coercivity phase resulted in incomplete demagnetization with ~10 to 15% of magnetization not affected by AF treatment. Some samples of modern soil have erratic demagnetization trajectories (Fig. 10a) or yielded anomalous (intermediate between normal and reversal) directions (Fig. 10b) or demonstrate anomalous direction (Fig. 10b).

Typical demagnetization results for Neopleistocene and Eopleistocene samples are show in (Fig. 10c, d) respectively. Although principal difference in samples' demagnetization behavior was not observed, some minor distinctions were noticed.

Median destructive fields of the lower two-thirds of Eopleistocene sediments were generally higher than in Neopleistocene samples (Fig. 11), which can be explained by hematite enhancement of Eopleistocene part of the section (see above). This hematite enhancement also affects NRM values (Fig. 11)—NRM mean directions of Neopleistocene statistically differs from Eopleistocene ones, while differences in Qn values are statistically insignificant (see Table 2).

Secondary, maximum angular deviation (MAD) values, which are larger for Neopleistocene samples (up to 12°) than for Eopleistocene (~5°) ones (Fig. 11). We interpret this observation as due to coarser-grained magnetic minerals in Neopleistocene sediments (sandy loam), which are less stable paleomagnetic recorders.

All Eopleistocene samples across the section yielded ChRM exclusively of normal polarity are (except for one Neopleistocene sample with southern declination).

Samples from the upper part of the weathering crust also characterized by normal polarity of ChRM directions (Fig. 10e). However, in the lower part of the weathering crust stable magnetizations are not preserved at all—magnetization vectors change chaotically during demagnetization (Fig. 10f).

The largest variations of the ChRM directions along the section observed in the modern soil, thick horizon of buried soils in the Neopleistocene part (layer 5), and in the weathering crust (Fig. 11). Anomalous (up to reversed polarity) characteristic directions in the modern soil most likely result from slope surface processes (solifluction). This interpretation is supported by AMS data (see above). Declination anomaly in the layer 5 is likely reflects pedogenic reworking of sediment or may represent a record of one of excursions in the Brunhes chron. ChRM directions in the weathering crust are evidently affected by chemical magnetization as seen from their high HIRM/Mrs ratios. The Bcr/Bcf and Mrs/Mf ratios (see Fig. 7) are also typical for chemical magnetization [22].

ChRM stereoplots for each stratigraphic units of the Ust-Obor section are shown in Fig. 12. Characteristic NRM directions of the modern soil are rather scattered (Fig. 12a) and obviously do not represent the direction of ancient geomagnetic field. ChRM direction distribution in the weathering crust (Fig. 12b) has an elongated shape and is not Fisher-distributed [23]. Most likely the acquisition of chemical magnetization occurred over a long time and thus ChRM directions of the weathering crust reflect different moment of geomagnetic field variations.

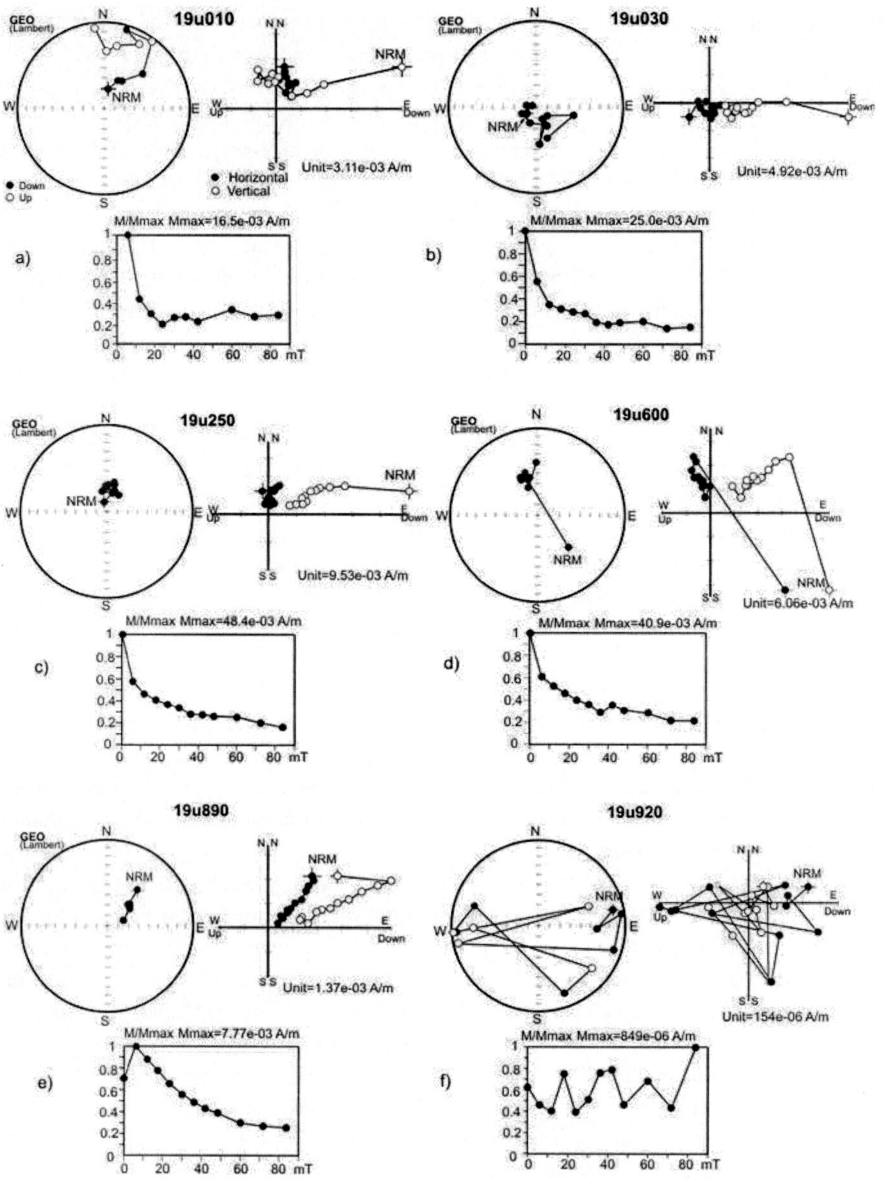

Fig. 10 AF demagnetization results for representative samples from the Ust-Obor section (stereoplots, orthogonal plots, and NRM decay pots). **a**, **b** Modern soil; **c** Neopleistocene; **d** Eopleistocene; **e** upper part of weathering crust; **f** lower part of weathering crust

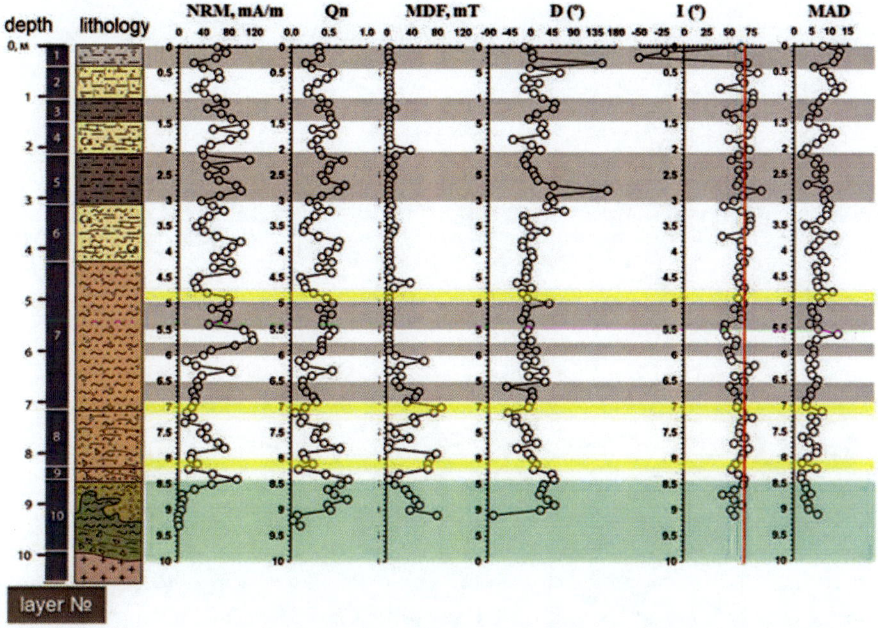

Fig. 11 Variations of paleomagnetic parameters in Ust-Obor section versus depth. MDF—median destructive field (50% of initial NRM remains); D—declination of characteristic component; I—inclination of characteristic component; MAD—maximum angular deviation of characteristic component. Red line marks inclination of GAD model

ChRM directions from Neopleistocene (Fig. 12c) and Eopleistocene (Fig. 12d) sediments are closer to Fisher distribution [23]. Mean directions for both units are given in (Table 4). Both units are characterized by normal polarity. The scatter of paleomagnetic directions for the Neopleistocene unit is slightly higher than for Eopleistocene even if we exclude directions of presumed excursion (Table 4). Mean paleomagnetic directions for Neopleistocene and Eopleistocene units are statistically different but difference become insignificant if we directions of presumed excursion (Table 4).

5 Discussion

Normal polarity of Eopleistocene part of the section does not agree with the generally accepted magnetic polarity during the Eopleistocene (Fig. 2). According to the Russian magnetic polarity time scale [24], predominant polarity during upper Eopleistocene was reversed (Matuyama chron), suggesting that the magnetization polarity of the strata containing the Ust-Oborian fauna should be of reversed (see Fig. 2).

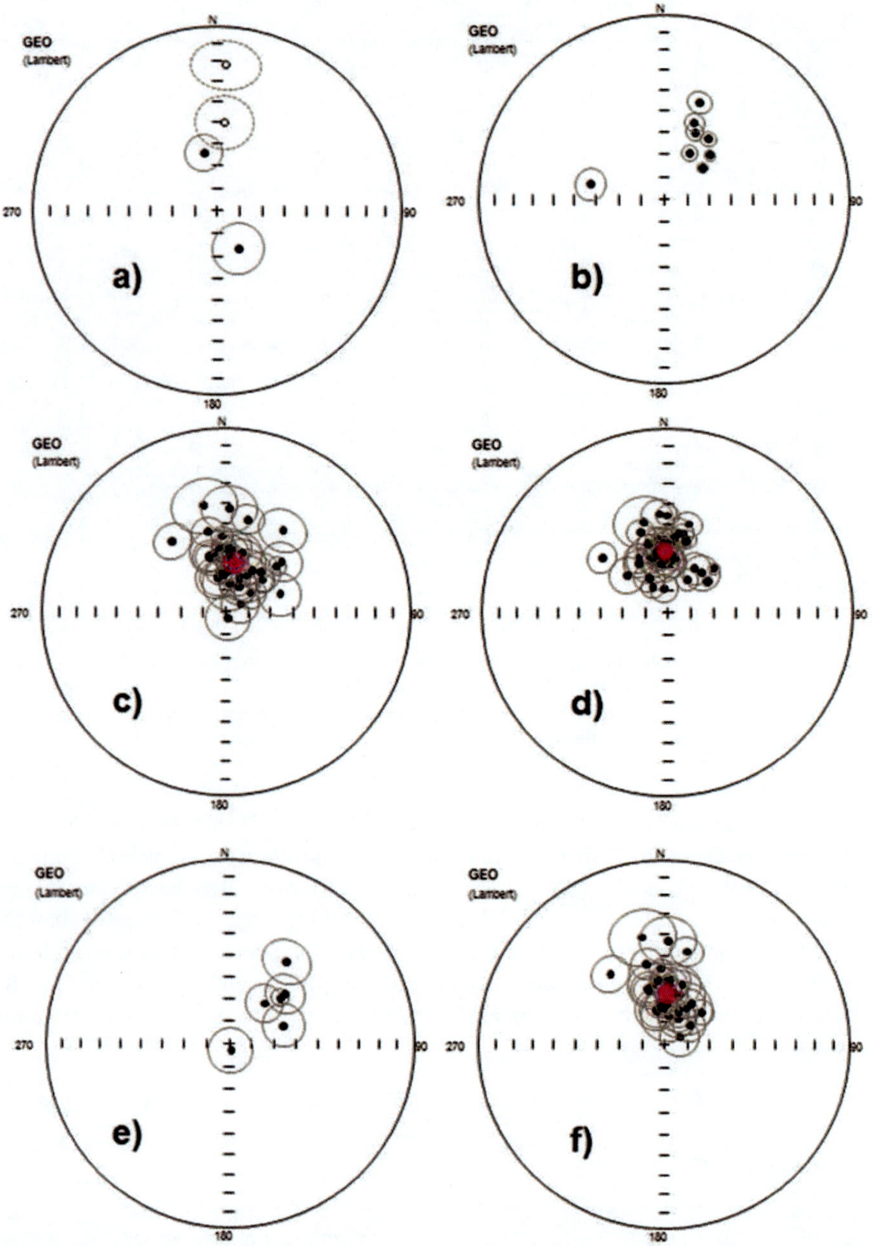

Fig. 12 Stereoplots of ChRM directions with confidence ellipses for different part of the Ust-Obor section. **a** Modern soil; **b** weathering crust; **c** Neopleistocene (layers 2–6) part; **d** Eopleistocene (layers 7–9) part; **e** interval of presumed excursion; **f** Neopleistocene (interval of excursion is excluded). Pink symbol—mean direction

Table 4 Mean ChRM directions of Neopleistocene and Eopleistocene units of the Ust-Obor section and their angular difference

Unit	N	D (°)	I (°)	k	a95 (°)
Neopleistocene	38	10.5	67.4	29.1	4.4
Eopleistocene	43	356.7	62.8	44.9	3.3
Neopleistocene (presumed excursion excluded)	32	3.2	66.9	38.7	4.2
Angular difference	7.1 ± 4.3				
Angular difference (presumed excursion excluded)	4.5 ± 4.1				

Note N—number of samples; D—mean declination; I—mean inclination; k—precision parameter; a95—confidence limit

Several possible scenarios can explain the observed discrepancy. The age of the Ust-Oborian fauna may require revision (for possible correlation of normal Ust-Obor section with Cobb Mountain or Olduvai subchrons) but paleontological methods alone can resolve this issue, which is beyond the scope of this article. Therefore, for the two potential explanations resolving the contradiction that follows, the age of the Ust-Oborian fauna is assumed to be correct.

5.1 Total Remagnetization of Eopleistocene Strata by Normal Field (Age of Magnetization Is Younger Than Fauna Age)

This assumption is supported by statistically insignificant differences in most of the rock magnetic and paleomagnetic parameters between the Neopleistocene and Eopleistocene parts of the Ust-Obor section very similar magnetic mineralogy in these sediments. Notwithstanding this, it is rather difficult to invoke a realistic remagnetization mechanism that could potentially be responsible for observed paleomagnetic properties. A remagnetization process of the Ust-Obor section should have resulted in acquisition of secondary chemical magnetization carried by coarse-grained multi-domain magnetite. Moreover this secondary magnetite should retain the primary sedimentary fabric. None of the widely known mechanisms described in [25] are suitable for remagnetization of sediments in the Ust-Obor section, since there are no required conditions for chemical reactions.

If we assume that hematite is a carrier of secondary magnetization (which concentration in the lower part of the section is higher than in the upper part), then the primary reverse magnetization should retain (at least partially) in samples in which the magnetization of magnetite prevails over the magnetization of hematite.

Postdepositional deformations of the sediments can also be excluded. The modern soil in Ust-Obor section can serve as an example of postdepositional deformations due to the observation of no stable demagnetization directions and isotropic AMS. Eopleistocene and Neopleistocene sediments in contrast to modern soil are characterized by noticeable AMS and stable ChRM directions which means do not coincide with maximal AMS axes. ChRM inclinations generally 5–20° lower than expected GAD inclinations (Fig. 11) which indirectly points at detrital sedimentary nature of magnetization.

An additional argument for the absence of remagnetization are the values of Qn less than unit. Such values are typical for detrital origin of magnetization in sediments rather than for chemical one [26]. In total all these arguments allow rejecting the remagnetization hypothesis.

5.2 *Redeposition of Ust-Oborian Fauna in Neopleistocene Strata (Age of Strata Is Younger Than Fauna Age)*

This assumption suggests that the Neopleistocene sandy loams and Eopleistocene loams constitute a single sequence accumulated in the Upper Neopleistocene, and the Ust-Oborian fauna was redeposited from older sediments. This hypothesis is also in good agreement with the consistent values of the rock magnetic parameters (concentration—dependent and AMS parameters) in both units. Statistically significant differences between parameters are characteristic mainly for a group of coercive parameters and can be easily explained by an increased concentration of hematite in the lower half of the section, which was redeposited due to erosion of the underlying weathering crust. Statistically significant differences between mean ChRM directions in the upper and lower deposits do not exceed the amplitude of secular variations and thus, cannot serve as a convincing basis for substantiating the different ages of upper and lower part of the section.

The similarity in sedimentary environment during deposition of the upper and lower units is also clear from almost identical compositions of magnetic minerals in the sediments, similar sedimentary magnetic fabric, and statistically insignificant differences in mean Qn values < 1.

Taking into account, that slope-wash and colluvial process played a significant role in the sedimentation process it is rather reasonable to assume the following scenario of sedimentation: Eopleistocene sediments were eroded from the upper parts of slopes and then the sediments containing Eopleistocene fauna were transported by gravity and water down the slopes where was deposited and depleted by aeolian material.

Eopleistocene climate in Baikal rift zone was warmer than Neopleistocene one [27], so the sedimentary environments should be widely different as well as magnetic properties of sediments. However, the magnetic properties of the lower and upper units in Ust-Obor section differ insignificantly. Moreover, magnetic properties of Eopleistocene reversely magnetized buried soils on the Olkhon Island (Lake Baikal) follows “Chinese (pedogenic)” mechanism [28] and differ dramatically from magnetic properties of normally magnetized buried soils in the Ust-Obor section with “Siberian mechanism”.

Summarizing all the arguments above, we arrive at a conclusion that the upper (layers 2–6) and lower (7–9) units of the Ust-Obor section represent deposits of a single Neopleistocene sedimentation stage, and the Eopleistocene fauna is highly likely redeposited.

6 Conclusions

All stratigraphic units of Ust-Obor section is characterized only by normal polarity, however the age of the Ust-Oborian fauna complex is reported to be of Eopleistocene thus within the Matuyama reversed polarity chron.

This contradiction cannot be explained by secondary magnetization overprinting the primary reversed magnetization, since rock magnetic and paleomagnetic data do not provide the necessary supporting evidence.

Close mean values of most of rock magnetic parameters, similar nature of the magnetic fabric, and the coincidence of mean ChRM directions within the limits of secular variations make strongly support a single sedimentation stage (Neopleistocene) for the lower and upper units. In this case, the Eopleistocene fauna is highly likely redeposited.

To finally resolve this controversy, it is necessary to perform additional studies of the Ust-Oborian Eopleistocene fauna to clarify the age and the possibility of redeposition.

Acknowledgements This work was supported by the Russian Science Foundation (project no. 19-17-00216, paleomagnetic study), RFBR (project no. 20-05-00247) and the Government of the Russian Federation (project no. 075-15-2019-866). We thank the reviewers France Lagroix and Boris Robert and the associate editor Evgeniy V. Kulakov for their useful comments on the manuscript.

References

1. Gnibidenko, Z.N., Erbaeva, M.A., Pospelova, G.A.: Paleomagnetism and biostratigraphy of some deposits of the Upper Cenozoic of Western Transbaikalia. In: Fotiadi, E.E. (ed.) Paleomagnetizm mezozoya i kainozoya Sibiri i Dal'nego Vostoka (Paleomagnetism of the Mesozoic and Cenozoic of Siberia and the Far East), pp. 75–95. IGiG SO RAN, Novosibirsk (1976) (in Russian)
2. Zudin, A.N.: Some problems of the Trans-Siberian paleomagnetic correlation of the key sections of the Quaternary and regional stratigraphy. In: Kochkovsky Horizon of Western Siberia and It's Age Analogies in Adjacent Areas. Proceedings of the Institute of Geology and Geophysics SBAS USSR, pp. 98–118. Nauka Press, Novosibirsk (1980) (in Russian)
3. Karasev, V.V.: Cenozoic of Transbaikalia, 128 p. Chita (2002) (in Russian)
4. Erbajeva, M.A., Shchetnikov, A.A., Kazansky, A.Yu., Matasova, G.G., Khenzykhenova, F.I., Filinov, I.A., Namzalova, O.D.-Ts., Nechaev, I.O.: New Pleistocene key section Ulan-Zhalga in Western Transbaikalia. Dokl. Earth Sci. **488**(3), 48–52 (2019). https://doi.org/10.31857/S0869-56524883277-281

5. Matasova, G.G., Kazansky, A.Yu., Shchetnikov, A.A., Erbajeva, M.A., Filinov, I.A.: New rock- and paleomagnetic data on quaternary deposits of the Tologoi key section, Western Transbaikalia, and their paleoclimatic implications. Izv. Phys. Solid Earth **56**(3), 392–412 (2020). https://doi.org/10.1134/S1069351320030052
6. Alekseyeva, N.V.: Evolyutsiya prirodnoi sredy Zapadnogo Zabaikal'ya v pozdnem kainozoe (po dannym fauny melkikh mlekopitayushchikh) (Environmental Evolution of Late Cenozoic of West Transbaikalia (Based on Small Mammal Fauna)). In: Dodonov, A.V. (ed.). GEOS, Moscow (2005) (in Russian)
7. Gradstein, F.M., Ogg, J.G., Schmit, M.D., Ogg, G.M.: Geologic Time Scale 2020. Elsevier (2020)
8. Karasev, V.V., Erbajeva, M.A.: Reference section of Quaternary sediments Ust-Obor (Transbaikalia). In: Quaternary Sediments of the Far East and Adjacent Territories of Khabarovsk, pp. 67–69. FSGGP Khabarovskgeologiya (2001) (in Russian)
9. Alekseyeva, N.V., Karasev, V.V., Erbajeva, M.A.: Pleistocene biostratigraphy of the Transbaikal area (South East Russia). Cour. Forsch.-Inst. Senckenberg **259**, 237–241 (2007)
10. Tauxe, L.: Essentials of Paleomagnetism. University of California Press, Berkeley (2010)
11. Zijderveld, J.D.A.: A.C. demagnetization of rocks: analysis of results. In: Collinson, D.W., Creer, K.M. (eds.) Methods in Paleomagnetism, pp. 254–286. Elsevier, Amsterdam (1967)
12. Kirschvink, J.L.: The least-squares line and plane and the analysis of palaeomagnetic data. Geophys. J. Int. **62**, 699–718 (1980)
13. Jasonov, P.G., Nurgaliev, D.K., Burov, B.V., Heller, F.: A modernized coercivity spectrometer. Geol. Carpath. **49**, 224–226 (1998)
14. Kosareva, L.R., Nourgaliev, D.K., Kuzina, D.M., Spassov, S., Fattakhov, A.V.: Ferromagnetic, dia/paramagnetic and superparamagnetic components of Aral Sea sediments: significance for environmental reconstruction. ARPN J. Earth Sci. **4**, 1–6 (2015)
15. Day, R., Fuller, M., Schmidt, V.A.: Hysteresis properties of titanomagnetites: grain-size and compositional dependence. Phys. Earth Planet. Inter. **13**, 260–267 (1977)
16. Evans, M.E., Heller, F.: Environmental Magnetism: Principles and Applications of Enviromagnetics. Academic Press, New York (2003)
17. Matasova, G.G., Kazansky, A.Yu., Zykina, V.S.: Superposition of "Alaskan" and "Chinese" models of paleoclimate records in magnetic properties of Upper and Middle Neopleistocene deposits in southern West Siberia. Russ. Geol. Geophys. **44**, 607–619 (2003)
18. Liu, Q., Roberts, A., Torrent, J., Horng, C.S.: What do the HIRM and S-ratio really measure in environmental magnetism? Geochem. Geophys. Geosyst. **8**, Q09011 (2007). https://doi.org/10.1029/2007GC001717
19. Dunlop, D.J.: Theory and application of the Day plot (Mrs/Ms versus Hcr/Hc). 1. Theoretical curves and tests using titanomagnetite data. J. Geophys. Res. Solid Earth **107**(B3) (2002). https://doi.org/10.1029/2001JB000486
20. Tarling, D.H., Hrouda, F.: The Magnetic Anisotropy of Rocks. Chapman and Hall, London (1993)
21. Tailor, S.N., Lagroix, F.: Magnetic anisotropy reveals the depositional and postdepositional history of a loess-paleosol sequence at Nussloch (Germany). J. Geophys. Res. Solid Earth **120**, 2859–2876 (2015). https://doi.org/10.1002/2014JB011803
22. McCabe, C., Channell, J.E.T.: Late Paleozoic remagnetization in limestone of the Craven Basin (northern England) and rock magnetic fingerprint of remagnetized sedimentary carbonate. J. Geophys. Res. **99**, 4603–4612 (1994)
23. Fisher, R.A.: Dispersion on a sphere. Proc. R. Soc. A **217**, 295–305 (1953)
24. Guzhikov, A.Yu., Shkatova, V.K.: About the implied changes in general magnetostratigraphic polarity scale of quaternary period. In: Decisions of the Interdepartmental Stratigraphic Committee of Russia and Its Regular Commissions, 68 p. Release № 44. VSEGEI publishing, St. Petersburg (2016) (in Russian)
25. Elmore, R.D., Muxworthy, A.R., Aldana, M.M., Mena, M.: Remagnetization and chemical alteration of sedimentary rocks. Geol. Soc. Lond. Spec. Publ. **371**, 1–21 (2012). https://doi.org/10.1144/SP371.15

26. Nagata, T.: Rock Magnetism, 350 pp. Maruzen Co., Tokyo (1961)
27. Mats, V.D., Ufimtsev, G.F., Mandelbaum, M.M., Alakshin, A.M., Pospeev, A.V., Shimaraev, M.N., Khlystov, O.M.: Cenozoic of Baikal Rift Depression: Structure and Geological History. SB RAS Publishing, Novosibirsk (2001).(in Russian)
28. Matasova, G.G., Kazansky, A.Y., Shchetnikov, A.A., Filinov, I.A.: Relationship between rock-magnetic and grain size data from early quaternary section "Togay" (Olkhon Island, Baikal Lake). In: *11th International Conference "Problems of Geocosmos"*, Book of Abstracts, pp. 156–157. St. Petersburg (2016)

Preliminary Rock Magnetic and Paleomagnetic Data from a 14.5 m Core of Lake Kotokel Sediments (Baikal Region)

A. A. Shchetnikov, A. Yu. Kazansky, I. A. Filinov, and G. G. Matasova

Abstract Lake Kotokel is a medium-size freshwater lake located close to the eastern coast of the Lake Baikal. Kotokel lake lacustrine sediments preserve a unique archive of vegetation and climate history for at least the last 46 kyr. Here we present the first preliminary rock magnetic and paleomagnetic data for Lake Kotokel. A 14.5 m long sediment core was retrieved from a water depth of 3.6 m in the southern part of the lake. Magnetic susceptibility (MS) and its anisotropy were measured from 385 oriented samples. The characteristic component of the natural remanent magnetization (ChRM) was obtained using stepwise alternating field (AF) demagnetization. Results of our rock magnetic analyses were interpreted in conjunction with available paleoclimatic and paleobotanical data. The response of magnetic properties to climatic changes is illustrated by the correlation of MS values with vegetation change in the region as indicated by spore-pollen and diatom data. The ancient geomagnetic field is poorly recorded in the studied sediments. Paleomagnetic directions in the upper part of the core are randomly distributed due both to very weak remanence as well as to mechanical perturbation of sediments, most likely caused by degassing processes in soft gyttja. Unstable and scattered ChRM directions result from secondary mineralogical modifications, such as dissolution of primary magnetic grains and growth of secondary magnetic minerals, caused, in turn, by environmental changes in the study area.

A. A. Shchetnikov · A. Yu. Kazansky (✉) · I. A. Filinov · G. G. Matasova
Geological Institute of Russian Academy of Sciences, Moscow, Russia

A. A. Shchetnikov · I. A. Filinov
Irkutsk State University, Irkutsk, Russia

Institute of the Earth's Crust, Siberian Branch, Russian Academy of Sciences, Irkutsk, Russia

A. A. Shchetnikov
Vinogradov Institute of Geochemistry, Siberian Branch, Russian Academy of Sciences, Irkutsk, Russia

A. Yu. Kazansky
Geological Department, Lomonosov Moscow State University, Moscow, Russia

G. G. Matasova
Trofimuk Institute of Petroleum Geology and Geophysics, Siberian Branch, Russian Academy of Sciences, Novosibirsk, Russia

A. Kosterov et al. (eds.), *Problems of Geocosmos—2020*, Springer Proceedings in Earth and Environmental Sciences, https://doi.org/10.1007/978-3-030-91467-7_12

Keywords Rock magnetism · Paleomagnetism · Lake sediments · Baikal region · Paleoclimate

1 Introduction

Lacustrine sediments of small isolated lakes with limited drainage are commonly composed of detrital material, originating from the surrounding catchment area. Small lakes respond quickly to environmental fluctuations, adapting to sudden changes in seasonal temperature, runoff catchment, and snowmelt. The ability of lakes to respond rapidly to environmental oscillations and to record such changes in their bottom sediments makes them important archives of paleoclimate history [1]. Magnetic properties of lake sediments are among the most widely used paleoclimatic proxies due to their high sensitivity to environmental and climatic fluctuations [2–4]. In addition, magnetic measurements are typically non-destructive and usually not very time-consuming. Rapidly accumulating sediments are traditionally used for studying the geomagnetic field behavior on geologically short time scales (secular variation and its extremes such as reversals and excursions).

Magnetic properties of lacustrine sediments, however, may be affected by geochemical processes (oxidation or reduction), and magnetic minerals can either precipitate, or dissolve in situ [5, 6]. Thus, for correct interpretation of the magnetic signal in such sediments, a full understanding of magnetization mechanisms and origin is required.

The Lake Baikal sediments represent the longest paleoclimatic record in the area [7], which makes them a natural laboratory for magnetic studies, with rock magnetic proxies serving as an important tool for paleoclimatic reconstructions and dating of sediments. At present, the magnetic susceptibility record, used for paleoclimatic interpretations of Lake Baikal sediments covers the last 6.7 Myr [8], while the magnetic polarity record spans the last 8.4 Myr [9].

The basis of paleoclimatic interpretation of rock magnetic data from Baikal sediments was originally proposed by Peck et al. [10]. It was shown that decreasing magnetic concentration and coercivity are linked to climatic variations through biogenic dilution during interglacial periods and to the increasing concentration of terrigenous high coercivity minerals during glacial periods. Later, biogenic magnetite was found in the upper part of the Lake Baikal sediments, which concentration decreased towards the bottom of the section due to diagenetic reduction [11]. Dearing et al. [12] proposed the following mechanism of the post-depositional transformation reductive diagenesis of ferrimagnetic minerals: authigenic formation of paramagnetic and canted antiferromagnetic oxyhydroxides (ferrihydrite and goethite); authigenic formation of greigite; and bacterial magnetosome production. High coercivity minerals in the sediments are represented by atmospherically derived particles and catchment-derived detrital material [12].

A more detailed scenario of magnetization processes was proposed by Demory et al. [6], who suggested that during glacial epochs magnetic properties of sediments are largely controlled by detrital magnetite. Detrital magnetite is usually associated with pronounced horizons of authigenic greigite, formed in a reducing environment. During interglacial stages, along with the dilution by biogenic silica and decreasing detrital input, the weakness of the rock magnetic signal was partly caused by reductive dissolution of magnetic particles. Small magnetite grains are typically dissolved, while larger low-coercivity grains with much greater volume-to-surface ratios are less affected by dissolution and remain largely in their initial multidomain state, leading to low stability of the paleomagnetic signal up the to loss of the primary NRM direction [13].

At low sedimentation rates, magnetite is preferentially dissolved after burial under the redox front and detrital hematite becomes a predominant mineral. High sedimentation rates and fast burial cause preservation of magnetite or its transformation into greigite under sulphate-reducing conditions. Formation of greigite ultimately affects initial magnetic properties of sediments [6]. A good example of such effects are sediments from Khubsugul (Hovsgul) Lake in Mongolia, a smaller rift-related sister-lake of Lake Baikal. Although Khubsugul Lake sediments carry a well-preserved record of the magnetic field for the last million of years [14], rock magnetic properties are rather complex. The reason is for this is the abundance of biogenic magnetite and authigenic and biogenic greigite in the sediments [15, 16]. Only detailed rock magnetic analyses allowed to determine terrigenous magnetic influx into the sediments and to reveal that variations in magnetite/greigite ratio reflect changes from suboxic (high ratio) to anoxic (low ratio) sedimentation conditions.

Overall, the relationship between rock magnetic properties of sediments and paleoclimate changes in the lakes of the Baikal area is rather complex and variable for different lakes and sedimentary environments.

Sedimentation rates in the Holocene-Upper Pleistocene interval of the Baikal sediments do not exceed 3.5–3.7 cm/kyr [9, 17] and 5 cm in Khubsugul Lake [14]. Such low rates resulted in limited sedimentary and paleoclimatic records. In contrast, small and medium-size lakes with more rapid sedimentation rates can provide more detailed information about the Upper Pleistocene–Holocene regional history. Therefore, such lakes are attracting an increasing attention for paleoclimatic reconstructions of the last 50–100 ka.

One of such small- to medium-sized lakes is the Kotokel Lake, the sediments of which provide unique archives of vegetation history, climate, and lake-internal bioproductivity (diatoms) for the Holocene und the Upper Pleistocene. Although available drill cores [18] represents the longest available sedimentary record, paleomagnetic and rock magnetic data on the Lake Kotokel sediments are still missing. Accordingly, the main purpose of this study was to obtain detailed records of magnetic susceptibility and geomagnetic polarity patterns from Lake Kotokel sediments in order to test their suitability for environmental and palaeomagnetic studies.

2 Geology and Sampling

Lake Kotokel (52.490° N, 108.090° E, 458 m a.s.l.) is a medium-sized freshwater lake located close to the eastern shore of Lake Baikal (Fig. 1). The lake has a surface area of 69 km^2 (15 km length on a SW–NE elongation and 6 km width), with an average depth of 4.5 m and maximum depth of 15 m. The catchment area was estimated as ~187 km^2 [19]. This lake was formed due to subsidence of the lithosphere in the Baikal rift zone. Lake Kotokel is located in coastal low-mountain massifs and is in contact with the lowland in the northeast (the Kotochik River area), where the ancient weathered crust is overlain by over 100 m thick middle Pleistocene–Holocene sediments [19]. The intricate shoreline of the lake mimics variable-size open bights and bays, peninsulas and islands, reflecting the topography of the surrounding area.

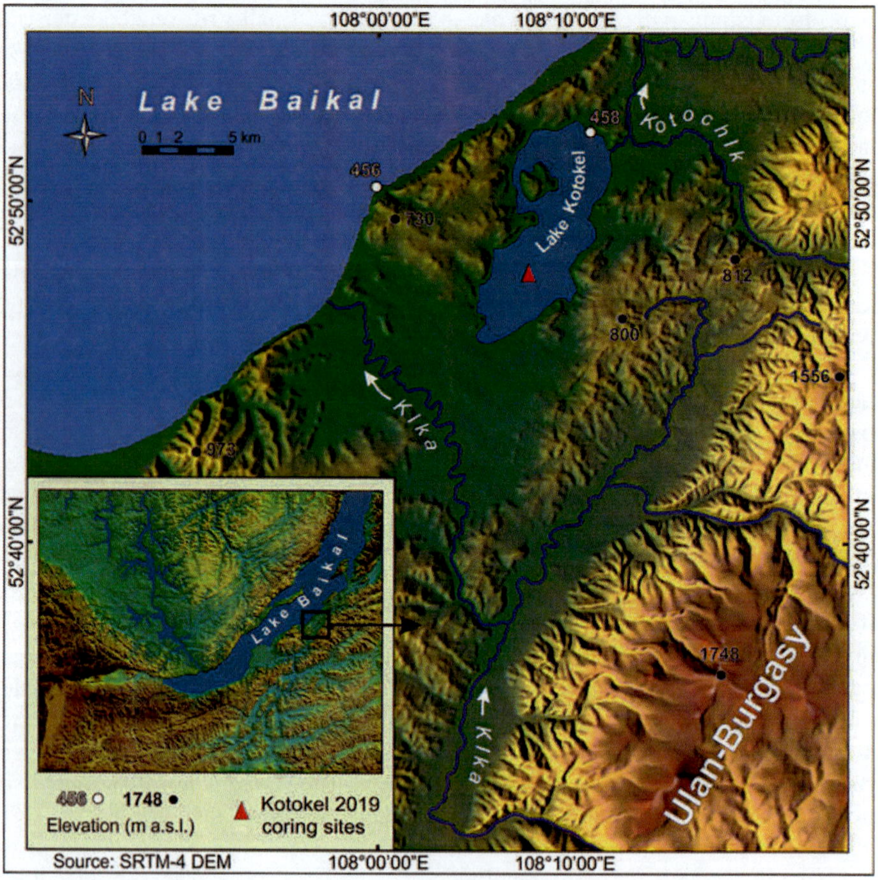

Fig. 1 Study area and location of the coring site. White elevation numbers mark lake shores, black ones mark all other relief

Three cores from the Kotokel Lake (KTK1 [20], KTK2 [18], and KOT-01/05 [21]) have been previously drilled and studied. In all cores the upper part of the sediments is composed by soft brownish gyttja with variable thickness ranging between 660 and 810 cm. The reported thickness variations likely result from the visual inspection of somewhat indistinct sedimentary boundaries of the same core by different workers [18, 22]. Sediments beneath the gyttja layer were described differently by different workers [18, 20, 21]. Here we use the description of the KTK2 core from [18] as it was used for interpretation of seismic stratigraphy of the lake sediments [23]. The seismic data reveals three depositional units distributed over the entire lake basin [23]. The upper unit 1 consists of the soft gyttja, which reaches a maximum thickness of 700–800 cm. The gyttja unit is underlain by a 300–400 cm thick, partly laminated clay and silty clay (Unit 2). The units 1 and 2 follow the pre-relief of the lake bottom and may have been formed as lake sediments. Unit 3 consists of black silty clay sediments of unknown thickness. The lake bottom stratigraphy as inferred from seismic data suggests that the same layers can be traced over the entire lake Kotokel basin. Thicknesses of these layers, however, may vary [23].

Unfortunately, the age of the Lake Kotokel sediments is ambiguous. The only reliable age model was established for the KTK2 core on the basis of 13 radiocarbon determinations [18]. Three age determinations available for the KTK1 core do not agree with each other and therefore do not provide additional age constraints [22]. There are no absolute age data for the KOT-01/05 core and its age model was constructed based on the correlation of pollen records in the core with those in cores KTK1 and KTK2 [21]. Here we use the age model proposed by [23], constructed on the basis of the KTK2 core age data. This model suggests that the accumulation of the Unit 1 (gyttja) began at the end of the Last Glacial (15 ka BP) and continued during the entire Holocene. The Unit 2 (partly laminated clay) was deposited during the Late Pleistocene approximately between 25 and 15 ka BP; The Unit III (silty clay) may have been formed during Marine isotope stage (MIS) 3. Those age constraints allow estimation of sedimentation rates in Lake Kotokel as ~47 cm/ka for Unit 1 and ~28 cm/kyr for Units 2 and 3 [18, 23].

The Kotokel core KTK-19/II (52° 47′ 20.94″ N, 108° 7′ 29.45″ E) was retrieved from a depth of 3.6 m in the southern part of Lake Kotokel in June 2019, at the site identified as having the most uniform sedimentation rates. A UWITEC piston corer platform (Austria) was used for coring. The UWITEC system consists of 2 m long 63 mm PVC liners in a stainless-steel liner, and the stationary piston system enables the operator to retrieve contiguous 2 m sections. The corer has a rubber sleeve core catcher, and operates using a 50 kg hammer. Totally 8 consecutive ~2-m sections (lots) of the core with a total length of 14.6 m were obtained. The core recovery was 95%. The core was hermetically packed and delivered to the Vinogradov Institute of Geochemistry of the Siberian Brunch, Russian Academy of Sciences, where, under laboratory conditions, it was divided along the length into two equal half-cores, one of which was archived. Samples were taken from the other half. The uppermost and lowermost 3–5 cm parts of each lot were excluded due to deformation of the sediments caused by the drilling process. The core consists of eight lots which represent a nearly continuous sediment succession up to 1450 cm.

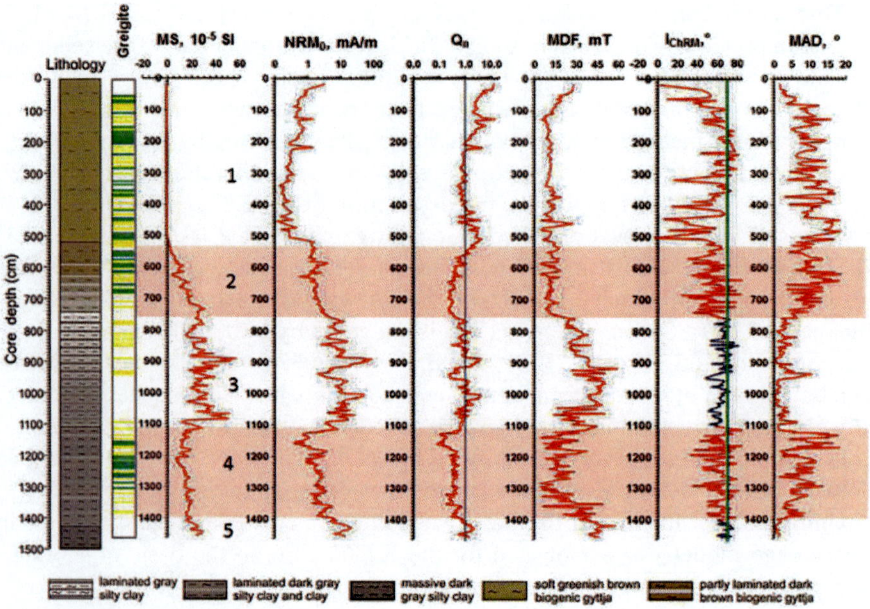

Fig. 2 Magnetic properties of the sediments from Lake Kotokel core KTK-19/II. MS—magnetic susceptibility; NRM_0—value of natural remanent magnetization before AF demagnetization; Q_n—Koenigsberger ratio (blue line marks $Q_n = 1$); MDF—median destructive field, I_{ChRM}—inclination of the characteristic component (green line marks the inclination from the Geocentric Axial Dipole (GAD) model): red curve—unstable behavior, dark blue curve—stable behavior (see text), MAD—maximal angular deviation. 1–5—Numbers of segments (see text). Pink bars mark even segments. "Greigite" column marks the presumed presence of greigite: green—most likely; yellow—probable, white—no greigite

The core sediments consist of soft greenish brown biogenic gyttja (0–520 cm), partly laminated dark biogenic gyttja (520–590 cm), intervening layers of dark biogenic gyttja with gray silty clay (590–740 cm), laminated and gray silty clay (740–1110 cm), laminated dark gray silty clay and clay (1110–1440 cm), and massive dark gray silty clay (below 1440 cm) (Fig. 2).

Totally, 385 samples were taken from undisturbed sediments on retention of their orientation with respect to the axis of the core. 7 cm^3 internal volume plastic boxes with rounded edges were pressed into the split surface of the U-channel approximately every 3–5 cm.

3 Methods

Laboratory studies were carried out in the Laboratory of the Main Geomagnetic Field and Petromagnetism of the Schmidt Institute of Physics of the Earth of the Russian

Academy of Sciences (RAS, Moscow) and the Laboratory of Geodynamics and Paleomagnetism of the Trofimuk Institute of Petroleum Geology and Geophysics of the Siberian Branch of RAS (Novosibirsk). Volume-normalized low field magnetic susceptibility (MS) and its anisotropy (AMS) were measured using an AGICO MFK-1 Kappabridge. Corrected anisotropy degree (Pj′) and shape factor (T) of AMS were calculated according to [24] using the AGICO Anisoft 4.2 AMS analytical software package.

Measurements of natural remanent magnetization (NRM) were made using a 2G Enterprises 755R cryogenic magnetometer. Because diamagnetic response overwhelmingly dominated in the gyttja layer samples, we calculated the Koenigsberger ratio $Q_n = \text{abs}(NRM_0/(MS * H_{amb}))$, where H_{amb} is the Earth magnetic field of at the sampling site (60 μT) and NRM_0 the NRM intensity before demagnetization. Specimens from each level were subjected to stepwise AF demagnetization, using a built-in automatic degaussing system Model 600 (2G) in 5–10 mT steps up to 70–120 mT. Demagnetization results were analyzed using vector end-point diagrams [25] and principal component analysis [26] implemented in the AGICO Remasoft 3.0 paleomagnetic analytical software package.

Greigite-bearing samples were identified by the increase NRM during AF demagnetization due to an acquisition of gyroremanent magnetization (GRM) [27] that is most pronounced in greigite [28]. However, such approach is not always definitive [29].

Because GRM is typically acquired above 40–70 mT demagnetizing AF, we used the ratio of the minimum NRM value on the decreasing path of demagnetization curve (above 40 mT demagnetization field) to the maximum NRM value on its increasing part, as a greigite indicator. AF peak amplitudes in our study vary from 70 to 120 mT, so we accepted the following provisory limits to rank the occurrence of greigite: (i) min/max ratio > 1.5—greigite is most likely present; (ii) 1.1 < min/max ratio < 1.5—greigite is probably present; (iii) min/max ration < 1.1—greigite is not present.

4 Results

Variations in rock magnetic parameters allowed subdividing the studied core into five segments with boundaries nearly perfectly coinciding with the lithological ones (Fig. 2).

The uppermost segment 1 of greenish brown biogenic gyttja (0–520 cm) is characterized by low (-0.3 to -0.8×10^{-5} SI) negative MS values. Low positive susceptibility values ($<1 \times 10^5$—SI) were also detected in some instances. The upper part of this segment (above 170 cm) shows NRM_0 values >0.5 mA/m and consequently $Q_n > 1$. Median destructive field (MDF) values in this part consistently decrease with depth from 28 to 8 mT. The lower part of the segment is characterized by the lowest NRM_0 values (<0.5 mA/m). Q_n in most cases is close to 1 and MDF values change from 5 to 15 mT.

The second segment 2 (520–740 cm), reveals a gradual increase of MS values (from 14 to 305 × 10^{-5} SI), a sharp (up to 4 mAm) increase of NRM_0, $Q_n > 1$ (above 600 cm) and $Q_n < 1$ (below 600 cm). The MDFs vary between 4.5 and 18.6 mT (Fig. 2).

Segment 3 (740–1110 cm) has higher MS values (116–566 × 10^{-5} SI), NRM_0 (2–89 mAm) and MDF (up to 51 mT). MS and NRM_0 demonstrate the maximum scatter in the entire section. Q_n varies around 1 but sometimes exceeds the value $Q_n = 3$ (Fig. 2).

Segment 4 (1100–1400 cm) is characterized by a sharp drop of MS (69–289 10^{-5} SI) and NRM_0 (0.48–7.9 mAm) accompanied by gradual decrease of Q_n and MDF values (Fig. 2). $Q_n < 1$ (0.01–0.8) within the whole segment while MFD values are scattered largely from 4.3 to 37 mT (Fig. 2).

The last segment 5 (1400–1454 cm) is composed of massive dark gray silty clay. It is characterized by slight increase of MS (155–291 10^{-5} SI) and NRM_0 (4.9–22 mAm), $Q_n > 1$ (up to 2.3) and enhanced MDF values 39–57 mT (Fig. 2).

The corrected anisotropy degree Pj′ changes from 1.05 to 1.4. Minimal Pj′ values correspond to the gyttja layer (segment 1). Several sharp peaks of Pj, detected for this segment are probably due to measurement errors, related to limited sensitivity and overall low values of measured parameters (Fig. 3). AMS in this segment is characterized by highly scattered shape factor values (T), which change sharply from prolate to oblate (Fig. 3). This change perfectly corresponds to lithological changes, i.e. sediments below 500 cm are mainly characterized by foliation fabric, whereas the uppermost sediments lack any foliation or preferred alignment. The T-factor distinctly marks the boundary between gyttja and clay units (Fig. 3). Maximum anisotropy values up to 1.3 correspond to the middle part of segment 4. For the rest of the core, Pj′ is relatively constant without sharp changes, except for a thin horizon between 1110 and 1116 cm with low degree of AMS, characterized by prolate AMS ellipsoids (Fig. 3).

Magnetic fabric in the two uppermost lots of the core is totally different from that in the parts below. The parts of the core (>552 cm, lots 4–8) have typical sedimentary fabrics with subvertical κ_{min} axes and subhorizontal κ_{max} axes evenly distributed within the bedding plane [24] (Fig. 4). We could not determine the actual declination of κ_{max} axes, because the core lots were not oriented horizontally; nevertheless, we can estimate their distribution within each lot. Figure 4 shows that each lot below 552 cm has preferred κ_{max} directions. In contrast, the upper part of the core (above 552 cm, lot 1–3, mainly gyttja) does not preserve a sedimentary fabric. Lot 1 has chaotic distribution of AMS axes, lot 2 has a partly inverse fabric (shallow κ_{min} inclinations, grouping of κ_{max} axes), lot 3 has a nearly sedimentary fabric but the anisotropy axes are more scattered than in the underlying sediments (Fig. 4).

AF demagnetization indicates the presence of greigite in all five segments of the core. Typical demagnetization results are shown in Fig. 5. The increase of NRM starts above 45–70 mT in different samples reaches values in a range between 1.1 and 5.5 times of the minimal NRM (Fig. 5a, c, d). Along with the intensity increase, an additional attribute of greigite bearing samples is systematical change of NRM direction towards the endpoint in a plane perpendicular to the last AF axis used

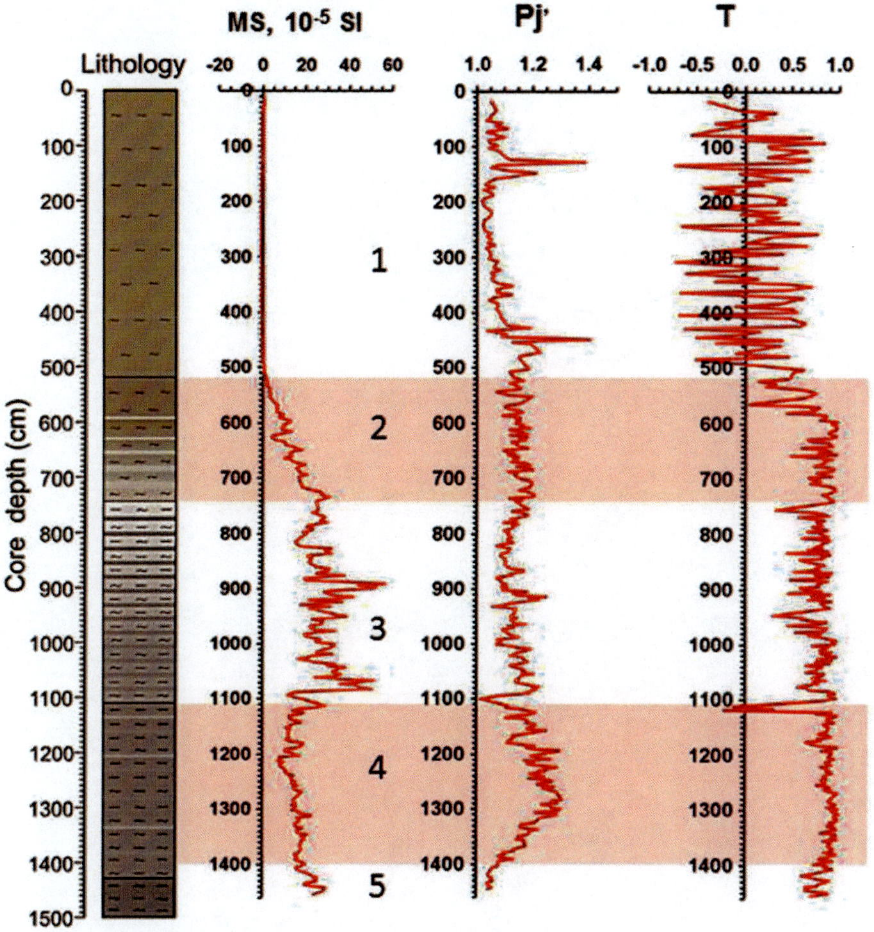

Fig. 3 AMS parameters of the sediments from Lake Kotokel core KTK-19/II. MS—magnetic susceptibility; Pj′—corrected anisotropy degree; T—shape parameter. 1–5—Numbers of segments (see text). Blank (not colored) and pink-colored parts of the plot mark odd and even segments of the core, respectively (see numbering in the text). Lithology legend is in Fig. 2

[27, 28]. Such a systematic manner of NRM changes was observed in all greigite-bearing samples (Fig. 5). Ranking the max/min ratios of samples (see Sect. 3) we've identified 73 samples as "most likely bearing greigite", 81 sample as "probably bearing greigite" and 231 samples as "greigite free" sediments. The distribution of greigite-bearing levels is shown in Fig. 2. Greigite is abundant in segments 1, 2 and 4 and has sporadic occurrence in sectors 3 and 5.

Stepwise AF demagnetization reveals three types of NRM behavior in greigite-free samples (Fig. 6). The first type corresponds to the upper part of gyttja (samples kot01 and kot05 in Fig. 6), which is characterized by relatively high coercivity and

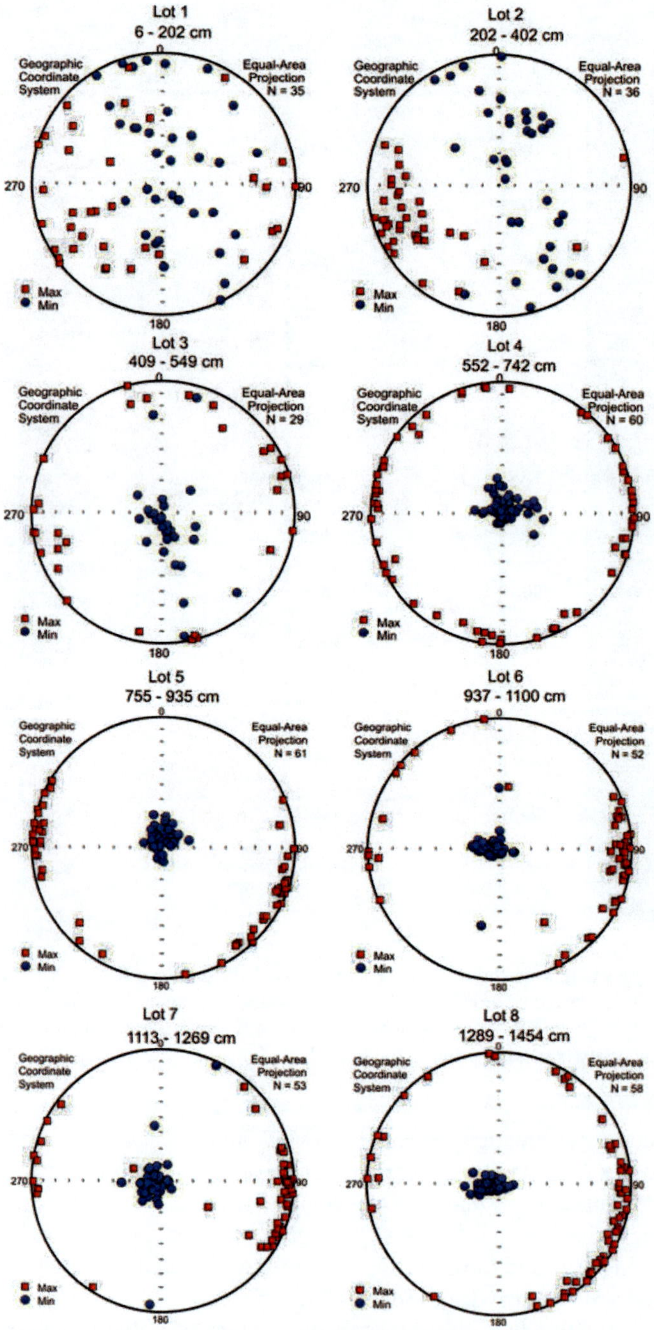

Fig. 4 Magnetic fabric of the Lake Kotokel sediments

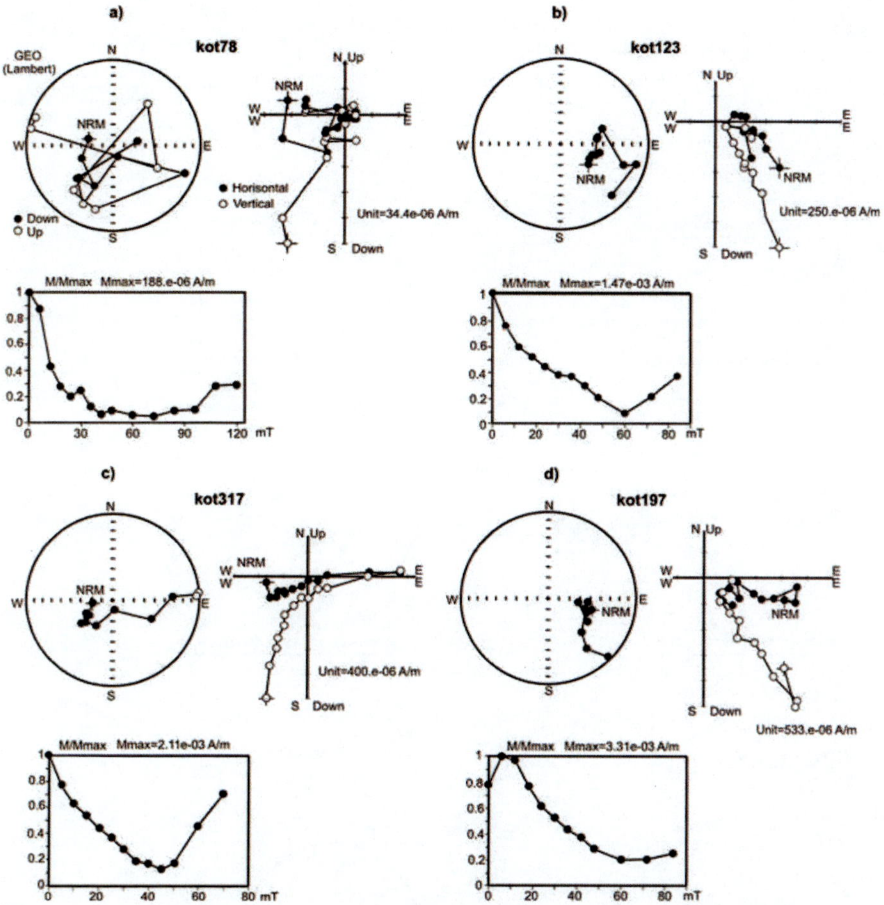

Fig. 5 Typical AF demagnetization results from greigite-bearing samples **a** segment 1; **b** segment 2; **c, d** segment 4

anomalously shallow inclination (the present geomagnetic field I = 72.7° and GAD model I = 69°) ChRM directions.

The second type of samples has poorly preserved stable NRM components (kot97 and kot298 in Fig. 6). Their orthogonal vector plots and stereoplots show largely scattered directions. Such a behavior is observed in all intervals of the core that are characterized by low NRM intensity and low MDF values (sectors 1, 2, 4).

The third type (segments 3 and 5) has the highest NRM and coercivity values (kot214 and kot372 on Fig. 6). Orthogonal vector plots of those samples display a nearly single component NRM. ChRM inclinations determined from orthogonal plots are shown in Fig. 2. However, the presence of greigite cannot be excluded in some samples due to the shift of NRM direction towards the plane perpendicular to the last AF axis used (Fig. 5, sample kot214, segment 3).

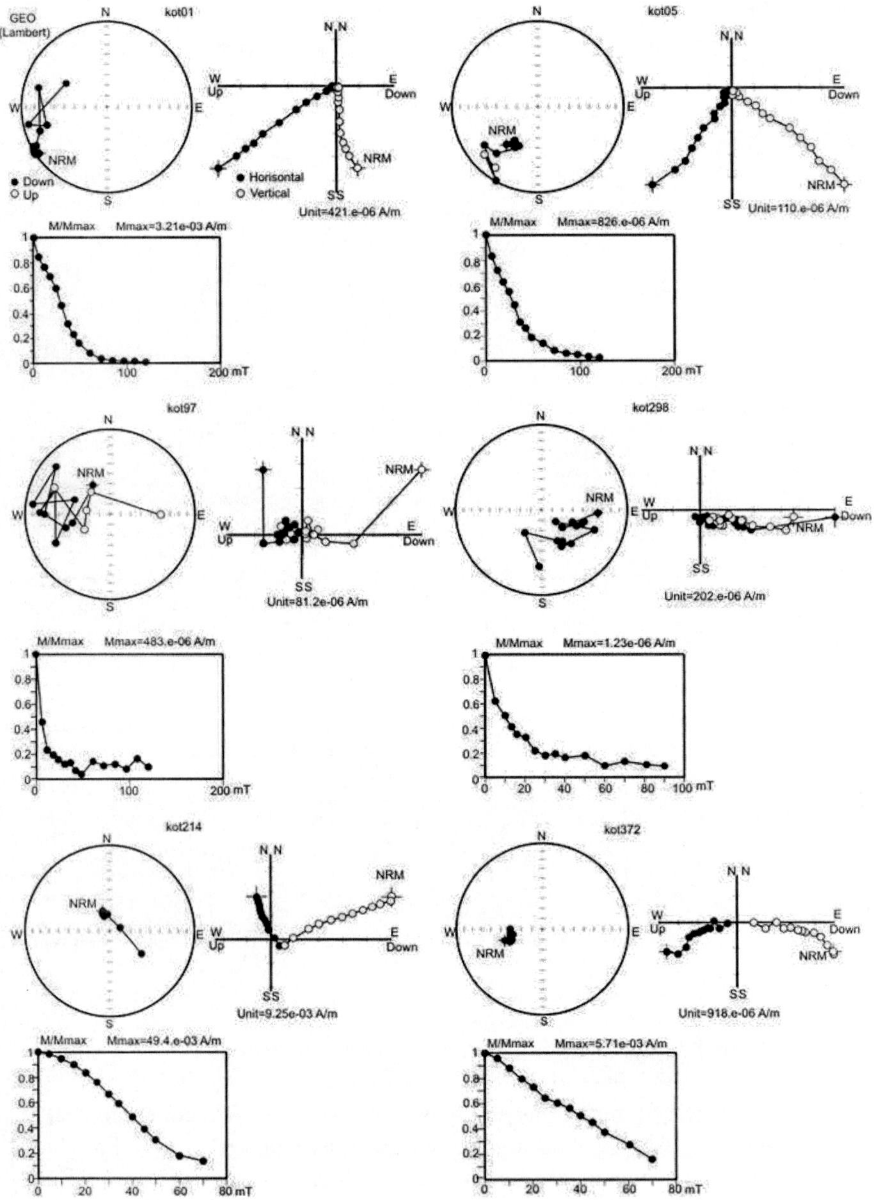

Fig. 6 Typical demagnetization results from Kotokel core. Depth of samples: kot01—5.6 cm, kot05—22.4 cm (both—segment 1), kot97—504 cm (segment 2), kot298—1176 cm (segment 4), kot214—983 cm (segment 3), kot372—1400 cm (segment 5)

Most of the studied samples are characterized by normal NRM polarity, but ChRM inclinations are generally 5–30° lower than expected GAD inclinations. Inclinations of some samples are 5–15° steeper than the GAD. Only a few samples have swallow to negative inclinations, which are concentrated in the upper and lower parts of the gyttja layer (Fig. 2). Those samples correspond to greigite-bearing levels. The last gyttja segment (400–550 cm) demonstrates maximal Pj′ values (see Fig. 3). Segments with low scatter of inclinations (close to GAD inclination) correspond to maximal MS and MDF values at 750–1100 cm (segment 3) and 1350–1450 cm (segments 5) (Fig. 2). As indicated above those segments we consider as predominantly greigite-free. All other segments (1, 2 and 4) demonstrate significant deviations from GAD inclinations that exceed 20°.

5 Discussion

The gyttja layer (segment 1) in lake Kotokel is a specific type of sediment which is absent in lakes Baikal and Khubsugul. Rock magnetic parameters including AMS of this layer differ markedly from uppermost Holocene sediments in Baikal and Khubsugul lakes as well as the underlying clayey sediments of Lake Kotokel. These differences are likely due to the size of Lake Kotokel and the specific sedimentary environment during the last stage of lake formation.

Magnetic properties of the clayey sediments from Lake Kotokel are comparable with those from Lake Baikal: MS and NRM_0 change within similar range [30, 31]. Their oblate magnetic fabric is identical to the sedimentary fabric of Lake Baikal sediments [17]. Such similarity suggests more or less similar sedimentation environments of clayey sediments in both lakes and probably similar magnetization processes.

Magnetic susceptibilities of sediments from Lake Khubsugul appear similar to those in Lake Kotokel. NRM_0 values, however, are doubled as compared to the most of Kotokel samples [32] and thus show the enhanced $Q_n > 1$. This phenomenon in Lake Khubsugul sediments may be explained by the presence of biogenic greigite [15]. Besides, abundant hydrotroilite (a product of bacterial sulfate reduction) in Khubsulul sediments as mottles, laminae, and small nodules cementing sand grains was reported [33]. It suggests a possible chemical origin of greigite. It should be noted, Kotokel sediments yielded very similar values, which indicate the cognate mechanism of NRM_0 enhancement, caused by formation of biogenic or chemical greigite in both lakes. Detailed rock magnetic and magnetic mineralogy studies of Kotokel sediments have not yet been carried out. Therefore, we can only tentatively assume the possible mechanisms of magnetic properties acquisition and their relation to the climate change. We use the climatic records of the nearby KTK2 core [18] for comparison with our magnetic records of KTK-19/II.

The structure of the KTK-19/II Kotokel core is very similar to that obtained by [18] for the KTK2 core located in close proximity (Fig. 7). Stratigraphic boundaries in both cores completely coincide above 750 cm. Below this depth, the thickness of the laminated grey silty clay in core KTK-19/II is one meter more than in core KTK2

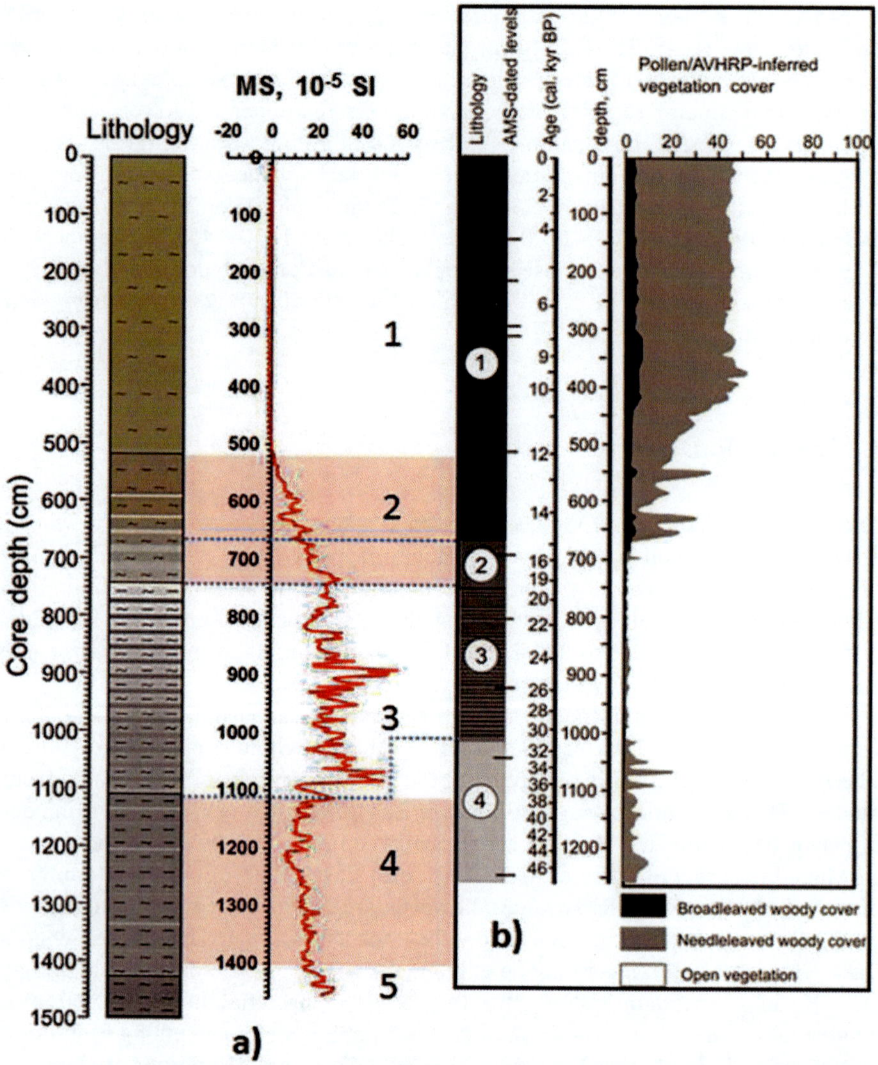

Fig. 7 Correlation of the sedimentary successions in Lake Kotokel cores: **a** lithology and magnetic susceptibility for KTK-19/II core (this paper); **b** lithology with Accelerator Mass Spectrometer-dated levels and age model, percentage changes in woody cover within a 21 × 21 km window for the KTK2 core (simplified from Bezrukova et al. [18]) AVHRR—of modern forest cover by Advanced Very High-Resolution Radiometer (see Tarasov et al. [20] for details). In column **a**: 1–5—Numbers of segments (see text). Pink bars mark even segments. Blue dotted possible correlation lines. Legend for lithology column **a** see in Fig. 2. Legend for lithology column **b**: 1—soft brownish black gyttja; 2—gray blackish silty clay; 3—laminated grey silty clay; 4—gray or dark gray silty clay

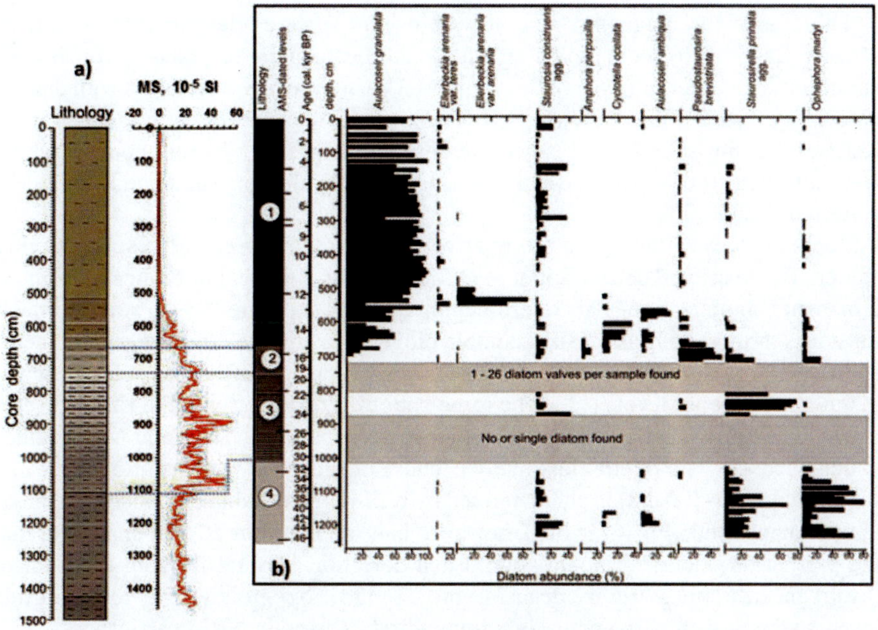

Fig. 8 Correlation of sedimentary successions in Kotokel cores: **a** lithology and magnetic susceptibility for the KTK-19/II core (this paper); **b** lithology, age model and simplified diatom percentage diagram for the KTK2 core (after [18]). For legend of column **b** see Fig. 6

(Figs. 7 and 8). Taking into account this difference, we can roughly estimate the age of the lowest level of core KTK-19/II to be older than 46 kyr.

MS variations in core KTK-19/II almost perfectly coincide with those of the vegetation type as recorded in core KTK2 (see Fig. 7). Maximum MS values correspond to open vegetation around the lake, while decrease of the magnetic signal depicts progressive changes in forest cover. Palynological data suggest well-established boreal forest during accumulation of the gyttja layer and a greater-than-present open landscape around Lake Kotokel [18]. We interpret these observations as being due to reduced erosion in the catchment due to vegetation, which diminished detrital input into the lake. The detrital input decrease in Baikal sediments during interglacial periods was discussed in [6]. They suggest that detrital input in Lake Baikal during interglacials was low compared to the glacials and magnetic signal was is mainly determined by in situ processes. In Lake Kotokel, however, it is probably the decisive factor since the gyttja layer has a biogenic origin [21] and contains almost no terrigenous material. This assumption is supported by gradual MS increase in the transitional segment between gyttja and clayey sediments (Segment 2) when wood cover gradually decreased (Fig. 7).

The absent sedimentary magnetic fabric in the gyttja layer also indicates its predominantly biogenic nature. It should be noted that the primary magnetic fabric in gyttja may have been reworked by degassing processes [23].

Dilution of the magnetic fraction by biogenic silica as described in sediments of Lake Baikal [10] seems to be an additional mechanism, responsible for the MS variations in Segment 1 (gyttja). It is supported by inverse MS correlation with diatom percentage in the KTK2 core (Fig. 7). These processes were likely weaker in Lake Kotokel than in Lake Baikal, since the abrupt changes in biogenic content which were revealed in the gyttja layer in KTK1 core [22] are not manifested in the MS pattern of KTK-19/II core.

NRM_0 values in the gyttja are relatively high with respect to the susceptibility values. The restricted detrital input to gyttja suggests chemical or biogenic origin of remanent magnetization. This interpretation is partly supported by $Q_n > 1$. Sediment top zones throughout Lake Baikal contain biogenic magnetite produced by magnetotactic bacteria postdepositional reduction, however, resulted in the loss of fine-grained magnetite at deeper levels [11]. The same pattern is observed in Lake Kotokel sediments. The uppermost gyttja layers demonstrate enhanced NRM_0 and MDF values which decrease with depth most likely because of secondary oxygenation (Fig. 2).

This layer (0–100 cm) is characterized by well-defined stable characteristic remanence components (Fig. 6), which, however, have inclinations 20–40° shallower that the present-day magnetic inclination. Such deviation may result from mechanical disturbance of soft gyttja by degasing process [23]. No greigite was found in this layer, so the biogenic magnetite is supposed to be the main NRM carrier.

Samples from the underlying gyttja yielded erratic demagnetization data (Fig. 6), typical for sediments which have been affected by reductive magnetite dissolution [6, 13]. Thus, the gyttja layer is characterized mainly by weak and unstable magnetization (large scatter of ChRM inclinations and large MAD values) so that no reliable paleomagnetic direction could be determined.

The possible reason is the occurrence of greigite. Greigite in lake sediments can be of chemical and biogenic origin. Chemical greigite forms by transformation of magnetite at constant sedimentation rate and fast burial at sulphate-reducing conditions [5]. Scanning electron microscope observations of sedimentary greigite demonstrate that it commonly grows within framboids and as larger aggregates that gives rise to strong magnetostatic interactions among magnetic particles [33], which can contaminate the primary magnetization. Thus, chemical greigite carries a secondary, remanent magnetization, which often obscures or even totally overwhelms the primary paleomagnetic signal [33].

Biogenic greigite is produced by magnetotactic bacteria that form greigite magnetosomes [34]. Those magnetosomes can contribute to the remanent magnetization of sediments due to their preservation as fossils after the death of magnetotactic bacteria [35].

Although we cannot confidently distinguish between chemical and biogenic greigite in the studied samples, most of the low inclinations in segment 1 correspond to greigite bearing layers likely suggesting its chemical origin.

According to the KTK2 climatic record [18], the transitional Segment 2 may correspond to the late-glacial–early Holocene climate shift. As mentioned above, this time interval is characterized by reduced forest vegetation, accompanied by decreasing detrital input into the lake. Probably biogenic dilution of magnetic minerals occurred due to slightly enhanced concentration of biogenic silica in this segment [22].

Our preferred interpretation is that continued reductive dissolution of detrital magnetite resulted in the disappearance of small grains and the overall coarsening of magnetic fraction therefore causing low stability of magnetization and relatively large scatter of ChRM inclinations (Fig. 2). Reduced vegetation and diatom content attest to a decrease of biogenic activity in the lake. The volume of biogenic material was insufficient for sulphate reduction and resulted in decrease of the chemical greigite generation. Low concentration of chemical greigite can partly explain the fact that the amplitudes of inclination changes are not so large than in the gyttja layer. In addition, MADs are somewhat smaller than in segment 1 (Fig. 2).

The segment 3 of the core KTK-19/II which most likely corresponds to the coldest period during sediments' accumulation? Of core KTK2 formed under steppe and tundra steppe vegetation conditions [18]. Open landscapes and the absence of diatom fauna during this period determined the intensive detrital input, marked by the highest MS, NRM_0 and MDF values. Detrital sedimentation is also supported by the typical sedimentary magnetic fabric. Stable demagnetization behavior and low MAD values indicate that reductive dissolution of detrital magnetite was low or absent. Nevertheless, $Q_n > 1$ suggest the presence of chemical or biogenic magnetization here. The possible biogenic magnetization may be governed by the formation of greigite. Sporadical greigite has been reported in glacial sediments of Lake Baikal [13] and its abundance in Lake Khubsugul [15, 16]. Demory et al. [6] suggested greigite formation in local microenvironments of sulphate reduction in Lake Baikal caused by presence of faecal pellets.

We explain the sporadic layers of greigite by its biogenic nature, i.e. formed due to magnetotactic bacteria. However, both detrital and biogenic magnetite are expected to be the main carriers of NRM. The NRM high stability during AF demagnetization and the small scatter in ChRM inclinations lends support to this interpretation (Fig. 2).

Segment 4 of Lake Kotokel core KTK-19/II probably corresponds to the lowest warm stage in core KTK2 which was characterized by partial appearance of woody vegetation [18] causing restricted input of detrital magnetite. Relatively low MS and NRM_0 values support this interpretation. Weak NRM stability, low MDF and large MAD values possibly indicate reductive dissolution of detrital magnetite and its transformation to greigite. Despite the presence of diatom fauna in lowermost horizons of core KTK2 [18], the role of biogenic dilution in this segment is not clear.

Segment 5. Analogues of this segment are not present in core KTK2. We suggest colder climate in this segment than in Segment 4. Increased values of MS, NRM_0 and MDF indicate more intensive detrital input of magnetic minerals. High NRM stability during AF demagnetization and low scatter of ChRM inclinations with small MAD values again suggest the absence of dissolution of detrital magnetite, while $Q_n > 1$ most likely reflects the presence of biogenic magnetite, since greigite was not found.

Thus, the acquisition of magnetic properties in Lake Kotokel sediments is controlled by the same four processes as in Baikal and Khubsugul Lake: detrital input, dilution of the magnetic minerals by biogenic silica, biogenic and chemical neoformation of magnetic minerals and reductive dissolution of magnetite. Detrital input which depends on the vegetation cover around the lake seems to be the main sedimentation controlling factor for Lake Kotokel. When forest vegetation predominates, the detrital input is reduced (almost to zero in the gyttja). It increases to maximum values with the reduction of forest vegetation. Detrital input in the segments 3 and 5 which correspond to cold intervals is accompanied by formation of biogenic greigite. Both processes are responsible for primary magnetization which may be used for paleomagnetic studies.

In the segments 1, 2 and 4, which represent warm climate intervals, the magnetization processes are slightly different. The primary magnetic minerals have different origin (biogenic and chemical in segments 1 and 2 and terrigenous in segment 4) but the main process here is reductive dissolution, which resulted in decreasing magnetic concentration and instability of magnetization.

Dilution of sediment by biogenic silica was detected only in Segment 1. Influence of this process is not evident in the other segments.

6 Conclusions

Magnetic properties of Lake Kotokel core KTK-19/II perfectly correlate with core lithology and environmental changes recorded in the cores from adjacent sites [18, 21, 22].

The nature of the magnetic signal in Lake Kotokel is complex. Similar to lakes Baikal and Khubsugul, the main process responsible for the observed rock magnetic pattern and NRM acquisition seems to be detrital input and reductive dissolution of magnetite. Magnetic dilution of sediment by biogenic silica is less pronounced. Magnetite—greigite transformation, associated with a reducing environment caused by high sedimentation rates and abundance of organic material was typical for interglacial periods. In addition, biogenic neoformation of greigite may have sporadically occurred during glacial periods.

Combination of those processes determines the magnetic properties of lake Kotokel sediments in total.

Magnetic susceptibility variations may reflect the following climate related changes: (i) forest vegetation in the surrounding area; (ii) intensity of detrital input and probably; (iii) bioproductivity of the lake (diatoms).

The record of geomagnetic field history is poorly preserved in core KTK-19/II due to dissolution of magnetite and neoformation of chemical greigite. Reliable ChRM directions are preserved only for cold intervals.

Degassing processes associated with tectonic activity [23] in the soft gyttja layer may have significantly affected the core structure.

Acknowledgements This work was supported by the Russian Science Foundation (project no. 19-17-00216—rock magnetic and paleomagnetic study), RFBR (project no. 20-05-00247) and the Government of the Russian Federation (project no. 075-15-2019-866).

We thank anonymous reviewers and associate editor Evgeniy V. Kulakov for their useful comments on the manuscript.

References

1. Bradley, R.S.: Paleoclimatology. Academic Press, Burlington (1999)
2. Evans, M.E., Heller, F.: Environmental Magnetism: Principles and Applications of Enviromagnetics. Academic Press, New York (2003)
3. Wang, H.Y., Liu, H.Y., Zhu, J.L., Yin, Y.: Holocene environmental changes as recorded by mineral magnetism of sediments from Anguli-nuur Lake, southeastern Inner Mongolia Plateau, China. Palaeogeogr. Palaeoclimatol. Palaeoecol. **285**, 30–49 (2010)
4. Duan, Z., Liu, Q., Yang, X., Gao, X., Su, Y.: Magnetism of the Huguangyan maar lake sediments, Southeast China and its paleoenvironmental implications. Palaeogeogr. Palaeoclimatol. Palaeoecol. **395**, 158–167 (2014)
5. Roberts, A.P.: Magnetic mineral diagenesis. Earth-Sci. Rev. **151**, 1–47 (2015)
6. Demory, F., Oberhänsli, H., Nowaczyk, N.R., Gottschalk, M., Wirth, R., Naumann, R.: Detrital input and early diagenesis in sediments from Lake Baikal revealed by rock magnetism. Glob. Planet. Change **46**(1–4), 145–166 (2005). https://doi.org/10.1016/j.gloplacha.2004.11.010
7. Colman, S.M., Peck, J.A., Karabanov, E.B., Carter, S.J., Bradbury, J.P., King, J.W., Williams, D.F.: Continental climate response to orbital forcing from biogenic silica records in Lake Baikal, Siberia. Nature **378**, 769–771 (1995)
8. Kravchinsky, V.A., Krainov, M.A., Evans, M.E., Peck, J.A., King, J.W., Kuzmin, M.A., Sakai, H., Takaoshi, K., Williams, D.F.: Magnetic record of Lake Baikal sediments: chronological and paleoclimatic implication for the last 6.7 Myr. Palaeogeogr. Palaeoclimatol. Palaeoecol. **195**, 281–298 (2003)
9. Kravchinsky, V.A.: Magnetostratigraphy of the Lake Baikal sediments: a unique record of 8.4 Ma of continuous sedimentation in the continental environment. Glob. Planet. Change **152**, 209–226 (2017)
10. Peck, J.A., King, J.W., Colman, S.M., Kravchinsky, V.A.: A rock-magnetic record from Lake Baikal, Siberia: evidence for Late Quaternary climate change. Earth Planet. Sci. Lett. **122**, 221–238 (1994)
11. Peck, J.A., King, J.W.: Magnetofossils in the sediment of Lake Baikal, Siberia. Earth Planet. Sci. Lett. **140**, 159–172 (1996)
12. Dearing, J.A., Boyle, F.L., Appleby, P.G., Mackay, A.W., Flower, J.R.: Magnetic properties of recent sediments in Lake Baikal, Siberia. J. Paleolimnol. **20**, 163–173 (1998)
13. Demory, F., Nowaczyk, N.R., Wirth, R., Oberhänsli, H.: High-resolution magnetostratigraphy of late quaternary sediments from Lake Baikal, Siberia: timing of intracontinental paleoclimatic responses. Glob. Planet. Change **46**, 167–186 (2005). https://doi.org/10.1016/j.gloplacha.2004.09.016
14. Fedotov, A., Kazansky, A.Y., Tomurhuu, D., Matasova, G., Ziborova, G., Zheleznyakova, T., Vorobyova, S., Phedorin, M., Goldberg, E., Oyunchimeg, T., Narantsetseg, T., Vologina, E., Yuldashev, A., Kalugin, I., Tomurtogoo, O., Grachev, M.: A 1-Myr record of paleoclimates from lake Khubsugul, Mongolia. EOS Trans. **85**, 387–390 (2004)
15. Nourgaliev, D.K., Iassonov, P.G., Kosareva, R.L., Kazanskii, A.Y., Fedotov, A.P.: The origin of magnetic minerals in the Lake Khubsugul sediments (Mongolia). Russ. J. Earth Sci. 7, ES3004 (2005)

16. Fabian, K., Shcherbakov, V.P., Kosareva, L., Nourgaliev, D.: Physical interpretation of isothermal remanent magnetization end-members: new insights into the environmental history of Lake Hovsgul, Mongolia. Geochem. Geophys. Geosyst. **17**, 4669–4683 (2016). https://doi.org/10.1002/2016GC006506
17. Sakai, H., Nomura, S., Horii, M., Kashiwaya, K., Tanaka, A., Kawai, T., Kravchinsky, V., Peck, J., King, J.: Paleomagnetic and rock-magnetic studies on Lake Baikal sediments: BDP96 borehole at academician ridge. In: Minoura, K. (ed.) Lake Baikal, pp. 35–52. Elsevier Science (2000)
18. Bezrukova, E.V., Tarasov, P.E., Solovieva, N., Krivonogov, S.K., Riedel, F.: Last glacial–interglacial vegetation and environmental dynamics in southern Siberia: chronology, forcing and feedbacks. Palaeogeogr. Palaeoclimatol. Palaeoecol. **296**, 185–198 (2010)
19. Shchetnikov, A.A.: Morphotectonics of lacustrine basins: The Baikal rift zone as an example. Russ. J. Pac. Geol. **1**, 120–129 (2007)
20. Tarasov, P.E., Bezrukova, E.V., Krivonogov, S.K.: Late Glacial and Holocene changes in vegetation cover and climate in southern Siberia derived from a 15 kyr long pollen record from Lake Kotokel. Clim. Past **5**, 285–295 (2009)
21. Fedotov, A.P., Vorobyeva, S.S., Vershinin, K.E., Nurgaliev, D.K., Enushchenko, I.V., Krapivina, S.M., Tarakanova, K.V., Ziborova, G.A., Yassonov, P.G., Borissov, A.S.: Climate changes in East Siberia (Russia) in the Holocene based on diatom, chironomid and pollen records from the sediments of Lake Kotokel. J. Paleolimnol. **47**, 617–630 (2012). https://doi.org/10.1007/s10933-012-9586-5
22. Shichi, K., Takahara, H., Krivonogov, S.K., Bezrukova, E.V., Kashiwaya, K., Takehara, A., Naka-mura, T.: Late Pleistocene and Holocene vegetation and climate records from Lake Kotokel, central Baikal region. Quatern. Int. **205**, 98–110 (2009)
23. Zhang, Y., Wünnemann, B., Bezrukova, E.V., Ivanov, E.V., Shchetnikov, A.A., Nourgaliev, D., Levina, O.V.: Basin morphology and seismic stratigraphy of Lake Kotokel, Baikal region, Russia. Quatern. Int. **290–291**, 57–67 (2013)
24. Tarling, D.H., Hrouda, F.: The Magnetic Anisotropy of Rocks. Chapman and Hall, London (1993)
25. Zijderveld, J.D.A.: A.C. demagnetization of rocks: analysis of results. In: Collinson, D.W., Creer, K.M. (eds.) Methods in Paleomagnetism, pp. 254–286. Elsevier, Amsterdam (1967)
26. Kirschvink, J.L.: The least squares line and plane and the analysis of paleomagnetic data. Geophys. J. R. Astron. Soc. **62**, 699–718 (1980)
27. Hu, S., Appel, E., Hoffmann, V., Schmahl, W.W., Wang, S.: Gyromagnetic remanence acquired by greigite (Fe_3S_4) during static three-axis alternating field demagnetization. Geophys. J. Int. **134**, 831–842 (1998). https://doi.org/10.1046/j.1365-246x.1998.00627.x
28. Stephenson, A., Snowball, I.F.: A large gyromagnetic effect in greigite. Geophys. J. Int. **145**, 570–575 (2001). https://doi.org/10.1046/j.0956-540x.2001.01434.x
29. Roberts, A.P., Chang, L., Rowan, C.J., Horng, C.S., Florindo, F.: Magnetic properties of sedimentary greigite (Fe_3S_4): an update. Rev. Geophys. **49**, RG1002 (2011). https://doi.org/10.1029/2010RG000336
30. Kravchinsky, V.A., Evans, M.E., Peck, J.A., Sakai, H., Krainov, M.A., King, J.W., Kuzmin, M.I.: A 640 kyr geomagnetic and paleoclimatic record from Lake Baikal sediments. Geophys. J. Int. **170**, 101–116 (2007). https://doi.org/10.1111/j.1365-246X.2007.03411.x
31. Sakai, H., Nomura, S., Horii, M., Kashiwaya, K., Kawai, T., Kravchinsky, V., Peck, J.: Paleomagnetism and paleoenvironmental magnetism studied on BDP-98 sedimentary cores from Lake Baikal. In: Kashiwaya, K. (ed.) Long Continental Records from Lake Baikal, pp. 233–243. Springer-Verlag, Tokyo (2003)
32. Kazansky, A.Y., Fedotov, A.P., Matasova, G.G., Yuldashev, A.A., Ziborova, G.A., Zheleznyakova, T.O., Vologina, E.G., Oyunchimeg, T., Narantsetseg, T., Tomurhu, D.: Results of paleo- and rock magnetic studies of bottom sediments from Lake Khubsugul (Mongolia). In: Paleomagnetism and Magnetism of Rocks: Theory, Practice and Experiment: Proceedings of International Seminar, pp. 301–305. Kazan University Publishing House, Kazan (2004) (in Russian)

33. Roberts, A.P.: Magnetic properties of sedimentary greigite (Fe_3S_4). Earth Planet. Sci. Lett. **134**(3–4), 227–236 (1995)
34. Bazylinski, D.A., Frankel, R.B., Heywood, B.R., Mann, S., King, J.W., Donaghay, P.L., Hanson, A.K.: Controlled biomineralization of magnetite (Fe_3O_4) and greigite (Fe_3S_4) in a magnetotactic bacterium. Appl. Environ. Microbiol. **61**, 3232–3239 (1995)
35. Vasiliev, I., Franke, C., Meeldijk, J.D., Dekkers, M.J., Langereis, C.G., Krijgsman, W.: Putative greigite magnetofossils from the Pliocene epoch. Nat. Geosci. **1**, 782–786 (2008). https://doi.org/10.1038/ngeo335

Determination of the Position of the South Magnetic Pole Based on Experimental Data Obtained During Russian Round-the-World Expeditions: 1820 (F. Bellingshausen) and 2020 (R/V "Admiral Vladimirsky")

I. M. Demina, V. G. Boyarskikh, S. A. Merkuryev, S. A. Ivanov, and V. A. Soldatov

Abstract During the round-the-world expedition of the R/V "Admiral Vladimirsky" (2019–2020), following closely the route of the voyage of Bellingshausen and Lazarev in 1819–1821, a large amount of magnetic data was obtained, including in the Antarctic region. One of the goals of this research has been to determine the position of the South Magnetic Pole (SMP) from experimental data. This was the reason to return to the declination data obtained during the Bellingshausen expedition and to determine the position of the SMP from these data. Several methods have been devised for this purpose. To determine the current position of the SMP, a small polygon survey in the area of its assumed position has been made. The polygon location was chosen based on the forecast of the IGRF-12 model. The three-component measurements on board the R/V "Admiral Vladimirsky" and scalar outboard measurements have been performed and estimates of the current position of the SMP obtained.

Keywords South magnetic pole · Component measurements

1 Introduction

In connection with the 200th anniversary of the First Russian Antarctic Expedition of F. Bellingshausen and M. Lazarev (1819–1821), the Russian Geographical Society organized a new round-the-world expedition of the R/V "Admiral Vladimirsky" (Dec. 2019–June 2020). The route of this expedition was planned to follow closely that of the Bellingshausen expedition (Fig. 1). One of the tasks of the expedition was to

I. M. Demina (✉) · V. G. Boyarskikh · S. A. Merkuryev · S. A. Ivanov · V. A. Soldatov
Pushkov Institute of Terrestrial Magnetism of the Russian Academy of Sciences, St. Petersburg Filial, Universitetskaya emb. 5, St. Petersburg, Russian Federation
e-mail: dim@izmiran.spb.ru

S. A. Merkuryev
St. Petersburg State University, Universitetskaya emb. 7-9, St. Petersburg, Russian Federation

A. Kosterov et al. (eds.), *Problems of Geocosmos—2020*, Springer Proceedings in Earth and Environmental Sciences, https://doi.org/10.1007/978-3-030-91467-7_13

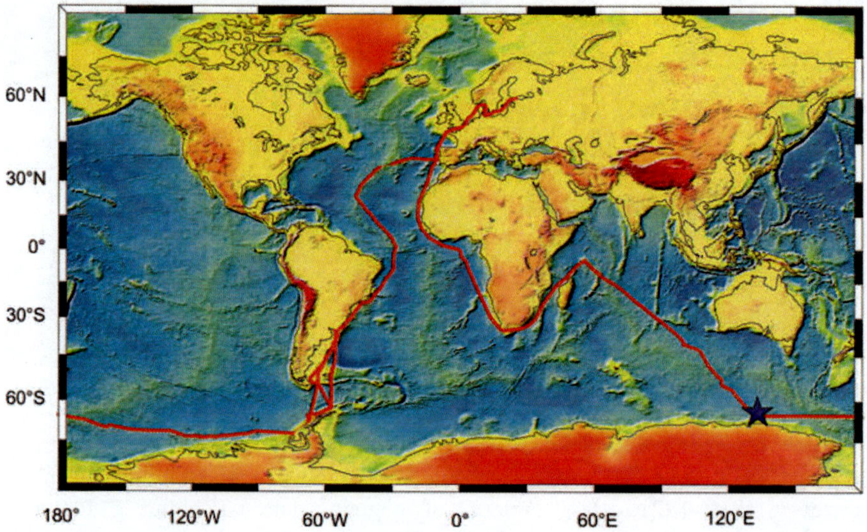

Fig. 1 Route map of the "Admiral Vladimirsky" expedition. The asterisk shows the position of the polygon for determining the location of the SMP

determine the position of the South Magnetic Pole (SMP). The problem of deriving the magnetic dip pole position using different methods has been studied in detail in [1].

The importance and relevance of determining the SMP position is related to the fact that this knowledge is necessary for understanding the nature and mechanism of the Earth's magnetic field generation [2]. In particular, it is assumed that an increase in the speed of the pole movement [3] may indicate both local processes in the core and associated jerks [4, 5], and global changes in the core, which can lead to field reversal [6]. The comprehensive analysis of the magnetic pole movements is given in [7].

In addition, it is important to assess the impact of the pole shift on the upper atmosphere [8], the Earth's thermosphere and ionosphere [9], climate [10] and biosphere [11]. To determine the current SMP position, it was planned to perform a small polygon survey in the area of its location predicted according to the IGRF-12 model [12].

2 The Results of Measuring the Declination *D* During the Bellingshausen Expedition

The first measurements of the declination *D* in the Antarctic region were carried out by the expedition of Captain J. Cook (1772–1775), but the task of determining the position of the SMP was not considered at this time. When preparing his expedition,

F. Bellingshausen took into account the results of J. Cook, and some points at which the measurements of *D* were to be carried out were chosen as close as possible to the points of measurement in Cook's expedition. Thus, the first measurements of the secular variation *D* in the Antarctic region were made. The locations of the points at which the *D* values were determined are shown in Fig. 2. The landing was not attempted by the Bellingshausen expedition, and inclination measurements were not possible at that time.

Only 52 points with measured declination values are mentioned in Bellingshausen's report [13] and only 46 of them in the Antarctic region. But in response to C. F. Gauss request, Bellingshausen sent him a list of 203 points with measured declination, which was gratefully published in [14]. Bellingshausen determined the position of the SMP using only 46 points, but we do not know how he obtained the position of the SMP from these points. We suggest that the most accessible method at that time would have been to find the SMP position from the intersections of great

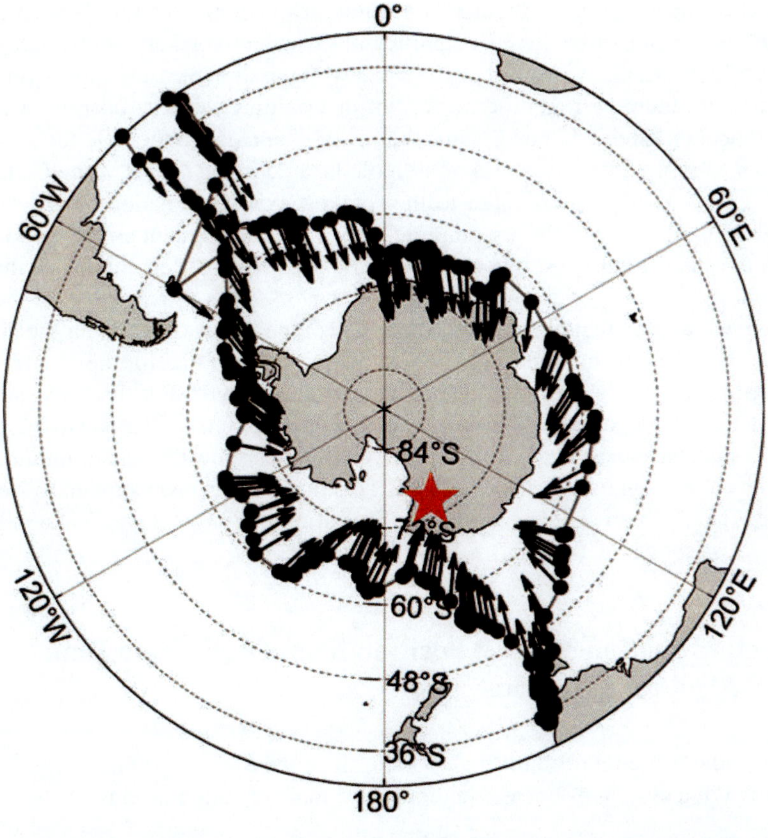

Fig. 2 Measurements of *D* by the expedition of F. Bellingshausen. The arrows show the direction of the vector of the horizontal component of the field *H*. The asterisk shows the position of the SMP according to Bellingshausen

circles. We therefore attempted to apply this method using 146 points above 50° S from the list published by Gauss.

3 Determination of the SMP Position According to Bellingshausen Data by the Method of Two Great Circles

To find the position of the SMP by the method of two great circles, a pair of points is selected at which the measurement of D is performed; great circles are drawn in the direction determined by the value D. The southern point of intersection of the circles gives an estimate of the SMP position, which we refer to as the virtual magnetic pole (VMP), see Fig. 3.

Let us consider main features of the method and the resulting errors. In the case of a purely dipole field, the method is accurate. Deviation from dipole is an uncorrectable systematic error that can significantly worsen the resulting estimate of the SMP position. Random errors are associated with possible measurement errors. The influence of random errors on the result of determining the pole position has been studied using a model. Let us assume that in an area close to the pole the magnetic field can be represented as a field of a dipole located in the center of the Earth with the magnetic moment M parallel to the rotation axis. Declination values D were calculated on the base of this assumption, and $\pm 2°$ measurement error was added to model D values. Figure 4 shows the effect of this error on the estimation of the pole position.

It is seen clearly that the area of possible SMP positions depends on the angular distance between the measurement points. Figure 4 shows that for the dipole model the smallest error is achieved at an angular distance of 90°. For the corresponding GUFM model the distances from the intersection points of large circles to the model's pole have been calculated. These calculations show that the minimum error is achieved at an angular distance of 60°. This method has been applied to Bellingshausen data, and an estimate of the SMP position at 73.01° S and 150.43° E was obtained, as shown in Table 1.

4 Determination of SMP Position from Declination Data by Algebraic Methods

These methods are referred to as algebraic, since the SMP position is calculated as a solution of a system of linear equations. The main equation used is

$$X \cdot \sin(D) = Y \cdot \cos(D), \tag{1}$$

where X and Y are the components of the geomagnetic field in the directions of geographical north and east, respectively.

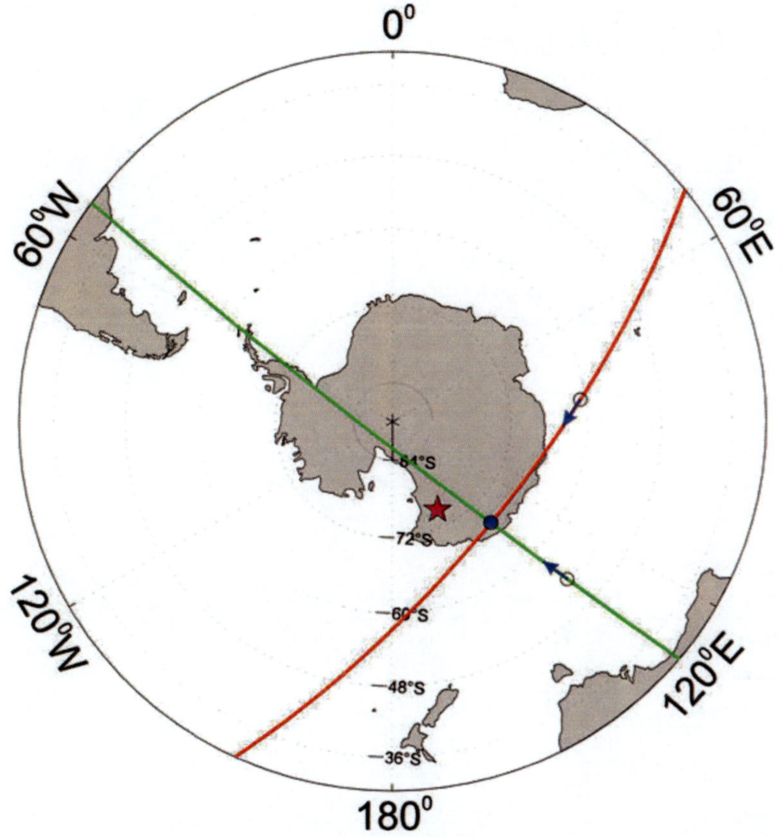

Fig. 3 Obtaining an estimate of the position of the SMP by the method of two great circles. The arrows show the azimuths corresponding to the D value at the measurement point. The asterisk is the position of the magnetic pole obtained by Bellingshausen

4.1 Equivalent Dipole Method

In this method, as in the method of great circles, the problem of determining the SMP position is solved in the dipole approximation. The dipole is located in the center of the Earth and has a magnetic moment M, which doesn't need to be parallel to the rotation axis. Having connected the parameters of this dipole with the declination values on the Earth's surface and setting the vertical component $M_z = 1$, we obtain a system of linear equations with respect to the vector components M_x and M_y. The solution of this system by least squares determines the direction of the vector M and thus the position of the SMP.

The method was tested on a model, for which the declination according to the GUFM model [15] was calculated at the points of measurements of D conducted by Bellingshausen. These values of D were used to solve the problem of determining the

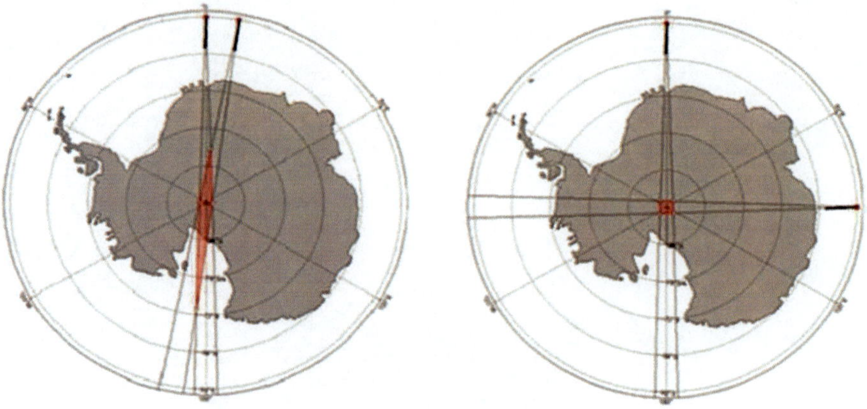

Fig. 4 Dependence of the error in determining the model SMP with an error in *D* equal to ±2° for two different angular distances between the measurement points

Table 1 Estimates of SMP coordinate

Method	Latitude, °	Longitude, °	Distance to Bellingshausen SMP, km
Two great circles	−73.01	150.43	406.7
Equivalent dipole	−73.2	154.8	476.8
Approximation	−72.0	151.5	522.1
Iteration	−74.8	153.0	322.6
GUFM for 1820	−74.39	152.93	345.5
Bellingshausen	−76	142.5	0

parameters of an equivalent dipole. The resulting pole position was compared with the one calculated from full GUFM representation. The difference between the two reached 150 km. SMP coordinates obtained by this method from the Bellingshausen data are shown in Table 1.

4.2 *Field Approximation by a Finite Sum of Spherical Harmonics*

This approach is also based on relation (1), but here X and Y are represented as a sum of spherical harmonics [16]. A system of equations is constructed for the coefficients of the series and the leading coefficient g_1^0 is set equal to 1. The obtained coefficients are used to calculate the horizontal component of the field inside the region. For this method, unlike the ones described above, the shape of the initial data assignment area is the main source of errors. Indeed, this area is practically a ring, and the problem of determining the coefficients of a series is solved for this ring, the result being used to extrapolate to the entire circle. This leads to an unpredictable "pole" daily variation as the order of the polynomial increases. Thus, raising the order of the polynomial above $n = 2$ does not make much sense. The result is shown in Fig. 5a.

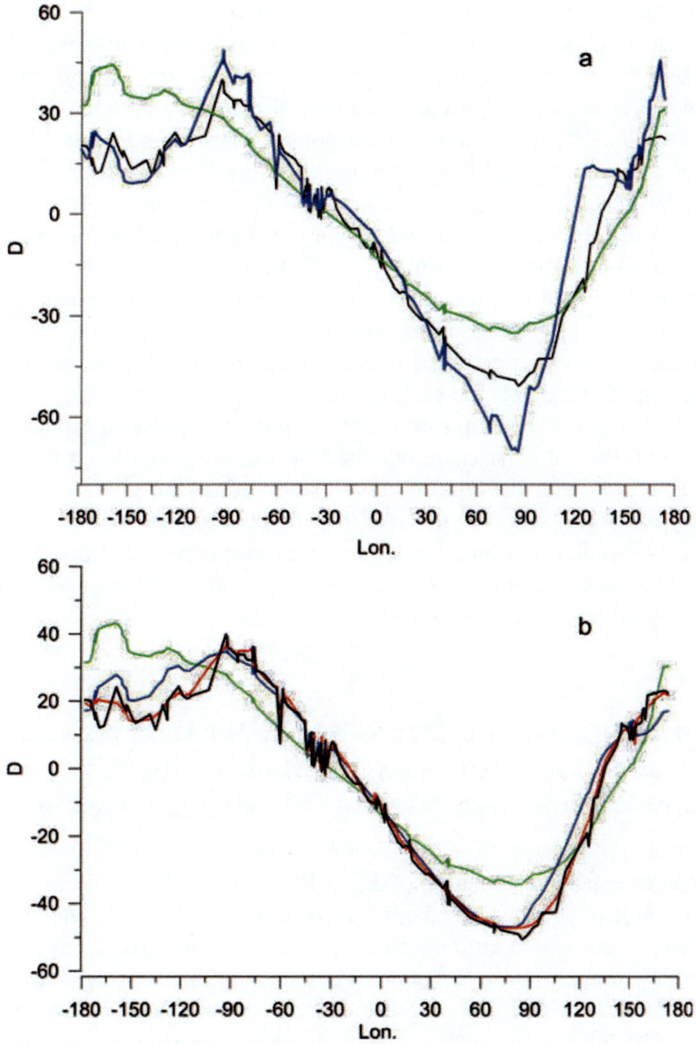

Fig. 5 Approximation of the measured *D* values by spherical functions. **a** Direct approximation, **b** iterative method; measured *D* values are shown in black, synthesized in green at $n = 1$, blue at $n = 2$, red at $n = 4$ (only for the iterative method)

4.3 Interpolation Approach

The configuration of the region suggests a possible modification of the above method, based on the transition from the approximation of D to the approximation of the X and Y components. Let the values of the horizontal component H at all points of measurement of D be the same and equal to an arbitrary value H_0. Then, from the measured D, one can calculate the estimates of X_0 and Y_0 and present these components as a series of spherical harmonics. We obtain a system of equations for determining the coefficients g and h at harmonics. These coefficients can be used to calculate H_i, X_i, and Y_i. This is how the iterative process is constructed.

For model D values, the process converged in 3–4 steps to the exact value at $n = 4$. From the calculations carried out, it can be concluded that even near the pole, the non-dipole part has a significant effect on determining the position of the SMP. The similar result was obtained in [1] for NMP. For Bellingshausen data, the number of steps was 5–7, depending on the order of the polynomial. There was no point in increasing the degree of the polynomial above $n = 4$, since the result of the approximation of D did not change (Fig. 5b).

All estimates of the position of the SMP obtained from Bellingshausen data are summarized in Table 1. The value obtained by Bellingshausen and that calculated for 1820 using the historical GUFM model is also added there.

The analysis of the presented results shows that the estimates of the SMP position obtained in the dipole approximation form a compact group. The coordinates obtained by the iterative method are closest to the historical model. In this case, all values differ from those obtained by Bellingshausen.

5 Determination of the SMP Based on the Data of Scalar and Component Measurements Made by the R/V “Admiral Vladimirsky” in the D’Urville Sea Area in 2020

For the experimental determination of the SMP, it was planned to perform a polygon survey in the area of its assumed position according to the IGRF-12 model. Measurements on the “Admiral Vladimirsky” were carried out with two sets of 3-component magnetometers: the MVC-2 magnetovariational complex, as well as a component magnetometer with three magnetoresistive sensors, both developed at SPbF IZMIRAN. The GPS sensor synchronized the operation of the entire system. Also, we had the outboard scalar magnetometer SeaSpy.

The polygon was carried out continuously for two days at April 6–7, 2020. According to the data collected at the Dumont d’Urville observatory, Terre Adelie, the geomagnetic field was quiet during the survey [17]. The polygon diagram is shown in Fig. 6.

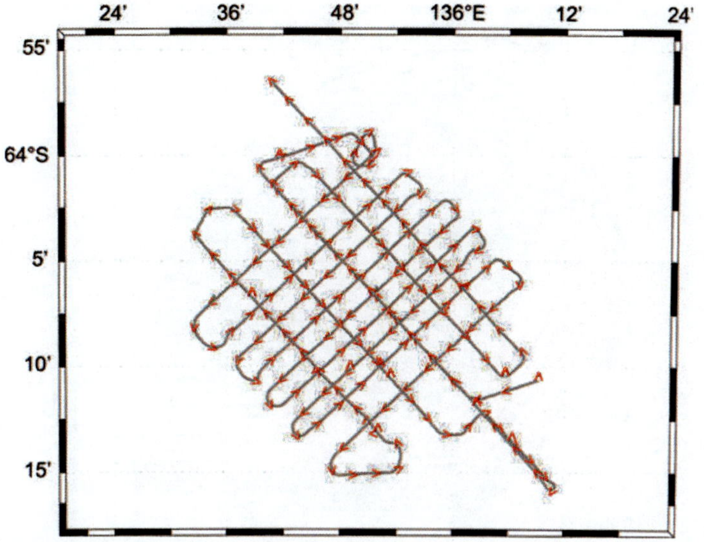

Fig. 6 Diagram of the polygon survey made by the R/V "Admiral Vladimirsky" in the area of the assumed SMP position

5.1 Estimate of the SMP Position by Scalar Data

The processing of scalar data was carried out following a conventional procedure including removal of outliers and filtering. According to nearby observatories the days were magnetically quiet. The correction for the diurnal variation was calculated from the residuals at the intersection points of the profiles. To obtain an estimate of the SMP position, the method of linearization of the expression for the total field intensity *T* through the components was used [18]

$$\delta \mathrm{T} = (\delta \mathrm{X} * \mathrm{X} + \delta \mathrm{Y} * \mathrm{Y} + \delta \mathrm{Z} * \mathrm{Z})/T.$$

Setting some initial values of the *X*, *Y* and *Z* components, we calculate *T* and find the differences with the measured values. After that, we present the correction as a sum of spherical harmonics and solve the resulting system of linear equations with respect to the coefficients by least squares. Using the coefficients, we calculate δX, δY, and δZ, which are introduced as corrections to the initial *X*, *Y*, and *Z*. Then the procedure is repeated until the minimum differences in successive steps are reached indicating procedure convergence. The final values of the components are used to calculate the horizontal component, the position of the minimum of which gives an estimate of the SMP position.

The spatial distribution of the synthesized *H* component for the order of the polynomial $n = 4$ is shown in Fig. 7 and in Table 2. The position of the minimum is marked in pink. The shape of isolines and the presence of the additional local

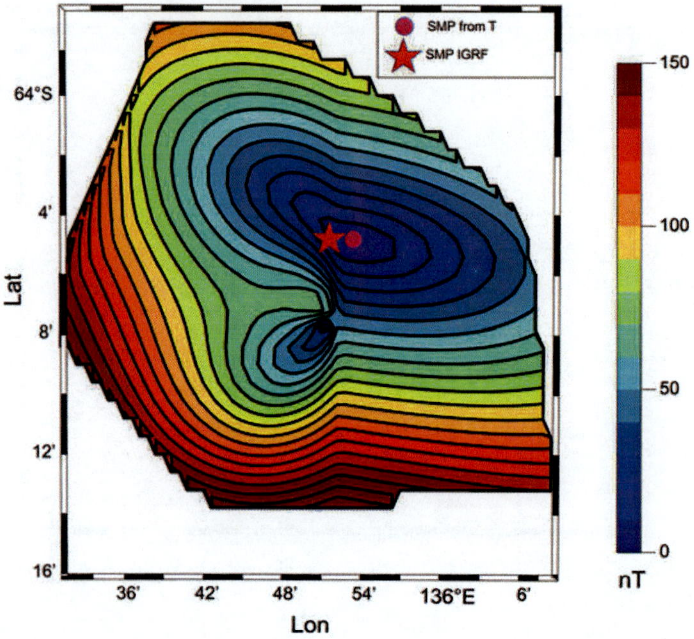

Fig. 7 Spatial structure of the synthesized horizontal component. The pink dot is the position of the global minimum. The red asterisk is the position of the SMP according to the IGRF model

Table 2 Estimations of current SMP position

Method	SMP coordinates		Distance to IGRF pole, km
	Latitude, °	Longitude, °	
Scalar data	64.08 S	135.90 E	4.045
Component data	64.08 S	135.80 E	1.477
IGRF-13	64.07 S	135.82 E	0

minimum indicate an influence of the anomalous field and diurnal variations on the structure of the horizontal component near the pole.

5.2 Determining the SMP Position from Component Data

The basis for determining the position of the pole from the component data is the results of processing the circulations, or 360° loops, performed during the cruise. Using the measured components, hodographs were constructed (the dependence of the *X* component on the *Y* component, for more details see [19]). The resulting ellipse allows one to determine the correction for "hard" iron, i.e., the center of the hodograph, and the calibration coefficients for the ratio of X and Y needed to reduce the ellipse to a circle.

Several circulations were performed and it was found that the "radius" of the hodograph decreases linearly with decreasing H, i.e., the radius of the hodograph is proportional to H and tends to 1000 nT as the point of circulation approaches the pole.

Figure 8 shows the points where the circulations were performed. The values of the horizontal component calculated from the IGRF-13 model at the points of circulation are plotted on the X axis, and the radius of the hodograph obtained from the measurement results is plotted on the Y-axis. The almost linear relationship is clearly observed.

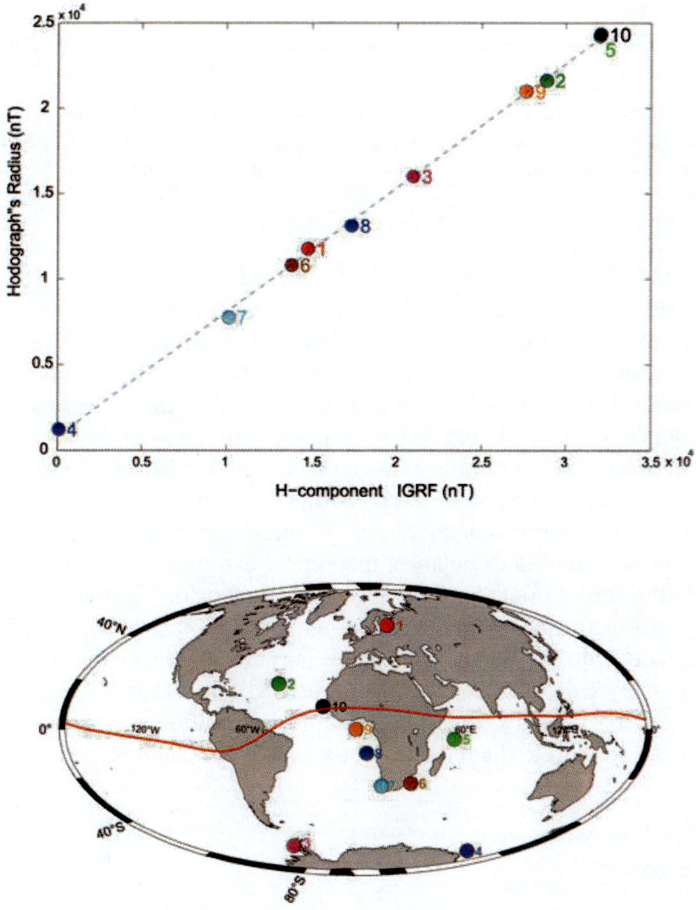

Fig. 8 Change in the hodograph radius depending on the value of the horizontal component H in the areas of circulation. The geographic location of the circulations is shown in the lower figure. The color and number in the top and bottom pictures are the same

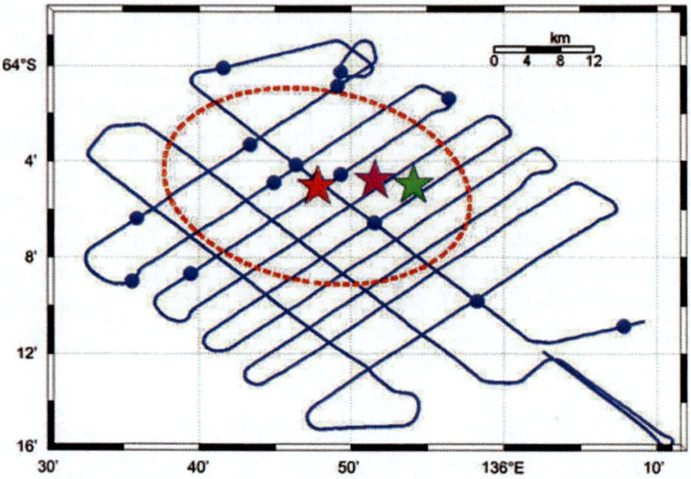

Fig. 9 The resulting estimates of the SMP position. The position of the VMP poles is shown by blue dots, the red star is the result of the VMP averaging, the lilac star is the SMP calculated from IGRF, and the green star is the SMP obtained from the scalar data. Red dashed line is 50%-confidence ellipse

Taking into account the parameters obtained from the hodographs, the value of the horizontal component was calculated. Its minima give the positions of the VMP. An estimate of the position of the SMP is obtained by averaging the VMP locations. The SMP position and the result of averaging are shown in Fig. 9 and in Table 2.

These methods give the average pole position over the time of the survey, which is 36 h (i.e. more than a day). In Fig. 9, the red dashed line shows the confidence ellipse with a 50% probability. Its semi-axes are 0.0671° and 0.1681°. These parameters are obtained by a standard method of the probability theory [20].

The probabilistic model corresponding to the scalar method appears to be very complex, not allowing us to obtain a theoretical estimate of the error. Numerical analysis of the method outcome, i.e. error calculation for hundreds of variants of randomly perturbed data, was not carried out. On the other hand, there have been diurnal variations during the survey, which may have led to a diurnal drift of the magnetic pole by tens of kilometers. Still, the SMP positions obtained by the two methods are fairly close to each other, and to the position calculated from IGRF, lending further support to the reliability of these determinations.

6 Conclusions

In the present study, estimates of the SMP position have been obtained from the data of declination measurements during the F. Bellingshausen 1819–1821 round-the-world expedition. It is shown that the influence of the non-dipole part of the field can have a significant effect on the SMP coordinates obtained by different methods, even if the declination was measured near the pole.

Based on the results of scalar and vector component measurements carried on board the R/V "Admiral Vladimirsky" at April 6–7, 2020, estimates of the SMP position were also obtained. The SMP coordinates obtained from the results of scalar and vector component measurements are close to each other and to the prediction of the IGRF-13 model. Thus, it is shown that the development of special methods for data processing of three-component magnetometers installed on board an iron carrier allows to solve a number of geophysical problems.

Acknowledgements Authors express their gratitude to all those who took part in the delivery of equipment for the expedition, promptly organized the transfer of information, provided communication and coordination along the route of the R/V "Admiral Vladimirsky":

Oleg Dmitrievich Osipov, Head of the expedition of the R/V "Admiral Vladimirsky", Deputy Head of the Department of Navigation and Oceanography of the Ministry of Defense of the Russian Federation.

Members of the research staff of the Department of Geomagnetic Research (SPbF IZMIRAN), Prof. Yu. A. Kopytenko, Dr. P. A. Sergushin, D. B. Zaitsev, V. A. Ivanenko, A. V. Petlenko.

Yegor Mikhailovich Krasinsky, Managing Director of Yuzhmorgeologiya JSC (Russian Geological Holding "Rosgeologiya").

Vladimir Timurovich Minligareev, Deputy Director for Research, Fedorov Institute of Applied Geophysics (IGA).

Evgeny Konstantinovich Grigoriev, Head of the Department of Gravimetric and Magnetometric Marine Research (Russian Geological Holding "Rosgeologiya").

Dr. Sergey Vladimirovich Protsaenko, employee of the 6th Atlantic Naval Expedition.

The results presented in this paper rely on the data collected at Dumont d'Urville, Terre Adelie, Antarctica. We thank EOST for supporting its operation and INTERMAGNET for promoting high standards of magnetic observatory practice (www.intermagnet.org).

The GMT program package has been used to produce the figures [21].

References

1. Newitt, L.R., Chulliat, A., Orgeval, J.-J.: Location of the north magnetic pole in April 2007. Earth Planet Space **61**, 703–710 (2009)
2. Merrill, R.T., McElhinny, M.W., McFadden, P.L.: The magnetic field of the earth: paleomagnetism, the core and the deep mantle, pp. 531. Academic Press, San Diego, Calif. (1996)
3. Korte, M., Mandea, M.: Magnetic poles and dipole tilt variation over the past decades to millennia. Earth Planet Space **60**, 937–948 (2008)
4. Newitt, L.R., Mandea, M., McKee, L.A, Orgeval, J.-J.: Recent acceleration of the north magnetic pole linked to magnetic jerks. Eos Trans. AGU **83**, 381 (2002)
5. Campbell, W.H.: Comments on "survey tracks current position of south magnetic pole" and "recent acceleration of north magnetic pole linked to magnetic jerks". Eos Trans. AGU **84**, 41 (2003)
6. Kuznetsov, V.V.: The north magnetic pole moves steadily according to the drift model. Vestn. KRAUNTS Phys.-Mat. Nauki **28**(3), 65–76 (2019) (in Russian)
7. Mandea, M., Dormy, E.: Asymmetric behavior of magnetic dip poles. Earth Planet Space **55**(3), 153–157 (2003)

8. Lyakhov, A., Zetser, Yu.I., Fuller-Rowell, T.: Possible consequences of the displacement of magnetic poles on the structure and dynamics of the earth's upper atmosphere. Dokl. Earth Sci. **409**(5), 688–690 (2006)
9. Namgaladze, A.A., Shapovalova, Yu.A., Knyazeva, M.A.: Investigation of the influence of the shift of geomagnetic poles on the thermosphere and ionosphere of the earth: problem statement. Proc. Kola Sci. Center Russ. Acad. Sci. Heliogeophysics **5**, 137–145 (2018) (in Russian)
10. Belikov, Yu.E., Burov, V.A., Dyshlevsky, S.V., Kotonaeva, N.G., Lapshin, V.B., Repin, A.Yu.: Possible connection between the movement of the magnetic pole and changes in solar activity with the Arctic climate. Part 3. Heliogeophys. Res. **19**, 25–31 (2018) (in Russian)
11. Reshetnyak, M.Yu., Pavlov, V.E.: Evolution of the dipole geomagnetic field. Observations and models. Geomagn. Aeron. **56**(1), 110–124 (2016)
12. Thébault, E., Finlay, C.C., Beggan, C.D., et al.: International geomagnetic reference field: the 12th generation. Earth Planet Space **67**, 79 (2015)
13. Bellingshausen, F.F.: Repeated surveys in the southern ocean and the 1819–1821 circumnavigation on Vostok and Mirnyi Sloops under the leadership of Captain Bellingshausen, the captain of Vostok. The captain of Mirnyi was lieutenant Lazarev, St. Petersburg, vols. 1–2 (1831) (in Russian)
14. Bellingshausen, F.F.: Abweichung der Magnetnadel, beobachtet vom Capitaine Bellingshausen in den Jahren 1819–1821. Resultate aus den Beobachtungen des Magnetischen Vereins im Jahre 1839, hrsg. von Carl Friedrich Gauß und Wilhelm Weber, pp. 117–119. Leipzig (1840) (in German)
15. Jonkers, A.R.T., Jackson, A., Murray, A.: Four centuries of geomagnetic data from historical records. Rev. Geophys. **41**(2), 1006 (2003)
16. Barraclough, D.R.: Spherical harmonic analyses of the geomagnetic field for eight epochs between 1600 and 1910. Geophys. J. R. Astron. Soc. **36**, 497–513 (1974)
17. https://www.intermagnet.org/
18. Cain, C.J., Hendricks, S.J., Langel, R.A., Hudson, W.V.: A proposed model for the international geomagnetic reference field-1965. J. Geomagn. Geoelectr. **19**(4), 335–355 (1965)
19. Ivanov, S., Merkuryev, S., Demina, I., Soldatov, V., Zaitsev, D.: Interpretation of component geomagnetic field measurements carried out on board a ferromagnetic vessel from the round-the-world expedition of the R/V "Admiral Vladimirsky" in 2019–2020. This issue
20. Tacq, J.: Multivariate normal distribution. In: Peterson, P., Baker, E., McGaw, B. (eds) International Encyclopedia of Education, 3rd edn, pp. 332–338. Elsevier Ltd. (2010)
21. Wessel, P., Luis, J.F., Uieda, L., Scharroo, R., Wobbe, F., Smith, W.H.F., & Tian, D. (2019). The generic mapping tools version 6. Geochem. Geophys. Geosyst. **20**, 5556–5564

Evolution and Statistics of the Geomagnetic Energy and Its Characteristic Timescales Since 1840

S. V. Starchenko and S. V. Yakovleva

Abstract The evolution of total (integrated from the core-mantle boundary to infinity) geomagnetic energy based on the COV-OBS.x1 geomagnetic field model is approximated by exponential functions with an error less than 3% in four time intervals between 1840 and 2020. Characteristic timescale is determined as a ratio of the energy to its time derivative ($T = E/P$, where $P = dE/dt$ has a meaning of power). Timescale values are statistically explored with annual resolution. Most of the timescales (87%) are negative, indicating decrease of energy with characteristic times of the order of thousand years. The remaining 13% indicate the energy increase with the minor timescales of about a few thousand years. The median timescale is − 1176 years, the arithmetic mean is + 1889 years and the most probable or mode Mo = − 483 years. The large standard deviation and RMS (44,571 and 44,400 years) indicate heavy tails which are clearly seen in the bimodal probability density distribution. We define a special geometric mean timescale (− 174 years) which is consistent with the known convective velocities ~ 0.3 mm/s and the observable magnetic heterogeneities drifts. The prevalent timescales are from ~ 500 years to a few thousand years. Corresponding characteristic velocities or the mean field alpha effects ensure a subcritical geodynamo regime. The results are also consistent with periodical and spectral estimates from geodynamo simulations and geomagnetic field models for both the modern era and the ancient geomagnetic field. We estimate roughly that a comparable timescale for magnetic field is about a factor of two longer than T, while the magnetic field periodicity may be several times longer.

Keywords Geodynamo · Geomagnetic energy · Geomagnetic variations · Timescales

---

S. V. Starchenko (✉) · S. V. Yakovleva
Pushkov Institute of Terrestrial Magnetism, Ionosphere and Radio Wave Propagation, Russian Academy of Sciences (IZMIRAN), Kaluzhskoe hwy 4, Troitsk, Moscow 108840, Russia

A. Kosterov et al. (eds.), *Problems of Geocosmos—2020*, Springer Proceedings in Earth and Environmental Sciences, https://doi.org/10.1007/978-3-030-91467-7_14

1 Introduction

The observed variations of the main magnetic field generated in the Earth's core have characteristic timescales that exceed several years, since shorter variations are almost completely shielded by the electrically conductive mantle [1–4]. Exceptions, relatively small in terms of their energy, are short-term phenomena such as torsional oscillations [5–9] with periods of a few years and jerks with durations of about a year [10–13], subject of active current study.

Changes of the Earth's internally generated magnetic field spanning the time scales from a few years to a few thousand years are classified as secular variation (SV) [14, 15]. The main characteristic of SV is its period; however, strictly speaking, it is not a period, but rather a characteristic time or timescale, since SV periodicity in the mathematical sense of the word has not been proved. This variation results from the effect of magnetic induction in the fluid outer core and from effects of magnetic diffusion in the core and the mantle [16, 17].

Temporal spectrum of the observed geomagnetic field has a semi-discrete structure described by the following three types of variations [18–23].

1. Variations of the first type are characterized by harmonic periods from a few tens up to a hundred years (in particular $T = 20$, 30 and 60 years). These variations are likely to be of internal origin, potentially being a manifestation of torsional oscillations in the fluid core [24]. The morphology of these variations and the high correlation with changes in the Earth's diurnal rotation also confirm their classification as torsional [25–28]. The physical nature of this class of variations as well as the mechanism of torsional waves generation is however still a matter of debate [5–9, 16, 29].
2. The main part of the SV spectrum includes periods of several hundred and several thousand years (in particular, 360, 600, 900, 1200, 1800, 2700, 3600, 5400 and 9000 $\pm$ 10% years). These variations occur in the upper part of the core (or at the core-mantle boundary) due to thermal and compositional convection, propagation of magnetohydrodynamic waves and short-scale diffusion processes [17, 30–35]. The variation of 9000 $\pm$ 1000 years is interpreted as the natural oscillation of the hydromagnetic dynamo [17, 19, 22, 32].
3. Fluctuations with timescales of tens of thousands of years and longer are primarily connected with decay modes, which are not oscillations. They result from magnetic diffusion processes in the entire core [14, 32–34]. Here we do not consider the longer variations, because they could hardly be covered by the model used in the present work, restricted by only < 200 years of observations.

Determination of the geomagnetic field frequency/time characteristics has always been an important problem, which has been approached using various mathematical methods: spectral Fourier analysis [28, 36], maximum entropy method [37, 38], autoregressive and correlation methods [39], and wavelet analysis [40, 41]. A detailed review of mathematical approaches used in the search of periodicities in geomagnetic data series is presented in [42, 43] and references therein.

One of the most significant limitations of the above-mentioned essentially harmonic methods [38–43] is the fundamental impossibility of studying a period exceeding the length of the time series. For example, according to tests performed by Jackson and Mound [24], the longest meaningful period would not exceed 75% of the time series length. Besides, an assumption about broadband continuous spectrum of field variability becomes more and more popular [17, 42–45]. Already in 1983 Barton [46] performed spectral analysis of declination and inclination time series, concluding that there was no evidence for discrete periods but, instead, for bands of preferred periods at 60–70, 400–600, 1000–3000 and 5000–8000 years. Still earlier, in 1968, Currie [47] argued that the temporal power spectrum of geomagnetic field observations was governed by a power law, i.e., f^k, where f is the frequency. More recently, Olson et al. [17] and Bouligand et al. [48, 49] carried out a detailed study of the frequency spectrum of dipole field variations from numerical geodynamo simulations, and found a broadband variability appropriately described by power laws. Their results agree well with the composite paleomagnetic dipole spectrum [50].

Against this background of numerous attempts of revealing any periodicity or characteristic times, we developed an original statistical approach to determining the long-term characteristics of time series [51–54]. For our studies, we chose the totally integrated energy E of the observed potential part of the main geomagnetic field [52–54]. This energy could be formally derived by means of the well-known work of Lowes [55], but apparently this was made only much later [56, 57]. This total energy seems to be the most adequate global variable for studying global temporal variations of the observed geomagnetic field.

The study of the energy contained in the potential part of the main geomagnetic field was initiated in [58]. Based on this work, Lowes [55, 59] determined the contribution of n-th spherical harmonic degree to the radial energy density normalized by the area of the sphere:

$$R_n = (n+1)\left(\frac{a}{r}\right)^{2n+4} \sum_{m}^{n} \left[(g_n^m)^2 + (h_n^m)^2\right] \tag{1}$$

Here g_n^m and h_n^m are standard Gauss coefficients, a is the radius of the Earth and r is the radius of a sphere that varies from the core radius r_c to infinity. The resulting R_n is expressed in (Tesla)^2.

Expression (1) is the so-called "spatial Lowes-Mauersberger power spectrum". We prefer (perhaps, more physically correct) to name it r-density spectrum of energy renormalizing it to J/m. This density varies with r, while the total energy E is spatially independent, compare with [43]. To derive an expression for energy, we first obtain the contribution of the n-th harmonic degree R_n (in J/m) to the radial energy density of the potential field as

$$\frac{4\pi r^2}{2\mu_0} R_n = 2\pi a^2 \frac{n+1}{\mu_0} \left(\frac{a}{r}\right)^{2(n+1)} \sum_{m=0}^{n} \left[(g_n^m)^2 + (h_n^m)^2\right] \tag{2}$$

Integrating (2) along the radius from the core-mantle boundary $r = r_c$ (where r_c is the radius of the Earth's core) to infinity we obtain the contribution of the n-th harmonic degree to the total energy (in J) as:

$$E_n = 2\pi a^{2n+4} \frac{n+1}{\mu_0} \sum_{m=0}^{n} [(g_n^m)^2 + (h_n^m)^2] \int_{r_c}^{+\infty} \frac{dr}{r^{2n+2}}$$
$$= \frac{2\pi a^3}{\mu_0} \left(\frac{a}{r_c}\right)^{2n+1} \frac{n+1}{2n+1} \sum_{m=0}^{n} [(g_n^m)^2 + (h_n^m)^2] \tag{3}$$

There is actually no need to integrate to infinity since even when integrating up to r equal to several times r_c, one obtains practically the same result with error less than 0.3%.

The integral energy E used in this study is simply the sum of all E_n terms from (3).

We use this E to obtain timescales $T = E/(dE/dt)$ which will be defined in detail in the next section. As an analogue to our timescales we consider the instantaneous correlation time of Hulot and Le Mouël [60] involving the ratio of the Gauss coefficients to their time-derivatives as

$$\tau_n = \left\{ \frac{\sum_{m=0}^{n} [(g_n^m)^2 + (h_n^m)^2]}{\sum_{m=0}^{n} [(\dot{g}_n^m)^2 + (\dot{h}_n^m)^2]} \right\}^{1/2}. \tag{4}$$

This timescale is usually linked to spectrum (1) by many authors [33, 44, 48, 49] studying the geomagnetic field variation and τ_n is one of the most popular characteristic timescale definitions. However, it is not so good for global investigations because it is a function of n. A more significant limitation of the correlation time (4) is that it is defined only on the Earth's surface. At the same time, the geomagnetic field energy is dominated by contributions from the regions located near the Earth's core, which is obvious from formulae (2) and (3). The integral energy E proposed here and all its derivatives are free from these disadvantages [52–54].

The primary goal of this paper is to explore global characteristic geomagnetic timescales. The detailed definition of these follows in Sect. 2 where we explore the 1840–2020 evolution of E and its timescales T. Section 3 deals with statistical and probability distribution descriptions of the annual energy timescales T. In Sect. 4 we outline the most statistically significant characteristic timescales for the geodynamo correlating them with modern/ancient geomagnetic field, trends, alpha effects, drifts and fluid velocities in the Earth's core. Final Sect. 5 presents a brief discussion and conclusions of this study.

2 Energy Evolution and Discrete Timescales

The central idea of this paper is to define and explore the characteristic instantaneous geomagnetic timescales T as the ratio of the global energy E, as defined by formula (3), to its time-derivative, referred to thereafter as power $P = dE/dt$. Thus, by definition

$$T \equiv E/(dE/dt) \quad (5)$$

An immediate bonus of this definition is introducing a positive/negative sign of the timescale T. The sign indicates rise/decay of the energy E. Definitions similar to (5), but without taking sign into account and in most cases local, has long been used [22, 33, 49, 60–62]. The innovation of our study is however to apply timescale definition (5) to a global evolutional and statistical analysis of geomagnetic energy as defined by (3).

The simplest time-process related to definition (5) is the one with $T = const$. Viewing (5) as a differential equation, we introduce its solution as an exponential dependence:

$$E = C \exp(t/T) \quad (6)$$

Therefore, a reasonable initial step would be to model the evolution of E by exponents (6) on some suitable time intervals where the timescale T could be regarded as constant.

We use the open-source (http://www.spacecenter.dk/files/magnetic-models/COV-OBSx1) geomagnetic model COV-OBS.x1 [63, 64]. It spans 1840–2020 and is based on annual initial data. The authors of the model successfully extrapolated it down to the half-year (yr/2) resolution. We used the results of this extrapolation for definition of the annual time-derivative dE/dt. It was calculated for each year as a ratio $[E(t - \text{yr}/2) - E(t + \text{yr}/2)]/\text{yr}$. Here t is the particular year and yr is 1 year. The corresponding time-values are transformed into seconds for calculating $P = dE/dt$ in Watts.

Figure 1 shows the evolution of the energy E and power $P = dE/dt$. At two instances, marked in Fig. 1, the power becomes zero, which corresponds to the local energy extrema. These naturally divide the plot of E into three monotonous parts that can be approximated by exponential functions (6) with an error less than 2% for two intervals from 1932 to 2020, whereas the first interval 1840–1931.5 was approximated with error about 5%. Dividing this interval into two, we reduce the approximation errors to their minimum values placing the dividing point near the first local maximum of P. We consider important to have an error sufficiently lower than the energy variation $2(E_{max} - E_{min})/(E_{max} + E_{min}) = 0.07$. Thus, we split the evolution of E for the given time interval into 4 epochs approximated with an error δ less than 3% (t is the time in years) as:

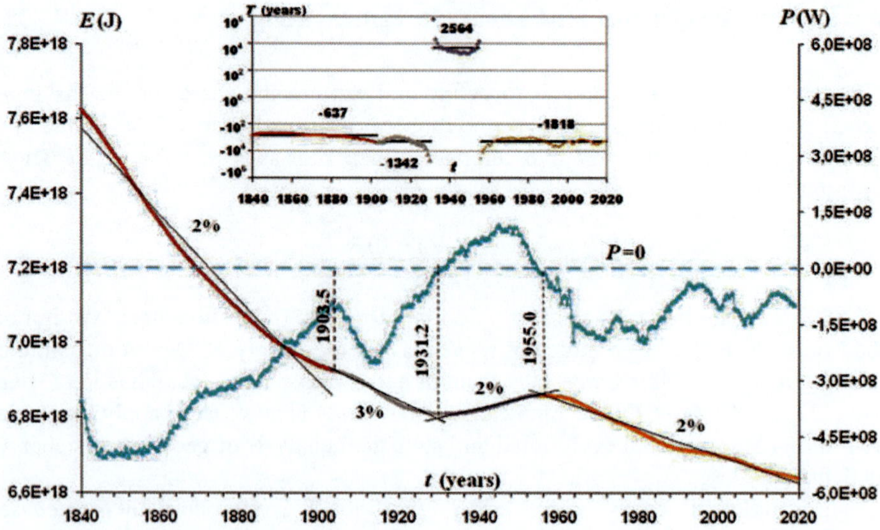

Fig. 1 Evolution of the global energy E (multicolor) and power P (blue) obtained from COV-OBS.x1 model [63, 64]. The left vertical axis is for E (Joules), the right for P (Watts). Exponential functions (7) approximate the energy E with approximation errors δ, indicated as a percentage. The horizontal dashed line corresponds to $P = 0$. The inset shows the continuous evolution of timescales T in years following (5). The relevant sections of the evolution of the energy and timescales are marked with the same colors. The corresponding discrete timescales (7–8) are displayed by horizontal lines

$$\begin{cases} E = 136 \cdot e^{t/T_a} \cdot 10^{18}\ \text{J}, & \text{where } 1840 < t < 1903.5 \text{ and } \delta = 2\% \\ E = 28.6 \cdot e^{t/T_b} \cdot 10^{18}\ \text{J}, & \text{where } 1904 < t < 1931.5 \text{ and } \delta = 3\% \\ E = 3.22 \cdot e^{t/T_c} \cdot 10^{18}\ \text{J}, & \text{where } 1932 < t < 1955.5 \text{ and } \delta = 2\% \\ E = 19.9 \cdot e^{t/T_d} \cdot 10^{18}\ \text{J}, & \text{where } 1956 < t < 2020 \text{ and } \delta = 2\% \end{cases} \tag{7}$$

At this level of accuracy those exponential functions are indistinguishable from linear fits. However, we can still extract timescales with errors of the order a few percents in a sense of (5–6). Accordingly, the ratio of energy to power ($T = E/P$) gives preliminary and rather rough discrete (in spectral sense) timescales for these four epochs (7) in years:

$$T_a = -637;\ T_b = -1342;\ T_c = 2564;\ T_d = -1818. \tag{8}$$

Now, we apply (5) and calculate the continuous evolution of the timescales for every year from 1840 to 2020. The result is shown in the inset to Fig. 1. The evolution is divided into relatively long, up to several decades, epochs alternating with short hyperbolic T-transitions. Hyperbolic T-transitions can have duration of several years and associated with the extrema of the energy where $P = 0$ and $|T|$ tends to infinity.

3 Statistics of the Characteristic Timescales

Following Fig. 1, we can reasonably split the energy evolution during 1840–2020 into four statistically different intervals (7). Each interval is described with a single spectral line or constant timescale T, (Eq. (8) and Fig. 1). The longest intervals are characterized with negative timescales indicating energy decrease. We observe three intervals with negative T and only one interval with positive T.

The intervals with negative T allow to estimate roughly the most probable (− 637 years) and median (about − 1500 years) timescales. At the same time, the shortest interval with positive timescales (2564 years) corresponds to a minor mode in a bimodal timescale distribution.

We now explore statistical and spectral features of annual timescales.

We apply formulae (5) and (3) to the Gauss coefficients provided by the COV-OBS.x1 model [64], and calculate 180 annual T values. The obtained timescales are then arranged in increasing order. This is essential for identifying extreme and median values, and for plotting probability distributions of T. The latter will be thereafter referred to as T_i, with i running from 1 to the maximum value $I = 180$.

Supposing that each T_i appears with the same probability, we plot the cumulative distribution function (thereafter CDF) for them as shown in Fig. 2.

The statistical parameters of the timescales are presented in a Table 1. All parameters were calculated from their standard definitions [65, 66] except for geometric mean (GM), which is explained at the end of this section.

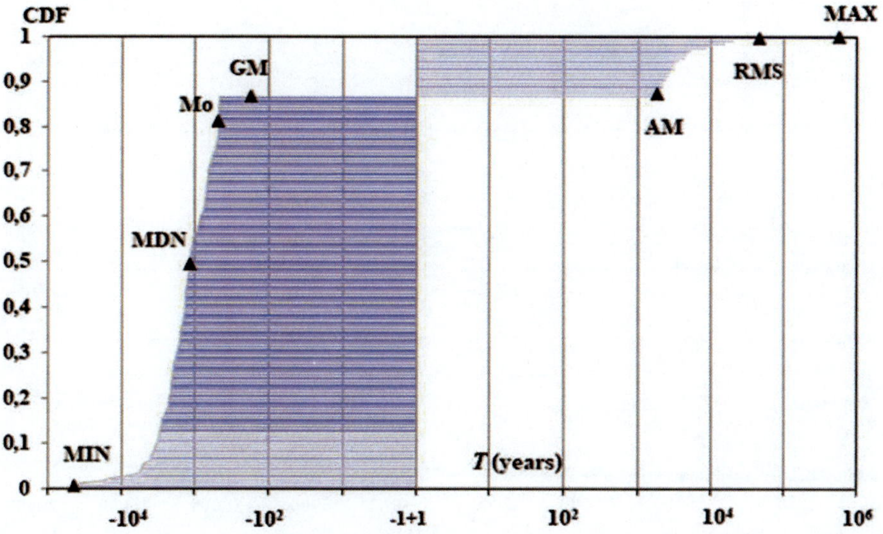

Fig. 2 The cumulative distribution function CFD for timescales T_i as defined in Sect. 2. Main statistical parameters of the timescales are shown (see Table 1 for details) with triangles. A confidence interval (75%) centered at MDN is shown in a darker color

Table 1 Statistical parameters of the characteristic timescales T_i, years

Range of T_i	Number (i)	Minimum (MIN), median (MDN), maximum (MAX)			Arithmetic Mean (AM), Root Mean Square (RMS), Standard Deviation (SD), Geometric Mean (GM)			
		MIN	MDN	MAX	AM	RMS	SD	GM
All	180	−43,790	−1176	592,198	1889	44,364	44,571	−174
$T_i > 0$	24	1855	3127	592,198	29,309	121,055	119,980	4571
$T_i < 0$	156	−43,790	−1347	−464.8	−2329	5393	4880	−1380

We start our analysis from the general (calculated from all T_i) median timescale MDN = − 1176 years (the first row of MDN column in Table 1). From this MDN we obtain a 75% confidence interval (based on the number of samples) bounded by timescales − 3400 and − 470 years that is visualized in Fig. 3. Therefore, the dominant trend is the decrease of energy with millennial characteristic timescales. This is similar to the timescales of drifts of the largest magnetic heterogeneities, and of the observed global geomagnetic variations [14, 35, 44, 61, 62, 67–70].

These 156 (87% of all T_i) T_i values with decreasing ($T < 0$) energy may be viewed as reflecting a major timescale of the order of 10^3 years (− 1347 years in the MDN column, last line of the Table 1) characterizing the modern geodynamo with diminishing energy of the observed geomagnetic field. This is in agreement with the characteristic timescale for the dipolar component of the geomagnetic field [15, 17, 32, 35, 60, 61].

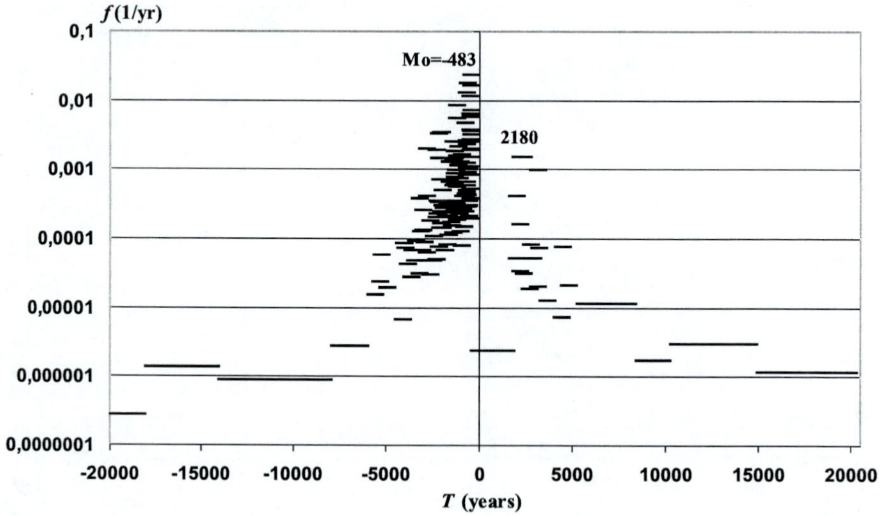

Fig. 3 The probability density function $f(T)$ as defined by formula (11). The most probable or mode value is Mo = − 483 years and the secondary positive mode is 2180 years

The remaining 13% or 24 positive T_i values indicate the energy increase of a timescales of about a few thousand years (3127 years in MDN column, second line of Table 1).

General arithmetic mean AM is 1889 years (the first line in the AM column of Table 1). This positive value does not seem appropriate bearing in mind the domination of negative timescales. An explanation to this apparent contradiction is the large standard deviations and RMS values. Indeed, we observe heavy tails which are linked to extreme minimal (−43,800 years) and maximal (592,000 years) timescales and are obviously overpopulated with respect to a Gaussian distribution. Thus, the true timescale distribution is not Gaussian but clearly bimodal. This is shown in Fig. 3 by a probability density function $f(T)$, which we defined as the simplest finite difference of CDF (see Fig. 2)

$$f_k = (I-1)^{-1}(T_{k+1} - T_k)^{-1}. \tag{9}$$

Here a density function $f(T) = d\mathrm{CDF}/dT$ is a constant f_k for $T_{k+1} < T < T_k$, and k runs from 1 to $I-1$. Outside the $[T_1, T_I]$ interval $f(T) = 0$. The integral of fdT over all T is obviously equal to one. The most probable, or mode, value Mo would then be − 483 years. It is respectively two and four times smaller than the median and mean values considered above. Besides, we observe extremely uneven f variation and very large range of timescales covering more than 9 orders of magnitude from a value lower than − 10^4 years up to over than 10^5 years.

In view of the above, a geometric mean should be a more appropriate measure of a characteristic mean timescale. The geometric mean is well-defined for positive-only values (see GM column in the Table 1 for $T_i > 0$), and can similarly be defined for $T_i < 0$ as follows:

$$GM_+ = \left(\prod_{i=157}^{180} T_i\right)^{1/24} = 4571 \text{ years}, \quad GM_- = -\left(\prod_{i=1}^{156} |T_i|\right)^{1/156} = -1380 \text{ years} \tag{10}$$

Introducing $S = \sum \mathrm{sgn}(T_i) \lg |T_i|$ we write $GM_\pm = \mathrm{sgn}(S) 10^{|S|/L}$, where L is 24 for positive and 156 for negative T_i. Combining these expressions for positive and negative T_i, we obtain for a general geometric mean (note the minus sign in the exponent to GM_+):

$$GM = -|GM_-|^{156/180} (GM_+)^{-24/180} = -174 \text{ years} \tag{11}$$

This value reflects a characteristic "rapid" change of the energy E in the modern era. Besides, $|GM|$ is close to the length (180 years) of the field annual means series. The corresponding geomagnetic field timescale is about $2GM = -348$ years, doubled due to the proportionality of the energy to the square of the field.

Let us round up this section by pointing out that there exists an extended gap containing no T_i's (Fig. 2). The range of missing values extends from the maximum negative T value (column MAX in Table 1) of − 465 years to the minimum positive T value (column MIN) of + 1855 years. A timescale within this gap might occur, but with very low probability, as seen in Fig. 3.

4 Physical Interpretation of Timescales Statistics

In order to make physical interpretation we need to compare our magnetic energy timescales (5) with known timescales of the magnetic field. Generally, such a comparison could be done by magnitude only, because there are rather different definitions of the timescales, e.g., compare expression (4) from Hulot and Le Mouël [60] with our expression (5) based on (3). Taking into consideration that axial dipole component is dominant everywhere, we could establish a rough relationship between our timescales T_0 and τ_0, similar to Hulot and Le Mouël's definition:

$$T_0 = |(g_1^0)^2/[d(g_1^0)^2/dt]| = |g_1^0/[dg_1^0/dt]|/2 = \tau_0/2. \tag{12}$$

We thus assume that our energy timescale is roughly a half of magnetic field timescale defined in a way similar to our definitions (3) and (5). This also follows from the simple consideration that the energy E is proportional to a square of the magnetic field intensity. Carrying out a Fourier transform on both parts of this relationship one sees immediately that an energy frequency is twice the magnetic field frequency.

An absolute value of the median energy timescale |MDN| = 1176 years corresponds to a geomagnetic field timescale of 2352 years. This is about 30% larger than the exponential decay time of the modern axial dipole, which is about 1700–1900 years and characterized by decreasing trend [32, 34, 35, 71]. Therefore, the magnetic field associated with energy E apparently decays slower even in comparison with the slowest decay mode of the axial dipole.

Dividing the radius of the Earth's core ($r_c = 3.5 \times 10^6$ m) by this characteristic time (2352 years), we obtain $\alpha = 0.04$ mm/s, which is an order of magnitude less than the known convective velocities [14, 35, 67, 72–77]. Therefore, α should roughly be an analogue of the alpha effect of the mean field theory [77–81]. However, it works against generation reducing the magnetic field intensity because all timescales MDN < 0. To consider the generation we use positive timescales (~ 4000 years, the second row of MDN/GM column in Table 1) giving $\alpha_+ = 0.02$ mm/s. The known molecular magnetic diffusion is about 1 m^2/s [80], while the plausible turbulent mean field diffusion η could be a few times larger [77–81]. Thus, a mean-field magnetic Reynolds number value Rm is:

$$\mathrm{Rm} = \alpha_+ r_c/\eta \approx 20 - 40. \tag{13}$$

This value is subcritical for a mean-field geodynamo [81–83]. The integral energy E decays due to prevalence of negative timescales over the positive, as shown above.

Let us find a correlation between the theoretically known [14, 35, 67, 72–77] convective velocities and corresponding observational drift velocities of magnetic heterogeneities [14, 35, 44, 61, 62, 67–70]. The convective velocities and drift velocities in the Earth's core are of the order of 0.3 mm/s. Dividing the radius of the Earth's core by these velocities, we obtain the values ~ 350 years that are approximately twice the absolute value of the geometric mean timescale $|GM|$.

This 348 years global magnetic field timescale is comparable with global "single time constant of secular variation" $\tau_{sec} = 535$ years defined by Christensen and Tilgner [33], who used a set of numerical dynamo models together with direct observations to derive scaling relations for magnetic energy. A similar value $\omega^{-1} = 415$ years was obtained by Bouligand et al. [49] where in their abstract they say: "based on dynamo simulations, the authors argue that a prior for the observational geomagnetic field over decennial to millennial periods can be constructed from the statistics of the field during the short satellite era". Adopting similar approach, we use bicentennial statistics to describe timescales from a half-thousand up to a half-million years.

5 Concluding Remarks

The central idea of this paper is to explore a characteristic timescale T which is defined as a ratio of the total geomagnetic energy E to its time derivative, or power $P = dE/dt$. The energy is defined as resulting from integrating from the core-mantle boundary to infinity. The major bonus of such definition is a possibility to determine from direct observation a number of timescales sufficiently larger than the length of the studied time series.

Thus, we can explore insufficiently studied millennia-long processes on the base of relatively short, but much better studied and detailed, time series covering less than several centuries. This is especially important for the geomagnetic variations of long, i.e., from 450 years for T in this study, and even much longer, up to half-million years, characteristic times, which could be obtained from rather short periods of direct measurements.

A novel feature we introduce is to consider the sign which appears at these timescales. Negative timescales $T < 0$ correspond to energy E decrease, while positive ones to energy E rise. The definitions of timescales similar to our have long been used [22, 33, 49, 60–62], but these studies did not take a sign into the account.

We investigated evolution of the power P and 180 annual timescales obtained from COV-OBS.x1 model [63, 64]. The evolution of timescales T is naturally divided into relatively long, up to several decades, epochs with almost constant T alternating with short hyperbolic T-transitions. Hyperbolic T-transitions can have duration of several years and are associated with the extrema of the energy where $P = 0$ and $|T|$ increases infinitely.

For the study period covering 1840–2020, evolution of energy E is approximated by exponents with an error of less than 3% if the period is divided into four intervals.

Each interval is described with a single spectral line or constant timescale T. We observe three intervals with negative T and only one interval with positive T. The intervals with negative T allow to estimate roughly the most probable (– 637 years) and median (about – 1500 years) timescales. At the same time, the shortest interval with positive timescales (2564 years) corresponds to the minor mode in a timescale bimodal distribution.

Statistical analysis of the annual T values determines their standard statistical characteristics. 87% of T values are negative and indicate the decreasing $T < 0$ trend in energy with the major timescale in the order of one thousand years attributable to modern geodynamo. The remaining 13% indicate the energy increase on the minor timescale that is about a few thousand years. The median timescale is – 1180 years, while the arithmetic mean is + 1890 years. Less probable heavy tails associated with the extreme minimal (– 43,800 years) and maximal (592,000 years) timescales are the likely reason for extremely large standard deviation and RMS values of 44,400 and 44,600 years respectively. There is a gap between – 465 and 1855 years with no T. However, any T could appear there with a probability similar to the probability for the revealed heavy tails.

We calculated and plotted the bimodal probability density function with the most probable value, or mode Mo $\approx -$ 500 years. The absolute value of |Mo| is two-four times smaller than absolute values of median and mean considered above. The minor mode is about 2000 years. Additionally, we observe extremely non-smooth variations in this function and a very wide range of time scales, covering more than 9 orders of magnitude, from below $-\ 10^4$ years to more than 10^5 years.

A geometric mean appears a more suitable measure to represent averaged timescales. The mean geometric timescale, $GM = -$ 174 years (Sect. 3), is consistent with the known convective velocities ~ 0.3 mm/s, the observable magnetic heterogeneities drifts [14, 35, 68–70] and with global geomagnetic timescales of about 500 years found previously [33, 49].

The prevalent timescales are from half a thousand years, where the maximum number of timescales is concentrated, and up to a few thousand years. Corresponding characteristic velocities or the mean field alpha effects ensure subcritical geodynamo. The obtained results are also consistent with periodical and spectral estimates from geodynamo simulations and geomagnetic field models for both the modern era and the ancient geomagnetic field.

To conclude, we consider how timescales T introduced here relate to long-term periods Q, obtained from studies of geomagnetic field spectrum in paleomagnetic and geodynamo models. Timescales T as defined here have physical meaning only on the interval where the change in energy E is monotonous. For the geomagnetic field, this interval is approximately $2T$ as follows from formula (12). A periodic process can then be subdivided into a minimum of two such intervals, one corresponding to increasing energy (field strength), and another to decreasing. Thus, in this case $Q = 4T$, but a large gap in the energy derivative $\mathrm{d}E/\mathrm{d}t$ arises when energy evolution changes from growth to decrease. This gap can be made smaller if the period Q is subdivided into four intervals, approximately corresponding to a sinusoidal dependence monotonously growing from zero to a maximum, decreasing to zero and then to a minimum, and finally growing back to zero, resulting in $Q = 8T$. Discontinuities can be further reduced by an approximation like one discussed in Sect. 2, Fig. 1 and formulae (7) and (8). This is of course only a rough estimate, and in order to reveal a relationship between Q and T, it would be necessary to compare the behavior of the same data series within the framework of the classical and our statistical spectral approach. However, for now, we restrict ourselves to the simplest relation

$$Q = AT \tag{14}$$

We assume an approximate value of $A \approx 6$, as an average resulting from the above estimates for Q. Bearing in mind a high uncertainty of this number, it can still be argued that the results of all long-term (covering > 10^4 years) studies of geomagnetic field evolution known to us [23, 43, 48, 49, 84–87] agree reasonably well with the results obtained in this work based only on a 200-year long series.

Acknowledgements We would like to express our deep gratitude to two anonymous reviewers for their important and useful comments that helped us significantly revise and improve the manuscript.

References

1. McDonald, K.L.: Penetration of the geomagnetic secular field through a mantle with variable conductivity. J. Geophys. Res. **62**, 117–141 (1957)
2. Braginsky, S.I., Fishman, V.M.: Magnetic field shielding in the mantle at conductivity concentrated near the boundary with the core. Geomagn. Aeron. **17**(5), 907–915 (1977)
3. Benton, E.R., Whaler, K.A.: Rapid diffusion of the poloidal geomagnetic field through the weakly conducting mantle: a perturbation solution. Geophys. J. R. Astr. Soc. **75**, 77–100 (1983)
4. Jault, D.: Illuminating the electrical conductivity of the lowermost mantle from below. Geophys. J. Int. **202**(1), 482–496 (2015)
5. Braginsky, S.I.: Torsional magnetohydrodynamic vibrations in the Earth's core and variations in day length. Geomagn. Aeron. **10**, 1–8 (1970)
6. Maffei, S., Jackson, A.: Propagation and reflection of diffusionless torsional waves in a sphere. Geophys. J. Int. **204**(3), 1477–1489 (2016)
7. Gillet, N., Jault, D., Canet, E., Fournier, A.: Fast torsional waves and strong magnetic field within the Earth's core. Nature **465**, 74–77 (2010)
8. Teed, R.J., Jones, C.A., Tobias, S.M.: Torsional waves driven by convection and jets in Earth's liquid core. Geophys. J. Int. **216**(1), 123–129 (2019)
9. Gillet, N., Jault, D., Canet, E.: Excitation of travelling torsional normal modes in an Earth's core model. Geophys. J. Int. **210**(3), 1503–1516 (2017)
10. Bloxham, J., Zatman, S., Dumberry, M.: The origin of geomagnetic jerks. Nature **420**, 65–68 (2002)
11. Mandea, M., Olsen, N.: Geomagnetic and archeomagnetic jerks: where do we stand? EOS Trans. Am. Geophys. Union **90**(24), 208–208 (2009)
12. Aubert, J., Finlay, C.C.: Geomagnetic jerks and rapid hydromagnetic waves focusing at Earth's core surface. Nat. Geosci. **12**(5), 393–398 (2019)
13. Kloss, C., Finlay, C.C.: Time-dependent low-latitude core flow and geomagnetic field acceleration pulses. Geophys. J. Int. **217**(1), 140–168 (2019)
14. Parkinson, W.D.: Introduction to Geomagnetism. Scottish Academic Press, Edinburgh (1983)

15. Bloxham, J., Gubbins, D., Jackson, A.: Geomagnetic secular variation. Phil. Trans. R. Soc. Lond. A **329**, 415–502 (1989)
16. Wardinski, I.: Geomagnetic secular variation. In: Gubbins, D., Herrero-Bervera, E. (eds.) Encyclopedia of Geomagnetism and Paleomagnetism. Springer, Dordrecht (2007)
17. Olson, P., Christensen, U., Driscoll, P.: From superchrons to secular variation: a broadband dynamo frequency spectrum for the geomagnetic dipole. Earth Planet. Sci. Lett. **319–320**, 75–82 (2012)
18. Pecherskij, D.M., Sokolov, D.D.: Paleomagnitology, Petromagnitology and Geology. Dictionary-reference for neighbors in the specialty. IFZ RAS, Moscow (2010). http://paleomag.ifz.ru (in Russian)
19. Braginsky, S.I.: On the spectrum of oscillations of the earth hydromagnetic dynamo. Geomagn. Aeron. **10**(2), 221–233 (1970)
20. Braginsky, S.I.: Origin of the Earth's magnetic field and its secular variations. Izvestiya Acad. Sci. USSR Phys. Solid Earth **10**, 3–14 (1972)
21. Burlatskaya, S.P.: Archaeomagnetism. The Structure and Evolution of the Earth's Magnetic Field. GEOS, Moscow (2007) (in Russian)
22. Nachasova, I.E., Pilipenko, O.V.: Archaeomagnetic studies at Schmidt institute of physics of the earth, Russian academy of sciences: history and main results. Izv. Phys. Solid Earth **55**(2), 298–310 (2019)
23. Petrova, G.N., Nechaeva, T.B., Pospelova, G.A.: Characteristic Variations in the Past Geomagnetic Field. Nauka, Moscow (1992).(in Russian)
24. Jackson, L.P., Mound, J.E.: Geomagnetic variation on decadal time scales: what can we learn from empirical mode decomposition? Geophys. Res. Lett. **37**(14), L14307 (2010)
25. Braginsky, S.I.: Short-period geomagnetic secular variation. Geophys. Astrophys. Fluid Dyn. **30**(1–2), 1–78 (1984)
26. Bondar, T.N., Golovkov, V.P., Yakovleva, S.V.: Secular variation in the geomagnetic field in 1980–2000. Geomagn. Aeron. **43**(6), 799–802 (2003)
27. Ivanov, V.V., Bondar, T.N.: Dynamics and physical model of secular variations within a range of up to 100 years. Geomagn. Aeron. **55**(3), 389–397 (2015)
28. Alldredge, L.R.: Geomagnetic variations with periods from 13 to 30 years. J. Geomagn. Geoelectr. **29**, 123–135 (1977)
29. Hulot, G., Finlay, C.C., Constable, C.G., Olsen, N., Mandea, M.: The magnetic field of planet Earth. Space Sci. Rev. **152**, 159–222 (2010)
30. Braginsky, S.I.: Magnetic waves in the core of the Earth. II Geophys. Astrophys. Fluid Dyn. **14**, 189–208 (1980)
31. Braginsky, S.I.: MAC-oscillations of the hidden ocean of the core. J. Geomagn. Geoelectr. **45**(11–12), 1517–1538 (1993)
32. Olson, H., Amit, Y.: Changes in Earth's dipole. Naturwissenschaften **93**, 519–542 (2006)
33. Christensen, U.R., Tilgner, A.: Power requirement of the geodynamo from ohmic losses in numerical and laboratory dynamos. Nature **429**, 169–171 (2004)
34. Starchenko, S.V.: Harmonic sources of the main geomagnetic field in the Earth's core. Geomagn. Aeron. **51**(3), 409–414 (2011)
35. Starchenko, S.V., Ivanov, V.V.: Nature of the diffusion, generation, and drift of the geomagnetic dipole from 1900 to 2010. Dokl. Earth Sci. **448**(1), 67–69 (2013)
36. Creer, K.M., Thouveny, N., Blunk, I.: Climatic and geomagnetic influences on the Lac du Bouchet palaeomagnetic SV record through the last 110 000 years. Phys. Earth Planet. Inter. **64**(2–4), 314–341 (1990)
37. Currie, R.G.: Geomagnetic line spectra – 2 to 70 years. Astrophys. Space Sci. **21**, 425–438 (1973)
38. Papitashvili, N.Y., Rotanova, N.M.: The method of maximum entropy and its application to the analysis of time series of the geomagnetic field. Geomagn. Aeron. **19**(3), 543–550 (1979)
39. Ivanov, V.V., Rotanova, N.M.: The Sompi spectral method and its application in the analysis of simulated and measured time series. Geomagn. Aeron. **36**(1), 89–94 (1996)

40. Burakov, K.S., Galyagin, D.K., Nachasova, I.E., Reshetnyak, M.Yu., Sokolov, D.D., Frick, P.G.: Wavelet analysis of geomagnetic field intensity for the past 4000 years. Izv. Phys. Solid Earth **34**(9), 773–778 (1998)
41. Rotanova, N.M., Bondar, T.N., Ivanov, V.V.: Wavelet analysis of secular geomagnetic variations. Geomagn. Aeron. **44**(2), 252–258 (2004)
42. Panovska, S., Finlay, C.C., Hirt, A.M.: Observed periodicities and the spectrum of field variations in Holocene magnetic records. Earth Planet. Sci. Lett. **379**, 88–94 (2013)
43. González-López, A., Campuzano, S.A., Molina-Cardína, A., Pavón-Carrasco, F.J., Osete, M.L.: Characteristic periods of the paleosecular variation of the Earth's magnetic field during the Holocene from global paleoreconstructions. Phys. Earth Planet. Inter. **312**, 106656 (2021)
44. Tanriverdi, V., Tilgner, A.: Global fluctuation in magnetohydrodynamics dynamos. New J. Phys. **13**, 033019 (2011)
45. Olson, P.: Core dynamics: an introduction and overview. In: Treatise on Geophysics, 2nd edn, vol. 8, pp. 1–25. Elsevier, Amsterdam (2015). https://doi.org/10.1016/B978-0-444-53802-4.00137-8
46. Barton, C.: Analysis of palaeomagnetic time series – technique and applications. Geophys. Surv. **5**, 335–368 (1983)
47. Currie, R.: Geomagnetic spectrum of internal origin and lower mantle conductivity. J. Geophys. Res. **73**, 2779–2790 (1968)
48. Bouligand, C., Hulot, G., Khokhlov, A., Glatzmaier, G.A.: Statistical palaeomagnetic field modelling and dynamo numerical simulation. Geophys. J. Int. **161**, 603–626 (2005)
49. Bouligand, C., Gillet, N., Jault, D., Schaeffer, N., Fournier, A., Aubert, J.: Frequency spectrum of the geomagnetic field harmonic coefficients from dynamo simulations. Geophys. J. Int. **207**, 1142–1157 (2016)
50. Constable, C., Johnson, C.: A paleomagnetic power spectrum. Phys. Earth Planet. Inter. **153**, 61–73 (2005)
51. Starchenko, S.V., Yakovleva, S.V.: MHD sources, 1600–2005 evolution and 1900–2005 probabilistic time analysis for logarithmic time-derivatives of geomagnetic spherical harmonics. In: Nurgaliev D., Shcherbakov V., Kosterov A., Spassov S. (eds) Recent Advances in Rock Magnetism, Environmental Magnetism and Paleomagnetism, pp. 513–522. Springer Geophysics. Springer, Cham. (2019)
52. Starchenko, S.V., Yakovleva, S.V.: Spatial-spectral evolution of the energy and power of the potential geomagnetic field since 1840. In: Bobrov, N.Yu., Zolotova, N.V., Kosterov, A.A., Yanovskaya, T.B. (eds.) Proceedings of the Conference "Problems of the Geocosmos", 08–12 Oct 2018, Peterhof, St. Petersburg, pp. 236–243. VVM Publishing House, St. Petersburg (2018) (in Russian)
53. Starchenko, S.V., Yakovleva, S.V.: Determination of specific time variations in the energy of the Earth's magnetic potential field from the IGRF model. Geomagn. Aeron. **59**(5), 606–611 (2019)
54. Starchenko, S.V., Yakovleva, S.V.: Energy and power spectra of the potential geomagnetic field since 1840. Geomagn. Aeron. **59**(2), 242–248 (2019)
55. Lowes, F.J.: Mean-square values on sphere of spherical harmonic vector fields. J. Geophys. Res. **71**, 2179 (1966)
56. Xu, W.: Unusual behavior of the IGRF during the 1945–1955 period. Earth Planets Space **52**, 1227–1233 (2000)
57. Bayanjargal, G.: The total energy of geomagnetic field. Geomech. Geophys. Geo-Energy Geo-Resour. **1**(1–2), 29–33 (2015). https://doi.org/10.1007/s40948-015-0006-y
58. Mauersberger, P.: Das Mittel der Energiedichte des geomagnetischen Hauptfeldes an der Erdoberflache und seine sakulare Anderung. Gerlands Beitr. Geophys. **65**, 207–215 (1956)
59. Lowes, F.J.: Spatial power spectrum of the main geomagnetic field, and extrapolation to the core. Geophys. J. R. Astr. Soc. **36**, 717–730 (1974)
60. Hulot, G., Le Mouël, J.L.: A statistical approach to the Earth's main magnetic field. Phys. Earth Planet. Inter. **82**, 167–183 (1994)

61. Hulot, G., Sabaka, T.J., Olsen, N. Fournier, A.: The present and future geomagnetic field. In: Treatise on Geophysics, 2nd edn, vol. 5, pp. 33–78 (2015). https://doi.org/10.1016/B978-0-444-53802-4.00096-8
62. Lhuillier, F., Fournier, A., Hulot, G., Aubert, J.: The geomagnetic secular variation timescale in observations and numerical dynamo models. Geophys. Res. Lett. **38**, L09306 (2011)
63. Gillet, N., Jault, D., Finlay, C.C., Olsen, N.: Stochastic modeling of the Earth's magnetic field: inversion for covariances over the observatory era. Geochem. Geophys. Geosyst. **14**(4), 766–786 (2013)
64. Gillet, N., Barrois, O., Finlay, C.C.: Stochastic forecasting of the geomagnetic field from the COV-OBS.x1 geomagnetic field model, and candidate models for IGRF-12. Earth Planet Space **67**, 71 (2015)
65. Bowley, A.L.: Elements of Statistics, 336 pp. P.S. King, London (1902)
66. Freund, R.J., Mohr, D.L., Wilson, W.J.: Statistical Methods, 3d edn, 800 pp. Academic Press, Salt Lake City USA (2010)
67. Braginsky, S.I., Roberts, P.H.: Equations governing convection in the Earth's core and the geodynamo. Geophys. Astrophys. Fluid Dyn. **79**, 1–97 (1995)
68. Bullard, E.C., Freeman, C., Gellman, H., Nixon, J.: The westward drift of the Earth's magnetic field. Philos. Trans. R. Soc. Lond. A **243**(6), 1–92 (1950)
69. Yukutake, T.: The westward drift of the Earth's magnetic field in historic times. J. Geomagn. Geoelectr. **19**, 103–116 (1967)
70. Holme, R., Whaler, K.: Steady core flow in an azimuthally drifting reference frame. Geophys. J. Int. **145**, 560–569 (2001)
71. Poirier, J.P.: Physical properties of the Earth's core. C. R. Acad. Sci. Paris, Sér. II **318**, 341–350 (1994)
72. Aubert, J.: Approaching Earth's core conditions in high-resolution geodynamo simulations. Geophys. J. Int. **219**, 137–151 (2019)
73. Starchenko, S.V.: Analytic scaling laws in planetary dynamo models. Geophys. Astrophys. Fluid Dyn. **113**(1–2), 71–79 (2019)
74. Christensen, U.: Geodynamo models with a stable layer and heterogeneous heat flow at the top of the core. Geophys. J. Int. **215**, 1338–1351 (2018)
75. Starchenko, S.V., Jones, C.A.: Typical velocity and magnetic field strengths in planetary interiors. Icarus **157**, 426–435 (2002)
76. Busse, F.H.: Generation of planetary magnetism by convection. Phys. Earth Planet. Inter. **12**, 350–358 (1976)
77. Moffatt, K., Dormy, E.: Self-Exiting Fluid Dynamos. Cambridge University Press, Cambridge (2019)
78. Kleeorin, N., Rogachevskii, I.: Magnetic helicity tensor for an anisotropic turbulence. Phys. Rev. E **59**, 6724–6729 (1999)
79. Ruzmaikin, A.A., Starchenko, S.V.: Kinematic turbulent mean-field geodynamo. Geomagn. Aeron. **28**, 475–480 (1988)
80. Driscoll, P.E., Du, Z.: Geodynamo conductivity limits. Geophys. Res. Lett. **46**, 7982–7989 (2019)
81. Shukurov, A.M., Sokolov, D.D., Ruzmaikin, A.A.: Oscillatory α^2–dynamo. Magnetohydrodynamics **21**(1), 6–10 (1985)
82. Starchenko, S.V.: Supercritical magnetoconvection in rapidly rotating planetary cores. Phys. Earth Planet. Inter. **117**(1–4), 225–235 (2000)
83. Starchenko, S.V.: Dynamo models with strong generation 1. Kinematic solution and axisymmetric $\alpha\omega$-dynamo. Geophys. Astrophys. Fluid Dyn. **77**, 55–77 (1994)
84. Bondar, T.N., Golovkov, V.P., Yakovleva, S.V.: Spatiotemporal model of the secular variations of the geomagnetic field in the time interval from 1500 through 2000. Geomagn. Aeron. **42**(6), 793–800 (2002)
85. Jackson, A., Jonkers, A.R.T., Walker, M.R.: Four centuries of geomagnetic secular variation from historical records. Philos. Trans. R. Soc. Lond. A **358**, 957–990 (2000)

86. Panovska, S., Constable, C.G., Korte, M.: Extending global continuous geomagnetic field reconstructions on timescales beyond human civilization. Geochem. Geophys. Geosyst. **19**(12), 4757–4772 (2018)
87. Morzfeld, M., Buffett, B.A.: A comprehensive model for the kyr and Myr timescales of Earth's axial magnetic dipole field. Nonlinear Process. Geophys. **26**(3), 123–142 (2019)

Heat Transfer Analysis of Tangentially Rotating Fluid Flow Past a Semi-infinite Vertical Cylinder Kept in Uniform Horizontal Magnetic Field Using Non-linear Regression and Back-Propagation Neural Network

Hari Ponnamma Rani, Koragoni Naresh, Yadagiri Rameshwar, and Sergey V. Starchenko

Abstract Understanding of the fluid flow over a vertical cylinder is important in many geophysical and technical applications. This paper addresses the numerical simulation of transient natural convective flow passing a uniformly heated semi-infinite vertical cylinder. The fluid is assumed to be viscous, incompressible and tangentially rotating. The cylinder is assumed to be kept in a uniform horizontal magnetic field, which is applied normal to the vertical axis of the cylinder. The closed-form solutions of governing non-dimensional transient coupled non-linear boundary layer equations are obtained numerically using the Crank-Nicolson implicit finite difference method. The momentum and heat transfer coefficients are expressed as average skin friction $\overline{\tau}$ and heat transfer rate $\overline{Nu}$, respectively, for distinct Ekman number (Ek) and magnetic parameter (M) values. The results obtained show that the flow and heat transfer characteristics are significantly influenced by these parameters. Based on the response surface methodology (RSM), the nonlinear regression equation of fitted line response data for $\overline{Nu}$ is obtained for different values of control parameters, such as Ek, M, Prandtl number (Pr) and Grashof number (Gr). The regression coefficients are considered until third order for achieving a more accurate solution and their significance is analyzed using variance (ANOVA) method. The maximum value of $\overline{Nu}$ is obtained by using the optimization technique. Artificial neural network (ANN) model is used to predict $\overline{Nu}$ value for a specified control parameters input. The multi-layer feed-forward network was trained by an algorithm of Back-propagation. It has been observed that the ANN model shows reasonably good agreement with numerical outcomes, so it is regarded as a trusted modeling

---

H. P. Rani (✉) · K. Naresh
Department of Mathematics, National Institute of Technology, Warangal, India
e-mail: hprani@nitw.ac.in

Y. Rameshwar
Department of Mathematics, University College of Engineering, Osmania University, Hyderabad, India

S. V. Starchenko
IZMIRAN, Troitsk, Russia

A. Kosterov et al. (eds.), *Problems of Geocosmos—2020*, Springer Proceedings in Earth and Environmental Sciences, https://doi.org/10.1007/978-3-030-91467-7_15

method. The Kappa coefficient, a quantitative measure of reliability for two raters that are rating the same thing, was calculated from the data obtained using the Crank-Nicolson method, non-linear regression equation and ANN model. The Kappa values obtained by all three methods are perfectly consistent.

Keywords Magnetohydrodynamic natural convection · Crank-Nicolson method · Regression analysis · Artificial neural network

1 Introduction

The boundary layer concept is a very useful and important aspect in understanding the transport processes occurring in geophysical external flows. For example, the effect of mountains on the Earth's atmosphere has long been recognized and studied [1–4]. Almost 50% of the Earth's land surface is covered by hilly and mountainous terrain [5] and the effect of mountains on weather and climate is often felt far away from the mountains. Passive effects involve momentum exchange between the surface and the atmosphere and occur when a flow is modified by the presence of mountains [6]. Examples include flow blocking, flow channeling, and lee waves. Also, examples include thermally driven wind systems generated by horizontal temperature and pressure differences [7]. These effects are most pronounced in cloudless conditions and weak synoptic pressure gradients. At any moment, passive and active effects play an important role in determining the dynamic and thermodynamic atmospheric structure. The atmospheric boundary layer is usually defined as the atmospheric layer that interacts directly with the Earth's surface on a time scale of a few hours or less (e.g., [8]). During daytime convective conditions, the atmospheric boundary layer is also called the convective boundary layer (CBL). Knowledge of CBL is useful or required in many applications such as air-quality studies, aerial spraying- and fire weather operations, regional climatology investigations, and weather forecasting [9]. It is also helpful in many industrial and technical processes including solar central receivers exposed to winds, electronic devices cooled by fans, nuclear reactors cooled during emergency shutdown, heat exchangers placed in a low-velocity environment, etc. [10]. For viscous liquid flows with heat transfer, the impact of linear dependence of density on dimensionless temperature, i.e., natural convection, appears to be highly pronounced in applications pertaining to geophysical flows and industrial manufacturing processes and cannot be ignored. Relation between temperature and density is nonlinear because of a large difference between the border and liquid temperatures, while nonlinear density temperature differences in the buoyancy force term would have a substantial effect on the circulation and energy transference features.

The laminar-convective boundary layer emerging on the outer surface of vertical circular cylinder is studied analytically and numerically by a number of researchers due to its importance in the field of science and technology [11–13]. The transient natural convection over heat generating vertical cylinders of different radii and thermal capacities has shown that as the radius of cylinder increases, the specific thermal capacity of the cylinder decreases and the surface heat flux increases [14]. Ganesan and Rani [15] investigated one such problem with respect to time along with mass transfer.

Studies of magnetohydrodynamic (MHD) flow have applications related to processes in solar wind, magnetosheath and magnetotail plasma sheet [16]. The reason is that the constructiveness of magnetic field fluid flow gives the best results in controlling the heat transfer. The impact of heat transfer and mass transfer on the magnetohydrodynamic flow passing a vertical cylinder is also paid attention in the literature [17, 18]. The increasing strength of magnetic field leads to a retardation in the velocity and rise in thermal boundary layer thickness, when the cylinder is subjected to thermal radiation [19]. For these problems the time required to reach the steady state is inversely proportional to the magnetic field intensity [20]. Nadeem and Akram [21] investigated the effect of inclined magnetic field on inclined symmetric or asymmetric channel with peristaltic flow of a Williamson fluid model.

In addition to MHD studies, numerous works deal with rotating boundary layer flows. Planetary boundary layer (PBL), also called atmospheric boundary layer, is the region of lower troposphere where the Earth's surface strongly influences temperature, moisture, and wind through the transfer of air mass. As a result of surface friction, winds in the PBL are usually weaker than above and tend to blow toward areas of low pressure. For this reason, PBL has also been called Ekman layer, after the Swedish oceanographer Vagn Walfrid Ekman, a pioneer in the study of the behaviour of wind-driven ocean currents. Greenspan and Howard [22] were the first to investigate the time-dependent motion of a rotating fluid. They found the presence of the Ekman layer and concluded that Ekman layer plays an important role in transient flow. In a series of papers, Chandrasekhar [23] emphasised the role of Coriolis force in problems of thermal instability and the stability of a viscous flow in the presence of an external magnetic field.

For the rotating stratified flow over a circular cylinder, it is observed that the wake asymmetries associated with the unstratified flows are destroyed by the stratification [24]. Experimental results on a similar problem with uniform free stream past a vertical oscillating circular cylinder showed that the rotation tended to induce rectified anticyclonic currents near the cylinder boundary [25]. For a flow passing the accelerated horizontal cylinder immersed in a rotating fluid, amplifying the rotation parameter results in the velocity profiles with low steep gradient in the axial direction and increasing gradient in the transverse direction [26]. Transient motions in a rotating cylinder containing stably stratified rotating fluid were studied by Holton [27] using Laplace transform method. The results show that the effect of viscosity is to force a meridional circulation in the interior of fluid as a result of convergence in Ekman boundary layers.

In the present study, a tangentially rotating fluid flow over an isothermal semi-infinite vertical cylinder in the presence of uniform horizontal magnetic field is considered. The physical problem under consideration is elaborated in Sect. 2. The dimensionless governing equations are listed in this section. In Sect. 3, the numerical method to solve the momentum and heat equations using the Crank-Nicolson implicit finite difference method is given. The outcome of our simulations has been confirmed with previous numerical results [13]. The velocity and temperature profiles are explained graphically in Sect. 4 for the control parameters arising in the system. Using the numerical data, the cubic regression equation for $\overline{Nu}$ is derived based on the response surface methodology. Also, the optimum $\overline{Nu}$ is obtained using the optimization technique for the considered range of control parameters. The inter-rater reliability of the results, i.e., the agreement between the results obtained by employing the finite difference method, the non-linear regression equation and Artificial Neural Network is calculated using Kappa coefficient. The concluding remarks are listed in Sect. 5.

2 Problem Description and Governing Equations

We consider the unsteady laminar fluid flow passing the semi-infinite vertical cylinder of radius r_0. The fluid is assumed to be homogeneous incompressible viscous and electrically conducting. The cylindrical coordinate system is assumed. The x-axis is directed along the axis of the cylinder vertically upward and the radial coordinate r is normal to the axis of cylinder. The cylinder is assumed to be kept in uniform magnetic field B_0, applied normal to the axis of the cylinder. The uniform magnetic field motivates the rotation. The fluid rotates about the tangential direction of the cylinder in a state of rigid body rotation with constant angular velocity, ω_0. In the entire flow field, uniform pressure is presumed. The axi-symmetric flow in the cylindrical coordinates can be derived in a similar manner as for two-dimensional flow in polar coordinates (x, r).

Initially, before starting experiment (time, $t' \leq 0$), both the cylinder and the surrounding fluid are presumed to be at the same ambient temperature, $T_\infty{}'$. Later ($t' > 0$), when the cylinder is maintained at the uniform temperature $T_w{}'(>T_\infty{}')$, there is a temperature difference between adjacent fluid layers present in the vicinity of the cylindrical wall. This process induces the density variation among the fluid layers. The heavier fluid moves down due to gravitational force and hence free convection occurs in a very thin layer, known as the boundary layer. Figure 1 shows the schematic representation of the above. The subplot in Fig. 1 shows the rotation angle, which is perpendicular to the axis of the cylinder. It can be noted that uniform magnetic field is applied in horizontal way.

It is assumed that magnetic Reynolds number is very small and therefore the induced magnetic field is negligible compared to the applied magnetic field. It is assumed that the impact of the Coriolis force is perpendicular to the axis of the cylinder. Also, the frictional boundary layer in the rotating fluid generates the Ekman layer. The Ekman layer thickness is given by $\delta = \sqrt{\nu/\omega}$ and it depends on the kinematic viscosity ν and the rotation speed ω. For very small viscosity, or rapid rotation, the thickness becomes progressively smaller.

The equations of motion for the above considered physical model are given by [17, 19, 20]:

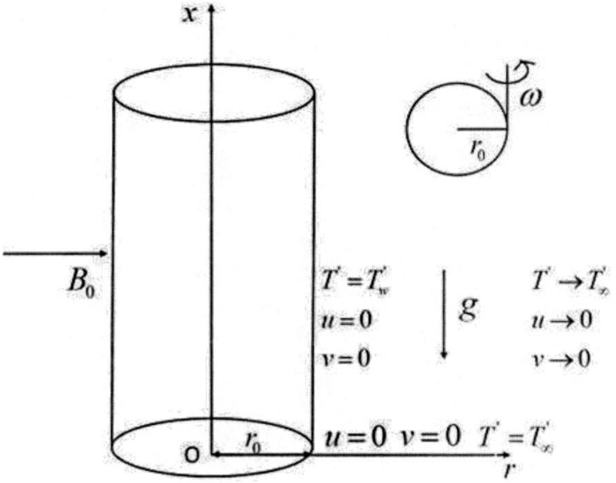

Fig. 1 Schematic of the investigated problem

Continuity equation

$$\frac{\partial(ur)}{\partial x} + \frac{\partial(vr)}{\partial r} = 0 \tag{1}$$

Momentum equation in axial direction

$$\frac{\partial u}{\partial t'} + u\frac{\partial u}{\partial x} + v\frac{\partial u}{\partial r} - 2\omega_0 v = g\beta\left(T' - T_\infty\right) + \frac{\nu}{r}\frac{\partial}{\partial r}\left(r\frac{\partial u}{\partial r}\right) - \frac{\sigma B_0^2}{\rho}u \tag{2}$$

Momentum equation in transverse direction

$$\frac{\partial v}{\partial t'} + u\frac{\partial v}{\partial x} + v\frac{\partial v}{\partial r} + 2\omega_0 u = \frac{\nu}{r}\frac{\partial}{\partial r}\left(r\frac{\partial v}{\partial r}\right) \tag{3}$$

Heat equation

$$\frac{\partial T'}{\partial t'} + u\frac{\partial T'}{\partial x} + v\frac{\partial T'}{\partial r} = \frac{\alpha}{\rho c_p}\frac{1}{r}\frac{\partial}{\partial r}\left(r\frac{\partial T'}{\partial r}\right) \tag{4}$$

The initial and boundary conditions are:

$$\left.\begin{array}{r} t' \le 0 : u = 0,\, v = 0,\, T' = T_\infty{}',\quad \forall x \; and \; r \\ t' > 0 : u = 0,\, v = 0,\, T' = T_w{}',\quad at\; r = r_0 \\ u \to 0,\, v \to 0,\, T' \to 0 \quad as\; r \to \infty \end{array}\right\} \tag{5}$$

Time (t) is made dimensionless by ω_0, the axial (X) and radial (R) coordinates by r_0, the corresponding velocity coordinates (U, V) by $\frac{r_0}{\nu}$ and temperature (T) by $\frac{T'-T_\infty{}'}{T_W{}'-T_\infty{}'}$ i.e.,

$$t = \omega_0 t', \ X = \frac{x}{r_0}, \ R = \frac{r}{r_0}, \ U = \frac{u r_0}{\nu}, \ V = \frac{v r_0}{\nu}, \ T = \frac{T' - T_\infty{}'}{T_W{}' - T_\infty{}'} \tag{6}$$

Equations (1), (2), (3) and (4) in dimensionless form are expressed as follows:

$$\frac{\partial (RU)}{\partial X} + \frac{\partial (RV)}{\partial R} = 0 \tag{7}$$

$$\frac{\partial U}{\partial t} + U\frac{\partial U}{\partial X} + V\frac{\partial U}{\partial R} - \frac{2}{Ek}V = GrT + \frac{1}{R}\frac{\partial U}{\partial R} + \frac{\partial^2 U}{\partial R^2} - MU \tag{8}$$

$$\frac{\partial V}{\partial t} + U\frac{\partial V}{\partial X} + V\frac{\partial V}{\partial R} + \frac{2}{Ek}U = \frac{1}{R}\frac{\partial V}{\partial R} + \frac{\partial^2 V}{\partial R^2} \tag{9}$$

$$\frac{\partial T}{\partial t} + U\frac{\partial T}{\partial X} + V\frac{\partial T}{\partial R} = \frac{1}{Pr}\left[\frac{1}{R}\frac{\partial T}{\partial R} + \frac{\partial^2 T}{\partial R^2}\right] \tag{10}$$

where magnetic parameter $M = \frac{\sigma B_0^2 r_0^2}{\rho \nu}$, Ekman number $Ek = \frac{\nu}{\omega_0 r_0^2}$, $Gr = \frac{g\beta r_0^3 (T_w{}' - T_\infty{}')}{\nu^2}$ and $Pr = \frac{\nu}{\alpha}$ are introduced. In these control parameters, ρ denotes density, σ is electrical conductivity, υ is kinematic viscosity, β is coefficient of volume expansion, g is acceleration due to gravity and α is thermal diffusivity. The axial and radial velocities are U and V along X and R directions.

The dimensionless form of initial and boundary conditions is given by:

$$\left.\begin{array}{r} t \le 0 : U = 0, V = 0, T = 0, \quad \forall X \ and \ R \\ t > 0 : U = 0, V = 0, T = 1, \quad at \ R = 1 \\ U = 0, V = 0, T = 1, \quad at \ X = 0 \\ U \to 0, V \to 0, T \to 0 \ \ as \ R \to \infty \end{array}\right\} \tag{11}$$

3 Numerical Method

In order to solve the unsteady non-linear coupled Eqs. (7)–(10) under the given conditions (11), an implicit finite difference scheme of Crank-Nicolson type has been employed as described in Richner and Hächler [6]. The solution domain is a rectangle limited by lines with Xmin = 0, Xmax = 1, Rmin = 1 and Rmax = 20, where Rmax assumed to lie outside the momentum and thermal boundary layers, so that the boundary conditions are satisfied. Using known initial and boundary values of U, V and T the next time step values are computed. Using these new set of values, the computations were repeated for several time steps until the absolute difference between values of velocity U as well as of temperature T at two consecutive time steps was less than 10^{-5} at all grid points.

Table 1 L2-relative errors for different grid size

Grid size	Velocity	Temperature
50 × 250 and 100 × 500	0.001329	0.000485
100 × 500 and 200 × 1000	0.000472	0.000441

The steady-state velocity U and temperature T values obtained with 100 × 500 grid system differ from those with 50 × 250 grid system in the second decimal place, and from those with 200 × 1000 grid system in the third decimal place. Thus, for all subsequent analyzes, the grid system of 100 × 500 was chosen, with mesh size in X and R directions being taken as 0.01 and 0.03, respectively. Further, from Table 1 it can be seen the L2-relative error of grid system of 100 × 500 and 200 × 1000 is better than that of 50 × 250 and 100 × 500. Hence the mesh size of 100 × 500 was considered appropriate for further calculations.

4 Results

The present study focuses on the flow variations in the boundary layer region that can be characterized as a specific region of the fluid in which steep inclinations of velocity as well as temperature can occur. The simulated results are presented to outline the general physics involved in the effects of M (= 1, 4, 25, 100) which is equivalent to the Hartmann number Ha (= 1, 2, 5, 10) and Ek (= 0.25, 0.5, 0.8, 1, 1.2, 3, 8, 10) on the transient velocity and temperature profiles with fixed Pr (= 0.71(air)) and Gr (= 20). Also, the profiles of dimensionless mean skin-friction and Nusselt number or rate of heat transfer are discussed in a graphical manner with respect to time and radial coordinates for different values of Ek and M.

The present results were corroborated with available numerical results in the literature, since there are no experimental or analytical studies on this problem. Hence the numerical results of Lee et al. [13] are used to corroborate the present simulations. Lee et al. [13] have used the cubic spline interpolation technique to solve the steady flow over vertical cylinder with variable surface temperature. Equations (7), (8) and (10) in the steady-state have been considered, without magnetic field and rotation. Resulting dimensionless equations were solved numerically for $Pr = 0.7$. In Table 2 the results obtained by Lee et al. [13] are listed along with ours and the agreement between the two is reasonably good.

4.1 Field Variables

The magnitude analysis shows that the Coriolis force is critical in comparison with inertial and viscous forces engaged with the fundamental field equations. The effect of fluid rotation in the tangential direction of the vertical cylinder is discussed using

Table 2 Comparison of the values of velocity and temperature

	Lee et al. [13]	Present results	Lee et al. [13]	Present results
R	*U*	*U*	*T*	*T*
1	0	0	1	1
1.5	0.3801372	0.33134	0.772893	0.71452
2	0.4659859	0.46921	0.6062271	0.58512
3	0.449322	0.43944	0.371794	0.34783
4	0.28275	0.2158	0.20146	0.24541
5	0.156428	0.17856	0.08424	0.08415
6	0.063973	0.05998	0.038461	0.03458
7	0.020694	0.02147	0.014652	0.01324
8	0.0040677	0.00854	0.00732	0.00851
9	0.00386354	0.00495	0.00183	0.00127
10	0.0000989	0.0000523	0.00004585	0.000088

graphs shown in Figs. 2 and 3 for fast and slow rotations and for a fixed value of $M = 4$. These figures plot curves corresponding to U and V against the radial coordinate R and time t. When Ek is smaller than one (Fig. 2) viscous forces are overwhelmed by the Coriolis force. The axial velocity U starts from zero at the leading edge of the cylinder, reaches its maximum value and then asymptotically approaches zero for large R. The transverse velocity V also starts from zero at the leading edge of the cylinder, reaches its negative value and then attains zero along the radial coordinate R, because the Coriolis and viscous forces are restricted in the transverse direction.

From Fig. 2b it can be seen that velocity varies/oscillates rapidly enough close to the wall since viscous effects play a vital role in the vicinity of boundary. Boundary

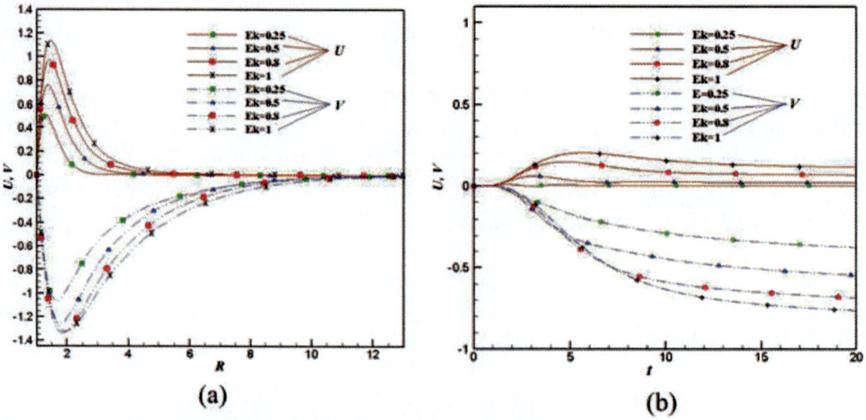

Fig. 2 Velocity profiles for $Ek < 1$ with $M = 4$ **a** against R at $t = 20$ and $X = 1$ and **b** against t at $(X, R) = (1.5, 1)$

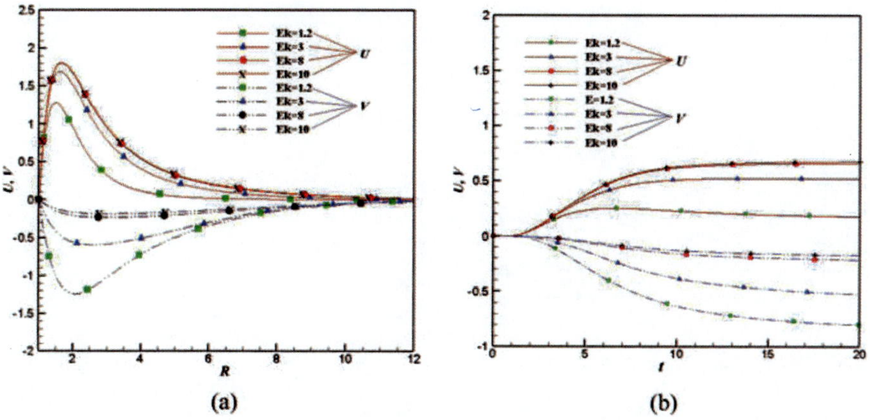

Fig. 3 Velocity profiles for large values of Ekman number with $M = 4$ a against R at $t = 20$ and $X = 1$ and b against t and $(X, R) = (1.5, 1)$

layers therefore transfer viscous effects only in the downstream direction. The velocity U attains the temporal maximum, then slightly decreases for increasing rotation rates and becomes asymptotically steady. For the plotted values of Ek, the transverse velocity increases and remains asymptotically constant as the impact of the viscous force is very small. The momentum layer thickness increases with Ek because rotation is reduced.

For $Ek > 1$, i.e. fluid rotating with low angular velocity, Fig. 3a, b plot dependences of U and V on R and t respectively. The axial momentum boundary layer thickness increases with Ek while the opposite effect is observed for transverse velocity profiles V. For the fast-rotating fluids, the initially arising small oscillations in the velocity profiles get vanished and they attained the temporal maximum quickly.

Figure 4a, b show the U and V profiles against the radial coordinate R and time t, respectively for different magnetic field strength values. From Fig. 4a it is observed that when the magnetic field strength is greater than the viscous forces and is increasing, the momentum boundary layer thickness decreases with respect to R. Also, in the considered range of M values, V starts from zero at the leading edge of the cylinder, reaches its negative value and then attains zero along R, since the impact of kinematic viscosity decreases. This type of flow is expected because the Lorentz force tends to inhibit the relative movement of fluid flow. Figure 4b shows that U attains maximum value after some time t and becomes asymptotically steady for increasing M, i.e., when the viscous forces are smaller than the magnetic forces. When the magnetic field strength is approximately comparable to viscous forces, i.e., $M = 1$ the axial velocity profile reaches its maximum and becomes oscillatory as time t increases. The transverse velocity profile V shows a similar oscillatory behavior for $M = 1$.

For different values of Ek and M, the temperature profiles/thermal boundary layer against R and t are shown in Fig. 5a–d, respectively. Figure 5a, b show that the

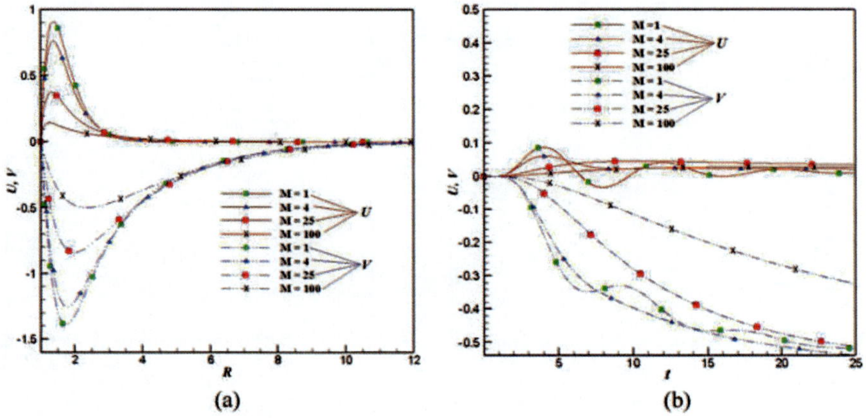

Fig. 4 Velocity profiles for different magnetic parameter M values with $Ek = 0.5$ **a** against R at $t = 20$ and $X = 1$ and **b** against t and $(X, R) = (1.5, 1)$

temperature profiles begin with the boundary condition of the hot wall temperature i.e., $T = 1$, and then monotonically decrease with increasing R, finally reaching the free stream temperature, $T = 0$. The main reason is that with increase in velocity cooling capacity of the fluid increases, the convective heat transfer effect between fluid and cylindrical wall becomes more intense, and the temperature of fluid reduces faster. It is also observed that as magnetic field strength (M) and Ek increase, the time required for T to reach the steady-state becomes longer. Figure 5a reveals that increasing magnetic field strength (M) yields highly intense time-independent temperature profiles. Also, the steady-state temperature profile grows with decreasing Ek. The most significant observation in Fig. 5a, b is that for the case of large values of Ek and M, i.e., the slow rotation and stronger magnetic field, the steady-state temperature profile appears to be more profound in the vicinity of the hot wall, especially, at the leading edge of the cylinder.

The unsteady temperature profiles as a function of t for selected M and Ek values are shown in Fig. 5c, d respectively. After reaching the peak value, the temperature attains a time-independent state. It can be noted that, as M and Ek amplifies, it takes longer for the temperature to reach its maximum. For smaller values of viscosity in comparison with the magnetic field strength and rotation rates, more pronounced thermal boundary layer profiles are observed.

4.2 Momentum and Heat Transfer Coefficients

The friction drag coefficient is an important parameter in heat transfer studies, since it is directly related to heat transfer coefficient. Heat transfer coefficient is a quantitative characteristic of convective heat transfer between the hot cylindrical wall and the

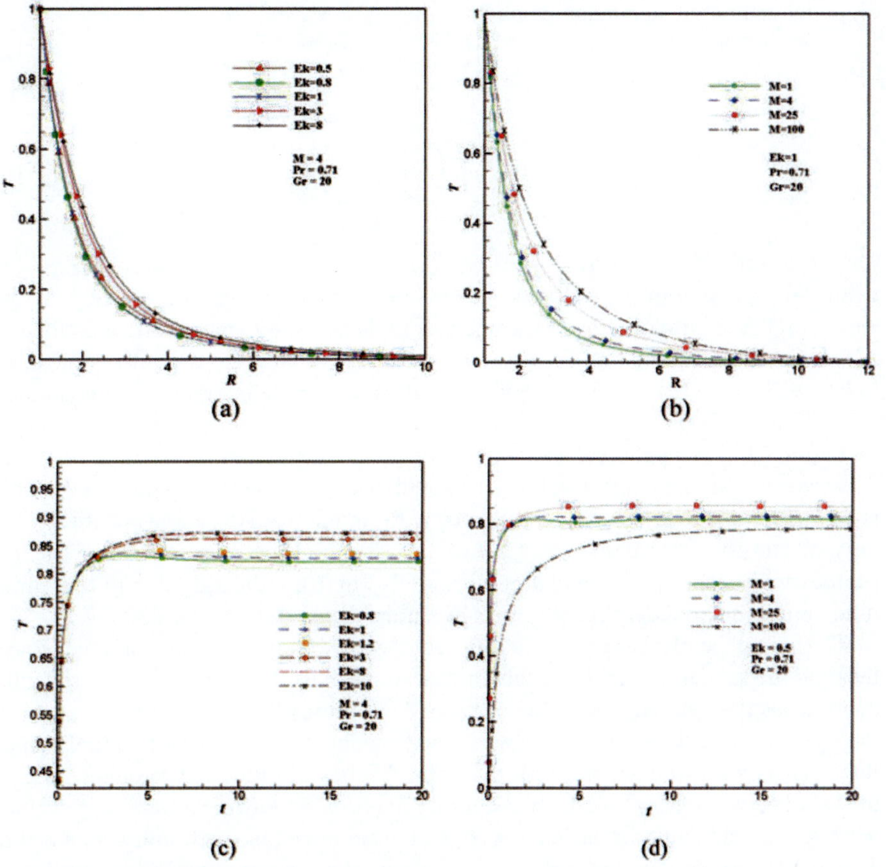

Fig. 5 Temperature profiles for different *Ek* and *M* **a**, **b** against *R* at $t = 20$ and $X = 1$ and **c**, **d** against *t* and $(X, R) = (1.5, 1)$

fluid. The skin friction drag characterizes the shear at the boundary. The increased skin friction is generally a disadvantage in technical applications. While the increased heat transfer can be exploited in some applications such as heat exchangers, it should be avoided in other cases such as gas turbine applications, for instance. Also, the calculation of heat transfer coefficient plays a key role in fractured media seepage theories and in heat transfer in fractured surrounding rock [28].

The instantaneous distribution of wall-shear stress (and hence its magnitude), $\overline{\tau}$, reflects the structure of the boundary layer close to the wall. The heat transfer coefficient is often calculated from the Nusselt number (rate of heat transfer). The non-dimensional average or mean skin-friction ($\overline{\tau}$) and heat transfer rate $\left(\overline{Nu}\right)$ are described as:

$$\overline{\tau} = \int_0^1 \left(\frac{\partial U}{\partial R} \right)_{R=1} dX \tag{12}$$

$$\overline{Nu} = -\int_0^1 \left(\frac{\partial T}{\partial R} \right)_{R=1} dX \tag{13}$$

The estimate for skin friction in Eq. (12) and rate of heat transfer in Eq. (13) were developed from the velocity and temperature gradients, respectively, at the wall with the assumption that the velocity and thermal boundary layers are laminar. A five-point approximation formula was used to evaluate the derivatives engaged in Eqs. (12) and (13) and then the numerical values of the integrals were computed using Newton-Cotes closed integration formula.

Figure 6a, b show the values of $\overline{\tau}$ for different Ek and M as a function of t. The $\overline{\tau}$ appears to rise with higher values of Ek and decline with higher values of M. This is due to that the flow velocity accelerates with rotation and decelerates with M (see Figs. 3b and 4b). For fixed values of Ek and M, an initial sharp rise in the skin friction profiles with respect to time can be observed (Fig. 6), indicating that in the initial time period the boundary layer flow is in a transitional state.

The time dependences of $\overline{Nu}$ for different Ek and M values are shown in Fig. 7. As the fluid in the heated boundary layer moves along an isothermal surface, it gradually approaches the temperature of the surface. This causes the temperature gradient in the fluid at the surface (and the rate of heat transfer) to decrease in the direction of flow. $\overline{Nu}$ is seen to decrease with time and reaches the time-independent state. It is also observed that at small t Nusselt numbers $\overline{Nu}$ for different parameter values overlap with each other at the leading edge of the cylinder and then split after a while. This is because conduction occurs at the early stage and dominates the transfer of

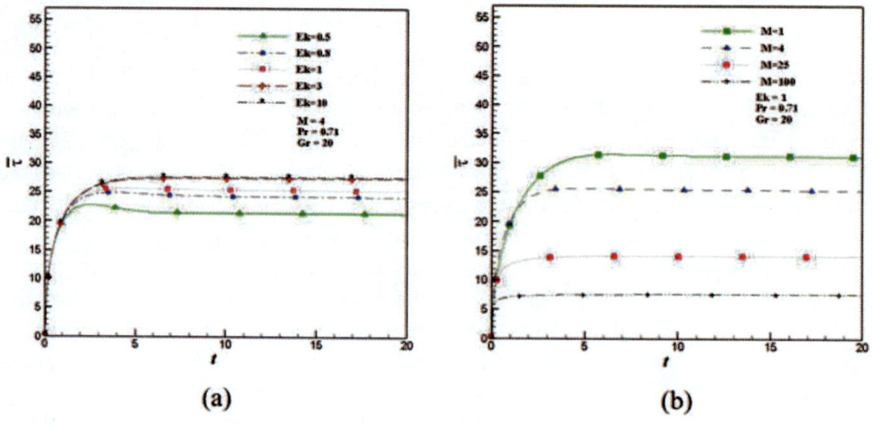

Fig. 6 Skin-friction coefficient ($\overline{\tau}$) for different values of **a** Ek and **b** M

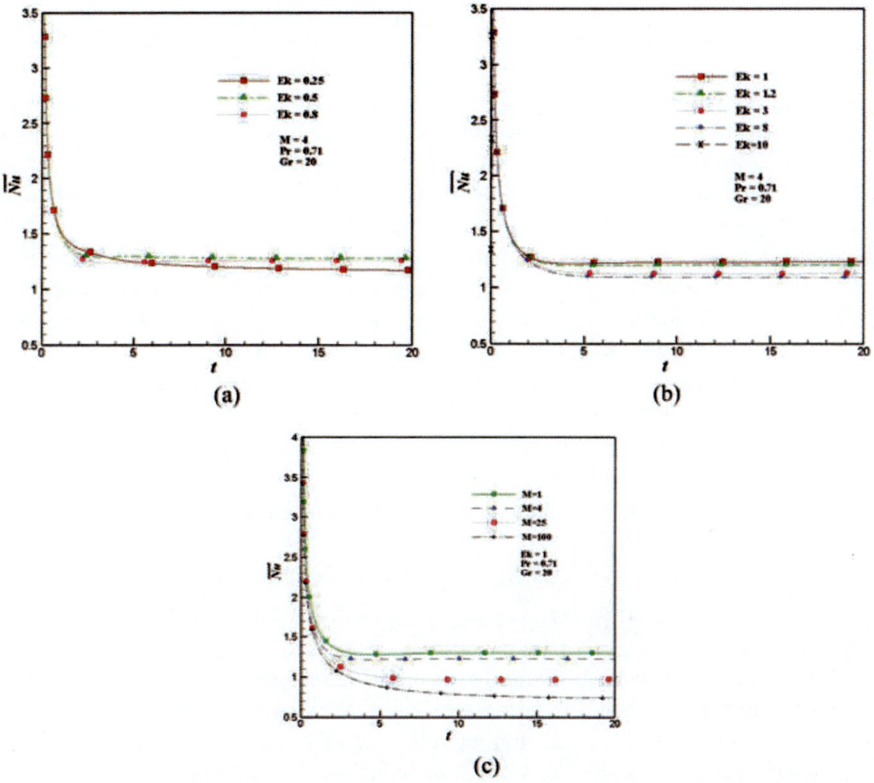

Fig. 7 Time dependences of $\overline{Nu}$ for different **a**, **b** *Ek* and **c** *M* values

convective heat. It is also recognized that there is a reduction in $\overline{Nu}$ for increasing *M* and *Ek* values. The reason is that thermal boundary layer increases with increase of the magnetic field strength or the rotation rate.

4.3 Response Surface Method (RSM)

Using the nonlinear regression based on RSM, the equation of fitted line response data for $\overline{Nu}$ was obtained for different values of control parameters, such as *Ek*, *Pr*, *M* and *Gr* arising in the present model. RSM was used for the existing data sequence of $\overline{Nu}$ for finding the optimal response.

The $\overline{Nu}$ data was generated using finite difference method for different combinations of parameters and the simulated values are shown in Appendix. This data was used to find the optimal $\overline{Nu}$ in the considered range of control parameters viz. $1 \leq M \leq 100$; $0.5 \leq Ek \leq 3$; $0.71 \leq Pr \leq 55$ and $10 \leq Gr \leq 30$.

The considered equation of fitted line response until third order is given below

$$y = \beta_c + \sum_{i=1}^{n} \beta_i x_i + \sum_{i=1}^{n} \beta_{ii} x_i x_i + + \sum_{i,j=1}^{n} \beta_{ij} x_i x_j + + \sum_{i,j,k=1}^{n} \beta_{ijk} x_i x_j x_k + \in \quad (14)$$

where $\in$ is the error. This equation includes β_c as the intercept, β_i as the i-th factor linear regression coefficient, β_{ii} as the i-th factor quadratic regression coefficients, β_{ij} as the i-th and j-th factor coefficient, and β_{ijk} as the i-th, j-th and k-th factor cubic regression coefficients. Also, x denotes the parameters M, Ek, Pr and Gr considered in the present study. It can be noted that Tahari et al. [29] used the quadratic regression coefficients, while cubic regression coefficients were used in the present study. For error assessment in Eq. (14), it is essential to consider certain requirements and analysis, such as, standardization of errors, uniforming the errors of variance, and establishing the correlation between the data. The excess and predicted values for all tested instances are shown in Appendix through regression Eq. (14). Analysis of Variance (ANOVA) was employed to decide the effect of every control parameter on the response function. For each of the non-linear regression coefficients β appearing in Eq. (14) the following hypothetical test was performed in ANOVA:

$$\begin{aligned} &\text{Null Hypothesis } \mathrm{H}_0 : \beta_1 = \beta_2 = \cdots = \beta_k = 0 \\ &\text{Alternative Hypothesis } \mathrm{H}_1 : \beta_j \ 0 \text{ for at least one } j \end{aligned} \quad (15)$$

The p-value signifies the probability of acquiring results by utilizing Eq. (15) with the assumption that the null hypothesis is right. Smaller p-values imply that there is more grounded proof for the alternative hypothesis. The p-values obtained from ANOVA results are tabulated in Table 3 showing the impact of every parameter on $\overline{Nu}$ and its value on the responses in Eq. (15).

In the present study we have considered the calculations with p-value ≤ 0.05 to be significant. Thus, the equation of fitted line response data for the present model is obtained as:

$$\begin{aligned} \overline{Nu}_{RSM} = {} & 0.81957 - 0.05147M + 0.17425Pr + 0.02667Gr + 0.00183M^2 \\ & - 0.00118MPr - 0.02384EkPr + 0.00278PrGr - 0.00001M^3 \\ & + 0.00001MPr^2 + 0.00307Ek^2Pr + 0.00018EkPr^2 \\ & - 0.00004Pr^2Gr - 0.00001Pr^2Gr \end{aligned} \quad (16)$$

Mathematica® software was used to determine the best instances among all computationally simulated and fitted response studies of $\overline{Nu}$. It was found that the maximum of $\overline{Nu}$ (= 10.83) occurs when $M = 1$, $Ek = 0.5$, $Pr = 55$ and $Gr = 30$.

Table 3 Using ANOVA analysis—the significance of coefficients (for clarity the significant parameters are marked in bold face)

Term	Coefficient	*p*-value	Significance
Constant	**0.81957**	0.00000	**Yes**
$\boldsymbol{M}$	**−0.05147**	0.00000	**Yes**
Ek	0.08386	0.18950	No
$\boldsymbol{Pr}$	**0.17425**	0.00000	**Yes**
$\boldsymbol{Gr}$	**0.02667**	0.04780	**Yes**
$\boldsymbol{M^2}$	**0.00183**	0.00003	**Yes**
Ek^2	−0.00823	0.90037	No
Pr^2	−0.00083	0.57593	No
Gr^2	0.00029	0.79511	No
$M\,Ek$	−0.00087	0.64381	No
$\boldsymbol{M\,Pr}$	**−0.00118**	0.00000	**Yes**
$M\,Gr$	−0.00014	0.50951	No
$\boldsymbol{EkPr}$	**−0.02384**	0.00000	**Yes**
$Ek\,Gr$	−0.01263	0.19258	No
$\boldsymbol{Pr\,Gr}$	**0.00278**	0.00000	**Yes**
$\boldsymbol{M^3}$	**−0.00001**	0.00009	**Yes**
M^2Ek	0.00001	0.58816	No
$\boldsymbol{M^2Pr}$	**0.00000**	0.00001	**Yes**
$M^2\,Gr$	0.00000	0.76195	No
$M\,Ek^2$	0.00000	0.99165	No
$M\,EkPr$	0.00000	0.57234	No
$M\,Ek\,Gr$	0.00002	0.38249	No
$\boldsymbol{M\,Pr^2}$	**0.00001**	0.00000	**Yes**
$M\,Pr\,Gr$	0.00000	0.35534	No
$M\,Gr^2$	0.00000	0.97726	No
Ek^3	−0.00416	0.86702	No
$\boldsymbol{Ek^2Pr}$	**0.00307**	0.00003	**Yes**
$Ek^2\,Gr$	0.00285	0.16903	No
$\boldsymbol{Ek\,Pr^2}$	**0.00018**	0.00100	**Yes**
$EkPrGr$	−0.00002	0.65524	No
$EkGr^2$	−0.00002	0.90797	No
Pr^3	−0.00003	0.33383	No
$\boldsymbol{Pr^2\,Gr}$	**−0.00004**	0.00000	**Yes**
$PrGr^2$	0.00000	0.88664	No
Gr^3	−0.00001	0.62400	No

4.4 Application of Artificial Neural Networks (ANN)

An ANN is defined as a network which includes a massive interconnected processors or neurons, parallelly operating and learning from experience. Before using the ANN model to provide desired output data, the model needs to be trained to identify the relationship between the inputs (*M*, *Ek*, *Pr* and *Gr*) and the desired output data i.e., in the present case, $\overline{Nu}$. The training or the learning process determines the weights. A set of input–output patterns are first prepared for conducting the training process. This multilayered ANN structure is shown in Fig. 8.

Back-propagation neural networks (BPNN) methodology was employed. In this methodology, of the numerically simulated data values, 70%, 15%, and 15% of the data was considered at the information level, to validate the model, and to test the advanced model, respectively. A network training feature, Trainlm-back propagation algorithm [30, 31] was used that updates the weight and bias values to Levenberg–Marquardt optimization. This algorithm is often the fastest and most preferred as a first-choice supervised algorithm, although it requires more memory compared to other algorithms. A total of 108 samples of information is provided to the algorithm, thus generating the 108 arrays.

Figure 9 shows the regression analysis between the neural network output and the respective training and testing target data of $\overline{Nu}$. The scatter plots in the figure reveal adequate output parameters, and data points are tightly clustered around the chosen optimal unit-slope line. The expected values show an outstanding agreement with the target values. The coefficient of correlation (R-value) between the targets and outputs describes the relation of targets that explain the variability in output. If the number is equal to 1, then there is a perfect correlation between the targets and outputs. The dotted line in Fig. 9 shows the best straight line fit while the solid line shows the perfect fit (output equals target). The simulated and predicted values are very close to each other, which means that for $\overline{Nu}$, the training and testing correlation factors are R = 0.99875 and R = 0.99924, respectively. From Fig. 9, it can be noted that the recommended ANN system has a strong correlation factor. It also indicates

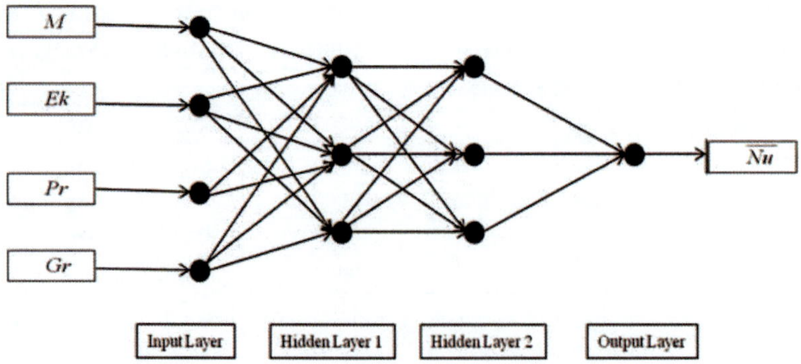

Fig. 8 Structure of multilayered neural network

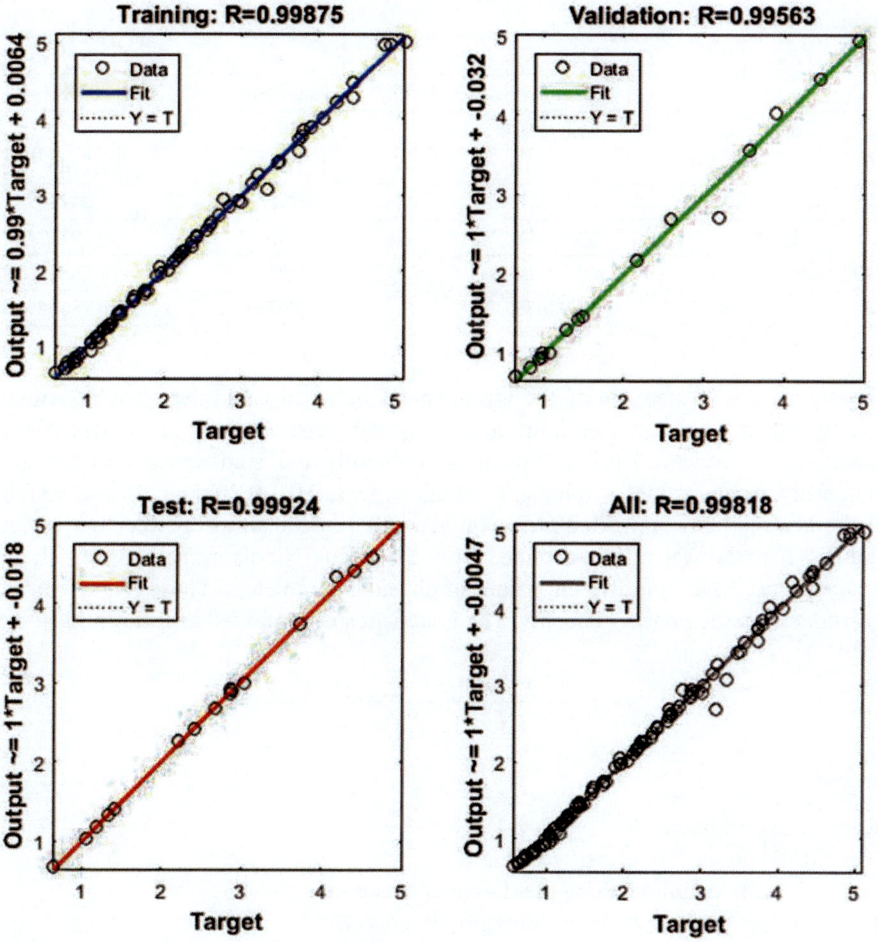

Fig. 9 Correlation factor (*R*) of $\overline{Nu}$ using ANN

that the practicing and testing correlation factors of $\overline{Nu}$ are R = 0.99563 and R = 0.99818, respectively.

4.5 *Kappa Coefficient* ($\hat{\kappa}$)

An increasing number of researches have used the agreement of Kappa coefficient as a measure of classification precision. The Kappa statistic was originally developed by Cohen [32]. The kappa statistic ($\hat{\kappa}$) is frequently used to test interrater reliability. The importance of rater (methods) reliability lies in the fact that it represents the extent to which the data collected in the study are correct representations of the variables

Table 4 Values from the data of $\overline{Nu}$

M	Ek	Pr	Gr	$\overline{Nu}$		
				Crank-Nicolson method	Regression analysis	ANN
1	0.5	0.71	10	1.09619	1.16939	1.1800
4	1	0.71	20	1.23121	1.20341	1.2286
25	1	7	20	1.94698	2.0495	2.04797
25	3	0.71	30	0.95815	0.9145	0.89787
100	3	55	30	2.77316	2.9381	2.76871
$\hat{\kappa}$				0.986732	0.986757	0.986789

measured. $\hat{\kappa}$ is a measure of the arrangement between two raters that determine which category belongs to a finite number of subjects, whereby agreement due to chance is factored out. $\hat{\kappa}$ is one of the most frequently used statistics to test inter-rater reliability. Kappa values ≤ 0 indicate no agreement, 0.01–0.20 poor, 0.21–0.40 fair, 0.41–0.60 moderate, 0.61–0.80 substantial and 0.81–1.00 almost perfect agreement. Table 4 lists the $\overline{Nu}$ values obtained from the Crank-Nicolson method, regression analysis, and ANN for different values of physical parameters. These $\overline{Nu}$ values are considered as the matrix elements. The $\hat{\kappa}$ statistics is computed as follows [32]:

$$\hat{\kappa} = \frac{N \sum_{i=1}^{n} m_{i,i} - \sum_{i=1}^{n}(G_i C_i)}{N^2 - \sum_{i=1}^{n}(G_i C_i)}, \tag{7}$$

where

- i class number;
- N total number of classified values;
- $m_{i,i}$ $\overline{Nu}$ values found along the diagonal of the matrix,
- C_i sum of each row values belonging to class i;
- G_i is the sum of each column values belonging to class i.

From Table 4 it can be seen that $\hat{\kappa}$ is approximately 0.986 for all three methods, namely, Crank-Nicolson method, non-linear regression and ANN. This shows that the inter-rater reliability is 98%. Hence there is almost perfect agreement between the data of $\overline{Nu}$ calculated from the numerical method, non-linear regression and ANN model.

5 Discussion and Conclusions

The natural convective boundary layer flow past the surface of vertical cylinder plays a central role in many applications, and has been extensively studied both analytically [20] and numerically [15]. However, few studies considered the time dependency of this non-linear problem. Thus, in this paper, we have numerically solved the unsteady non-linear coupled partial differential equations governing the tangentially rotating fluid flow passing a vertical cylinder kept in uniform constant horizontal magnetic field. Firstly, the mathematical model has, in this and other studies, been carefully verified against non-rotating and non-magnetic flows. Based on the numerical results and scaling analysis, the basic flow structure, wall shear stress, and heat transfer rate have been discussed. Due to the presence of Coriolis force, in addition to thermal boundary layers velocity boundary layers, called Ekman layers, exist in the heat transfer regime. Some important observations appear below:

1. The faster rotation and stronger magnetic field reduce the thickness of momentum boundary layer.
2. The mean heat transfer rate decreases, i.e., the heat transfer by convection is deteriorated with increasing rotation rate or with stronger magnetic field.
3. Response surface method is used to find the maximum average heat transfer rate, $\overline{Nu}$ with the cubic regression coefficients. Using this method, the maximum $\overline{Nu}$ obtained was 10.8275 for $M = 1$, $Ek = 0.5$, $Pr = 55$ and $Gr = 30$.
4. Artificial Neural Network (ANN) was applied for the prediction of data of $\overline{Nu}$ using 108 simulated training samples. It can be concluded that ANN modeling allows a considerable time-saving when predicting $\overline{Nu}$.
5. Kappa coefficient calculated from the numerical, non-linear regression equation and ANN model is in perfect agreement with data obtained from the methods and models described in the paper.

With the help of numerically simulated results the general regression equation was obtained to find the optimum rate of heat transfer. To save time, ANN methodology was adopted to predict this optimum value. Thus, the methodology adopted in the present study is a valid method for different natural convective problems in predicting the rate of heat transfer in less time.

Appendix

$\overline{Nu}$ data for 108 different combinations of control parameters

S. No.	M	Ek	Pr	Gr	$\overline{Nu}$	Estimation (regression analysis)	Errors
1	1	0.5	0.71	10	1.09619	1.113973	−0.01778
2	1	0.5	0.71	20	1.34937	1.372415	−0.02305
3	1	0.5	0.71	30	1.49701	1.521827	−0.02482
4	1	0.5	7	10	2.27816	2.255666	0.022494
5	1	0.5	7	20	2.72638	2.674523	0.051856
6	1	0.5	7	30	3.01714	2.970041	0.047097
7	1	0.5	55	10	3.90157	3.915322	−0.01375
8	1	0.5	55	20	4.62905	4.633879	−0.00483

(continued)

(continued)

S. No.	M	Ek	Pr	Gr	$\overline{Nu}$	Estimation (regression analysis)	Errors
9	1	0.5	55	30	5.11515	5.119903	−0.00475
10	1	1	0.71	10	1.14792	1.129813	0.018107
11	1	1	0.71	20	1.29958	1.352800	−0.05322
12	1	1	0.71	30	1.38961	1.473127	−0.08352
13	1	1	7	10	2.23456	2.226095	0.008465
14	1	1	7	20	2.61718	2.609012	0.008168
15	1	1	7	30	2.85963	2.874960	−0.01533
16	1	1	55	10	3.74114	3.782435	−0.0413
17	1	1	55	20	4.40947	4.461343	−0.05187
18	1	1	55	30	4.85415	4.914087	−0.05994
19	1	3	0.71	10	1.06231	1.056324	0.005985
20	1	3	0.71	20	1.19655	1.213219	−0.01667
21	1	3	0.71	30	1.28485	1.292936	−0.00809
22	1	3	7	10	2.05953	2.005754	0.053776
23	1	3	7	20	2.36524	2.320634	0.044605
24	1	3	7	30	2.56461	2.544027	0.020582
25	1	3	55	10	3.46685	3.414292	0.052557
26	1	3	55	20	4.05698	4.010327	0.046652
27	1	3	55	30	4.44425	4.405681	0.038569
28	4	0.5	0.71	10	1.03774	1.055083	−0.01734
29	4	0.5	0.71	20	1.2786	1.307410	−0.02881
30	4	0.5	0.71	30	1.42465	1.452646	−0.028
31	4	0.5	7	10	2.15867	2.175472	−0.0168
32	4	0.5	7	20	2.59523	2.588051	0.007178
33	4	0.5	7	30	2.88076	2.879231	0.001529
34	4	0.5	55	10	3.74017	3.786001	−0.04583
35	4	0.5	55	20	4.45107	4.497030	−0.04596
36	4	0.5	55	30	4.92585	4.977465	−0.05162
37	4	1	0.71	10	1.06571	1.067570	−0.02983
38	4	1	0.71	20	1.23121	1.284772	−0.00617
39	4	1	0.71	30	1.32885	1.401254	0.023396
40	4	1	7	10	2.07905	2.142635	0.016035
41	4	1	7	20	2.46299	2.519602	0.075627
42	4	1	7	30	2.70833	2.781541	0.099218
43	4	1	55	10	3.53075	3.650497	0.089672
44	4	1	55	20	3.53075	4.322206	0.128863
45	4	1	55	30	4.61155	4.769692	0.156158

(continued)

(continued)

S. No.	M	Ek	Pr	Gr	$\overline{Nu}$	Estimation (regression analysis)	Errors
46	4	3	0.71	10	0.98141	0.995126	−0.01372
47	4	3	0.71	20	1.12703	1.147555	−0.02053
48	4	3	0.71	30	1.22177	1.224746	−0.00298
49	4	3	7	10	1.89709	1.923678	−0.02659
50	4	3	7	20	2.22391	2.233929	−0.01002
51	4	3	7	30	2.43447	2.454632	−0.02016
52	4	3	55	10	3.21517	3.286339	−0.07117
53	4	3	55	20	3.7968	3.876494	−0.07969
54	4	3	55	30	4.18545	4.267909	−0.08246
55	25	0.5	0.71	10	0.86086	0.786233	0.074626
56	25	0.5	0.71	20	1.04862	0.999039	0.04958
57	25	0.5	0.71	30	1.17287	1.118334	0.054535
58	25	0.5	7	10	1.74179	1.764967	−0.02318
59	25	0.5	7	20	2.11981	2.136878	−0.01707
60	25	0.5	7	30	2.37314	2.400971	−0.02783
61	25	0.5	55	10	3.14411	3.088580	0.05553
62	25	0.5	55	20	3.79423	3.750194	0.044035
63	25	0.5	55	30	4.23055	4.194796	0.035754
64	25	1	0.71	10	0.83209	0.775357	0.056732
65	25	1	0.71	20	0.98553	0.955347	0.030183
66	25	1	0.71	30	1.08487	1.048196	0.036673
67	25	1	7	10	1.61535	1.709362	−0.09401
68	25	1	7	20	1.94698	2.047972	−0.10099
69	25	1	7	30	2.19133	2.285132	−0.0938
70	25	1	55	10	2.896	2.934858	−0.03886
71	25	1	55	20	3.49249	3.559461	−0.06697
72	25	1	55	30	3.89365	3.973422	−0.07977
73	25	3	0.71	10	0.74115	0.710629	0.030521
74	25	3	0.71	20	0.86867	0.835081	0.033589
75	25	3	0.71	30	0.95815	0.897874	0.060275
76	25	3	7	10	1.44267	1.500505	−0.05784
77	25	3	7	20	1.747	1.781633	−0.03463
78	25	3	7	30	1.95579	1.986793	−0.031
79	25	3	55	10	2.62652	2.598995	0.027524
80	25	3	55	20	3.18517	3.151281	0.033889
81	25	3	55	30	3.56092	3.518405	0.042515
82	100	0.5	0.71	10	0.69213	0.731428	−0.0393

(continued)

(continued)

S. No.	M	Ek	Pr	Gr	$\overline{Nu}$	Estimation (regression analysis)	Errors
83	100	0.5	0.71	20	0.81858	0.850035	−0.03146
84	100	0.5	0.71	30	0.90881	0.923631	−0.01482
85	100	0.5	7	10	1.3195	1.310908	0.008591
86	100	0.5	7	20	1.59458	1.584527	0.010053
87	100	0.5	7	30	1.78765	1.798826	−0.01118
88	100	0.5	55	10	2.42768	2.423757	0.003923
89	100	0.5	55	20	2.96196	2.955840	0.00612
90	100	0.5	55	30	3.33468	3.319409	0.015271
91	100	1	0.71	10	0.64393	0.638560	0.005369
92	100	1	0.71	20	0.74759	0.732597	0.014993
93	100	1	0.71	30	0.83357	0.787993	0.045577
94	100	1	7	10	1.19907	1.175441	0.023629
95	100	1	7	20	1.43943	1.424003	0.015426
96	100	1	7	30	1.61126	1.619616	−0.00836
97	100	1	55	10	2.20344	2.206418	−0.00298
98	100	1	55	20	2.69364	2.709736	−0.0161
99	100	1	55	30	3.04164	3.050910	−0.00927
100	100	3	0.71	10	0.57126	0.607178	−0.03592
101	100	3	0.71	20	0.65532	0.678660	−0.02334
102	100	3	0.71	30	0.71619	0.736983	−0.02079
103	100	3	7	10	1.05152	1.008446	0.043073
104	100	3	7	20	1.26702	1.232510	0.034509
105	100	3	7	30	1.42599	1.429106	−0.00312
106	100	3	55	10	1.96699	1.977404	−0.01041
107	100	3	55	20	2.43744	2.441387	−0.00395
108	100	3	55	30	2.77316	2.768708	0.004451

References

1. Atkinson, B.W.: Meso-scale Atmospheric Circulations. Academic Press, London (1981)
2. Barry, R.G.: Mountain Weather and Climate, 2nd edn. Routledge, London (1992)
3. Whiteman, C.D.: Mountain Meteorology - Fundamentals and Applications. Oxford University Press, New York, NY (2000)
4. Chow, F.K., De Wekker, S.F.J., Snyder, B. (eds.): Mountain Weather Research and Forecasting. Recent Progress and Current Challenges. Springer, Berlin (2013)
5. Meybeck, M., Green, P., Vörösmarty, C.J.: A new typology for mountains and other relief classes: an application to global continental water resources and population distribution. Mt. Res. Dev. **21**, 34–45 (2001)

6. Richner, H., Hächler, P.: Understanding and forecasting alpine foehn. In: Chow, F., De Wekker, S.F.J., Synder, B. (eds.) Mountain Weather Research and Forecasting: Recent Progress and Current Challenges, pp. 219–260. Springer, New York, NY (2013)
7. Zardi, D., Whiteman, C.D.: Diurnal mountain wind systems. In: Chow, F., De Wekker, S.F.J., Synder, B. (eds.), pp. 35–119. Springer, New York, NY (2013)
8. Stull, R.B.: An Introduction to Boundary Layer Meteorology. Kluwer Academic Publishers, Dordrecht (1988)
9. De Wekker, S.F.J., Kossmann, M.: Convective boundary layer heights over mountainous terrain—a review of concepts. Front. Earth Sci. **3**(76), 1–17 (2015)
10. German, M., Francisco, V., VaZquez, L.: Characterization of the Martian convective boundary layer. J. Atmos. Sci. **66**, 2044–2058 (2009)
11. Sparrow, E.M., Gregg, J.L.: Laminar free convection heat transfer from the outer surface of a vertical circular cylinder. Trans. ASME **78**, 1823–1829 (1956)
12. Goldstein, R.J., Briggs, D.G.: Transient free convection about vertical plates and circular cylinders. Trans. ASME C J. Heat Mass Transf. **86**, 490–500 (1964)
13. Lee, H.R., Chen, T.S., Armaly, B.F.: Natural convection along slender vertical cylinders with variable surface temperature. J. Heat Mass Transf. **110**, 103–108 (1988)
14. Velusamy, K., Garg, V.K.: Transient natural convection over a heat generating vertical cylinder. J. Heat Mass Transf. **35**, 1293–1306 (1992)
15. Ganesan, P., Rani, H.P.: Transient natural convection along vertical cylinder with heat and mass transfer. Heat Mass Transf. **33**, 449–455 (1998)
16. Karimabadi, H., Roytershteyn, V., Wan, M., Matthaeus, W.H., Daughton, W., Wu, P.: Coherent structures, intermittent turbulence, and dissipation in high-temperature plasmas. Phys. Plasmas **20**(1), 012303 (2013)
17. Ganesan, P., Rani, H.P.: Unsteady free convection MHD flow past a vertical cylinder with heat and mass transfer. Int. J. Therm. Sci. **39**, 265–272 (2000)
18. Ganesan, P., Loganathan, P.: Magnetic field effect on a moving vertical cylinder with constant heat flux. Heat Mass Transf. **39**, 381–386 (2003)
19. Loganathan, P., Kannan, M., Ganesan, P.: Thermal radiation effects on MHD flow over a moving semi-infinite vertical cylinder. Int. J. Math. Anal. **5**(6), 257–274 (2011)
20. Deka, R.K., Ashish, P.: Transient free convective MHD flow past an infinite vertical cylinder. Theor. Appl. Mech. Lett. **40**(3), 385–402 (2013)
21. Nadeem, S., Akram, S.: Influence of inclined magnetic field on peristaltic flow of a Williamson fluid model in an inclined symmetric or asymmetric channel. Math. Comput. Model. **52**, 107–119 (2010)
22. Greenspan, H.P., Howard, L.N.: On a time–dependent motion of a rotating fluid. J. Fluid Mech. **17**, 385–404 (1963)
23. Chandrasekhar, S.: Hydrodynamic and Hydromagnetic Stability. England, Oxford (1961)
24. Davies, P.A., Davis, R.G., Foster, M.R.: Flow past a circular cylinder in a rotating stratified fluid. Philos. Trans. R. Soc. Lond. **A331**, 245–286 (1990)
25. Xu, Y., Boyer, D.L., Zhang, X.: Rotating oscillatory flow past a cylinder. Phys. Fluids **5**(4), 868–880 (1993)
26. Ashish, P.: Flow past an accelerated horizontal cylinder in a rotating fluid. J. Appl. Fundam. Sci. **2**(2) (2016). ISSN 2395-5562
27. Holton, J.R.: The influence of viscous boundary layers on transient motions in a stratified rotating fluid: Part I. J. Atmos. Sci. **22**, 402–411 (1965)
28. Chen, L., Chu, Y., Zhang, Y., Han, F., Zhang, J.: Analysis of heat transfer characteristics of fractured surrounding rock in deep underground spaces. Math. Probl. Eng. **2019**, Article ID 1926728, 1–11 (2019)
29. Tahari, M., Ghorbanian, A., Hatami, M., Jing, D.: Physical effect of a variable magnetic field on the heat transfer of a nanofluid-based concentrating parabolic solar collector. Eur. Phys. J. Plus **132**, 549 (2017)
30. Ali, S., Smith, K.A.: On learning algorithm selection for classification. Appl. Soft Comput. **6**(2), 119–138 (2006)

31. Pham, D., Sagiroglu, S.: Training multilayered perceptrons for pattern recognition: a comparative study of four training algorithms. Int. J. Mach. Tools Manuf. **41**, 419–430 (2001)
32. Cohen, J.: A coefficient of agreement for nominal scales. Educ. Psychol. Measur. **20**(1), 37–46 (1960)

Numerical Flow Analysis in Γ Shaped Enclosure: Energy Streamlines and Field Synergy

Hari Ponnamma Rani, Vekamulla Narayana, Yadagiri Rameshwar, and Sergey V. Starchenko

Abstract Understanding heat transfer becomes more important as the severity of climate, either hot or cold, increases. Managing heat flows is critical to occupant's thermal comfort, durability, energy efficiency, and, increasingly, thermal resilience during periods of extended power outages. In the present study, the convective air flow in the differentially heated gamma (Γ) shaped enclosure is simulated numerically. Isothermal temperature conditions are assumed at the vertical walls of the enclosure in which the temperature of the step wall is higher than that of other vertical walls. Top and bottom walls of the enclosure are considered to be adiabatic. The governing equations of the problem are discretized using the finite volume approach. To accelerate these simulations, the message passing interface (MPI) protocols are employed using the OpenMPI standard library. High-resolution simulation results are presented. We compare our results with those obtained by state-of-the-art methods to validate the performance of employed numerical methods. The flow behavior is elucidated with the aid of streamlines, isotherms, energy streamlines and synergy between the velocity and temperature gradient vectors by varying the control parameter, Rayleigh number, in the laminar regime.

Keywords Convection · Rayleigh number · Streamlines · Field synergy

H. P. Rani (✉)
Department of Mathematics, National Institute of Technology, Warangal, India
e-mail: hprani@nitw.ac.in

V. Narayana
Department of Mathematics, Balaji Institute of Technology and Science, Narsampet, Warangal, India

Y. Rameshwar
Department of Mathematics, University College of Engineering, Osmania University, Hyderabad, India

S. V. Starchenko
IZMIRAN, Troitsk, Russia

A. Kosterov et al. (eds.), *Problems of Geocosmos—2020*, Springer Proceedings in Earth and Environmental Sciences, https://doi.org/10.1007/978-3-030-91467-7_16

1 Introduction

Natural convective heat transfer is the self-sustained field since there is no external force needed for the fluid motion. Natural convection is basically well-known as an energy efficient process, and is important in several natural and industrial applications such as geothermal reservoirs [1], flow and heat transfer in solar ponds [2], float glass production [3], etc. When these thermodynamic systems are exposed to thermal gradients and friction effects, they are subject to the energy flow from one region to another due to conduction or convection. Analysis of cooling of a convective flow in enclosed spaces that trap the heat by greenhouse effect, remains a challenge. Heat transfer through opaque enclosure/cavity components is primarily due to conduction and convection within wall and roof cavities. Air leakage and ventilation represent a significant proportion of the total heat flows across a building enclosure unless the effective air barrier system and ventilation heat recovery are deployed. Understanding the mechanisms driving heat flows enable the designers to develop more efficient and cost-effective enclosures by allocating budgets to components that are most critical to high-performance buildings.

There has been a renewed interest in convective flow in cavities owing to its importance in environmental and energy management problems in current scientific and geo-political context [4–9]. Numerous researchers have also considered different configurations of cavities to understand the performance of thermosfluid-flow systems by flow structures and geometries, as well as to investigate the process of self-organization and self-optimization in nature [10]. A major area of recent investigation has been the reconciliation of the several mathematical models currently in use. Of particular interest are the boundary conditions at the walls and the prediction of the heat transfer coefficients when the convective currents are driven by different temperature distributions. The main objective of the present study is therefore to analyze the natural convection in a Γ shaped cavity, filled with air, with the walls kept at different temperatures. Also, to examine the energy flow in this cavity with respect to the streamlines, visualization of total energy transfer in the enclosure is carried using the energy streamlines [11].

A number of previous studies [12–17] have analysed energy streamline concept for different physical situations. It was observed that, by increasing the curvature radius of the corners, the irreversibility ratio (ratio between the viscous and thermal irreversibility) decreases while the Bejan number (ratio of heat transfer irreversibility to total irreversibility due to heat transfer and fluid friction) increases. The Γ-shaped enclosure has several applications related e.g., to building corners. Heat transfer inside this type of enclosures is governed by several common influential factors, like interwall spacing, enclosure height, sharpness of the corners, multicellular flow pattern at high Grashof numbers, which is the ratio of the buoyancy to viscous force acting on a fluid. Due to the importance of these geometries in convective flow the extensive literature exists on L-shaped enclosures [18–20]. In our previous work [21–23] convective heat transfer characteristics in a cavity have been analysed numerically, while the present work deals with these flow characteristics in a Γ-shaped enclosure. Rani et al. [21, 23] studied and visualized streamlines, isotherms, field synergy and energy streamlines on natural and mixed convective flows in 2D and 3D cavity enclosures. These visualisation concepts also employed in the present study.

Kimura and Bejan [24] and Bejan [25] proposed "heatline" concept as a powerful alternative way to visualize thermal energy flow. Heatlines consider both conduction and convection thermal energies simultaneously. In the limit of vanishingly small convective flow, the heatline reduces to the heat flux line. The total flow of energy across each heatline is zero [26]. Heatfunction is derived such that it satisfies the net energy balance. A number of previous studies [27–31] have analyzed the heatline concept for various physical situations.

Plotting of isotherms and streamlines is a common visualization tool for convection heat transfer problems to analyse the flow field and thermal characteristics. Hooman [32] developed energy flux vectors which discover the gap between the heatlines and energy streamlines. With help of field synergy principle (FSP), Guo et al. [33, 34] revealed the intensity of convective heat transfer is related to temperature gradient. Reduction of heat loss in natural convection problem attributes to weakening synergy between velocity vector and temperature gradient [35]. Nusselt number is calculated from heat transfer coefficient and is the ratio of convective to conductive heat transfer at a boundary in an incompressible fluid or gas. Convection includes both advection (fluid motion) and diffusion (conduction). Heat transfer coefficient is a quantitative characteristic of convective heat transfer between the fluid and the solid wall. Also, the calculation of heat transfer coefficient plays a key role in fractured media seepage theories and in heat transfer in fractured surrounding rock [36]. The heat transfer improvement can all be attributed to the better synergy between the velocity field and temperature field, which may offer some guidance in the design of electronic devices. The underlying mechanism of FSP will be discussed in detail in Sect. 4.4.

The paper is organized as follows. The next section deals with the considered physical system and corresponding governing equations and boundary conditions. The equations are nonlinear and coupled. The higher-order differencing schemes are used to linearize the convection terms. Attention has been paid to selecting the appropriate convective discretization schemes. The simulations are done for different Rayleigh number Ra ranging from 10^3 to 10^6, and the results are shown in Sect. 3. The coordination between the velocity vector and temperature gradient is shown with respect to Ra. The observations and the conclusions are summarised in the last section.

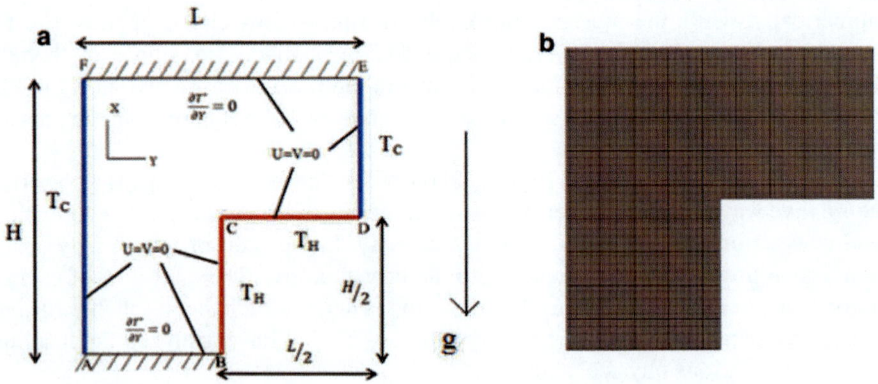

Fig. 1 **a** Schematic diagram and **b** simulated grid 80 × 80

2 Mathematical Formulation

2.1 Problem Description

The flow in the Γ-shaped enclosure is investigated with respect to its significant geometrical and boundary wall characteristics as shown in Fig. 1a. Uniform grid 80 × 80 is considered in the geometry shown in Fig. 1b. The characteristic length 'L' and height 'H' of the cavity are assumed to be the same, i.e., $L = H$. In thermodynamics, an isothermal process is by definition a process when the temperature of the system remains constant. Isothermal or constant temperature is assumed at all vertical walls. The step wall, BCD, is presumed to be kept at a higher temperature (T_h) than the temperature (T_c) of other vertical walls (AF and DE). The remaining borders/walls are assumed as perfectly insulated (adiabatic). The step wall is of height $H/2$ such that the aspect ratio of the step, i.e., ratio between the height and the length of the step is 1.

2.2 Governing Equations

The working fluid, air, filled inside the cavity is assumed to be an incompressible Newtonian fluid with constant properties. The Mach number is assumed to be inside the limits of the incompressible circulation (i.e., less than 0.3). Also, the following assumptions are made: the flow is laminar; the viscous dissipation is absent; radiation heat transfer is negligible; the gravitational force, g, acts in vertical direction. The fluid flow inside the cavity is governed by the following steady state dimensionless equations in the Boussinesq approximation:

$$\nabla.\boldsymbol{V} = 0 \tag{1}$$

$$\boldsymbol{V}.\nabla\boldsymbol{V} = -\nabla P + Pr\nabla^2\boldsymbol{V} + RaPrT^* \tag{2}$$

$$\boldsymbol{V}.\nabla T^* = \nabla^2 T^* \tag{3}$$

In the above equations the fluid flow (field) variables $\boldsymbol{V}$, P and T^* denote the velocity vector, pressure and temperature, respectively, and are made dimensionless by L/α, $L^2/(\rho\alpha^2)$ and ΔT, respectively. The symbols α, ρ and ΔT represent thermal diffusivity, density and temperature difference between hot and cold walls, respectively. Vector product is denoted with divergence (.) operator. The control parameters of the system are Rayleigh number *Ra* and Prandtl number *Pr*. The value of *Pr* is considered as 0.7 since the working fluid is air. Equations (1)–(3) are closed with the following boundary conditions:

No-slip conditions are imposed on all the bounding walls i.e.,

$$\boldsymbol{V} = 0 \text{ when } \left\{ \begin{array}{c} X = 0, 0 \le Y \le 1 \\ X = 1, 0.5 \le Y \le 1 \\ Y = 0, 0 \le X \le 0.5 \\ X = 0.5, 0 \le Y \le 0.5 \\ Y = 0.5, 0.5 \le X \le 1 \end{array} \right\}$$

The imposed isothermal (constant) temperature and adiabatic conditions on the walls are given by

$$T^* = 1 \text{ when } \left\{ \begin{array}{c} X = 0.5, 0 \le Y \le 0.5 \\ Y = 0.5, 0.5 \le X \le 1 \end{array} \right\}$$

$$T^* = 0 \text{ when } \left\{ \begin{array}{c} X = 0, 0 \le Y \le 1 \\ X = 1, 0.5 \le Y \le 1 \end{array} \right\}$$

$$\frac{\partial T^*}{\partial Y} = 0 \text{ at } Y = 0, 1.$$

3 Numerical Methodology

The dimensionless governing Eqs. (1)–(3) along with the boundary conditions are solved using the open source software, OpenFOAM [37]. The employed system configuration consists of Intel Core i5-5200U CPU @ 2.20 GHz 2.20 GHz processor with 8 GB RAM, 64-bit Operating system in the x64 based processor.

The spatial derivatives are discretized by employing the second order upwind linearization technique. The Laplacian and Divergence terms are discretized by the Gauss linear and Self-Filtered Central Difference (SFCD) schemes, respectively. The Conjugate Gradient Scheme (CGS) was employed to accelerate the convergence. The nonlinear advection terms are discretized by using Upwind Difference Scheme (UDS), QUICK, SUPERBEE and Self-Filtered Central Difference (SFCD) schemes [37].

Central differencing approximation has been used to discretize the diffusion terms and linear interpolation is employed to compute the cell face values for the convective terms. For a uniform grid the cell face values of the field variables are expressed as the average of the neighbouring nodal values. The central differencing scheme is unable to identify the flow direction, which is one of the major inadequacies of this scheme. To overcome this difficulty, an Upwind Difference Scheme (UDS) considers the flow direction, i.e., whether the flow is in the downstream (positive) or upstream (negative) direction. Thus, the field variable value at a cell face considered to be equal to the value at the downstream side node, when the flow is in the positive direction, otherwise value of field variable at a cell face is considered to be equal to the value at the upstream side node. In the higher order (HO) quadratic upstream interpolation for convective kinetics (QUICK) scheme the face value of field variable is calculated from a quadratic function passing through two neighbouring nodes (on each side of the face) and the upstream side node.

The first order scheme UDS is one of the simplest and most stable discretization schemes, however, it is more dissipative. The central differencing scheme (CDS) is more accurate than the first-order UDS, but it leads to oscillations in the solution or divergence if the local Peclet number, Pe, defined as the ratio between convective and diffusive transport, is larger than 2. It is then common to switch to first order upwind scheme in cells where Pe > 2. Such an approach is called a hybrid scheme. The QUICK scheme is generally very accurate but in regions with strong gradients, overshoots and undershoots can occur. This can lead to stability problems in the calculation.

To overcome these difficulties, the higher-order (HO) differencing scheme SUPERBEE [37] is expressed as a sum of UDS and a "limited" HO scheme, as shown below:

$$\Phi = \Phi_{UDS} + \psi[\Phi_{HO} - \Phi_{UDS}] \tag{4}$$

where Φ HO designates the higher-order face value of the field variable Φ. If ψ denotes the flux limiter and

$$r = \frac{\Phi_C - \Phi_U}{\Phi_D - \Phi_C} \tag{5}$$

Then $\psi = \psi(r)$, takes maximum [0, min(2r, 1), min(r, 2)] [38]. The subscripts U, C and D in the above Eq. (5) represent upstream, centre and downstream locations, respectively, for Φ.

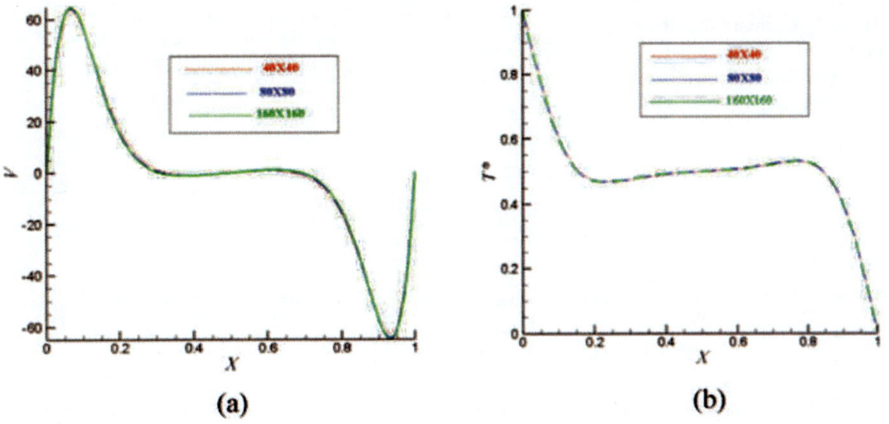

Fig. 2 Grid independence test for $Ra = 10^5$. **a** Profiles of velocity (V) component at mid height of cavity, **b** temperature (T^*) distribution at mid height of cavity

The SFCD is the combination of CDS and UDS with a blending factor γ and is defined as

$$\Phi = \gamma \Phi^{CDS} + (1 - \gamma)\Phi^{UDS} \tag{6}$$

By definition, parameter γ takes the values between 0 and 1.

Figure 2 shows the $\boldsymbol{V}$ and $\boldsymbol{T}^*$ solutions calculated using different meshes, namely, 40^2, 80^2 and 160^2 for $Ra = 10^5$. Figure 2a, b show profiles of velocity and temperature distributions at mid height of the cavity, respectively. The grid independence test is a process used to find the optimal grid. For the grid twice as fine, the solution did not vary much and the discrepancy is observed only at the fourth decimal place. In contrast, when the grid is roughened by the same factor, the solution changed already in the second decimal place. This ensures that the solutions obtained with 80^2cells represent real flow physics. Hence, in all calculations, 80^2 cells are used.

To accelerate these simulations, the message passing interface (MPI) protocols are employed using the OpenMPI standard library. The simulations are terminated when differences between two consecutive iterations of solutions of field variables are less than 10^{-5}. To validate the present simulated results, they are corroborated with the experimental/numerical results available in the literature, which are expressed with respect to mean Nusselt number, $\overline{Nu}$.

The employed different convection differencing schemes, like UDS, QUICK, SUPERBEE and SFCD are compared with respect to their CPU time. Table 1 lists the simulated $\overline{Nu}$ at the step (hot) wall. When MPI code is implemented, CPU time of SFCD is substantially reduced for all employed differencing schemes.

We compare our results with the state-of-the-art methods to validate the performance of employed numerical methods. Table 2 shows the comparison between the present results and the results from the literature with respect to $\overline{Nu}$. It can be

Table 1 Computational cost of convective differencing schemes when $Ra = 10^5$ with and without MPI

Scheme	$\overline{Nu}$	CPU time with MPI (s)	CPU time without MPI (s)
UDS	3.248	88,245	352,941
QUICK	4.126	84,568	338,472
SUPERBEE	4.372	84,498	337,882
SFCD	4.527	82,589	330,354

Table 2 Validation of present work with literature results in terms of $\overline{Nu}$ and Ra

Grid size	80^2	80^2	80^2	160^2
Ra	10^3	10^4	10^5	10^6
de Vahl Davis and Jones [39]	1.118	2.243	4.519	8.799
Markatos and Pericleous [40]	1.108	2.201	4.430	8.754
Rincon-Casado et al. [41]	1.118	2.241	4.522	8.819
Rani et al. [21]	1.068	2.039	4.057	6.558
Present work	1.127	2.238	4.527	8.858

observed that as Ra increases the present MPI results agree reasonably well with literature results. Comparing Tables 1 and 2, it can be noted the results obtained by SFCD scheme are closest to those found in the literature.

4 Results

For different values of Ra (= 10^3, 10^4, 10^5, 10^6) corresponding to laminar flow, the flow velocity and thermal field are depicted in terms of streamlines and isotherms in Figs. 3 and 4, respectively.

4.1 Streamlines

Figure 3a–d show the simulated streamlines for $Ra = 10^3, 10^4, 10^5, 10^6$, respectively. All figures exhibit two isothermal circular cores in the cavity. The main (larger) circulation cell is located on the left side cavity. This vortex occupies the region between the left cold vertical wall and the front face of the step, and re-circulates in the anti-clockwise direction. On the other hand, the secondary, weaker recirculation vortex occupies the region between the forward step and the right cold wall and rotates in the clockwise direction. Only at $Ra = 10^3$ the vortex in the left (larger) area penetrates considerably into the right (smaller) area, as shown in Fig. 3a.

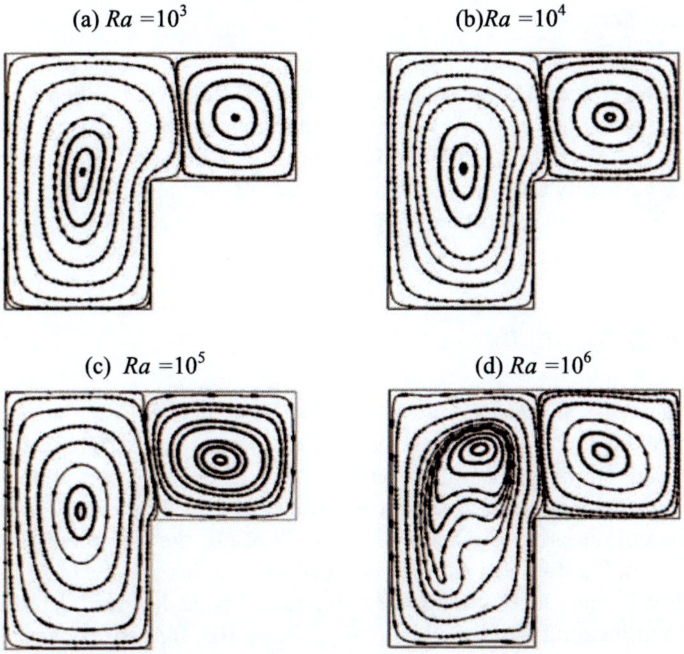

Fig. 3 Simulated streamlines for different *Ra*

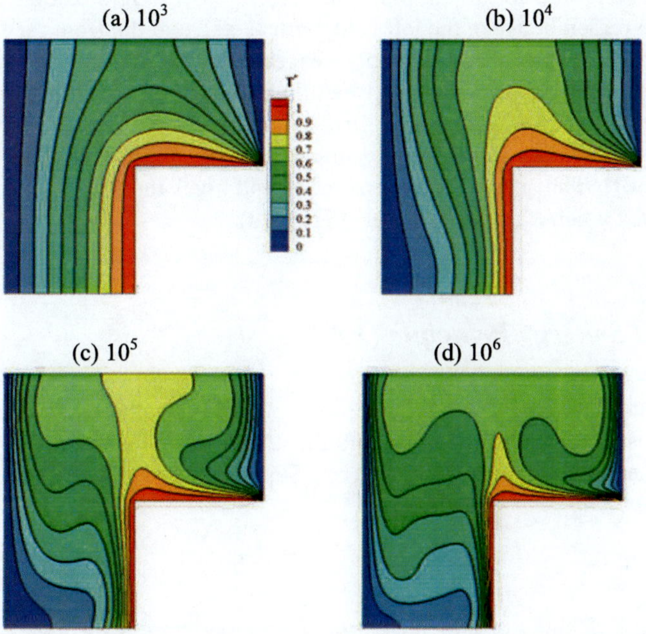

Fig. 4 Simulated isotherms for different *Ra*

4.2 Isotherms

Figure 4 shows isotherm contours in the cavity for different *Ra*. When $Ra = 10^3$, the flow, starting at the hot wall (step), forms isotherms which are smooth and vertical, due to presence of conduction mode for small *Ra*. When *Ra* further increases, the asymmetric vertically clustered isotherms appear in the vicinity of the walls. At the corners of the cavity, weaker thermal gradients are observed.

4.3 Energy Streamlines

Consider the Poisson equation $\nabla^2\Phi = (\nabla \times \boldsymbol{E}).\boldsymbol{k}$, where $\boldsymbol{E}$ denotes energy flux density vector, $\boldsymbol{k}$ denotes unit vector and Φ is known as energy streamfunction. By solving this equation, the energy streamlines are obtained. Roe et al. [38] used the energy flux density vectors to sketch imaginary energy flow lines. Two different types of streamlines are observed in the cavity, namely, the free energy streamlines, which represent the flow that starts at hot wall, moves through the cavity and closes at the cold wall, and closed streamlines originating at the hot wall, passing through the cavity and forming circular loops. When *Ra* is 10^3, free energy streamlines are observed due to conduction mode of heat transfer (Fig. 5a). When *Ra* is greater than 10^3, closed energy streamlines form (Fig. 5b–d). The main and larger circulation cell is on the left side and re-circulates in the anti-clockwise direction. This vortex occupies the region between the left cold vertical wall and the front face of the step. On the other hand, the secondary and weaker recirculation cell on the right side of the cavity occupies the region between the forward step and the right cold wall circulating in the clockwise direction. In the vicinity of vertical walls, the free energy streamlines form, while closed energy streamlines form in middle part of the cavity when *Ra* is 10^4. The size of vortex increases and gives the space for formation of another vortex when *Ra* grows further (Fig. 5c, d).

4.4 Field Synergy Principle (FSP)

The FSP is used to express the enhancement of convective heat transfer. The better synergy of velocity and temperature gradient/heat flow fields leads to the higher convective heat transfer rate [33, 34]. The FSP is calculated using the formula

$$\beta_f = \cos^{-1}\left(\frac{V.\nabla T}{|V||\nabla T|}\right) \tag{7}$$

where V and T denote the velocity vector and temperature, respectively.

For better synergy of the two vector fields, V and ∇T

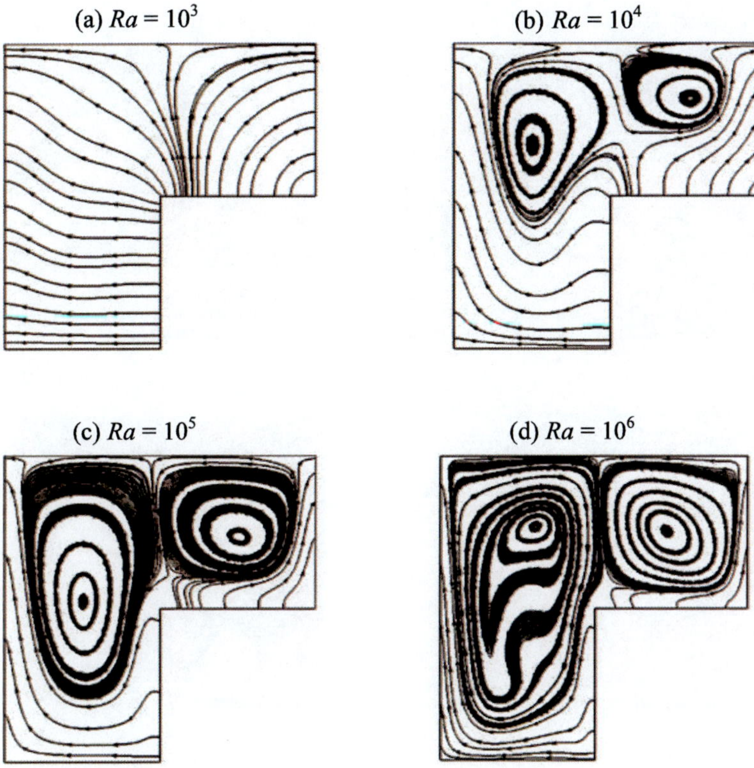

Fig. 5 Simulated energy streamlines for different *Ra*

- the angle between the velocity and the temperature gradient should be as small as possible, i.e., the velocity and the temperature gradient should be as parallel as possible;
- the local values of the three scalar fields, $|\boldsymbol{V}|, |\nabla T|$ and cos β_f, should all be simultaneously large, i.e., larger (~1) values of cos β should correspond to larger values of the velocity and the temperature gradient;
- the velocity and temperature profiles at each cross-section should be as uniform as possible.

Therefore, the strength of the convective heat transfer depends not only on the velocity and the temperature gradient, but also on their synergy. Figure 6 shows the simulated field synergy for different Ra. Synergy angle is zero when $\boldsymbol{V}$ and ∇T are parallel. When the synergy angle is 180°, $\boldsymbol{V}$ and the ∇T are in the opposite direction. When Ra is small, the angle between $\boldsymbol{V}$ and ∇T is small, i.e., both are in same direction (Fig. 6a shown in blue colour). As Ra increases this region (blue colour contour) starts decreasing. This is due to that the average Nusselt number is increasing with Ra.

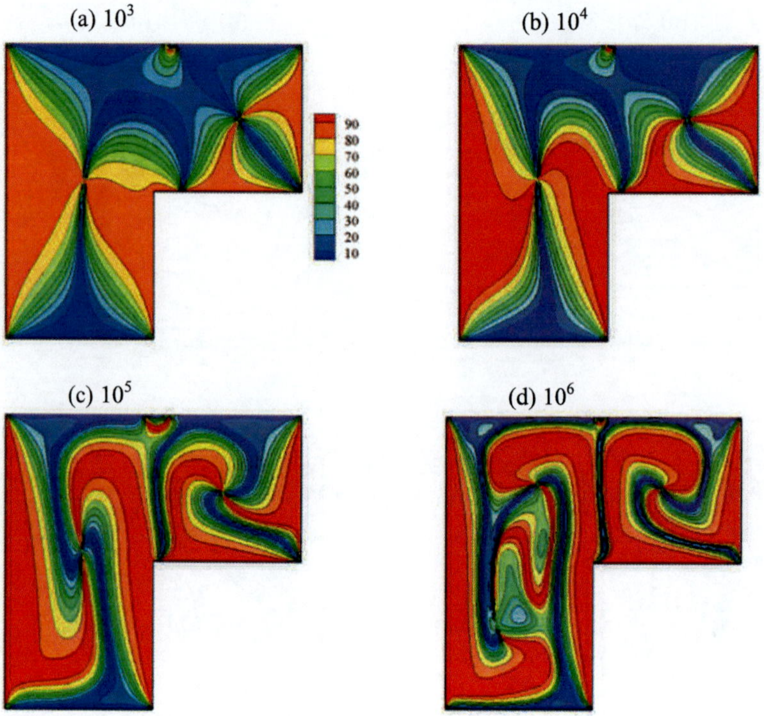

Fig. 6 Simulated field synergy for different *Ra*

5 Discussion and Conclusions

There are several studies in the literature on natural convection in cavities filled with viscous fluid kept at different temperature conditions that are imposed on the bottom or side walls owing to their importance in science and technology. The cavities with protrusions play a vital role in understanding the flow dynamics in the environmental related problems. For example, the studies related to the convective flow in Γ-shaped enclosure are useful in understanding the movement of heat and flow in building corners. Even though several studies are carried to understand the fluid flow inside the Γ-shaped enclosure, to date, the energy streamlines-based analysis of heat transfer evaluation is yet to appear in literature, so the reported results are new and original. Thus, in the present study the fluid flow is simulated inside the gamma (Γ) shaped enclosure by assuming isothermal temperature at the vertical walls and the step wall temperature higher than the temperature of other vertical walls. The top and bottom walls of the cavity are considered as adiabatic. The convective terms present in the governing equations are discretized and solved by using finite volume approach. The present results are corroborated by comparison with those available in the literature, validating the performance of employed numerical methods. The convection schemes are compared based on their accuracy and computational time. The streamlines show that the vortex occupies the region between left cold vertical wall and the front face of the step. This vortex re-circulates in the anti-clockwise direction. At the same time, the secondary and weaker recirculation vortex located on the right side of the cavity occupies the region between the forward step and the right cold wall and moves in the clockwise direction. Only at $Ra = 10^3$ vortex in the left (larger) area penetrates into the right (smaller) area. The circular shape of the vortex corners results in weaker thermal gradients at the boundary of the cavity walls. In the vicinity of vertical walls free energy streamlines form, while closed energy streamlines form in middle part of the cavity when Ra is 10^4. The size of circular region increases and gives space for forming another region for larger Ra. The synergy between velocity vector and temperature gradient decreases with increasing Ra.

References

1. Jue, T.C.: Analysis of flows driven by a torsionally-oscillatory lid in a fluid-saturated porous enclosure with thermal stable stratification. Int. J. Therm. Sci. **41**, 795–804 (2002). https://doi.org/10.1016/S1290-0729(02)01373-X
2. Mansour, R.B., Nguyen, C.T., Galanis, N.: Numerical study of transient heat and mass transfer and stability in a salt-gradient solar pond. Int. J. Therm. Sci. **43**, 779–790 (2004). https://doi.org/10.1016/j.ijthermalsci.2004.02.018
3. Prieto, M., Diaz, J., Egusquiza, E.: Analysis of the fluid-dynamic and thermal behaviour of a tin bath in float glass manufacturing. Int. J. Therm. Sci. **41**, 348–359 (2002). https://doi.org/10.1016/S1290-0729(02)01325-X
4. Bagchi, A., Kulacki, F.A.: Natural Convection in Superposed Fluid-Porous Layers. Springer, New York (2014)
5. Jansen, J.D.: A Systems Description of Flow Through Porous Media. Springer, New York (2013)
6. Poulikakos, D.: Natural convection in a confined fluid-filled space driven by a single vertical wall with warm and cold regions. ASME J. Heat Transf. **107**, 867–876 (1985). https://doi.org/10.1115/1.3247515
7. Lage, J.I., Bejan, A.: The resonance of natural convection in an enclosure heated periodically from the side. Int. J. Heat Mass Transf. **36**, 2027–2038 (1993). https://doi.org/10.1016/S0017-9310(05)80134-6
8. Bilgen, E., Wang, X., Vasseur, P., Meng, F., Robillard, L.: Periodic conditions to simulate mixed convection heat transfer in horizontal channels. Numer. Heat Transf. A **27**, 461–472 (1995). https://doi.org/10.1080/10407789508913712
9. Sarris, I.E., Lekakis, I., Vlachos, N.S.: Natural convection in a 2D enclosure with sinusoidal upper wall temperature. Numer. Heat Transf. A **42**, 513–530 (2002). https://doi.org/10.1080/10407780290059675
10. Mohebbi, R., Izadi, M., Sajjadi, H., Delouei, A.A., Sheremet, M.A.: Examining of nanofluid natural convection heat transfer in a Γ-shaped enclosure including a rectangular hot obstacle using the lattice Boltzmann method. Physica A **526**, 120831 (2019). https://doi.org/10.1016/j.physa.2019.04.067
11. Mahmud, S., Fraser, R.: Visualizing energy flows through energy streamlines and pathlines. Int. J. Heat Mass Transf. **50**, 3990–4002 (2007). https://doi.org/10.1016/j.ijheatmasstransfer.2007.01.032

12. Tasnim, S.H., Collins, M.R.: Suppressing natural convection in a differentially heated square cavity with an arc shaped baffle. Int. Commun. Heat Mass Transf. **32**, 94–106 (2005). https://doi.org/10.1016/j.icheatmasstransfer.2004.05.022
13. Tasnim, S.H., Fraser, R.A.: Flow and energy fields in a resonant channel. Int. Commun. Heat. Mass Transf. **36**, 539–546 (2009). https://doi.org/10.1016/j.icheatmasstransfer.2009.03.016
14. Tasnim, S.H., Fraser, R.A.: Computations of the flow and thermal fields in a thermosacoustic refrigerator. Int. Commun. Heat Mass Transf. **37**, 748–755 (2010). https://doi.org/10.1016/J.ICHEATMASSTRANSFER.2010.04.006
15. Cao, N., Olson, J.R., Swift, G.W., Chen, S.: Energy flux density in a thermoacoustic couple. J. Acoust. Soc. Am. **99**, 3456–3464 (1996). https://doi.org/10.1121/1.414992
16. Ishikawa, H., Mee, D.J.: Numerical investigation of flow and energy fields near a thermosacoustic couple. J. Acoust. Soc. Am. **111**, 831–839 (2002). https://doi.org/10.1121/1.1430687
17. Ziapour, B.M., Dehnavi, R.: Finite-volume method for solving the entropy generation due to air natural convection in γ-shaped enclosure with circular corners. Math. Comput. Model. **54**(5–6), 1286–1299 (2011). https://doi.org/10.1016/j.mcm.2011.03.039
18. Mahmoodi, M.: Numerical simulation of free convection of a nanofluid in L-shaped cavities. Int. J. Therm. Sci. **50**, 1731–1740 (2011). https://doi.org/10.1016/j.ijthermalsci.2011.04.009
19. Mahmud, S.: Free convection inside an L-shaped enclosure. Int. Commun. Heat Mass Transf. **29**(7), 1005–1013 (2002). https://doi.org/10.1016/S0735-1933(02)00420-7
20. Kalteh, M., Hasani, H.: Lattice Boltzmann simulation of nanofluid free convection heat transfer in an L-shaped enclosure. Superlattices Microstruct. **66**, 112–128 (2014). https://doi.org/10.1016/j.spmi.2013.12.004
21. Rani, H.P., Narayana, V., Rameshwar, Y., Starchenko, S.V.: Aspect ratio effects on bottom heated 2D cavity using energy streamlines and field synergy principle. Latin Am. Appl. Res. **50**, 41–46 (2020). https://doi.org/10.52292/j.laar.2020.164
22. Narayana, V., Rani, H.P.: Analysis of visualization techniques of bottom heated lid-driven square cavity. Heat Transf. **49**, 3549–3559 (2020). https://doi.org/10.1002/htj.21787
23. Rani, H.P., Naresh, K., Narayana, V.: Differentially heated cubical cavity using energy pathlines and field synergy. Heat Transf. **49**, 3683–3701 (2020). https://doi.org/10.1002/htj.21795
24. Kimura, S., Bejan, A.: The boundary layer natural convection regime in a rectangular cavity with uniform heat flux from the side. J. Heat Transf. **106**(1), 98–103 (1984). https://doi.org/10.1115/1.3246666
25. Bejan, A.: Convection Heat Transfer. Wiley, New York (1984)
26. Eckert, E.R., Drake, G.R.M., Jr.: Analysis of Heat and Mass Transfer. McGraw-Hill, New York (1972)
27. Bello-Ochende, F.L.: A heat function formulation for thermal convection in a square cavity. Int. Commun. Heat Mass Transf. **15**, 193–202 (1988). https://doi.org/10.1016/0735-1933(88)90065-6
28. Morega, A.M., Bejan, A.: Heatline visualization of forced convection laminar boundary layers. Int. J. Heat Mass Transf. **36**, 3957–3966 (1993). https://doi.org/10.1016/0017-9310(93)90146-W
29. Zhao, F.Y., Liu, D., Tang, G.F.: Application issues of the streamline, heatline and massline for conjugate heat and mass transfer. Int. J. Heat Mass Transf. **50**, 320–334 (2007). https://doi.org/10.1016/j.ijheatmasstransfer.2006.06.026
30. Costa, V.A.F.: Heatline and massline visualization of laminar natural convection boundary layers near a vertical wall. Int. J. Heat Mass Transf. **43**, 3765–3774 (2000). https://doi.org/10.1016/S0017-9310(00)00028-4
31. Costa, V.A.F.: Unification of the streamline, heatline and massline methods for the visualization of two-dimensional heat and mass transfer in anisotropic media. Int. J. Heat Mass Transf. **46**, 1309–1320 (2003). https://doi.org/10.1016/S0017-9310(02)00404-0
32. Hooman, K.: Energy flux vectors as a new tool for convection visualization. Int. J. Numer. Methods Heat Fluid Flow **20**, 240–249 (2010). https://doi.org/10.1108/09615531011016984

33. Guo, G., Li, D., Wang, B.: A novel concept for convective heat transfer enhancement. Int. J. Heat Mass Transf. **41**, 2221–2225 (1998). https://doi.org/10.1016/S0017-9310(97)00272-X
34. Guo, Z., Tao, W., Shah, R.: The field synergy (coordination) principle and its applications in enhancing single phase convective heat transfer. Int. J. Heat Mass Transf. **48**, 1797–1807 (2005). https://doi.org/10.1016/j.ijheatmasstransfer.2004.11.007
35. Li, Y., Liu, G., Rao, Z., Liao, S.: Field synergy principle analysis for reducing natural convection heat loss of a solar cavity receiver. Renew. Energy **75**, 257–265 (2015). https://doi.org/10.1016/j.renene.2014.09.055
36. Chen, L., Chu, Y., Zhang, Y., Han, F., Zhang, J.: Analysis of heat transfer characteristics of fractured surrounding rock in deep underground spaces. Math. Probl. Eng. Article ID 1926728, 1–11(2019). https://doi.org/10.1155/2019/1926728
37. https://www.openfoam.com/
38. Roe, P.L.: Large scale computations in fluid mechanics. Part 2 Lect. Appl. Math. **22**, 163–193 (1985)
39. de Vahl Davis, G., Jones, I.P.: Natural convection in a square cavity: a comparison exercise. Int. J. Numer. Meth. Fluids **3**, 227–248 (1983). https://doi.org/10.1002/fld.1650030304
40. Markatos, N.C., Pericleous, K.A.: Laminar and turbulent natural convection in an enclosed cavity. Int. J. Heat Mass Transf. **27**, 755–772 (1984). https://doi.org/10.1016/0017-9310(84)90145-5
41. Rincon-Casado, A., Sanchez de la Flor, Chacon Vera, E., Sanchez Ramos, J.: New natural convection heat transfer correlations in enclosures for building performance simulation. Eng. Appl. Comput. Fluid Mech. **11**, 340–356 (2017). https://doi.org/10.1080/19942060.2017.1300107

Seismology

The Lake Baikal Unified Scaling Law for Earthquake Regional Coefficients

Anastasiya Nekrasova and Vladimir Kossobokov

Abstract Seismic hazard assessment requires an adequate understanding the earthquake distribution in magnitude, space, and time ranges. Lacking data for a period of several thousand years makes probabilistic approach to estimating the recurrence time of earthquakes unreliable. Nevertheless, the probabilistic seismic hazard assessment (PSHA) maps keep actively refined both at global and national scales. At the same time alternative models and methods are being developed to improve the accuracy and reliability of reproducible seismic hazard maps that pass intensive testing of historical evidence and simulated scenario earthquakes. One of those, the neo-deterministic seismic hazard assessment (NDSHA) provides reliable and effective tools for understanding and mitigating object-oriented earthquake risks. Consider the unified scaling law for earthquakes (USLE) as a part of NDSHA. USLE is generalizing the Gutenberg-Richter relationship (GRL) as $\log_{10} N(M, L) = A + B(6 - M) + C\log_{10} L, M_{-} \leq M \leq M^{-}$, where $N(M, L)$ is the number of magnitude M earthquakes in an area of linear size L, A characterizes the average level of seismic activity in terms of the annual rate of earthquakes of magnitude $M = 6$, B is the ratio $N(M)/N(M - 1)$, C estimates fractal dimension of the set of epicenters, and $[M_{-}, M^{-}]$ is the range where the relationship holds. Naturally, C complements to A and B (analogous to a and b of GRL) showing how the number of earthquakes changes with an area linear size. Recent studies have shown that using the USLE provides adequate estimation of seismic hazard at national or even regional scale when the existing earthquake data are a few decades long and rich enough in magnitude determinations.

Keywords Unified scaling law for earthquakes · Neo-deterministic seismic hazard assessment · Regional scale

---

A. Nekrasova (✉) · V. Kossobokov
Institute of Earthquake Prediction Theory and Mathematical Geophysics RAS, Profsoyuznaya Street, 84/32, Moscow 117997, Russian Federation
e-mail: nastia@mitp.ru

A. Kosterov et al. (eds.), *Problems of Geocosmos—2020*, Springer Proceedings in Earth and Environmental Sciences, https://doi.org/10.1007/978-3-030-91467-7_17

1 Introduction

Lake Baikal region is one of the most seismoactive regions of Russian Federation with relatively high population. We have already carried out analysis of the unified scaling law for earthquakes [1, 2] aimed at the seismic hazard estimation for this region [3, 4]. In this paper, we present results of a new analysis and mapping of the USLE coefficients at the more detailed space scale based on the uniform regional catalog of earthquake data from 1994 to 2019.

2 Method

We apply the Scaling Coefficient Estimation algorithm (SCE) [4] for compiling the maps of the USLE coefficients in Lake Baikal and surroundings.

Also, the Zaliapin and Ben-Zion algorithm [5] was used for seismic cluster identification with $b = B = 0.92$, and $d = C = 1.02$ as input parameters. These values are the averages of the corresponding USLE coefficients from the global coarse estimation available at the ISC Seismological Dataset Repository [6].

3 Data

3.1 Regional Catalog

In the Baikal region earthquake determinations by the Baikal Division of the Geophysical Survey, Federal Research Center of the Russian Academy of Sciences (48°–58° N and 99°–122° E) in 1994–2019 [7] are sufficiently complete by events with $M = 2.6$ (corresponding to the energy class $K = 8.6$ accepted in the Soviet Union). We used the original online version of the catalog with man-made seismic events excluded and the Rautian relation between energy class and magnitude, i.e., $K = 4 + 1.8 \times M$ at $K \leq 14$ and $K = 8 + 1.1 \times M$, if $K > 14$. Figure 1 presents the map of epicenters, the Gutenberg-Richter and magnitude versus time plots for all 8130 earthquakes.

The regional network configuration around Lake Baikal for the time interval considered was not uniform in space and time. For example, as shown in Fig. 2 the easternmost part of the region (i.e., longitudes more than 118° E) is probably represented with the completeness comparable to the other territories from 2005 to 2015 only, when a temporary station was installed in the region [8].

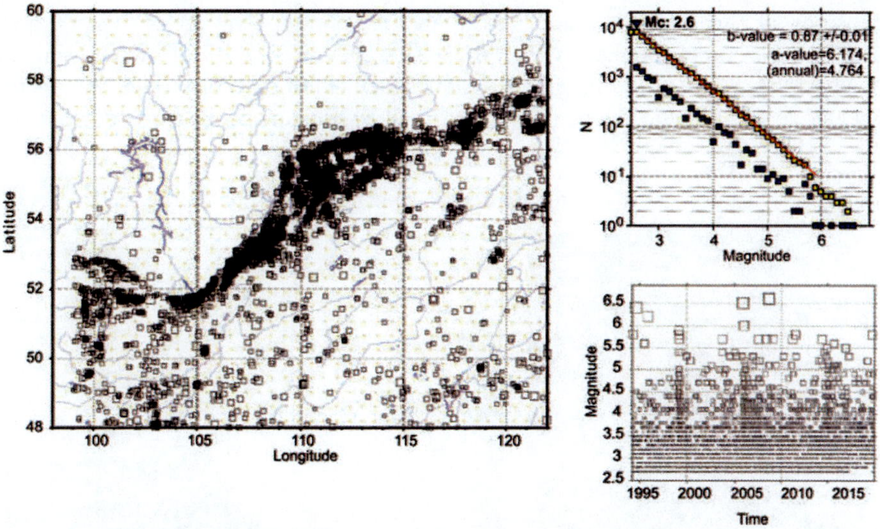

Fig. 1 The regional Lake Baikal catalog of earthquakes from 1/1/1994 to 12/31/2019. Left panel—the map of epicenters. Top right panel—the Gutenberg-Richter plots with the best linear fit *a*- and *b*-values for cumulative distribution (red line). Bottom right panel—magnitude versus time distribution of earthquake magnitude

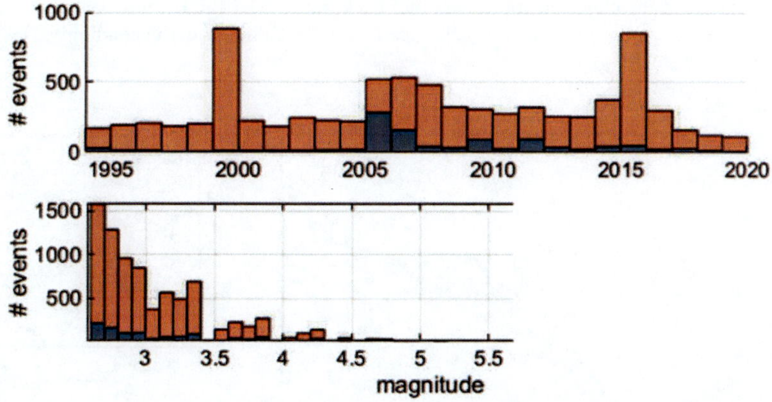

Fig. 2 The time (top panel) and magnitude (bottom panel) distributions of the latitude dependent part of the regional catalog 1/1/1994 to 12/31/2019 (red—all territories, blue—longitudes more than 118° E)

3.2 Declustering Catalog

The nearest neighbour method [5] applied to earthquakes in the Lake Baikal region with $p = q = 0.5$ parameters provides two joint distributions of rescaled space and time components of η *(R, T)*. The density distribution of η with the best fit Gaussian densities for clustered (blue) and background (red) components are shown in Fig. 3. The initial scaling parameters $b = B = 0.92$ and $d = C = 1.02$ were used for declustering. Same as in Fig. 1, Fig. 4 shows the map of epicenters, the Gutenberg-Richter and magnitude versus time plots for all 3989 mainshocks of the declustered regional catalog.

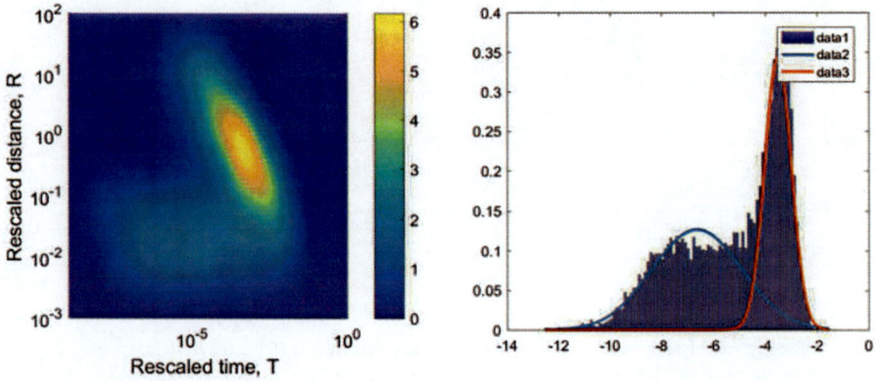

Fig. 3 Nearest-neighbour method application to earthquakes in Lake Baikal region. Left panel—the 2D joint distribution of rescaled space and time components of $\eta = R \times T$, with $p = q = 0.5$. Right panel—the 1D density distribution of the USLE control parameter η, with estimated Gaussian densities for clustered (blue line) and background (red line) components. The scaling parameters used are: $b = 0.92$ and $d = 1.02$ (i.e., B and C parameters correspondingly, Nekrasova and Kossobokov [6])

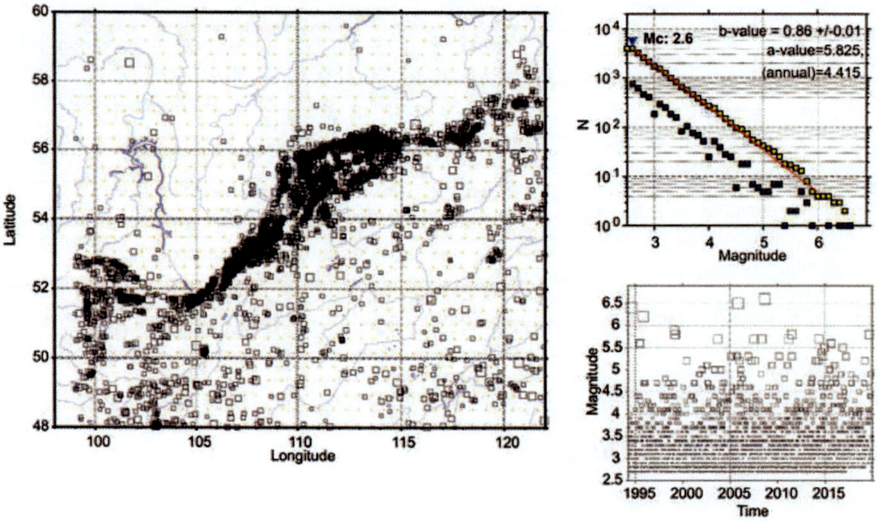

Fig. 4 The regional catalogue of mainshock from 1/1/1994 to 12/31/2019. Notes same as in Fig. 1

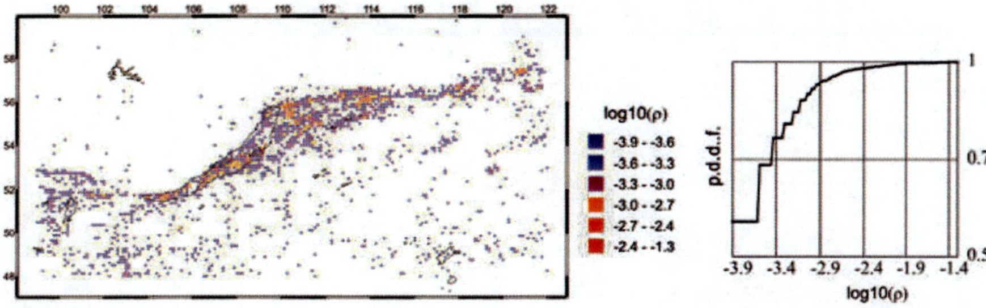

Fig. 5 The regional Lake Baikal catalogue of earthquakes from 1/1/1994 to 12/31/2019. Left panel—the active cells of a regular grid map. Right panel—the empirical probability function of seismic events ρ

To quantify the scaling parameters of earthquake occurrence in the Lake Baikal region the SCE algorithm was applied to the complete and declustered regional catalogs. Parameters of the USLE are evaluated in the seismically active cells of a regular grid 1/8° × 1/8°, considering a hierarchy of cascading areas with linear size of 1/8°, 1/4°, 1/2°, 1°, 2°, and 4°. The map of the 1813 active cells of a regular grid and empirical probability function of seismic events ρ ($\log_{10} \rho \geq -3.9$) is shown in Fig. 5. The reliable estimations of the USLE parameters were obtained for 1460 cells. The standard errors of the coefficients *A*, *B*, and *C* do not exceed 0.1, thus supporting the accuracy of parameter estimates and the assumption of self-similarity.

The resulting maps of the USLE coefficients *A*, *B* and *C* are shown in Fig. 6. The easternmost part of the region with the highest value of *A* and low values *B* and *C*, might be a result of the incompleteness of data in time. Therefore, in this part of the region the USLE parameters need to be critically considered in case of a seismic hazard map evaluation.

The parameter *A*, which corresponds to the logarithmic estimate of seismic activity at magnitude $M_0 = 6.0$, normalized to a unit area of 1° × 1° and unit time of one year, ranges between − 3.3 and − 1.4, i.e., from one earthquake in 2000 years to one earthquake in 25 years, correspondingly. The coefficient of magnitude balance, *B* (analogous to the *b-value* of GRL) concentrates in the range from 0.4 to 1.0, while the fractal dimension of the earthquake epicenters *C* varies from 0.3 to 1.2. The correlation dimension $D_2 = 0.92$ obtained from detailed observation (from 5 to 100 km spacing) for the South Baikal Basin [8] corresponds the maximum of density distribution of the coefficient *C*.

The grey areas in Fig. 7 mark the regional empirical density distributions of the USLE coefficients estimated based on the data from 1964 to 2008 and more robust hierarchy of square boxes with linear size from 8° down to 1/2° [4]. One can see that distributions of *A* and *C* based on different hierarchy are mainly within the same ranges. The distribution of *B* at more detailed scale shows up the triple-humped shape with maxima at 0.6, 0.8, and 1.0, which differs from the one based on the robust scale with the maximum about 0.75.

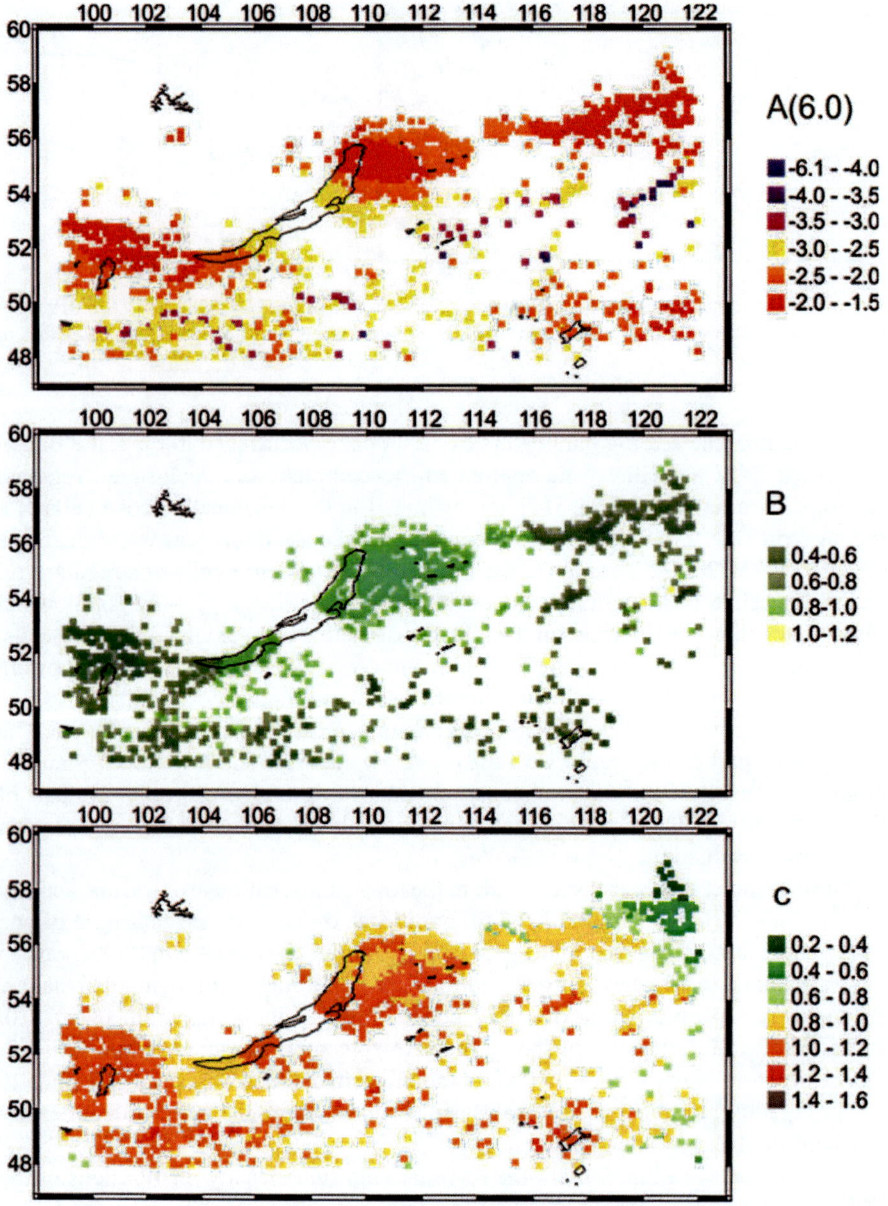

Fig. 6 The spatial distributions of the USLE regional coefficients *A*, *B* and *C*

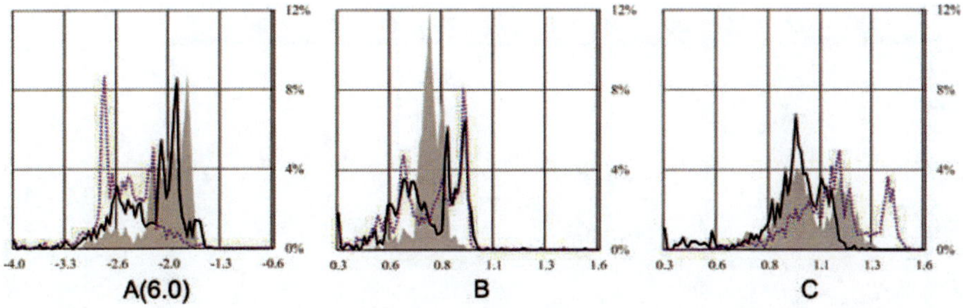

Fig. 7 The empirical probability density distribution functions of *A*, *B*, and *C* coefficients for complete (black line) and declustered (dashed line) catalogs on top of gray areas of the corresponding distributions obtained with more robust hierarchy of square boxes [4]

Additionally, the USLE coefficient estimation for the same 1/8° × 1/8° grid cells was performed for the declustered catalogue of mainshocks. See Fig. 7 for the empirical probability density distribution functions (dashed lines) and Fig. 8 for the spatial distributions *A*, *B* and *C* obtained for the catalog mainshocks. Note that there is no territory of missing estimates in the center of Lake Baikal, which is present in Fig. 6. Apparently, the complete catalog for this territory discloses a problem for SCE algorithm; presumably, the inadequacy of assumptions in case of more complex multi-fractal distribution, e.g., like in the area where the general strike of Baikal rift zone changes for about 45°.

The declustered catalogue, as expected, has lower range of the seismic activity, with maximum *A* about − 2.8, which corresponds to one mainshock in 600 years. The *B* coefficient has mainly the same distribution, while the range of *C* coefficient distribution shifted up for about 0.3 or more from 0.8 to 1.5. The maximum of *C* values for mainshocks is observed on the northernmost edge of Baikal Lake (Fig. 8).

4 Conclusion

The reliability USLE coefficients assessment is of special importance when aimed to provide an adequate estimation of seismic hazard at national or regional scale. We find it achievable when the existing earthquake data are just a few decades long but rich enough in magnitude determinations. In such a case, the maximum magnitude, $M_{\max}$, expected in T years at significance level *p%* can be obtained from $N(M_{\max}, L) = p\%$, then used for mapping ground shaking parameters by means of NDSHA algorithms [9]. Based on this study, the reliable USLE based seismic hazard maps could be obtained for a grid of about 1/8° (~ 14 km) or larger. Evidently, seismic hazard estimates may not rely exclusively on earthquake instrumental data and require the neo-deterministic assessment based on solid geological evidence, pattern recognition, and earthquake scenario simulations [10]. So far, it is early for definitive conclusions about the predictive power of the USLE based maps and

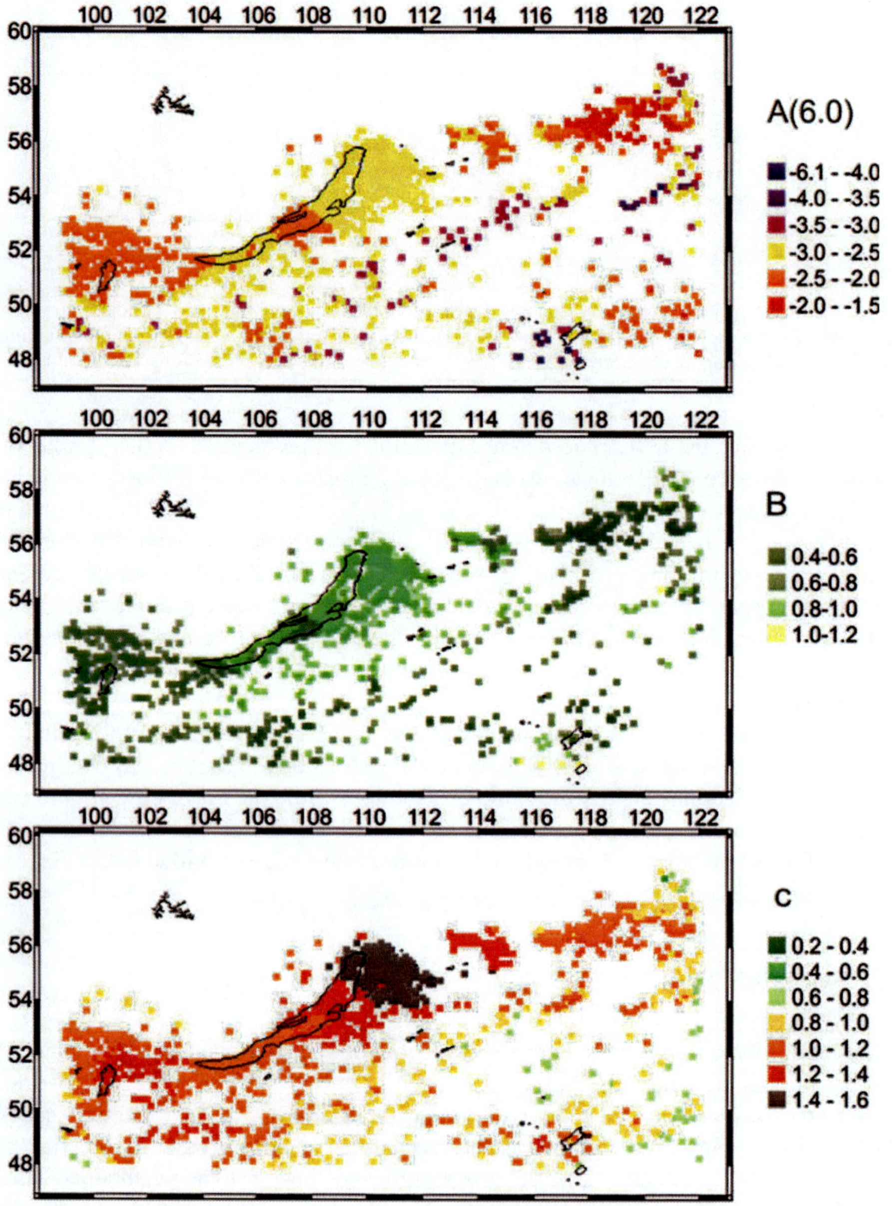

Fig. 8 Spatial distributions USLE regional coefficients *A*, *B* and *C* estimated for declustered catalogue of mainshocks

patterns. Numerous systematic studies and statistical testing of emerging patterns are needed and will follow in the near future.

Acknowledgements We are indebted to I. Zaliapin for providing the code for nearest-neighbor analysis. The research presented in this paper benefited from funding provided by RSF project №20-17-00180.

References

1. Kossobokov, V., Mazhkenov, S.: On similarity in the spatial distribution of seismicity. In: Chowdhury, D.K. (ed.) Computational Seismology and Geodynamics/AGU, 1st edn, pp 6–15. The Union, Washington (1994)
2. Nekrasova, A., Kossobokov, V.: Generalizing the Gutenberg–Richter scaling law. EOS Trans AGU **83**(47), NG62B–0958 (2002)
3. Nekrasova, A., Kossobokov, V.: General law of similarity for earthquakes: evidence from the Baikal Region. Dokl. Earth Sci. **407A**(3), 484–485 (2006)
4. Nekrasova, A., Kossobokov, V., Parvez, I., Tao, X.: Seismic hazard and risk assessment based on the unified scaling law for earthquakes. Acta Geod. Geophys. **50**(1), 21–37 (2015)
5. Zaliapin, I., Ben-Zion, Y.: Earthquake clusters in southern California I: identification and stability. J. Geophys. Res. **118**(6), 2847–2864 (2013)
6. Nekrasova, A., Kossobokov, V.: Unified Scaling Law for Earthquakes: Global Map of Parameters, ISC's Seismological Dataset Repository (2019)
7. Baikal Division of the Geophysical Survey, Federal Research Center of the Russian Academy of Sciences. Homepage. http://www.seis-bykl.ru/modules.php?name=Data&da=1. Last accessed 2020/12/10
8. Radziminovich, N., Miroshnichenko, A., Zuev, F.: Magnitude of completeness, b-value, and spatial correlation dimension of earthquakes in the South Baikal Basin, Baikal Rift System. Tectonophysics **759**, 44–57 (2019)
9. Nekrasova A., Kossobokov, V., Parvez, I.; Tao, X.: Unified scaling law for earthquakes as applied to assessment of seismic hazard and associate risks. Izv. Phys. Solid Earth **56**(1), 83–94 (2020)
10. Kossobokov, V.: Hazard, risks, and prediction. In: Panza, G.F., Kossobokov, V.G., Laor, E., De Vivo, B. (eds.) Earthquakes and Sustainable Infrastructure, pp. 1–25. Elsevier, Amsterdam (2022)

Clustering as One of Scenario of Development of Instability: An Earthquake Case

M. V. Rodkin, M. Yu. Andreeva, and E. V. Liperovskaya

Abstract The scenarios of development of bifurcations in different natural systems are believed to be similar. One of the examples of the development of bifurcation is the case of earthquakes, perhaps the mostly studied due to its practical importance and the amount of available data. However, even in this case the scenario of development of instability remains to be unclear. The method of examination of generalized vicinity of large earthquake is applied to reveal the typical scenario of development of seismic instability. Besides of the fore- and aftershock cascades two other types of precursor behavior were found. The time clustering of the main events and a moderate increase in background seismicity since several years before the large earthquake occurrence were found. Taking into account the uniformity of scenarios for the development of instability in systems of different nature, similar modes of instability development probably can origin also in other systems.

Keywords Generalized vicinity of large earthquake · Precursor behavior · Earthquake clustering · Modes of development of instability

1 Introduction

Significant uniformity of scenarios of the development of instabilities (bifurcations) in systems of very different physical nature is a widespread opinion. The signs of the

M. V. Rodkin
Institute of Earthquake Prediction Theory and Mathematical Geophysics, Russian Academy of Sciences (IEPT RAS), Profsoyuznaya Street, 84/32, Moscow 117997, Russian Federation
e-mail: rodkin@mitp.ru

M. V. Rodkin · M. Yu. Andreeva
Institute of Marine Geology and Geophysics Far Eastern Branch Russian Academy of Sciences (IMGG FEB RAS), Nauki Street, 1B, Yuzhno Sakhalinsk, Sakhalin Region 693022, Russian Federation

E. V. Liperovskaya (✉)
Schmidt Institute of Physics of the Earth of the Russian Academy of Sciences (IPE RAS), B. Gruzinskaya Street, 10, Build. 1, Moscow 123242, Russian Federation
e-mail: liper@ifz.ru

A. Kosterov et al. (eds.), *Problems of Geocosmos—2020*, Springer Proceedings in Earth and Environmental Sciences, https://doi.org/10.1007/978-3-030-91467-7_18

approaching bifurcation are usually assumed to be an increase in energy release, the development of long-range correlations, an increase in the low-frequency component in the spectrum of oscillations of the given system [1–7, and etc.]. One of the examples of the development of instability is the case of large earthquake occurrence, maybe the mostly studied because of its evident practical importance and an availability of statistical data. From here it seems natural that earthquake's precursors were applied for instability forecasting in other fields, even in such an unexpected area as forecasting the results of presidential elections in the United States [8, 9]. On the contrary, the features of critical state approaching, known in the physics of critical phenomena, were applied in earthquake forecasting [2, 7].

However, despite the fact that the case of earthquakes is one of the most studied, the scenario of development of a typical seismic instability remains controversial. This is due to the large individual (random) variability in change of flow of events in the spatial-temporal vicinity of each individual strong earthquake. This variability is so large that doubts are periodically expressed about the validity of even such well-known law of seismicity as the Omori-Utsu law [10]. The doubts are even more grounded for the case of validity of existence of foreshock cascades [11]. Hence, the detailed description of the mean typical scenario of precursor and post-shock behavior including possibly different modes of such behavior is fairly actual.

2 Typical Scenario of Instability Development Based on Seismological Data

In order to clarify the typical general features of the fore- and aftershock process a method of constructing and analyzing the generalized vicinity of a strong earthquake (GVLE) was proposed and implemented [6, 12–15]. The GVLE method combines data from many separate fore- and aftershock sequences into a single fore-aftershock sequence. This approach is close to the overlapping epochs method. The GVLE method differs from the works of other authors by the large (up to 1000) number of combined fore- and aftershock sequences and by normalization of the distance between events to the source size of the corresponding main earthquake. The GVLE method has provided huge growth in statistics. As a result, it turned out to be possible to describe the average typical features of the fore- and aftershock sequence in detail inaccessible in any other way.

As an example, Fig. 1 presents the change of fore- and aftershock numbers in GVLE [14]. In GVLE constructing the ISC world catalog was used; earthquakes with a magnitude Mb $\geq$ 5.0 and a depth H $\leq$ 70 km were taken into account. In this GVLE the vicinities of 500 strongest earthquakes were summarized; and the events located at a distance R (km) less than 5 source sizes of the main event from the hypocenter of the corresponding main event were used. To calculate the main event's source size the relations from [16, 17] were used. The difference in the formulas used for calculation of the source size of the main event does not affect the

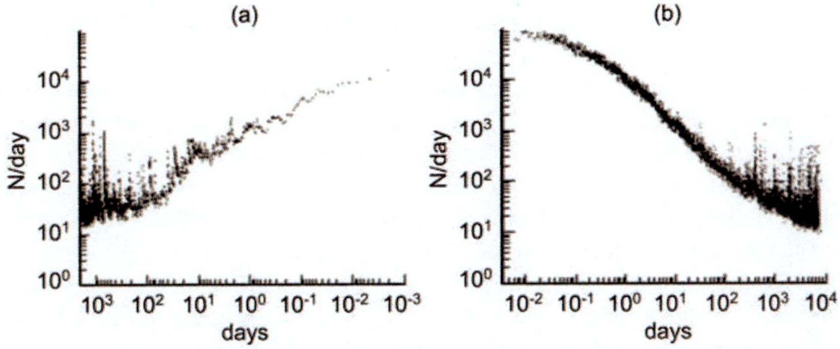

Fig. 1 Fore- (**a**) and aftershock (**b**) sequences in GVLE, ISC catalog, see details in the text

results qualitatively. Both the power-law foreshock cascade growing to the moment of the generalized main event and the generalized Omori-Utsu law of change in aftershocks' number are seen clearly in Fig. 1.

Let us now present a spatial-temporal generalized vicinity. Figure 2 shows the mean typical change in intensity of number of small earthquakes in the generalized vicinity of 500 strongest shallow (H < 70 km) earthquakes; the world GCMT catalog was used here, time's scale at the x-axis is close to logarithmic, along the y-axis the distance from the epicenter of the main earthquake at a scale of the source size of the corresponding main event is given. Removals of no more than 6 sizes of the focus of the main event are considered. Foreshock and aftershock cascades are clearly visible,

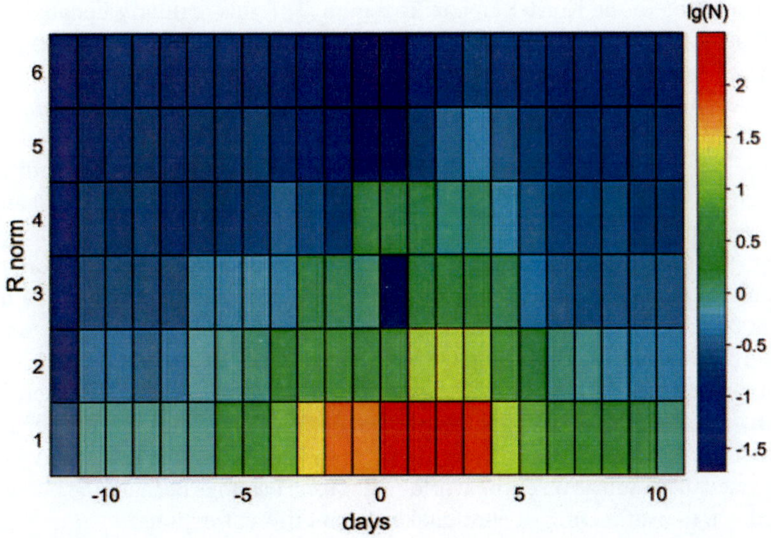

Fig. 2 Spatial-temporal vicinity of the generalized main event, see details in the text

they are located from the epicenter of the main event at a distance of 3–4 main event source size.

Besides the fore- and aftershock cascades, the moderate but clearly seen increase in a mean number of events that begins 5–10 years before the occurrence of a generalized strong earthquake can be seen. The increase takes place at distances 2–3 source sizes of the main event from the generalized main event. This increase in the earthquake number can be considered as another kind of precursor behavior, in addition to the power-law cascade of increase in foreshocks' number.

Note also that in a time vicinity of about one day from the generalized large event (GLE), at distances of 4–5 sizes of the focus of the main event, a tendency towards a decrease in the number of events is revealed. This effect can be associated with unloading the vicinity of the upcoming large event. The seismicity minimum immediately after the GLE occurrence at a distance of 3 times the source size can have a random character, due to the small number of events in this cell.

3 Regional Analysis of the Generalized Vicinity of a Large Earthquake

Applying the GVLE method to regional data allows new results to be obtained. Further, we discuss some features of the precursor behavior which had been masked when using the world earthquake catalogs because of summing up the earthquakes from different regions.

One of the suitable catalogs of the active continent-ocean transition zone is the regional catalog of the Kuril-Kamchatka region [18]; this regularly updated catalog is used below. As can be seen from Fig. 3b, the completeness of the catalog can be considered satisfactory for events with a magnitude mb $\geq$ 4.0.

The seismicity of the Kuril-Kamchatka region can be subdivided into two depth ranges, separated by a well-identified minimum of seismic activity at depths 300–400 km. This allows us to consider separately the seismicity for events with depths less than 300 km and more than 400. Fore- and aftershock sequences in Fig. 4 are seen clearly despite much fewer number of strong events in the Kuril-Kamchatka regional catalog. The GVLE for Kuril-Kamchatka region was constructed using vicinities of 72 events with mb $\geq$ 6.3 and depth $H < 300$ km. The generalized vicinity of deeper ($H \geq 400$ km) earthquakes was built for 12 strong (mb $\geq$ 6.3) earthquakes only.

At Fig. 4, besides the fore- and aftershock cascades, a few strong bursts of seismicity can be seen in the foreshocks area (and to a lesser extent, in the aftershocks' domain). These bursts in earthquakes number appear to be caused by the occurrence of strong nearby events. Similar strong bursts in fore- and aftershock activity were not detected in examination of the world earthquake catalogs because the bursts were masked due to overlapping of earthquakes from different regions.

Result Fig. 4 can be generated by strong earthquakes clustering in time. The clustering does take place. Figure 5 shows the time row of occurrences of strong

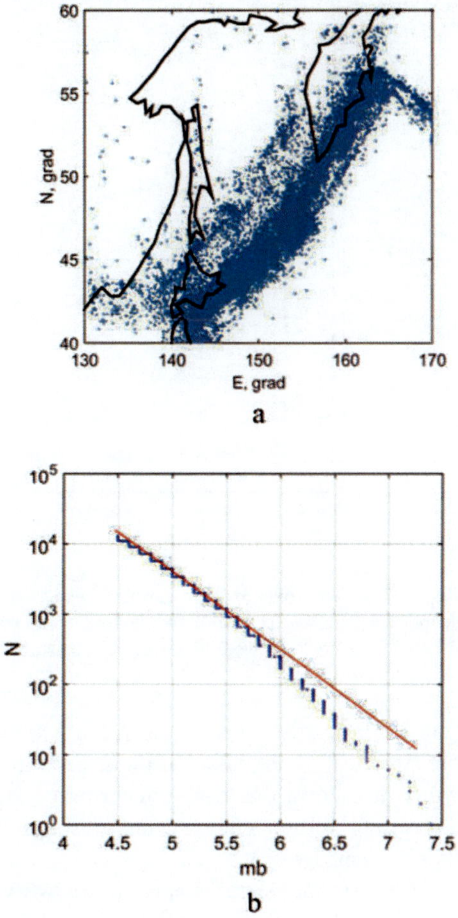

Fig. 3 The earthquakes in Kuril-Kamchatka region, blue dots (**a**), and the earthquakes' recurrence plot (**b**)

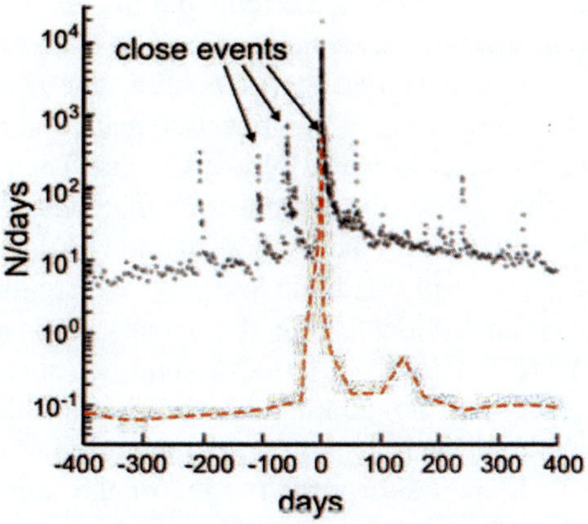

Fig. 4 The foreshock and aftershock sequences of the generalized vicinity of a large earthquake; black dots represent the case of H < 300 km main events, the red dashed line represent the H > 400 km events. Arrows mark some events that are close to each other over time

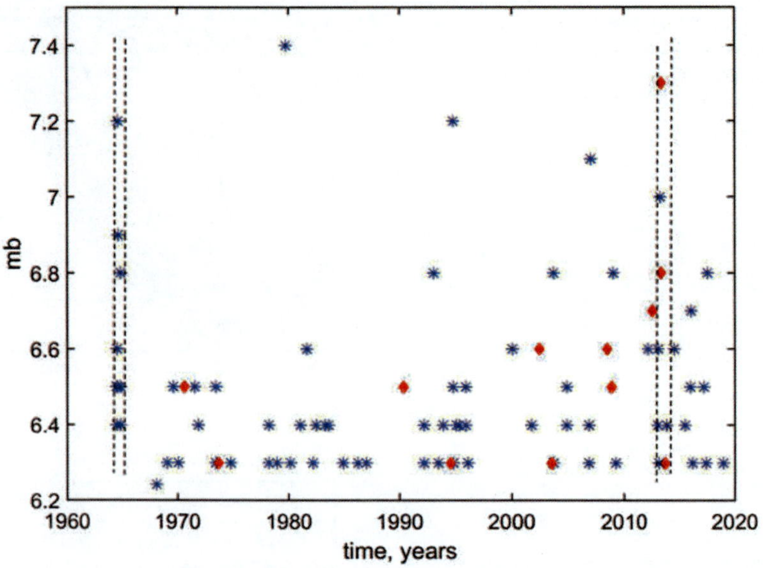

Fig. 5 The time rows of large earthquakes that were used to make the generalized vicinity; less deeper earthquakes (H < 300) are shown as blue stars, the deeper ones (H > 400) as red diamonds. Dashed lines enclose two clusters of high seismicity

(mb > 6.3) earthquakes in the Kuril-Kamchatka region, 84 events in total. Two of the strongest earthquakes of them have magnitudes mb = 7.4 and mb = 7.3, from which it follows that such clustering cannot be caused by aftershocks occurrence. According to Bath's law, a typical strongest aftershock is 1.0 magnitude and weaker than its main event.

The fact that cluster events, in the overwhelming majority, are not aftershocks is also supported by their location, Fig. 6. It can be seen that the events from one cluster are widely dispersed throughout the territory of the region, and therefore cannot be interpreted as aftershocks of one of the events of the cluster [19].

Let us now consider the entire time row of earthquakes, not limited to only the strongest ones. A significant contribution, from 30 to 50% of the total number of earthquakes is made by aftershocks. The aftershocks essentially mask the seismic regime. There are a number of methods of identification of aftershocks. The most widely used are the window method [20] and the method [21] in the software implementation of Vladimir Smirnov. We supplemented these two approaches with the method for identifying aftershocks proposed in [22]. The method [20] essentially overestimates the aftershocks number, since all independent events occurring in the space-time neighborhood of the main events fall into the number of aftershocks. The method [22] is a simplified formulation of the closest ancestor method [23, 24].

Figure 7 shows the change of the number of main events in the region under consideration when using the three mentioned above algorithms of aftershocks identification. One can see both significant differences in the number of main events

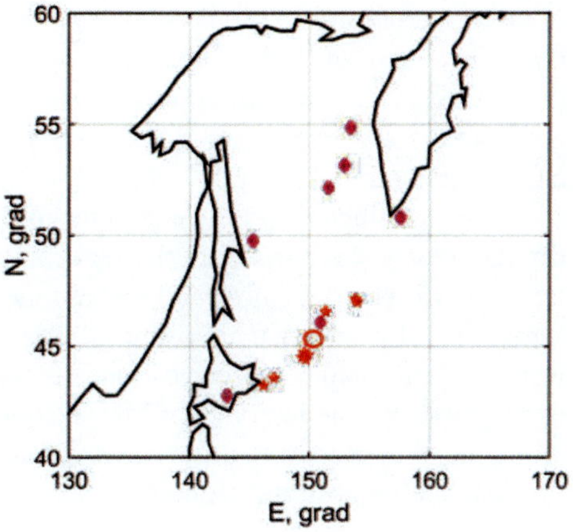

Fig. 6 The large earthquakes that compose two clusters of seismicity from Fig. 5 are shown by stars and diamonds. The size of the source of mb = 7.0 event is shown by circle

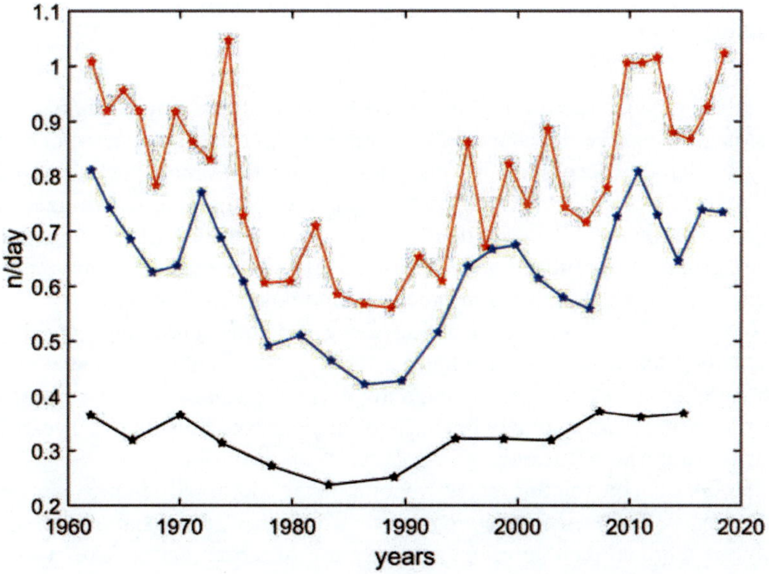

Fig. 7 Average density of the number of main events (M $\geq$ 4.0) for time-sequential groups of 500 earthquakes; black—by the window method [20], blue—by the algorithm [21], red—by the method [22]

identified by different methods, and the similarity of change of the number of the main events with time.

Below we discuss the results obtained with the use of the aftershocks identification method [21]. We had checked that other methods give similar tendencies. Events mb $\geq$ 4.0 were taken into account, and the earthquakes with mb $\geq$ 5.0 were considered as possibly accompanied by aftershocks. The distributions of the number of earthquakes in time intervals of 15, 30 and 60 days were compared for the actual distribution of the main events and for artificial catalogs, with the same number of events randomly distributed in the same time interval. In all cases 10 artificial catalogs were used to estimate the statistical deviations (Fig. 8). Time intervals 15, 30 and 60 days in all cases were counted from the moment of each next event both real and artificial.

In all examined cases the character of distributions of real and artificial catalogs differs greatly (Fig. 8a–c). In the real catalog, the probability of occurrence of time intervals with the very high and low number of events always were found to be much higher than for the artificial catalogs. The abnormally high and low seismic activity time intervals appear to correspond to time clusters of high seismic activity and compensating time intervals of low seismic activity. Note, that in linear coordinates along the ordinate, these differences would be less noticeable than at Fig. 8 where at "y" axis the logarithmic scale is used.

4 Discussion

The method of constructing a generalized vicinity of a strong earthquake GVLE allows one to describe the character of averaged precursor and post-shock seismic activity in detail inaccessible in other ways. The foreshock power-law cascade and aftershock sequence relaxation, proceeding according to the generalized Omori-Utsu law, are seen in GVLE excellently. The results of the GVLE examination give, however, grounds to assume that the fore- and aftershock sequences are not the only elements of the typical scenario of precursor and post-shock activity.

In result of the GVLE constructing using the most numerous world data, a moderate long-term increase in background activity was found. The increase in earthquakes number takes place around the main event at distance up to 2–3 source sizes of the main event. This anomaly begins several years, possibly up to 10 years before the main earthquake occurrence.

When GVLE constructing using regional data, additional features are revealed that were masked when using heterogeneous world data. In an examination of Kuril-Kamchatka regional data an effect of clustering of strong earthquakes was found. Note, that similar feature was found earlier. In [25], the seismicity of the whole Pacific region was considered, and it was concluded that the observed temporal clustering of strong (approximately M8.5+) seismicity can occur randomly with a probability less than 1%. In [25], it was suggested, that, in terms of the occurrence of strong earthquakes, the Earth's system behaves to a large extent as a self-synchronizing system.

The analysis of non randomness of the time distribution of the strongest earthquakes was confirmed by results of examination of the sequence of the main events (after the exclusion of aftershocks). It is shown that the time distribution of the main events differs significantly from the random distribution fitting the Poisson's law. For real seismicity a clustering of the main events in some time intervals takes place.

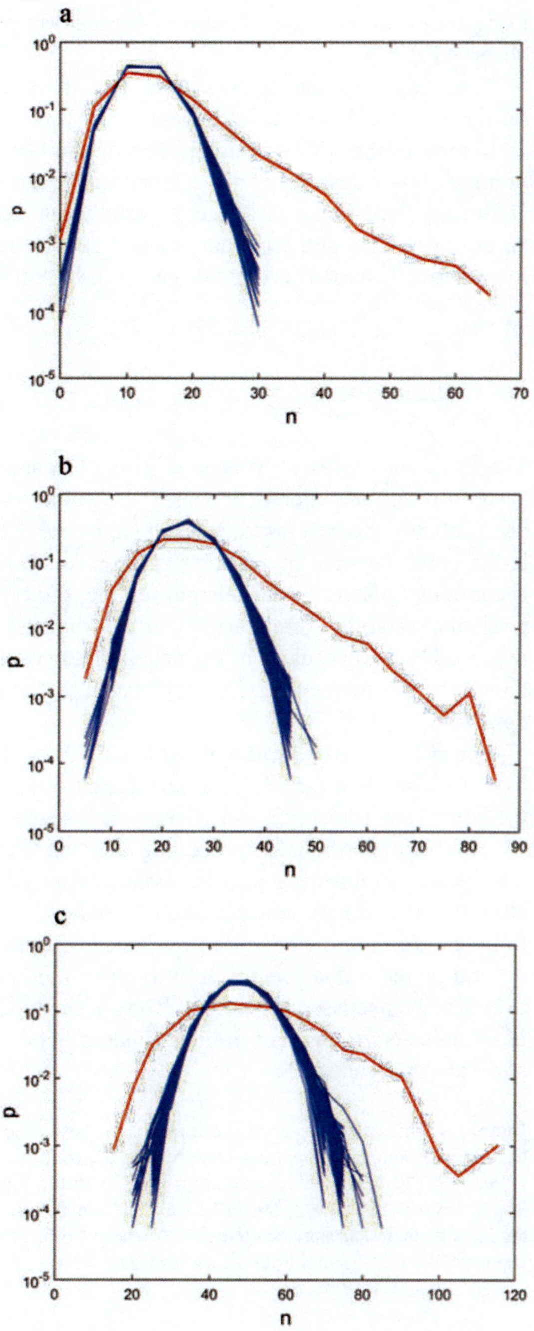

Fig. 8 Comparison of the distribution densities of the number of real (red line) and artificial (blue lines) main events falling into intervals of 15 (**a**), 30 (**b**), and 60 (**c**) days

Long-term (on the scale of the first decades) fluctuations in the seismic activity are observed also.

Thus, it seems can be concluded that the scenarios of seismic activity are not limited to the effect of development of the foreshock and aftershock cascades.

In view of the wide-spread opinion on the similarity of scenarios of development of instability in systems of different nature, it can be suggested that such phenomena as the main earthquakes clustering can occur also in systems of other nature. As an example, such political phenomenon as the Arab Spring of 2011 maybe could be hypothetically treated as an analogue of the effect of clustering of strong earthquakes.

5 Conclusions

The essential universality of scenarios of development of instability occurring in systems of different nature appears to be a common opinion. One of the most studied and probably the best provided with statistical data cases of instability development is the case of origin of a strong earthquake. But even in this case, due to a great random component in the seismic regime, the typical scenario of a development of a seismic instability remains unclear. For the purpose of a detailed description of the averaged typical scenario of the development of a seismic instability, the method of constructing a generalized vicinity of a large earthquake, GVLE, was suggested and applied.

The GVLE method was applied to data from world and regional earthquake catalogs. As a result of the GVLE application, in addition to the mode of development of the power-law foreshock and aftershock cascades, a long-term moderate increase in the number of earthquakes preceding an occurrence of a large earthquake was found. The effect of clustering of main events (after aftershocks excluding) was revealed also. Whereas the existence of the power-law fore- and aftershock cascades appears to be common knowledge, the existence of clustering of main events and a background increase in a seismic activity prior the large earthquake occurrence appears to be hardly discussed previously. Thus, it seems can be concluded that the scenarios of seismic activity are not limited by an effect of development of the foreshock and aftershock cascades.

Funding This work was carried out for a state assignment of the Institute of Earthquake Prediction Theory and Mathematical Geophysics, Russian Academy of Sciences (topic AAAAA19-119011490129-0), Schmidt Institute of Physics of the Earth of the Russian Academy of Sciences (topic AAAA-A19-119110690110-6) and of the Institute of Marine Geology and Geophysics, Far East Branch, Russian Academy of Sciences (topic AAAA-A18-118012290125-2.2), and was supported by the Russian Foundation for Basic Research, project no. 19-05-00466.

References

1. Bak, P.: How Nature Works. Springer, New York (1996)
2. Sornette, D.: Critical Phenomena in Natural Sciences. Springer, Berlin (2000)
3. Sobolev, G.A.: Evolution of periodic oscillations of seismic activity before strong earthquakes. Izv. Russ. Acad. Sci. Phys. Solid Earth Engl. Transl. **11**, 3–15 (2003)
4. Sobolev, G.A., Lyubushin, A.A.: Microseismic impulses as earthquake precursors. Izv. Russ. Acad. Sci. Phys. Solid Earth Engl. Transl. **42**(9), 721–733 (2006)
5. Pruessner, G.: Self-Organised Criticality: Theory, Models and Characterisation. University Press, Cambridge (2012). https://doi.org/10.1017/CBO9780511977671
6. Rodkin, M.V.: Patterns of seismicity found in the generalized vicinity of a strong earthquake: agreement with common scenarios of instability development. In: Sharma, A.S., et al. (eds.) Extreme Events and Natural Hazards. The Complexity Perspective. Geophysical Monograph Series, pp. 27–39. AGU, Washington (2012). https://doi.org/10.1029/2011GM001060
7. Shebalin, P.N., Zaliapin, I., Keilis-Borok, V.I.: Premonitory rise of the earthquakes' correlation range: Lesser Antilles. Phys. Earth Planet. Inter. **122**, 241–249 (2000)
8. Keilis-Borok, V.I., Lichtman, A.J.: Pattern recognition applied to presidential elections in the United States, 1860—1980: the role of integral social, economic, and political traits. Proc. Natl. Acad. Sci. U.S.A. **78**, 7230—7234 (1981)
9. Lichtman, A.J.: Predicting the next president. In: The Keys to the White House. Rowman & Littlefield, USA (2020)
10. Romashkova, L.L., Kosobokov, V.G.: The dynamics of seismic activity before and after great earthquakes of the world, 1985–2000. Vychisl. Seismol. **32**, 162–189 (2001) (in Russian)
11. Hardebeck, J.L., Felzer, K.R., Michael, A.J.: Improved tests reveal that the accelerating moment release hypothesis is statistically insignificant. J. Geophys. Res. **113**, B08310 (2008). https://doi.org/10.1029/2007JB005410
12. Rodkin, M.V.: Seismicity in the generalized vicinity of large earthquakes. J. Volcanol. Seismol. **2**(6), 435–445 (2008)
13. Rodkin, M.V., Tikhonov, I.N.: The typical seismic behavior in the vicinity of a large earthquake. Phys. Chem. Earth **95**, 73–84 (2016)
14. Rodkin, M.V.: A typical foreshock and aftershock anomaly: observations, interpretation, and applications. J. Volcanol. Seismol. **14**(1), 58–69 (2020)
15. Rodkin, M.V.: Earthquake prediction: old expectation and new results. Seismic Instrum. **57**, 438–455 (2021)
16. Sobolev, G.A.: Fundamentals of Earthquake Prediction. Nauka, Moscow (1993) (in Russian)
17. Wells, D.L., Coppersmith, K.J.: New empirical relationships among magnitude, rupture length, rupture width, rupture area, and surface displacement. BSSA **84**(4), 974–1002 (1994)
18. Andreeva, M.Yu., Kim, Ch.U.: Earthquakes of Kuril-Kamchatka Region (1737–2009). IG&G FEB RAS, Yuzhno-Sakhalinsk (2012)
19. Rodkin, M.V., Andreeva, M.Yu., Grigorieva, O.O.: An analysis of the generalized vicinity of a large earthquake using regional data: the Kuril–Kamchatka region. J. Volcanol. Seismol. **14**(6), 410–419 (2020)
20. Gardner, J.K., Knopoff, L.: Is the sequence of earthquakes in Southern California, with aftershocks removed, Poissonian? Bull. Seismol. Soc. Am. **64**(5), 1363–1367 (1974)
21. Molchan, G., Dmitrieva, O.: Aftershock identification and new approaches. Geophys. J. Int. **109**, 501–516 (1992)
22. Pisarenko, V.F., Rodkin, M.V.: Declustering of seismicity flow: statistical analysis. Izv. Phys. Solid Earth **55**(5), 733–745 (2019)
23. Zaliapin, I.A., Gabrielov, V., Keilis-Borok, V.I., Wong, H.: Clustering analysis of seismicity and aftershock identification. Phys. Rev. Lett. **101**(1), 1–4 (2008)
24. Zalyapin, I., Ben-Zion, Y.: Earthquake clusters in Southern California I: identification and stability. J. Geophys. Res. **118**, 2847–2864 (2013)
25. Bufe, Ch., Perkins, D.: Evidence for a global seismic-moment release sequence. Bull. Seismol. Soc. Am. **95**(3), 833–843 (2005)

Possible Relationship of Some Weak Earthquakes in Turkey with Industrial Explosions

V. I. Zhuravlev, A. Ya. Sidorin, and A. A. Lukk

Abstract Possible industrial explosion effects on weak earthquake seismicity in Turkey are considered. The spatial and temporal relationships between the studied events within short periods of time and over short distances, as well as their probable long-term effects, are investigated. Potential correlation between normalized distributions of daily and weekly earthquake and explosion numbers is estimated. Possible changes in the fractal dimension of the spatial seismicity distribution in connection with industrial explosions are studied. To search for spatially compact regions of explosions and local earthquake groups, the method of identifying clusters was applied. Clusters of explosions and earthquakes that are closely spaced have been found. In these clusters, the indicated time series behavior was compared. The significance of the observed temporal and spatial changes is assessed by the method of creating artificial catalogs, as well as by constructing inverse statistics of the earthquake-explosion.

Keywords Spatiotemporal analysis · Periodicities · Earthquake-explosion statistics · Turkey

1 Subject of the Study

The trigger mechanism of the earthquake occurrence under the industrial factors influence on the seismic regime is quite often investigated in scientific works. These factors include, in particular, the mode of filling reservoirs [1]. Some authors assumed the influence of water reservoirs on the daily and seasonal frequencies [2]. In addition, explosions are considered as one of the factors influencing seismic activity [3]. To analyze the relationship between earthquakes and explosions, it is necessary to have at least some separate catalogs of these events, which are often not easy to find. Recently, Turkey science organization made obtaining such data possible. There is a free downloadable resource for scientific investigation that we used when conducting this small research (http://udim.koeri.boun.edu.tr/zeqdb/indexeng.asp). Date source

V. I. Zhuravlev (✉) · A. Ya. Sidorin · A. A. Lukk
Schmidt Institute of Physics of the Earth, Russian Academy of Sciences, Moscow 123242, Russia

A. Kosterov et al. (eds.), *Problems of Geocosmos—2020*, Springer Proceedings in Earth and Environmental Sciences, https://doi.org/10.1007/978-3-030-91467-7_19

is provided by Boğaziçi University Kandilli Observatory and Earthquake Research Institute—Regional Earthquake—Tsunami Monitoring Center. This resource allows us to download earthquake and explosion catalogs. The source is intended only for scientific research purposes and mandatory citation is required. In undertaking this work, among other tools, we used the GMT6 package for working with maps (version for windows) and Cookbook-GMT 6.1.1 documentation (generic-mapping-tools.org). We have previously studied the seismic regime properties in Turkey and the possible connection between the specific diurnal periodicity and the industrial explosions, and we identified one, as a matter of fact [4]. Note that the catalog used in the mentioned study contained both earthquakes and, apparently, explosions. In the aforementioned work on the analysis of the differences in weekday and weekend seismicity maps, as well as the differences in night and daytime maps, the possible blasting operations places were clearly identified, which was also confirmed by the literature analysis. Furthermore, we obtained similar results when analyzing the seismicity of Iran [3]. This fact is not only a positive stimulus for further research, but also raises the question of whether earthquake catalogs are free of explosions, and vice versa, whether explosion catalogs are free of earthquakes. Several works have been devoted to searching for explosions in earthquake catalogs, particularly in the area under study [5–10].

2 Data Features

Let us present the parameters of the earthquake and explosion catalogs obtained by us from the indicated Internet resource. The earthquake catalog includes data on 244,491 events from 1995 to October 2020. The explosion catalog includes data on 25,387 events from 2005 to October 2020 as well. These are quite large-scale and representative statistical data.

When comparing the two processes, we will use only the interval from 2005, the volume of the earthquake catalog with 210,197 events. Before starting the analysis, we checked whether the same event was mistakenly included in both catalogs. To achieve this objective, we have the total time in the earthquake source for the same time in the explosion catalog, and vice versa, the explosion time from the explosion catalog was searched in the earthquake catalog. We did not find such matches in the catalogs. Figure 1 shows histograms of earthquakes and explosions magnitude distributions for these catalogs.

It is obvious from the figure that the relative number of strong events in explosions is much less than for earthquakes. Events that are competitive in magnitude are up to $M < 1.8$. Even the strongest explosions are weaker than many earthquakes. It also casts doubt on the fact that explosions can seriously affect seismicity. Thus, this influence may not be readily apparent. The following hypotheses of such influence are possible. Explosions can slightly change a strong earthquake timing. Setting off intensive explosions in the same area can relieve the earthquake preparation zone, weaken or prevent the earthquake, and it is possible to change the place of its

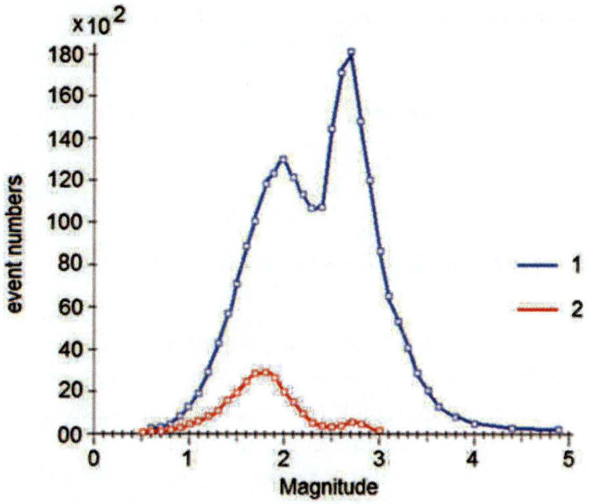

Fig. 1 Histograms of earthquake (1) and explosion (2) magnitudes

occurrence. Strong isolated explosions can affect weak seismicity. Figure 2 shows the earthquake and explosion general map. The boundary coordinates for the earthquake catalog are as follows:—26.0000–44.9992 E, 35.0000–41.9983 N. For explosions: 26.1443–44.0447 E, 35.0523–41.9803 N. All maps in GMT soft were plotted with "–RTR" option, which is Turkey.

A high density of explosions is observed only in one place where the stations are concentrated. For a preliminary assessment of the possible connection between explosions and earthquakes, visible on the given map in numerical rather than visual representation, we did the following: divided the entire map into 40 divisions from south to north and 80 from west to east. In this case, we get cells with a size of 0.174° by 0.225° (19.45 * 19.32 km). This cell size is larger than the coordinate

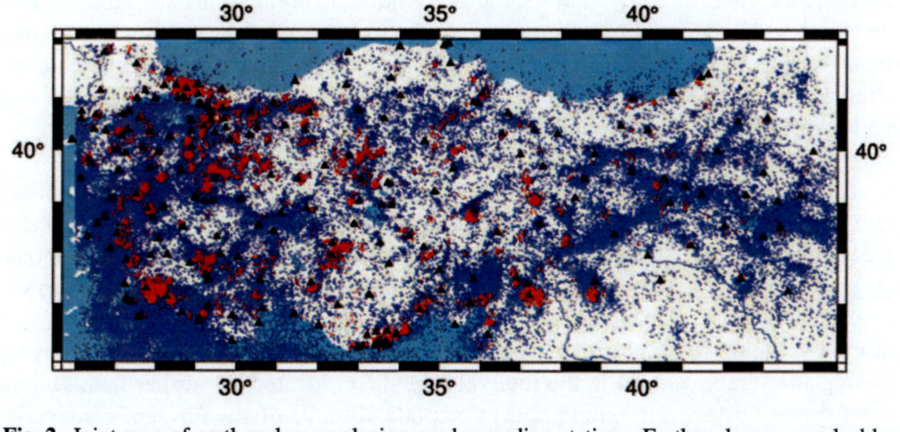

Fig. 2 Joint map of earthquakes, explosions and recording stations. Earthquakes are marked by blue circles, explosions—by red circles, stations—by black triangles

measurement error and makes it possible to have good event statistics in many cells. Let us count the number of explosions and the number of earthquakes in each of these cells, and arrange the result according to the number of explosions from minimum to maximum. The result gave low numbers of cells, where numbers of earthquakes and blasts were big enough.

3 Daily and Weekly Periodicities

We calculated the spectra of the earthquake and explosion series for periods of up to two days. We considered each event as a delta function with time from the catalog. The common feature of both spectra is the main extremum at the 24 h period and the maxima derived from it at multiple periods of 12, 8, 6 h and other periods. This type of spectrum indicates a strong deviation of the daily course from the harmonic in favor of the degree. However, there are two more periods on the explosion spectrum, 18 and 28 h, which are a consequence of the nonlinear modulation of the diurnal harmonic with a period of a week. This is precisely not an additive inclusion of the week period, but nonlinear modulation. This is a natural consequence of the fact, that the number of explosions carried out, has a weekly periodicity. An analysis of the weekly average blasting progress showed that the number of explosions conducted on Saturday is about 15% less than the number of explosions in the middle of the week, and the number of explosions conducted on Sunday is about 58 percent of the number of explosions on any workday. For a weekly series of earthquakes, the day-to-day contrast is insignificant. The number of events on Sunday is only 4% less than on weekdays. However, if we consider only those earthquakes that occurred at distances less than 10 km from the explosions, the weekly course for them has a noticeable contrast, and the number of earthquakes on weekends turns out to be noticeably larger than on weekdays. One of the following reasons can explain the results. First of all, noise hypothesis. Industrial noise, in the areas of explosions, is less on weekends, and therefore, recorded against a background of less noise. As a result, more earthquakes are recorded. The second possible reason is that some tectonic stresses are released by explosions on weekdays, but the accumulated stresses are unleashed on weekends, when there are no such explosions. This is a long-term cumulative effect and not an effect within a particular week.

The analysis of the daily explosion frequency showed that the overwhelming number of explosions occurs in the daytime from 6 to 9 a.m. During the rest of the day, there are only isolated events. There are two extremums on the daily course at 11 a.m. and 4 p.m. Although diurnal earthquakes have the maximum number of events at night, during the day there are also extrema similar to those in the diurnal variation for explosions at about 11 a.m. and 4 p.m. When we plot maps of earthquakes that occurred in the interval of 1–12 h, located no further than 10 km from the explosion, we found almost complete absence of such in the seismically active area in the southwest of the region and places with many blasts. Thus, the

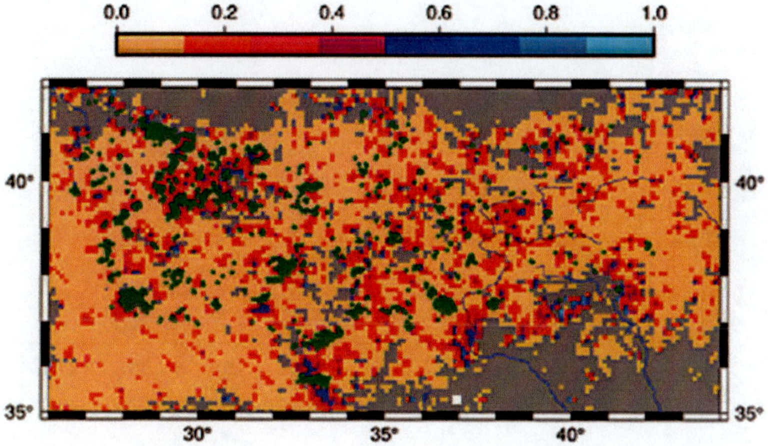

Fig. 3 Distribution of earthquake number occurred in the 11–12 h interval is presented with color map. Explosions are marked by green. Areas of blue color are overwhelmingly characterized by high activity from 11 to 12 h

effluent explosion features are detected in the seismicity properties only in a part of the region and cannot explain day effects on them (Fig. 3).

4 Spatial Data Analysis

Let us analyze the difference in distances for various time delays. Consider the explosion statistics ahead of the earthquake, the earthquake ahead of the explosion, and the chosen earthquakes ahead of the remaining earthquakes. Figure 4 shows the frequencies histograms of the distances between explosions and earthquakes. One to ten days separates them in time. Although the visual difference between plots 1 and 2 is small, it is quite significant in terms of the confidence interval sqrt(n) to n. The confidence interval for assessing the significance of the contrast is about 6 km about 30 events and about 13 km about 38 events, which is several times less than the observed difference in the curves. The maximum at 12–13 km is absent in the explosion-earthquake graph. Therefore, we can apparently talk about some kind of unloading of earthquake activity after explosions at a specified distance. Undoubtedly, this is a rather insignificant effect. The dependences with a time difference of up to a day show, on the contrary, the effect of some activation in the number of earthquakes after explosions at distances of about 9 km.

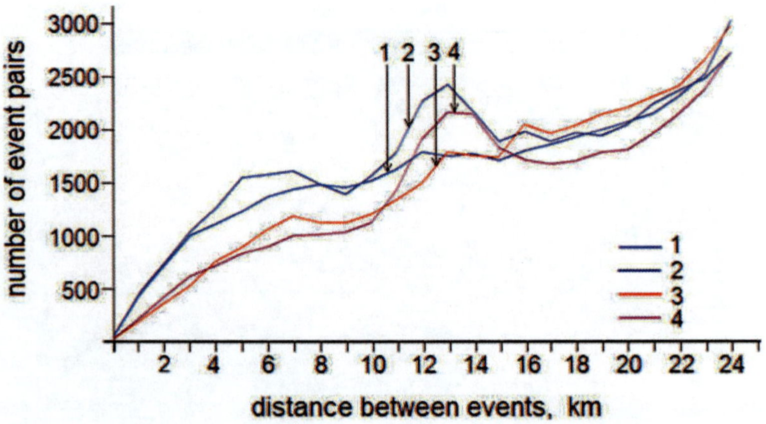

Fig. 4 1—The distribution of the distances from the explosion to the earthquake, the explosion is ahead of the earthquake; 2—same as 1, but the earthquake is ahead of the explosion; 3—distance between the sampled earthquake catalog and the remaining earthquake catalog; 4—same as 3, but the time difference between events is less than a day

5 Time-Distance Relations

Let us now consider the spatial-temporal domain interaction of explosions and earthquakes. Let us calculate two maps where earthquakes are ahead of explosions in time and explosions are ahead of earthquakes, if the difference in the distance between the explosion and the earthquake is less than 20 km (variants with less than 7 and 10 km are also calculated). We took the time difference between events and explosions to be no more than 10 days. A derivative result of the construction of such maps was also the distribution of distances and explosions from each other. To our surprise, on a completely chaotic schedule, there was a section where several earthquakes and explosions are located with a time delay of 100–200 and 500–550 s and are very close in distance from each other (less than 8 km). The following variants of the indicated phenomenon are assumed: (1) Trigger of seismic event after explosion; (2) Explosions occurring with a short delay one after the other. In this case, the recording network and the program for determining the hypocenters as a seismic event, not an explosion, interpret the recording of the second event; (3) After the explosion, due to the specifics of the place, a certain reflected wave appears with a short time delay, which is also mistakenly interpreted as a seismic event, and not as an explosion. There are 32 events satisfying the time condition from 65 to 250 s and at distances less than or equal to 8 km.

Pairs of explosions and earthquakes are marked in red, where the time difference is from 60 to 250 s, and the distance does not exceed 8 km. The remaining events are marked in blue, when the time periods between explosions and earthquakes are up to 20 days, and the distance is up to 10 km. Close events from one of the above-mentioned artificial catalogs and the remaining earthquakes are shown in green.

Table 1 Statistics of explosion-earthquake events, separated from each other up to 600 s and 8 km. There are in the columns pairs number (No), earthquake magnitude (Mag_{eq}), explosion magnitude (Mag_{exp}), time difference for "explosion-earthquake" (T_{diff}) in sec, distance between them (Dist) in km

No	Mag_{eq}	Mag_{exp}	T_{diff}	Dist	No	Mag_{eq}	Mag_{exp}	T_{diff}	Dist
1	1.9	1.7	14	5.499	17	1.8	2.4	117	2.160
2	1.6	1.9	15	7.790	18	1.9	2.0	140	2.997
3	1.6	1.5	24	2.898	19	2.9	2.2	145	6.959
4	1.5	1.6	25	6.193	20	1.6	1.6	165	7.815
5	2.9	1.8	36	6.889	21	1.9	1.7	177	4.562
6	3.4	1.7	39	2.622	22	1.5	1.4	178	5.311
7	1.7	2	46	1.169	23	1.3	1.8	180	5.198
8	1.9	1.9	47	3.975	24	2.1	2.1	186	0.000
9	1.5	1.2	67	6.783	25	1.8	2.0	186	5.424
10	1.6	1.6	80	4.796	26	1.4	1.0	193	6.953
11	1.9	1.9	85	2.183	27	3.8	2.2	202	0.678
12	1.5	1.5	90	5.543	28	1.7	1.8	203	4.694
13	1.6	2	92	1.051	29	2.3	1.5	209	5.894
14	2.1	2.2	99	6.680	30	2.5	2.7	212	6.124
15	1.3	1	101	2.818	31	2.6	2.7	219	7.432
16	1.5	1.7	113	4.182	32	1.7	1.6	222	3.693

These pairs show space distributed pattern and the fact, that the relation of the blasts and earthquakes is significant. The events marked in red fall under the possible definition of triggered events. However, one should exclude those that fall under the criterion that the earthquake is ahead of the explosion by a comparable time interval. If any of the events marked in red falls into the green area, its triggering nature is called into question. But the green domains are outside red. And so, judging by the small area with red color, potential pairs of events, which can be interpreted as triggers, are removed from each other for short periods of time. These are pair events "explosion-earthquake" where time difference between them is less than 600 s and distance is less or equal 8 km. Pair number 24 with distance 0 can be a mistake of event type detection (Table 1).

6 Strange Clusters

Let us define a spatial cluster as follows. This is an area in the plane where the number of events in a 6-km radius circle is twice as large as in the next ring after the circle with an area equal to the area of the circle. We looked at all cells of the event map with a step of 6 km. We did not expect that the clusters found in the earthquakes would coincide with the clusters from the explosions. The earthquake distribution is clearly fractal in nature, while the explosion distribution is less random. On a qualitative level, this was seen in Fig. 5. We analyzed spatial clusters of earthquakes and explosions and found two strange objects. These samples were close in space and were characterized by high numbers of earthquakes and explosions. For explosions, the sample center has the following coordinates 37.1773 N 27.8943 E, for earthquakes it has coordinates 37.2398 N 28.2068 E. The sample had a 10-km radius. The time series were plotted without any time averaging (Fig. 6).

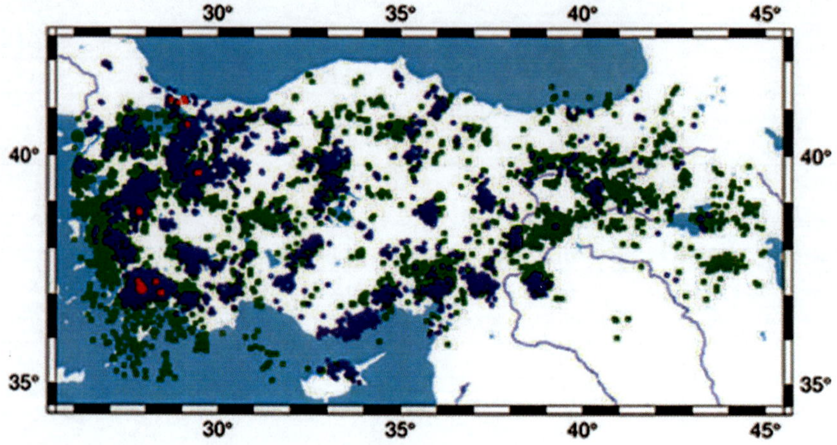

Fig. 5 Earthquakes and explosions close in time and distance

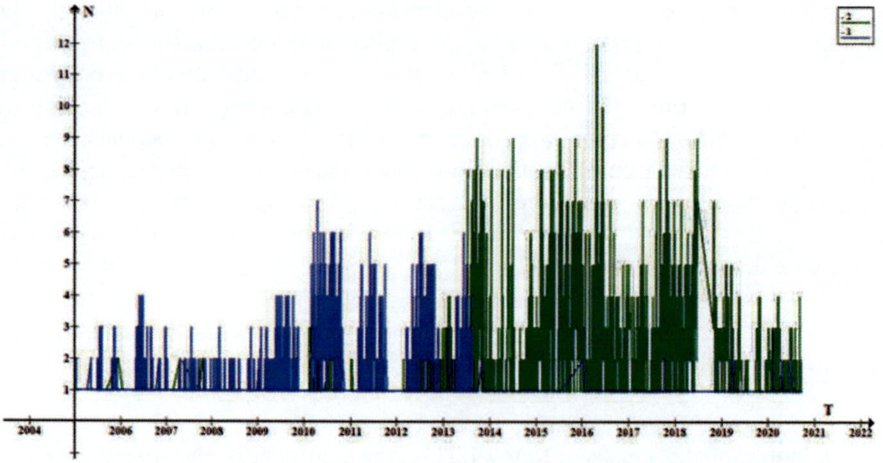

Fig. 6 Time series of earthquakes (blue) and explosions (green) in closely spaced samples with the highest numbers of events

We can see from the figure that the series actually overlap in time a little. At first, we observe, only earthquakes, then almost only explosions. Analysis of other spatial samples nowhere gave even a remotely similar result. We can make possible conclusions from the rows in the figure. A cataloging error occurred when explosions prior to mid-2012 were mistakenly interpreted as earthquakes. Active and constant local explosions practically reduce seismic activity.

7 Conclusions

To assess the significance of the explosion-earthquake relationship we applied two counter-statistics: (1) earthquakes followed by explosions, and (2) explosions followed by earthquakes. Similar statistics from randomly selected earthquakes and remaining earthquakes were also used.

Using earthquake and explosion catalogs with a large number of events, which, generally, makes it possible to obtain greater confidence in the conclusions, also creates a problem in assessing confidence. Undoubtedly, high seismic activity over almost the entire area always increases the seismic event occurring after an explosion. On the other hand, it is still problematic to assess the significance of the explosion-earthquake causal relationship. All conclusions drawn from the study are tentative. Presumably, about 30 earthquakes out of more than two hundred thousand are somehow connected with the explosion that happened within a short distance before. It seems possible that the 200 earthquakes that occurred within a day after the explosion and at a short distance from it are somehow related to it. However, these conclusions are less significant since earthquakes swarms with similar spatial and temporal properties are observed in the same or nearby areas, perhaps, in the long term, the active explosion areas somewhat change the spatial distribution of earthquakes around them, creating areas of unloading and activation, thereby slightly affecting the fractal distribution of earthquakes. Using primary seismogram material would probably make this possible and allow for a more accurate assessment. Then we can identify the possible explosion-earthquake relationship. Nevertheless, such an analysis is likely to be fragmentary because it is difficult to comprehensively analyze a seismogram's intricate structure. Owing to the fact that, on average, the energy class of explosions is much weaker than that of earthquakes, the physics behind the possible impact of an explosion-earthquake is controversial.

References

1. Smirnov, V., Ponomarev, A.V., Kartseva, T.I., Aydarov, F.: Dynamics of induced seismicity during filling of the Nurek reservoir. Phys. Earth **54**(4), 641–651 (2018)
2. Smirnov, V., Mikhailov, V.O., Ponomarev, A.V., Potanina, M.: On the dynamics of seasonal components of induced seismicity in the Koyna-Varna region, Western India. Phys. Earth **54**(4), 632–640 (2018)

3. Zhuravlev, V.I., Lukk, A.A.: Features of the daily periodicity of weak earthquakes in Iran. Phys. Earth **48**(1), 61–77 (2012)
4. Zhuravlev, V.I., Lukk, A.A.: Midday activation of seismicity in Turkey and a number of other regions of the world. Geophys. Res. **12**(4), 31–57 (2011)
5. Deshherevsky, A.V., Sidorin, A.Ya.: Estimation of the number of explosions in the Dushanbe-Vakhsh catalog of earthquakes. Questions Eng. Seismol. **43**(4), 61–76 (2016)
6. Yılmaz, S., Bayrak, Y., Cinar, H.: Discrimination of earthquakes and quarry blasts in the eastern Black Sea region of Turkey. J. Seismol. **17**(2), 14 (2013)
7. Horasan, G., Boztepe-Güney, A., Küsmezer, A., Bekler, F., Öğütçü, Z., Musaoğlu, N.: Contamination of seismicity catalogs by quarry blasts: an example from Istanbul and its vicinity, northwestern Turkey. J. Asian Earth Sci. **34**, 90–99 (2009)
8. Arrowsmith, M.D.: Mining explosion identification as an application to treaty verification. Ph.D. dissertation, Southern Methodist University, p. 519 (2009)
9. Kartal, Ö.F., Horasan, G.: Trabzon ve civarındaki deprem ve patlatma verilerinin birbirinden ayırt edilmesi (Discrimination of quarry blasts and earthquakes in Trabzon and its vicinity). SAÜ Fen Bilimleri Dergisi **15**, 68–74 (2011)
10. Wiemer, S., Baer, M. : Mapping and removing quarry blast events from seismicity catalogs, short notes. Bull. Seismol. Soc. Am. **90**, 525–530 (2000)

Horizontal Velocity Inhomogeneities of the Mantle Under Central Asia from Rayleigh Wave Phase Velocities

Oxana A. Solovey

Abstract Deep velocity structure of the upper mantle of the continent of Asia was investigated on the basis of a representative dataset of phase velocity dispersion curves of Rayleigh waves. A method of surface-wave tomography developed for a spherical surface was applied to obtain phase velocity distributions in the period range 10–200 s. Average dispersion curves were calculated for various tectonic regions and then inverted to vertical S-wave velocity sections up to the depth of about 600 km. The obtained results allow us to explore main features of the deep velocity structure of the upper mantle under the considered region.

Keywords Rayleigh waves · Phase velocities · Surface-wave tomography

1 Introduction

In this paper, we study main tectonic structures of Central Asia (Fig. 1), which is a composition of ancient blocks of various sizes (platforms and microcontinents), separated by fold belts [1]. The stable Paleozoic West-Siberian Plate and the Precambrian Siberian Platform are located at the north of the study region. At the south and southeast, the area includes large folded structures of Central and Eastern Mongolia and Transbaikalia, where the main part of the relief was formed in the Late Cenozoic [2, 3]. Currently, these regions are characterized by moderate mountain building. The southwestern part of the study area is represented by extensive regions of intense mountain building in Southern Siberia and Western Mongolia (Altai, Sayany, Khangai uplift), characterized by high seismic activity.

The degree of knowledge by different geophysical methods varies for all structures. The studies of the all territory were carried out using gravimetric methods [4, 5], geothermics [6–9] and various seismological methods [10–17].

O. A. Solovey (✉)
Institute of the Earth's Crust SB RAS, 128, Lermontov St., Irkutsk, Russia 664033

A. Kosterov et al. (eds.), *Problems of Geocosmos—2020*, Springer Proceedings in Earth and Environmental Sciences, https://doi.org/10.1007/978-3-030-91467-7_20

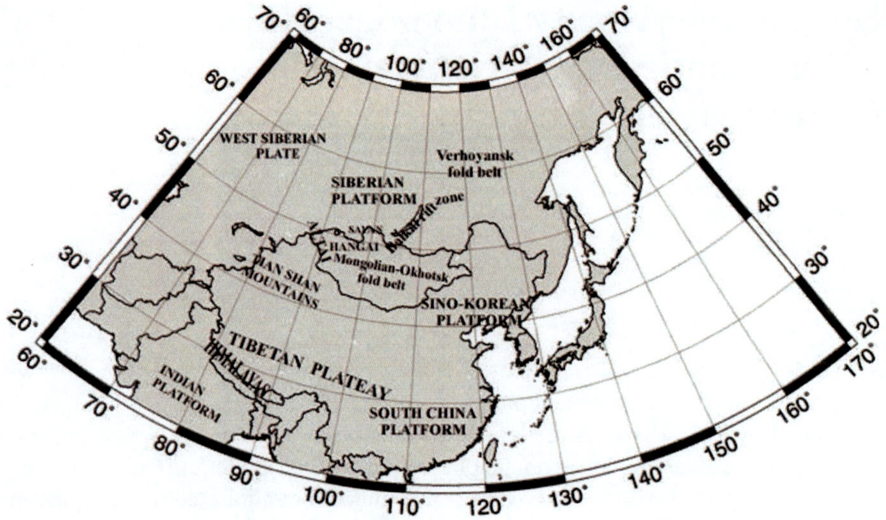

Fig. 1 Main geological structures of the study areas

2 Data and Methods

Initial data for this study were records of Rayleigh waves from earthquakes at broadband digital seismic stations of the IRIS networks (LHZ channels) located in the Asian continent. Quality of surface wave records (signal-to-noise ratio) and position of earthquake epicenters were adopted as selection criteria for the data. Seismic events were chosen in such a way that their epicenters were located at the same great circle path connecting a pair of seismic stations. The deviation in azimuth from the inter-station profiles did not exceed 5°. In total, 69 earthquakes with magnitudes M $\geq$ 6.0 were used. The range of epicentral distances was 8–15 thousand km, which allowed us to calculate the dispersion curves for periods 10–200 s. In this case, the depth of the method for a period of 200 s is about to 650–700 km [18]. As a result, a sample of 198 observed dispersion curves of phase velocities was obtained for 51 pairs of seismic stations.

Then, the obtained dispersion curves for each pair of seismic stations were averaged. The standard deviations of the velocities at the corresponding periods from their averaged values were calculated [19]. The doubled values of these deviations were taken as the determination errors. The analysis of errors in determining the phase velocities showed that the smallest errors (0.5–2%) are confined to the interval of periods of 30–120 s. For periods from 120 to 200 s. their values increase and are in the range of 1–3%. The largest determination errors (5–7%) were noted at periods of 10–15 s. Thus, the phase velocities at periods of 10–15 s are unreliable. However, their use for calculating velocity sections practically does not affect the mantle velocities of S-waves, which are the subject of research in this work. It should be noted that for each pair of stations, several dispersion curves were calculated for earthquakes which epicenters were located on opposite sides of the profile formed by these stations.

To obtain the inter-station curves of phase velocities, the processing of surface wave records was carried out by the method of frequency-time analysis (FTAN). This method is based on replacing signals given in the time domain or in the frequency domain with two-dimensional frequency-time representations. This is achieved by repeatedly filtering the seismic signal using narrow-band filters [19–21]. This allows to effectively separate the analyzed signals into their component harmonics and to obtain correct estimates of phase velocities.

The study of the upper mantle structure was carried out by mapping the distributions of surface wave velocities using a surface-wave tomography. The method is based on the Backus-Gilbert formalism. The program and the calculation methodology were developed by the professor of St. Petersburg University T. B. Yanovskaya for the case of a spherical surface [22, 23]. The method consists of solving a two-dimensional inverse kinematic problem, by calculation the function $V(\phi, \lambda)$–the distribution of the phase velocities of Rayleigh waves, followed by presentation of phase velocity maps distribution for each of the specified periods. The computation of maps of the distribution of phase velocities was carried out for steps of 5, 10 and 25 s for periods of 10–30 s, 30–100 s and 100–200 s, respectively. The minimum number of traces (127, 134) was obtained for periods of 10, 15 s, respectively, the maximum – 198 traces is typical for a period of 200 s. The calculations were carried out with a step of 3° in latitude and 4° in longitude. The input data are dispersion curves of Rayleigh wave phase velocities calculated for seismic paths crossing the study area in different directions. In addition to calculating the function $V(\phi, \lambda)$, the program also provides estimates of resolution of the data by calculating an effective averaging radius (R, ϕ, λ). The best resolution of the results of mapping phase velocities was obtained for the central part of the study area, where the effective averaging radius varies within 400–500 km. This is due to the fact that its central regions are covered by the largest number of seismic traces. The peripheral regions are characterized by a sharp increase in R values up to more than 1000 km. Based on this fact, as the boundary of the study area (Fig. 2), an area was identified where the effective smoothing radius does not exceed 1000 km. The results of calculating the averaging radius can also be presented in the form of maps.

Analysis of the obtained distributions of phase velocities, presented in [24], allows us to conclude that there are large-scale horizontal inhomogeneities in the mantle of the study region. For more accurate representation of their shapes and depths, an inversion of the Rayleigh wave phase velocities was carried out. For these purposes, local dispersion curves were inverted by minimizing the standard deviations of the calculated and observed phase velocities dispersion curves [24]. The standard PREM model [25] was assumed as a starting model. It consisted of twenty horizontal layers and 2000 km half space. The choice of velocity and density parameters is based on the results of studies by the DSS method and other geophysical and seismological methods [26–28]. The variable parameters in the calculations were the thickness of the layers and the values of the velocities of S-waves in these layers. The calculations were carried out in two stages. At the first stage, only S-wave velocities served as

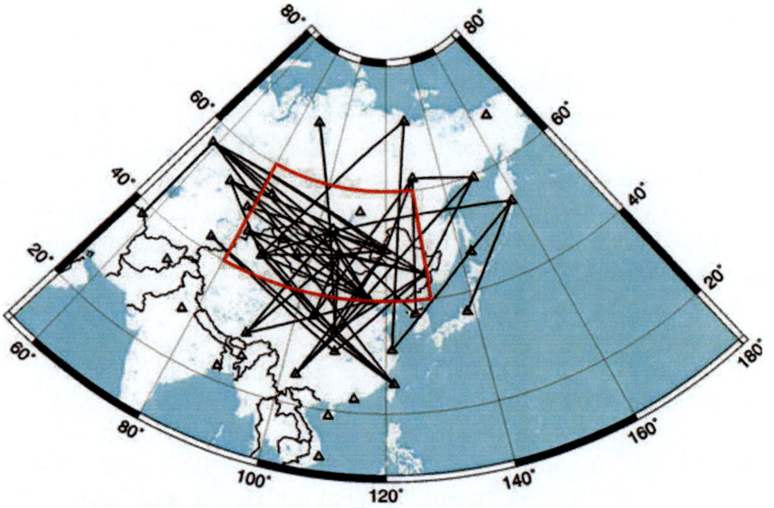

Fig. 2 Coverage of the area with seismic traces, where the triangles are seismic stations. The research area is highlighted in red

variable parameters. At the second stage, these velocities were fixed and the thicknesses of the layers were varied. The program used in the calculations is based on the technique of solving inverse problems for the cases of models of the environment with plane-parallel layers. Therefore, to eliminate errors associated with the sphericity of the study area, a correction for sphericity was introduced [29]. As a result, a set of 51 cross-section of seismic waves was obtained, characterizing the deep velocity structure in the areas between pairs of seismic stations (Fig. 2).

Maps of the distributions of S-wave velocities in the mantle for a set of the specified depths from 50 to 600 km were calculated separately with 50 km depth step and were based on selection wave velocities for each of these depths. Mapping the phase velocity distributions, the Earth's radius was assumed to be constant, but the mapping of S-wave velocities was made taking into account the depths of the corresponding spherical surfaces. As a result, a three-dimensional model of the distribution of S-wave velocities in the mantle of the study region was obtained and two-dimensional velocity sections were constructed for 5 profiles intersecting the main tectonic structures of the study area (Fig. 3).

3 Results

The general trend in the distribution of S-wave velocities in the mantle of the study region is shown on the vertical velocity cross-sections in Fig. 4.

Profile I–I′ is oriented from the northwest to southeast. It crosses the Siberian Platform, Baikal rift zone, and structures of the Mongol-Okhotsk fold belt in eastern

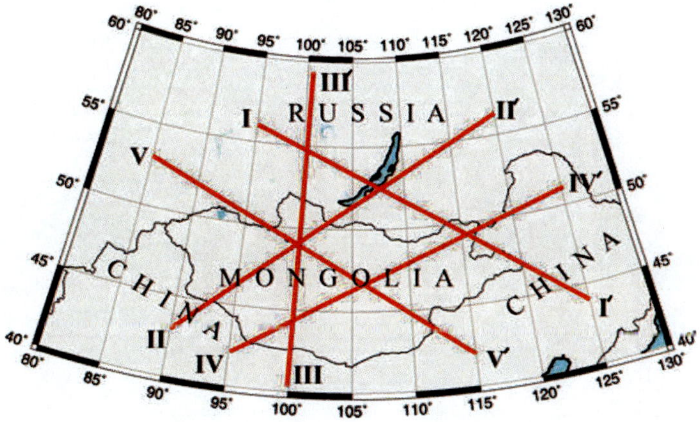

Fig. 3 Profiles, intersecting the study area in different directions

Mongolia and northeastern China. A pronounced feature of the profile is a thick asthenospheric layer under the structures of the Mongol-Okhotsk fold belt. Its top is at the depth of about 150 km, and its base is at the depths of 210–230 km. S-wave velocities within this layer vary within 4.30–4.38 km/s. The layer is significantly thinned under the Siberian Platform, where its thickness does not exceed 50 km, and the top is confined to the depth of about 200 km. In this case, S-wave velocities are slightly increased and vary in the range of 4.35–4.40 km/s. S-waves velocities above this layer are typical velocities for platformic structures and vary in the range 4.55–4.60 km/s. Below the asthenospheric layer, at the depths of 250–350 km, the mantle is also characterized by increased velocities from 4.60 to 4.70 km/s.

The velocity distribution at the southeast differs from the one at the northwest. Here, the velocities of S-waves above the asthenospheric layer are lower and vary within the range 4.40–4.50 km/s. Only at the northern periphery of the Sino-Korean Platform Vs increases up to values of 4.55–4.60 km/s and these velocities are observed in a thin layer located directly under the Moho. From the base of this layer to the top of the asthenosphere, S-wave velocities decrease from 4.50 to 4.40 km/s. Below the asthenospheric layer, contrary to the general tendency of the velocity increase with depth, two low-velocity layers are observed. One of them is attributed to the depths of approximately 250–300 km, the second one—to the depths of 350–380 km. S-wave velocities in these layers are reduced 4.30–4.35 km/s. Vs above and below these layers, as well as in the depth range between them, are typical for these depths in the mantle.

Depths of 440–480 km are characterized by a significant increase in S-wave velocities from about 4.80 to 5.30 km/s. Apparently, this is a so-called transition zone or zone of transition from the upper to lower mantle. It should be noted that this velocity distribution is typical for all five profiles.

The velocities of S-waves below the bottom of the transition zone at the north-western and central sections of the profile reach 5.30–5.45 km/s, with their subsequent

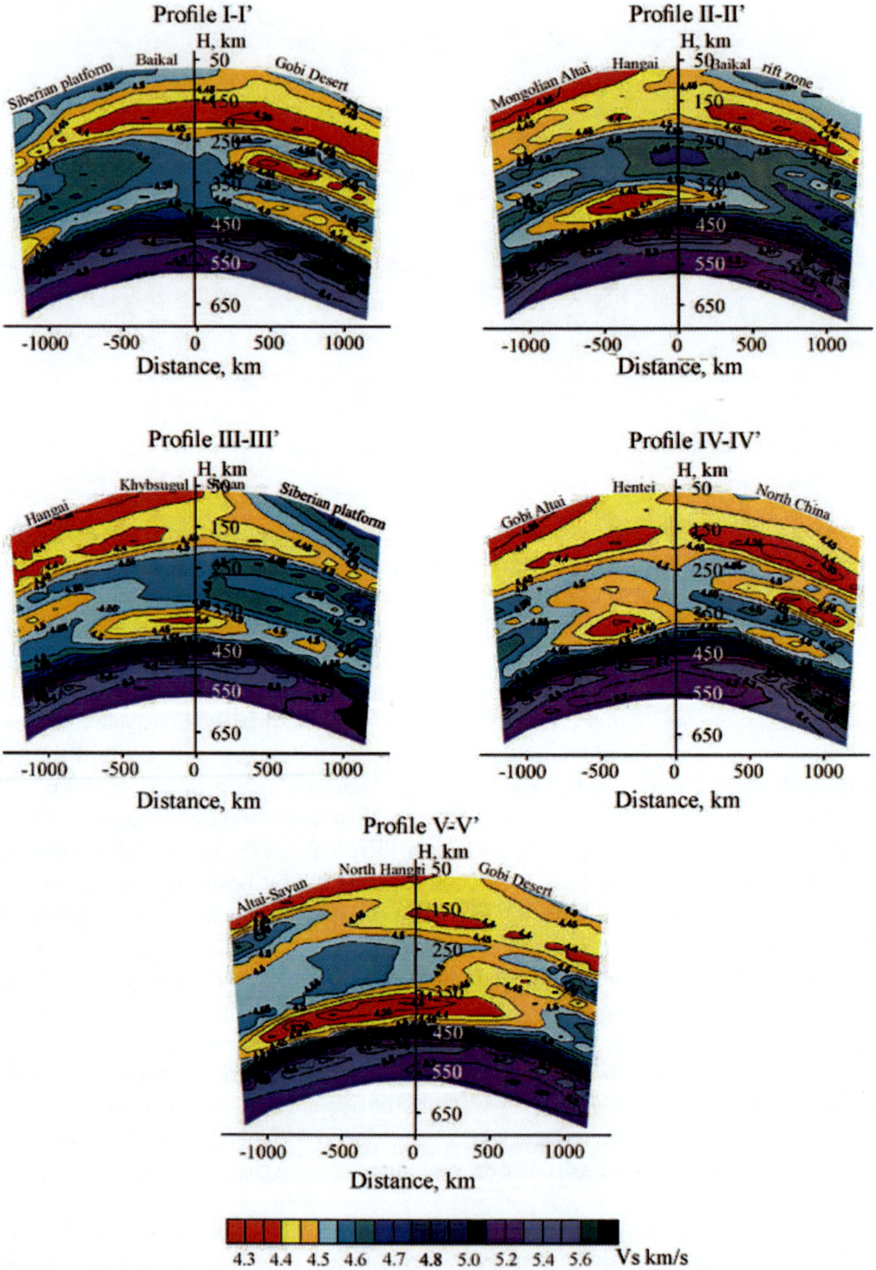

Fig. 4 The distribution of S-wave velocities with the depth along five profiles of the study area

insignificant decrease by 0.10 km/s at the depths of approximately 550–600 km. A different pattern of S-wave velocities is observed at the southeastern flank below the transit zone. Here, at the depths of 450–500 km, a high-velocity layer appears with a local velocity maximum of 5.50–5.60 km/s. Above and below this layer, the velocities vary within 5.30–5.45 km/s.

Profile II–II′ is oriented from the southwest to northeast. The profile is completely located within the various structures of the Mongol-Okhotsk fold belt. It originates at the foothills of the eastern Tien Shan, crosses the mountain structures of the Gobi Altai and the Khangai uplift, runs almost along the axial line of the Baikal rift zone up to the Stanovoy ridge. A significant difference from the previous profile is that the upper mantle to the depth of about 200 km is generally characterized by reduced S-wave velocities. Moreover, under the major part of the profile, the velocities vary within 4.30–4.45 km/s an exception is the Baikal rift zone. Here, the velocities in the upper 100 km layer reach values of 4.50–4.55 km/s, which is more typical for platformic and subplatformic structures than for rift systems. Below this high-velocity layer, down to depths of about 200 km, Vs decreases up to the values of ~4.30–4.35 km/s.

The depths of 200–400 km are characterized by typical S-wave velocities for these mantle depths, varying between 4.50 and 4.60 km/s. However, two regions of high velocities (4.60–4.65 km/s) are observed at these depths. One region is located to the southeast of Lake Baikal at depths of 250–300 km, the other one—under the northeastern part of the profile at depths of 300–350 km. At the southeastern part, at the same depth, a layer with a relatively decreased velocities (by 0.10 km/s) is observed. At the depths of 350–450 km, another low-velocity area appears. Its center is located under the high-mountain structures of Western Mongolia (Khangai region), Vs vary within 4.35–4.50 km/s.

For depths below the transition zone, as in the case of the first profile, insignificant velocity variations are observed. But, unlike the first profile, at the depth of 500 km it is characterized by the presence of a high-velocity layer of 50 km thickness with S-wave velocities of 5.40–5.45 km/s. Below it, a velocity decrease is observed up to the values of 5.25–5.30 km/s. Moreover, this decrease is observed almost throughout the whole profile at the depths of about 500–570 km, with an exception of its northeastern outskirts.

Profile III–III′ is oriented from the south to north. It crosses the eastern outskirts of the Gobi Altai, the Khangai uplift, the Sayan Mountains and ends at the Siberian Platform. At the depths, starting from the base of the crust and up to approximately 170 km under the high-mountain structures of Western Mongolia, an area of reduced velocities (about 4.40 km/s) appears. Within its limits, three inclusions with anomalously low S-wave velocities are revealed. One of them, occurs in the form of a layer, stretches from the Gobi Altai, through Khangai and Lake Khubsugul to the northern foothills of the Eastern Sayan. Two other are located at the depths from 240 to 270 km under the structures of the Gobi Altai and Khangai, respectively. At the north, the considered region of low velocities is significantly thinned up to 40 50 km and gradually sinks under the Siberian Platform at the depths of 200–250 km. The overlying mantle under the Siberian Platform is characterized by S-wave velocities within 4.50–4.65 km/s. As in the case of the first profile, there is a trend [26, 27] of a smooth decrease in velocities as they approach the thinned asthenospheric layer.

The structure of the mantle below the asthenospheric layer and down to the depth of about 430 km is identical to the structure of the mantle under the profile II–II′ at the same depths. Here, at the depths of 250–300 km, a relative velocity decrease is seen at the southern part of the profile. The second low-velocity area, confined to the Khangai uplift, occurs in the intersection area with profile II–II′ at the depths of 350–400 km. Some similarity is also observed at the deeper layers of the mantle below the transition zone, the high-velocity layer is located above the layer with relatively low velocities.

Profile IV–IV′ is oriented from the southwest to northeast. It passes through the southern regions of the Gobi Altai, crosses the central and northeastern regions of Mongolia, including the Khentei Mountain structures, the extreme northern regions of China, and ends at the Russian-China border. The S-wave velocities in the upper part of the mantle, at the depths of 50–200 km, along the entire profile, are relatively low and do not exceed the values of 4.40–4.45 km/s. Three regions with anomalously low velocities typical for asthenospheric material (approximately 4.30–4.35 km/s) are observed. One of these areas is localized at the southwestern flank of the profile under the Gobi Altai. The other two form a single layer of reduced velocities with a connection in the central part. The top of this layer is located at the depths of about 140–150 km, and the base—at the 180–200 km depths.

At the northeastern part of the profile, at the depths of about 100 km, the velocity increase up to 4.45–4.50 km/s is observed with a local maximum of 4.55 km/s, confined to the subplatformic region of northern China.

The mantle at the depths of 250–400 km differs significantly in velocity characteristics from the mantle at the same depths under previous three profiles. Here, despite the high Vs in the initial model, low-velocity inclusions are observed, encompassing about a half of the area at the considered depths. One of them is located under the Khentei uplift. This inclusion is an area with a complex configuration and occupies a depth range from about 250 to 400 km. The S-wave velocities within depths vary in the range 4.35–4.50 km/s. The second inclusion with anomalously low velocities of 4.30–4.35 km/s is confined to the northeastern flank of the profile and is a relatively thin layer at the depths of approximately 250–300 km, connected by two vertical channels with the overlying asthenosphere. The third layer of low velocities (4.45–4.50 km/s) is located in next to the top of the transition zone. Its thickness does not exceed 50 km. This layer gravitates to the previously considered first region of lower wave velocities under Khentei, where a wedge-shaped "ejection" in its direction occurs at similar depths.

The S-wave velocities in the mantle containing the low-velocity inclusions at the depths of 200–400 km are lower than the velocities of three previously considered profiles. In absolute terms, they are close to 4.50 km/s with a local maximum of 4.60 km/s at depths of about 250 km and 350 km. A similar one is also seen at the northeastern part of the profile at the depths of 300–350 km.

At the depths below the transition zone, a 100 km thick layer in form of a wedge with significantly higher Vs than typical mantle velocities under the northeastern part of the profile, with velocities of 5.40–5.60 km/s, is observed. A similar layer at the same depths appears on I–I′ profile.

V–V′ profile is oriented from the northwest to southeast. Its beginning is located at the southwestern outskirts of the West-Siberian Plate, crosses the mountain structures of the Altai and Sayan Mountains, the northern outskirts of the Khangai uplift, the sub-platformic regions of Central Mongolia, the northern regions of the Gobi Desert, and ends in China near the Mongolian border.

Almost throughout the entire length of the profile, the upper levels of the mantle, down to the depth of about 250 km, represent an area of low S-wave velocities. The velocities at these depths are less than 4.40 km/s. A layer with Vs typical for asthenospheric material (4.35 km/s) appears under the high-mountain structures of the Khangai, Altai and Sayan Mountains; the top is located near the base of the crust, and its bottom lies at the depth of about 90 km.

The upper mantle under the southeastern edges of the profile is characterized by higher S-wave velocities. The top of the mantle is a thin 30-km layer with velocities of about 4.50–4.55 km/s. At the depths of 70–100 km, the velocities decrease to 4.40 km/s. A high-velocity layer also appears under the northwestern edges of the profile at the depths of approximately 100–150 km. The velocities in this layer vary within 4.55–4.65 km/s. The layer under consideration is confined to the foothills of the Altai and Sayan Mountains.

The mantle at the depths of 200–320 km beneath the northwestern and central regions of the profile contrasts sharply with the southeastern flank of the profile in its velocity characteristics. The velocities vary in the range of 4.50–4.55 km/s. At the same time, an extensive high-velocity inclusion with velocities of 4.55–4.60 km/s is revealed in the mantle under the northern Khangai. Almost everywhere at the southeast of the profile, the mantle is characterized by low velocities, within the range of 4.40–4.55 km/s. However, there are also two high-velocity inclusions. One of them, with velocities of 4.55–4.70 km/s, is located at the depths of approximately 250–280 km, and the second one, with velocities not exceeding 4.65 km/s, is confined to the depths of 330–450 km. It should be noted that Vs low at the southeastern section of the profile occurs as a channel connecting the overlying asthenosphere with a layer of anomalously low velocities under the northwestern and central sections of the profile. The top of this layer lies at the depth of about 350 km, and the base—at about 400 km, that is, in the immediate vicinity of the transition zone.

Below the transition zone, S-wave velocities vary within 5.20–5.40 km/s. At the depths of about 480–600 km, a pronounced inversion of velocities is observed: a layer with velocities close to 5.40 km/s is underlain by a layer with velocities of about 5.25 km/s. A similar pattern is typical for all the previously considered profiles.

4 Conclusions

The results of this study indicate significant variations in S-wave velocity structure under pronounced tectonic structures in Central Asia. These differences are well expressed for the upper 400 km. Thus, the upper mantle to the depths of about 200 km beneath the regions of intense mountain building in Southern Siberia, Western Mongolia, and the Baikal region are characterized by lower S-wave velocities than Vs in the platformic and subplatformic structures of Siberia, Eastern Mongolia, and China. Moreover, contrary to the general trend of the velocity increase with depth, layers and inclusions with anomalously low velocities are observed. The maximum Vs velocities are observed under the Siberian Platform, with an exception of a thin, not more than 50 km, asthenospheric layer at the depths of 200–250 km. This finding does not contradict the previously obtained results on the thickness of the lithosphere under this region [11, 28, 30].

Thus, based on the presented profiles, it can be concluded that the mantle of Central Asia is distinguished by a complex structure over the entire range of the considered depths. Its inhomogeneities occurred both in horizontal and vertical directions. The above-mentioned alternation of low-velocity layers with high-velocity layers, the presence of regions of local minima and maxima opposite to general trend of an increase in Vs with depth indicates complex tectonic and convection processes of heat and mass transfer. The most contrasting inhomogeneities are revealed for the depths up to 400 km, where the velocity variations reach 0.40–0.50 km/s, while at the depths of 500–600 km, Vs vary, with rare exceptions, within 0.20 km/s.

Another prominent feature of the deep structure of the region is a zone of high gradient of velocity in the mantle at the depths of about 440–480 km. S-wave velocities vary in the range 4.80–5.30 km/s. This zone is revealed at the same depths for all five considered profiles.

References

1. Zonenshain, L.P., Kuzmin, M.I., Natapov, L.M.: Tectonics of Lithospheric Plates in the Territory of the USSR (in 2 books). Nauka, Moscow (1990).(in Russian)
2. Devyatkin, E.V.: Structures and formational complexes of the Cenozoic activation stage. In: Tectonics of the Mongolian People's Republic. Nauka, Moscow, pp. 182–195 (1974). (in Russian).
3. Devyatkin, E.V., Shuvalov, V.F.: Continental Mesozoic and Cenozoic Mongolia (stratigraphy, geochronology, paleogeography). In: Evolution of Geological Processes and Metallogeny of Mongolia. Nauka, Moscow, pp. 165–177 (1990). (in Russian)
4. Buyantogtokh, B., Turutanov, E.K., Kanaykin, V.S.: The structure of the earth's crust in the Ulan Bator region of Mongolia according to gravimetric data. Geodyn. Tectonophys. **10**(3), 585–602 (2019). (in Russian) https://doi.org/10.5800/GT-2019-10-3-0428
5. Turutanov, E.K.: Gravity Anomalies, Deep Structure and Geodynamics of the Mongolian-Siberian Region. Publishing House IRNITU, Irkutsk (2018).(in Russian)
6. Golubev, V.A.: Heat flow through the bottom of Lake Khubsugul and its mountainous surroundings (Mongolia). Phys. Earth **1**, 48–60 (1992). (in Russian)
7. Golubev, V.A.: Conductive and Convective Heat Output in the Baikal Rift Zone. Academic Publishing House "Geo", Novosibirsk (2007). (in Russian)
8. Khutorskoy, M.D., Golubev, V.A., Kozlovtseva, S.V., Mitnik, M.M., Yarmolyuk, V.V.: Thermal Regime of the Subsoil of the Mongolian Republic. Nauka, Moscow (1991).(in Russian)
9. Lysak, S.V., Dorofeeva, R.P.: Thermal state of the lithosphere in Mongolia. Geol. Geophys. **44**(9), 929–941 (2003). (in Russian)

10. Koulakov, I., Bushenkova, N.: Upper mantle structure beneath the Siberian Craton and surround areas based on regional tomographic inversion of P and PP travel times. Tectonophysics **486**, 81–100 (2010). https://doi.org/10.1016/j.tecto.2010.02.011
11. Kozhevnikov, V.M., Seredkina, A.I., Solovei, O.A.: 3D mantle structure of Central Asia from Rayleigh wave group velocity dispersion. Russ. Geol. Geophys. **55**(10), 1239–1248 (2014). (in Russian). https://doi.org/10.1016/j.rgg.2014.09.010
12. Mordvinova, V.V., Deschamps, A., Dugarmaa, T., Deverscher, J., Uzliybat, M., Sankov, V.A., Artemiev, A.A., Perrault, J.: Study of the velocity structure of the lithosphere on the Mongol-Baikal transect 2003 by exchange SV-waves. Phys. Earth **2**, 21–32 (2007). (in Russian)
13. Mordvinova, V.V., Treusov, A.V., Turutanov, E.K.: On the nature of the mantle plume beneath Khangai (Mongolia) according to seismogravimetric data. Doklady Earth Sci. **460**(3), 334–338 (2015). (in Russian)
14. Pandey, S., Yuan, X., Debayle, E., Priestley, K., Kind, R., Tilmann, F., Li, X.: A 3D shear-wave velocity model of the upper mantle beneath China and the surrounding areas. Tectonophysics **633**, 193–210 (2014). https://doi.org/10.1016/j.tecto.2014.07.011
15. Solovei, O.A., Kozhevnikov, V.M.: Three-dimensional velocity model of the upper mantle of Central Asia. Dokl. Earth Sci. **421**(6), 934–936 (2008). (in Russian)
16. Zhao, D.: Global tomographic images of mantle plumes and subducting slabs: insight into deep Earth dynamics. Phys. Earth Planet. Int. **146**, 3–34 (2004)
17. Zhou, Y., Nolet, G., Dahlen, F.A., Laske, G.: Global upper-mantle structure from finite-frequency surface-wave tomography. J. Geophys. Res. **111**, B04304 (2006). https://doi.org/10.1029/2005JB003677
18. Ritzwoller, M.N., Levshin, A.L.: Eurasian surface wave tomography: group velocities. J. Geophys. Res. **108**(B3), 4839–4878 (1998)
19. Levshin, A.L.: Surface and channel seismic waves. Nauka, Moscow (1973).(in Russian)
20. Bath, M.: Spectral Analysis in Geophysics. Nedra, Moscow (1980).(in Russian)
21. Levshin, A.L., Pisarenko, V.F., Pogrebinsky, G.A.: On a frequency-time analysis of oscillations. Annal. Géophys. **28**(2), 211–218 (1972)
22. Yanovskaya, T.B.: A surface wave tomography method based on the Backus-Gilbert approach. Comput. Seismol. **32**, 10–27 (2001). (in Russian)
23. Yanovskaya, T.B.: Surface Wave Tomography in Seismological Studies. Saint Petersburg: Nauka, Saint Petersburg (2015). (in Russian)
24. Kozhevnikov, V.M., Solovey, O.A.: Three-dimensional model of Central Asian mantle based on the dispersion of the phase velocities of Rayleigh waves. Volcanol. Seismol. **4**, 30–39 (2010). (in Russian)
25. Dziewonski, A.M., Anderson, D.L.: Preliminary reference Earth model. Phys. Earth Planet. Int. **25**, 297–356 (1981)
26. Pavlenkova, G.A., Priestly, K., Cipar, J.: 2D model of the crust and uppermost mantle along rift profile. Siberian cruton. Tectonophysics **355**, 171–186 (2002)
27. Pavlenkova, N.I.: The structure of the upper mantle of the Siberian platform according to data obtained on superlong seismic profiles. Geol. Geophys. **47**(5), 630–645 (2006). (in Russian)
28. Zorin, Y.A., Novoselova, M.R., Turutanov, E.K., Kozhevnikov, V.M.: Structure of the lithosphere of the Mongolian-Siberian Mountainous Province. J. Geodyn. **11**, 327–342 (1990). https://doi.org/10.1016/0264-3707(90)90015-M
29. Bolt, B.A., Dorman, J.: Phase and group velocities of Rayleigh waves in a spherical, gravitating Earth. J. Geophys. Res. **66**, 2965–2981 (1961)
30. Yanovskaya, T.B., Kozhevnikov, V.M.: 3D S-wave velocity pattern in the upper mantle beneath the continent of Asia from Rayleigh wave data. Phys. Earth Planet. Int. **138**, 263–278 (2003)

Various Scale PS ECWM-CDP Seismic Prospecting for Crustal Structure Studies

E. V. Isanina, N. V. Sharov, and S. I. Yuzhaninova

Abstract We construct depth sections of deep crust in the range from 0.8 to 1.2 km (low velocity zone) to 40–60 km (the upper mantle) by jointly using the active seismic data and the results of the PS converted wave method (PS ECWM). Crustal blocks and the boundary zones of different areas including the Onega-Ladoga Block, South Karelia were identified and delineated. Their geomorphological and geodynamic characteristics were described. Comparison of the deep sections of various mineralogenic provinces shows that in spite of the minor differences in evolution and the modern geodynamic settings, they have a lot in common, particularly: (1) All sections reveal tectonic (mantle) zones that extend through the crust and control the velocity contrast at the Moho discontinuity. These zones were commonly rejuvenated during various activation epochs. (2) Presence of strong waves converted at the velocity contact of the Moho boundary (disintegration and elevated geodynamic tension zones). (3) There is a wedge-like zone in the upper part of the earth crust, where no converted waves occur («silence zones»). (4) This domain always adjoins the tectonic (mantle) zone in the middle crust at 10–20 km depth. Similarities of the ore zones makes possible using of analogy in the prognostic studies for various mineralogenic provinces. Additionally, the RWM-CDP sections from the Salmi and Svir-Oyat prospects were constructed for depth varying from 1–2 to 300 m. Seismic microzoning (SMZ) was carried out as a part of the integrated areal seismic and seismological studies. It led to the assessment of the current state of the grounds and to delineation of areas of unconsolidated ground together with the geodynamically weakened (consequently, less stable) zones of possible residual deformation. Thus, the SMZ maps show seismic hazard differentiation for this area.

Keywords Various scale seismic prospecting · Earth crust · Deep sections · Moho · Seismic microzoning

---

E. V. Isanina · S. I. Yuzhaninova
Regional Geoecological Centre (RGEC) Urangeo JSC Branch, St. Petersburg, Russia

N. V. Sharov (✉)
Institute of Geology, KarRC, RAS, Petrozavodsk, Russia
e-mail: sharov@krc.karelia.ru

A. Kosterov et al. (eds.), *Problems of Geocosmos—2020*, Springer Proceedings in Earth and Environmental Sciences, https://doi.org/10.1007/978-3-030-91467-7_21

1 Introduction

The geological and geophysical surveys carried out on the Fennoscandian Shield in recent years aim at solving the fundamental problems of tectonics and geodynamics of the Early Precambrian, including petrology, evolution of the crustal processes, deep geology and physics of the tectonosphere. In 2002, a set of digital maps was published by the Finnish researchers: a geological map of the Pre-Quaternary formations [1], a map of anomalous magnetic field $(\Delta T)_a$ [2] and a map of the gravity field (s = 2.67 g/cm^3) of with a scale of 1:2,000,000 [3].

In Southern Finland, the Fennoscandian Shield occurs as the Ladoga-Bothnian geoblock; its southeastern partis covered by the Ladoga Lake (Priladozhye). The structural-geological position of the geoblock is ruled by surrounding Svecofennian Folded Province, which can be traced in latitudinal and northwestern directions from Finland into Sweden and in southern and southeastern directions towards the Russian Plate. In the northeast, the geoblock rests against the Karelian Megablock via the Ladoga-Bothnian (Raahe-Ladoga) Suture Zone (LBZ), which can be traced on satellite images from the central Russian Plate to the northeastern shore of the Gulf of Bothnia. In the northwest, south and southeast, directions the geoblock faces the Lower Middle Riphean aulacogens (Gulf of Bothnia, Gulf of Finland and Ladoga (Pasha-Ladoga) within the Baltic-White Sea rift-related system.

2 Results and Discussion

We used the Delta-Geon-02 3-component recorders and the Laccolite X-M2 digital recorder (Figs. 1 and 2). To image the crustal structure with the PS converted wave method (PS ECWM) based on the waves from teleseimic and regional earthquakes and with the seismic prospecting method (PSM), which was based on registration of the earthquakes and blasts from pneumatic emitters and other sources. The refraction wave (RWM) method and common depth point (CDP) methods were also utilized. The research conducted by the Nevskgeologia FGUGP in the Onega-Ladoga geoblock in 1980–2010 (Fig. 3) revealed relation of the deep structure and geodynamic state of the crustal blocks to their minerogenic specialization [4–10]. However, an emphasis was made on the connection of the middle and lower crustal parts (5–45 km) to the upper zone of the mantle (40–120 km) [11–13].

We know much less about the geological sequences at (0–1.5–3 km) than about the sequences at 3–5 km depth. However, the structure, the current geodynamic processes and the origin of the earth crust are essential for re-determining the main lines of the mineral prospecting and exploration in the study area.

The deep seismological section along profile XI is shown in Fig. 4. In its central portion, over a distance of 50 km (PC 1037–1050), in an elevated-intensity converted wave field, is a block in which the effect of the Pitkäranta tectonic dislocation system is felt. High values of the exchange capacity (P-S reflection coefficient) clearly

Fig. 1 Equipment for the 3-component PSM-PS ECWM survey (3-component seismic station)

Fig. 2 Equipment for seismic prospecting (cable of geophones attached to the data storage device)

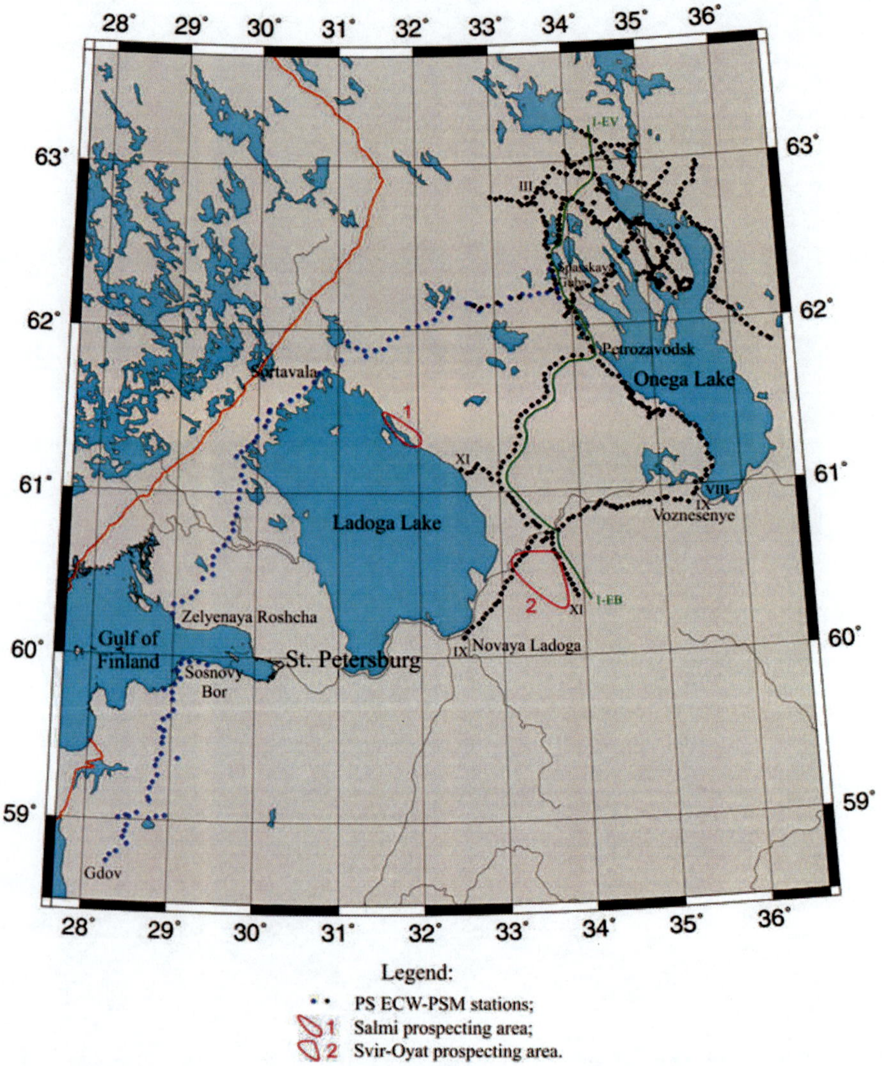

Fig. 3 Exploration of the southern slope of the Baltic Shield by the PSM-PS ECWM and RWM-CDP methods

indicate the mantle origin of this rift-forming structure [14]. This domain leans south at depths exceeding 40 km. Analysis of the exchange capacity in the plane of the fault system points out to the complex internal structure of the planar section of the Pitkäranta fault system. We observe high exchange capacity zones and the low exchange capacity zones as well. Moreover, zones free of the converted waves are occasionally observed. Horizontal boundaries are scarce and short, except for a boundary at 9–11 km depth. This indicates diversity in tension and intensity of

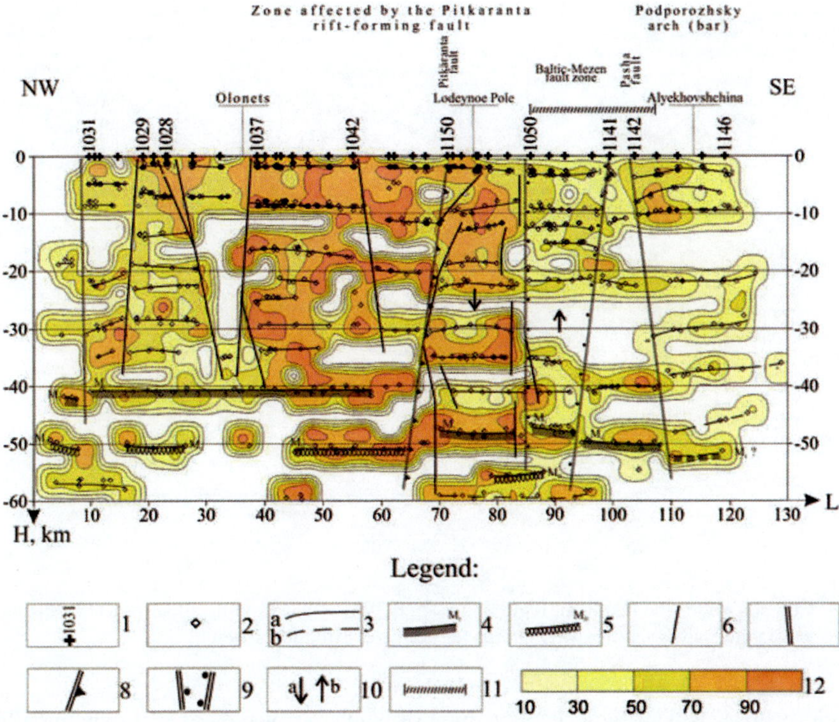

Fig. 4 Deep section and medium P-S reflection coefficient estimated by the PS ECWM. **Profile XI.** *Legend*: 1. Station location and its number. 2. Converted wave site. 3. Velocity contrast boundaries: (a) reliable, (b) less reliable. 4. Top of the crust-mantle transition zone. 5. Bottom of the crust-mantle transition zone. 6. Intracrustal faults. 7. Mantle faults breaking M_I and M_{II}. 8. Pitkäranta rift-forming fault (northern marginal fault of the Pasha graben of the Pasha-Ladoga aulacogen). 9. Ladoga-Onega fragment of the Baltic-Mezen zone of the Vendian-Paleozoic activation. 10. Descending (a) and ascending (b) blocks. 11. Potentially uraniniferous zones. 12. Exchange capacity of the medium, %

the movements along the entire plane of the Pitkäranta rift-forming fault zone. The Moho was localized at a depth of about 41 km.

The deep PS converted wave-CDP sections were used to delineate promising areas for a subsequent detailed ECWM-CDP seismic prospecting. The structural-tectonic characteristics of near surface (2–5 m) were determined on the ECWM-CDP sections obtained in the Salmi and Svir-Oyat prospects (Fig. 5). Proterozoic basement together with the fracturing and dislocation zones were delineated.

A three-dimensional model of the deep structure, based on intersecting profiles, was developed (Fig. 6). It confirms the hypothesis of the tectonic significance of the Pitkäranta Fault. The block between the Janisjärvi and Kashkan faults displays the low exchange capacity of the earth crust. Low-intensity converted waves are observed both at the Moho and at the contacts of dislocation zones. It turns out to be rigid, fairly stable, slightly tense massif, in which all tensions are relaxed. The southern

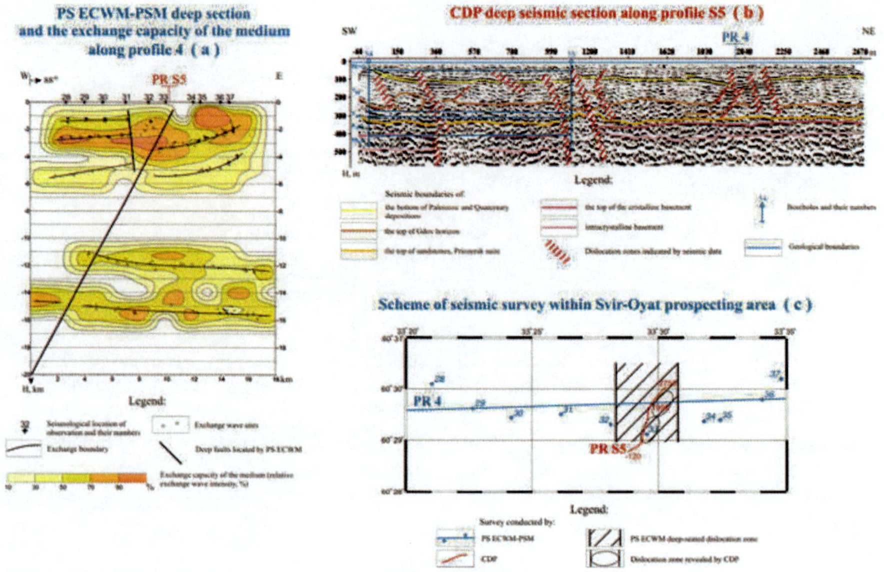

Fig. 5 Comparison of the results of regional (a. PS ECWM-PSM) and detailed (b. CDP) seismic studies within the Svir-Oyat prospecting area (c. profile distribution scheme)

branch of the Kashkan fault displays intense converted waves. It is a dynamically tense zone controlling the western flank of the Onega rift-related structure.

The block structure of the earth crust (Fig. 6) is indicated by the presence of tectonic dislocation zones of various ranks, weak and strong exchange capacity and layering. The earth crust was found to be considerably deformed at the intersection of northwestern (Pitkäranta and Janisjärvi) and northeastern (Baltic-Mezen and Voznesensk) zones. The main structures of the area, Janisjärvi and Baltic-Mezen, were found to be 10–50 km wide mantle fault systems. The internal structure of these units is complicated by the longitudinal, cross-cutting, and transverse faults of crustal origin that make up a high tectonic tension and permeability system. This system favours the ascent of mantle and crustal ore-forming fluids during activation periods.

Identifying a deep tectonic zone (Fig. 5), which extends into the inner (deeper than 20 km) crust, in the PS earthquake converted wave section and other features (the appearance of a wedge-shaped «silence zone» in the upper portion (0–5 km) allow to identify the PC-31-34 zone in the PS ECWM profile 4 for launching the detailed CDP studies (profile 5) followed by the location of a borehole.

To assess the mineralogical potential of the blocks and to compile seismic microzoning (SMD) maps, the structure of the upper part of the section (2–3 to 300–500 m) was interpreted with the help of the shallow-depth CDP sections and the seismological velocity sections (Fig. 7).

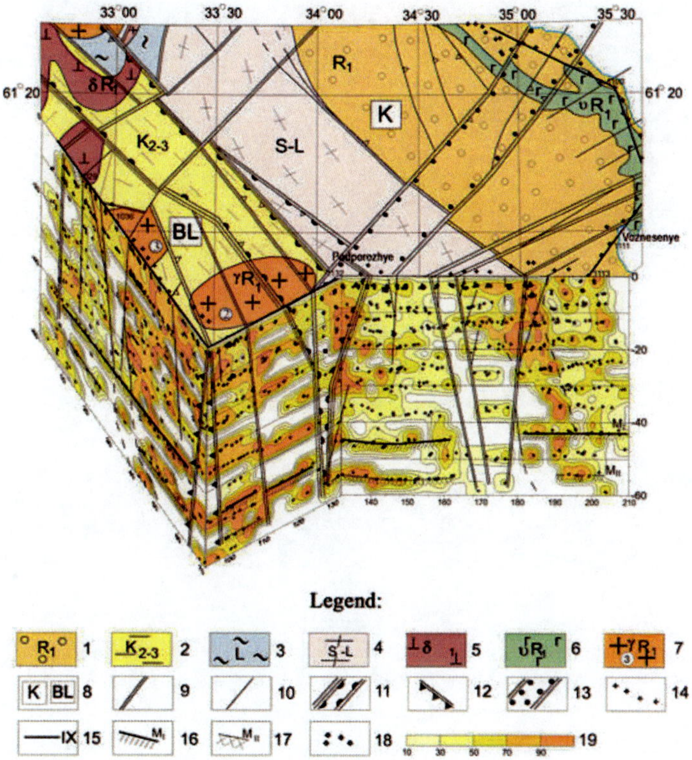

Fig. 6 3-D model of the deep structure of the Onega-Ladoga prospecting area. *Legend*: 1. Early Riphean Petrozavodsk-Shoksha sequence. Sandstones, siltstones, quartzitic sandstones, phthanites. 2. Kalevian Ladoga series, Ludocovian Sortavala series. Epidote-biotitic, amphibole-biotitic graphite-bearing schists and gneisses. 3. Lopian Gimoly series. Volcanic-sedimentary rocks of the Girvas-Manga granite-greenstone belt. 4. Undivided Saamian-Lopian, schist-gneiss sequence. 5. Early Riphean norites, gabbro-norites and gabbro-anorthosites. 6. Early Riphean gabbro-diabases and gabbro-dolerites of the Ropruchei sill, Prionezhsky Depression. 7. Early Riphean macroporphyroblastic microcline granites of the Olonets (1) and Podporozhsky (2) massifs. 8. Megablocks: K—Karelian, BL—Baltic-Ladoga. 9. Mantle faults. 10. Intracrustal faults. 11. Janisjärvi interblock fault system. 12. Pitkäranta rift-forming fault (northern marginal fault of the Pasha graben of the Pasha-Ladoga aulacogen). 13. Ladoga-Onega fragment of the Baltic-Mezen (BMZ) Vendian-Paleozoic activation zone. 14. Seismological monitoring sites. 15. PS ECWM profiles and their numbers. 16. Top of the crust-mantle transition zone. 17. Bottom of the crust-mantle transition zone. 18. Converted wave sites. 19. Exchange capacity, %

3 Conclusions

1. Deep crustal sections from 0.8 to 1.2 km to the top of the mantle (40–60 km) were constructed using the seismic PS-earthquake converted wave data from the Onega-Ladoga geoblock. Blocks and their border zones were delineated and

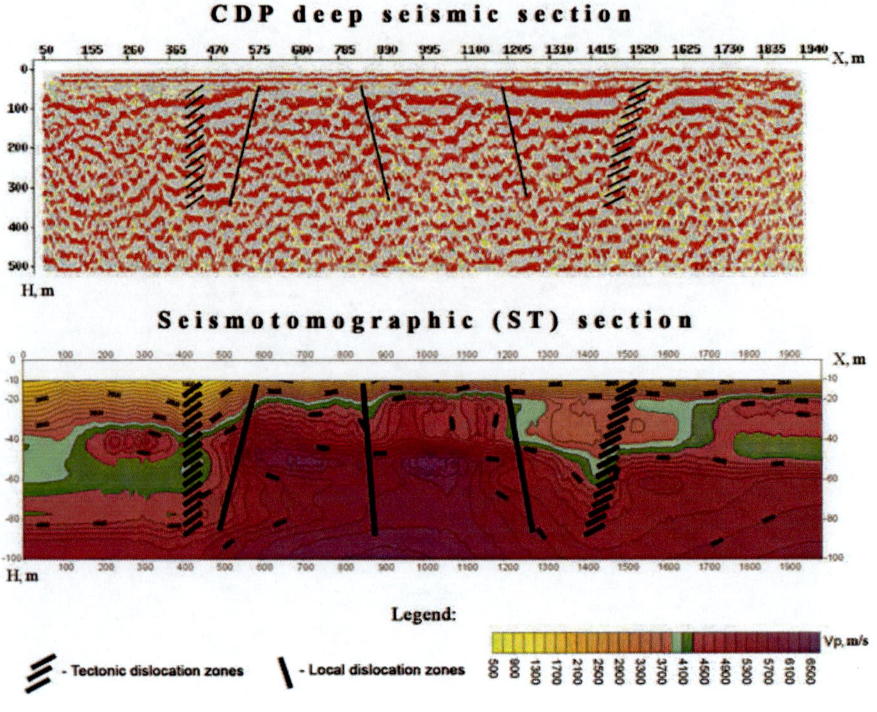

Fig. 7 Comparison of seismic sections, South Karelia

their geodynamic characteristics (exchange capacity of the earth crust) were estimated.

2. Comparison of the deep sections of various mineralogenic provinces showed deep structure similarities despite different evolution and the current geodynamic state:

The main promising mineralogenic crustal blocks lie either at or between the contacts of deep-seated dislocation zones and in regions of high exchange capacity values at block contacts.

There is a wedge-shaped region free of converted waves on the top of the earth crust («silence zone») at the depth of 5–7 km.

The contrasting zones always adjoin the tectonic (mantle) zone extending through the crust at the middle crust depth of 7–20 km.

There are tectonic (mantle) zones extending through the crust and controlling the velocity discontinuity at the Moho. They were commonly rejuvenated during various activation epochs.

Converted waves at the Moho, near the contact with a mantle fault zone are strong (in the regions of high geodynamic tension) (Fig. 8).

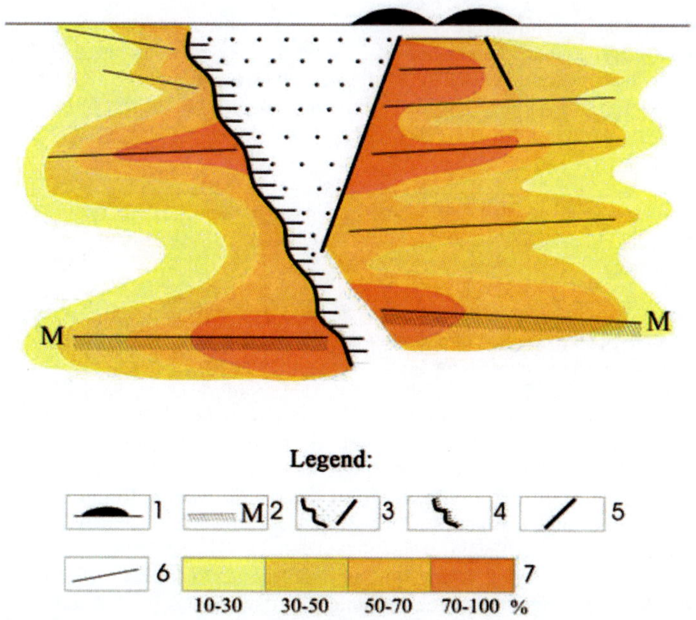

Fig. 8 Model of the uraniniferous province, as shown by PSM-PS ECWM data. *Legend*: 1. Deposits; 2. Earth crust bottom. 3. Silence zone. 4. Mantle fault zones. 5. Intracrustal fault zones. 6. Boundaries in the earth crust. 7. Exchange intensity, %

3. The mentioned above similarities allows using an analogy method for earthquake prediction.
4. The seismic prospecting RWM-CDP sections obtained in the Salmi and Svir-Oyat prospects were constructed for the depth varying from 2–3 to 300–350 m.
5. The data suggests that the deep crustal structure of the southeastern Fennoscandian Shield should be studied in more detail by seismic methods to compile a well-based map of tectonic zoning with metallogenic forecast. For this purpose, additional seismic surveys of the deep "land-sea" structure should be carried out in the Ladoga block employing the pneumatic emitters in the Ladoga and Onega lakes and the profile-areal recordings of miscellaneous source types (earthquakes, explosions and pneumatic emitters).
6. In order to prevent disasters and to make the geotechnical decisions more reliable, the current geodynamic state in active quarries should be controlled during the seismogeodynamic monitoring and seismic microzoning (SMZ).
7. We propose simultaneous integrated seismic studies at a depth of 2–5 m (RCWM-CDP) to 40–60 km PS ECWM-PSM):

   (a) to construct 0.8–3 to 40–60 km deep crustal sections;
   (b) to construct 2–3 to 400–500 m deep crustal sections;
   (c) to study the physical properties of rocks and to compile the SMD maps.

Acknowledgements The authors are grateful to Grigory Sokolov, Artem Lebedev and the reviewers for valuable improvements of the manuscript. This work was partly supported by the Institute of Geology Research Programme AAAA-A18-118020290086-1 and funded by RFBR as a part of the Research Project 19-05-00481 «Lithospheric structure and dynamics of the White Sea Region».

References

1. Koistinen, T., Stephens, M.B., Bogatchev, V., Nordgulen, O., Wenneström, M., Korhonen, J. (Comps.): Geological map of the Fennoscandian Shield, scale 1:2,000,000. Espoo/Trondheim/Upsala/Moscow (2001)
2. Korhonen, J.V., Aaro, S., All, T., Nevanlinnce, H., Skilbrei, J.R., Säävuori, H., Vaher, R., Zhdanova, L., Koistinen, T. (Comps.): Magnetic Anomaly Map of the Fennoscandian Shield, scale 1:2,000,000. Geological Surveys of Finland, Norway and Sweden and Ministry of Natural Resources of Russian Federation (2002)
3. Korhonen, J.V., Aaro, S., All, T., Elo, S., Haller, L.Ä., Kääriäinen, J., Kulinich, A., Skilbrei, J.R, Solheim, D., Säävuori, H, Vaher, R., Zhdanova, L., Koistinen, T. (Comps.): Bouguer Anomaly Map of the Fenoscandian Shield, scale 1:2,000,000. Geological Surveys of Finland, Norway and Sweden and Ministry of Natural Resources of Russian Federation (2002)
4. Sharov, N.V. (ed.): Deep structure and seismicity of the Karelian Region and its surroundings. KarRC RAS, Petrozavodsk, p. 353 (2004). (in Russian)
5. Sharov, N.V., Isanina, E.V., Pozhilenko, V.I., Stupak, V.M.: Increasing the validity of the geological models of the earth crust in the Sredny-Rybachy peninsula area by complexing CDP, DSS and PSM seismic methods. Uralski geofizicheski vestnik **4**, 98–105 (2007). (in Russian)
6. Glushanin, L.V., Sharov, N.V., Shchiptsov, V.V. (eds.): Paleoproterozoic Onega structure: geology, tectonics, deep structure and mineral genesis. KarRC RAS, Petrozavodsk, p. 431 (2011) (in Russian)
7. Sharov, N.V., Isanina, E.V., Zamozhnyaya, N.G., Krupnova, N.A., Suleimanov, A.K.: Results of the CDP, DSS and ECWM deep seismic study of the Onega Ladoga area, Republic of Karelia, Russia. Geofizicheskiy zhurnal **33**, 23–39 (2011). (in Russian)
8. Sharov, N.V.: Superdeep Drilling and Its Impact on the Seismic Models of the Fennoscandian Shield. In: Yanovskaya, T., Kosterov, A., Bobrov, N., Divin, A., Saraev, A., Zolotova, N. (eds.) Problems of Geocosmos–2018. Springer Proceedings in Earth and Environmental Sciences, pp. 253–261. Springer, Cham (2020). https://doi.org/10.1007/978-3-030-21788-4_21
9. Sharov, N.V.: Lithosphere of North Europe based on seismic data. KarRC RAS, Petrozavodsk, p. 173 (2017) (in Russian)
10. Sharov, N.V. (ed.): Proterozoic Ladoga structure geology, deep structure and mineral genesis. KarRC RAS, Petrozavodsk, p. 435 (2019) (in Russian)
11. Thybo, H., Janik, T., Omelchenko, V.D., Grad, M., Garetsky, R.G., Belinsky, A.A., Karataev, G.I., Zlotski, G., Knudsen, U.E., Sand, R., Yliniemi, J., Tiira, T., Luosto, U., Komminaho, K., Giese, R., Guterch, A., Lund, C.-E., Kharitonov, O.M., Ilchenko, T., Lysynchuk, D.V., Skobelev, V.M., Doody, J.J.: Upper lithosperic seismic velocity structure across the Pripyat Trough and Ukrainian Shield along the EUROBRIDGE '97 profile. Tectonophysics **371**, 41–79 (2003). https://doi.org/10.1016/S0040-1951(03)00200-2
12. Janik, T., Kozlovskaya, E., Heikkinen, P., Yliniemi, J., Silvennoinen, H.: Evidence for preservation of crustal root beneath the Proterozoic Lapland-Kola orogen (northern Fennoscandian shield) derived from P and S wave velocity models of POLAR and HUKKA wide-angle reflection and refraction profiles and FIRE4 reflection transect. J. Geophys. Res. **114**(B06308), 1–34 (2009). https://doi.org/10.1029/2008JB00568
13. Starostenko, V., Janik, T., Mocanu, V., Stephenson, R., Yegorova, T., Amashukeli, T., Czuba, W., Sroda, P., Murovskaya, A., Kolomiyets, K., Lysynchuk, D.: RomUkrSeis: seismic model

of the crust and upper mantle across the Eastern Carpathians-From the Apuseni Mountains to the Ukrainian Shield. Tectonophysics **794**, 228620 (2020). https://doi.org/10.1016/j.tecto.2020.228620

14. Yanovskaya, T.B., Dmitrieva, L.A.: Influence of non-rigidity of contact of elastic media on the coefficient of reflection, refraction and exchange. Izvestiya of the Academy of Sciences of the USSR. Phys. Solid Earth **2**, 17–22 (1991). (in Russian)

Evaluation of Hypocenters Distribution Based on the Geoelectric Models in the Tien Shan Earthquake-Prone Areas

Kseniia Nepeina and Elena Bataleva

Abstract This work is aimed to develop a new approach to study the relationships between the spatial distribution of the electrical conductivity and earthquake hypocenters in seismically active regions such as the Tien Shan. The basis of a complex deep complex model (up to 30 km depth) of the Tien Shan lithosphere is a fault-block model. The joint analysis of the geoelectric model and seismicity provides important information for understanding the distribution of deep electrical conductivity, which is an indicator of active geodynamic processes and provides indirect estimates of the stress-strain state of the geological media. The purpose of our investigation is to extend the study of the relationship between hypocenter depth distribution and geoelectric boundaries on a 2D cross-sections in the Central Tien Shan. This paper shows complex models, built for several meridional magnetotelluric soundings (MTS) profiles located (74.3° E and 75.2° E). We use seismic bulletins for 5 years (2015–2019) and select earthquakes nearest to the MTS profile lines. Based on our previous studies, we are interested in the behavior of seismicity with energy classes $K < 8$ and $K > 8$ separately. We calculate relative coordinates (latitudes) to the kilometers from the starting point of the profile. As a result, the graphs present colored 2D cross-section with profile distance (R) and earthquake hypocenters into the depth marked by dots. The analysis of this study indicates the presence of a certain relationship: the earthquakes hypocenters clustering attach to the boundaries of objects and outline geological bodies, or are located inside certain zones.

Keywords Crustal conducting layer · Resistivity heterogeneities · Geoelectric model

K. Nepeina (✉) · E. Bataleva
Research Station RAS, 49, 720049 Bishkek, Kyrgyzstan
e-mail: neks@gdirc.ru

A. Kosterov et al. (eds.), *Problems of Geocosmos—2020*, Springer Proceedings in Earth and Environmental Sciences, https://doi.org/10.1007/978-3-030-91467-7_22

1 Introduction

The occurrence of earthquake hypocenters is usually the result of a stress-strain state or the presence of fractured zones associated with faults. In this case, the concentration of bulk of seismic events in some spatial clusters can be reviewed as an indicator of modern geodynamic activity, and their depth distribution as a characteristic of rheological or geological boundaries. To assess such boundaries, it seems possible to compare deep geoelectric models and the distribution of earthquake hypocenters. Moreover, the overwhelming majority of earthquakes depth correlate with the top of the conducting layer, and are verified by zones of high gradients of physical parameters—electromagnetic, porosity, etc. [1, 2]. The crustal conductive layer is characterized by low seismic velocities and practically does not contain earthquake hypocenters, which indicates its fluid saturation and plasticity, i.e., inability to accumulate elastic energy sufficient for the occurrence of earthquakes. The crustal conductive layer could be very likely defined by deep magnetotelluric sounding (MTS) studies. Referring to [3], a high correlation between seismic activity, the day-surface deformations, and the morphology of the crustal conducting layer was established.

In [4], we attempted to analyze how seismic events correlate with the geoelectric heterogeneities of the geoelectric model built along the NARYN transect (total length 370 km along $76° \pm 0.2°$ E). Previous studies [4–6] have established that seismic events are crustal in terms of the depth of occurrence—earthquake hypocenters in the basement of the Tarim block lie at depths of up to 40 km, while the overwhelming majority of the Tien Shan earthquake hypocenters are localized at depths of up to 20 km [7]. Upon research [8] an analysis of the features of the focal mechanisms of earthquake sources concludes that each depth level (0–15 km, 15–30 km, and over 30 km) is distinct from others. This indicates a change in the stress-strain state of the medium with depth, as well as tectonic layering of the Earth's crust [8]. Based on the sample, which included 167 events with $K > 10.5$ for the period from 1987 to 2015, it was concluded that seismicity is confined to high-resistivity blocks of the geoelectric model ($\rho \approx 10^3$–$10^4\ \Omega \cdot m$) and their boundaries. Therefore, for this study, we decide to use the MTS results and the earthquake bulletins (2015–2019) to find any relationships between hypocenter distribution within the Central Tien Shan and geoelectric anomalies on cross-sections such as the crustal conducting layer.

2 Methods and Modeling Results

According to our technique used for the study in [4], we decided to repeat the same complex of methods (geoelectric models and seismic bulletins) and apply to the broad profiles of deep MTS and analyze the features of the lithosphere of the Central Tien Shan. We carried out three profiles (MTS points 900–906, 909–914, 915–920 + 810–827) in the Central Tien Shan in 2011 and 2018 and received all data with

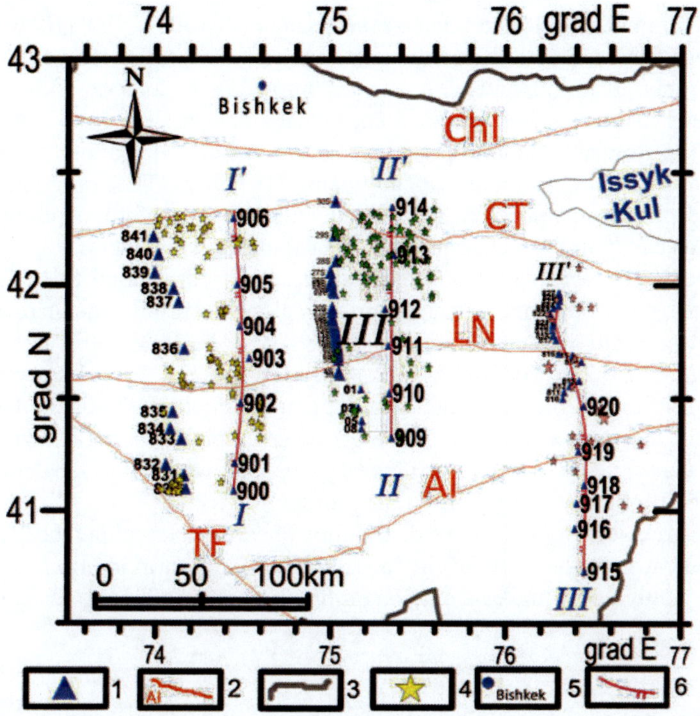

Fig. 1 Map of the Central Tien Shan with highlighted meridional profiles of deep magnetotelluric sounding (MTS) and selected earthquakes: 1—MTS points; 2—main faults (ChI—Chu-Ily, CT—Central Terskey, LN—Nikolaev Line, AI—Atbashy-Inylchek, TF—Talas-Fergana); 3—borders of the Kyrgyz Republic; 4—earthquakes (2015–2019) according to the seismic bulletin: only for adjacent territories (±25 km), 5—city, 6—main profile line for the geoelectric model (I—74.3° ± 0.2°; II—75.2° ± 0.2°; III—76.3° ± 0.2°)

modern Phoenix MTU-5 equipment (Fig. 1). The detailed geoelectric models for meridians (I—74.3° ± 0.2° E; II—75.2° ± 0.2° E, III—76.3° ± 0.2° E) after MTS data processing are created. Two models (profiles I and II) are presented for the first time. The total length for each profile is (km): I—145, II—120, III—150. So the sounding depth is more than 30 km. Using some assumptions for the true geoelectric models, we use the new MTS data from adjacent profiles (e.g., «Kokemeren» MTS points 810–827, «Son-Kul» MTS points 03–29 s, «Karabuk» - MTS points 01–08). The geoelectric models are built with the standard processing—for 2D inversion with number of iterations using Rodi-Mackie software [9]. Geoelectric model is a 2D cross-section: depth and profile line in kilometers. However, this software does not contain the possibility to add any other information, e.g. earthquake hypocenters. For this reason, we need to consider using another program to form a graph, e.g. Excel.

As we were looking before [4] in the time interval 1967–2015, we are moving forward to look at next 5 years (2015–2019). During the period under consideration (2015–2019), more than 300 earthquakes of energy class K from 6 to 14 were registered in the Central Tien Shan by the seismic network KNET [10]. According to statistics from the bulletin, earthquake hypocenters are distributed in the depth interval from 0 to 25 km, and are most concentrated in the interval of 5–15 km.

We need to transfer geographical coordinates of hypocenters to relative coordinates. The Research station operators compute the earthquake parameters (coordinates, focal depth, etc.) automatically using HYPOCENTER program [11]. We take the depth parameters from the seismic bulletins. Therefore, depth hypocenter distribution easily lies in 2D graph on a vertical axis in kilometers.

The next step is to select only earthquakes, that correspond to MTS profiles coordinates. We need to put them appropriately along the profile line. In this case, we limit three areas for each profile line: I—41.08°–42.4° N, 74.3° ± 0.2° E; II—41.3°–42.4° N, 75.2° ± 0.2° E; III—40.74°–42.4° N, 76.3° ± 0.2° E. In the interactive mode, operator select only the corresponding geographic coordinates of the earthquake bulletin for the further analysis. These earthquake epicenters are shown in Fig. 1 with the stars. Finally, we have various latitudes, that we need to recalculate to the kilometers from the starting point of the profile.

To construct joint models of the distribution of earthquakes along the 2D geoelectric models, we use the depth of the focus in its original form in kilometers and we calculate relative coordinates along the profile, using the following formulas:

$$\varphi_r = \varphi_{eq} - \varphi_{MT} \quad \text{and} \quad R = \Delta^\circ \cdot \varphi_r \tag{1}$$

where φ_r—relative latitude of the epicenter projection point, φ_{eq}—true latitude of the epicenter from the seismic bulletin (longitudes Λ_{eq}: I—74.3° ± 0.2° E; II—75.2° ± 0.2° E; III—76.3° ± 0.2° E), φ_{MT}—latitude of the starting MTS point: I—41.08° N, II—41.3° N, III—40.74° N; R—the distance along the profile in kilometers, where $\Delta^\circ = 110$ km.

Afterward, we can form a graph, where we have profile distance (R) along axis X and along axis Y—depth. Then we superpose the graph with the transparent plot area into Excel and keep only earthquake hypocenters marked by dots. The detailed geoelectric models for profiles I and II with the depth distribution of the hypocenters are considered in Figs. 2 and 3 correspondingly. Based on our previous studies, we are interested in the behavior of seismicity with energy classes K < 8 and K > 8 separately.

3 Results

According to the analysis of the distribution of hypocenters that occurred near the 74.30° meridian, an interesting pattern is observed (Fig. 2). Events with energy classes less than 8 often fall on conducting objects ($2 < \rho_{app} < 20$ Ω m) or are confined to the gradient boundaries between high- and low-resistivity bodies. The hypocenters are scattered from almost near-surface to significant depths of 25 km. However, earthquakes with K < 8 are not closely clustered. They spread everywhere

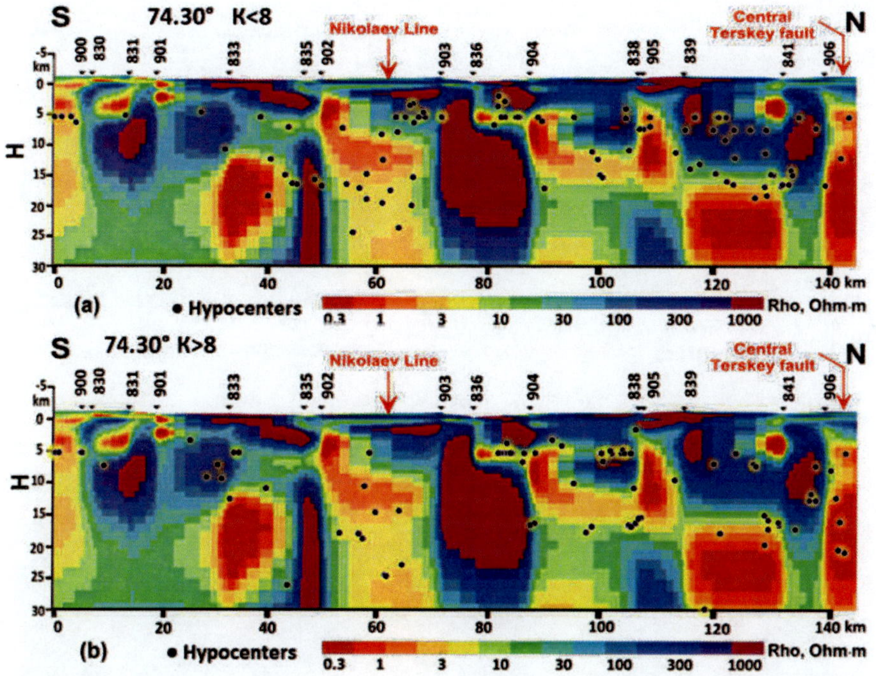

Fig. 2 Geoelectric model along 74.30° with earthquake hypocenters distribution from the KNET bulletin for 2015–2019: **a** K < 8, **b** K > 8

inside the range 5–25 km (MTS points 902–903) and 5–20 km (905–906) (Fig. 2a). As for the earthquakes with K < 8, they lined up in a horizontal line at a depth of ~ 5 km (MTS points 836–906 in Fig. 2a) and for K > 8 near 5 km and 15 km (MTS points 836–838 in Fig. 2b). This may be due to the peculiarities and errors in automatic depth calculations, as well as the limited sensitivity of the local seismic networks (e.g. KNET) for the western part of the Central Tien Shan. On the other hand, this fact of linear distribution of hypocenters may indicate layering and rheological heterogeneity of the crust of the Central Tien Shan. We also observe, that the stronger earthquakes (K > 8) the deeper locate than smaller earthquakes (K < 8).

For the meridian 75.2° ± 0.2° (Fig. 3), pattern of the hypocenters' appearance differs from the previous situation, and is very similar for earthquakes with K > 8 and K < 8. We think, the most part of them trace the borders between isolators and conductors. However, visually, the southern and northern parts of the profile correspond to different seismic regimes. Thus, for the southern part, between MTS points 909–910, rare seismicity is observed, while the northern part is distinguished by a greater number of earthquakes. At the same time, in the southern part, the hypocenters are completely inside the isolators, for the central part of the profile (MTS points 911–913); the hypocenters are located along the boundaries of the conducting bodies. In the northern region (MTS points 913–914), earthquakes at

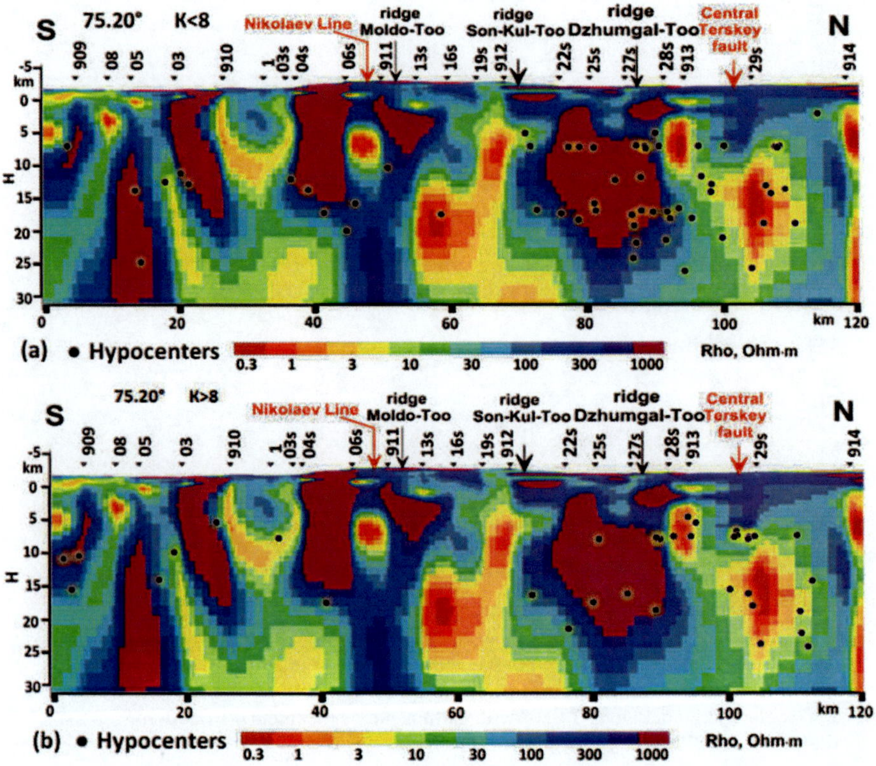

Fig. 3 Geoelectric model along 75.20° with earthquake hypocenters distribution from the KNET bulletin for 2015–2019: **a** K < 8, **b** K > 8

depths of 10–25 km are inscribed in massive bodies with low values of geoelectric anomalies. The zone under Central Terskey fault characterizes high hypocenters' density on both graphs Fig. 3a, b. Near MTS points 03–06 s the hypocenters clearly outline isolator boundaries at a depth of 10–20 km. Unfortunately, profile III (76.3° ± 0.2° E) is not so representative in order to conclude the distribution of earthquakes in deep cross-sections. Despite it is the longest regional profile between selected MTS profiles, it has less similarity with profiles I and II, due to the lack of number of earthquakes nearby for 2015–2019. Only 26 earthquakes occurred at 5 km depth, mostly in the northern part of the profile. In this regard, we conclude the seismicity on this profile decreases from North to South.

4 Conclusions

Further analysis of such plots shows that the boundaries of geoelectric heterogeneities correlate well with the spatial distribution of earthquake hypocenters. This fact indicates the presence of a certain relationship: the hypocenters of earthquakes are confined to the boundaries of geoelectric objects or are located within certain zones. The nature of the distribution of geoelectric heterogeneities and earthquake hypocenters along the profile indicates the existence of mechanically weakened layers in the Tien Shan lithosphere, e.g. Central Terskey fault and main ridges zones. Also we assume that the earthquake hypocenters discontinuously contour the weakened layer ~15 to 20 km depth—the bottom edge of the isolator bodies (e.g. 75.20° cross-section in Fig. 3).

The construction of reliable two-dimensional geoelectric models for the regime (indicator) points located on the regional profiles of the MTS opens up an additional opportunity to control a seismically active region. Definitely, for a more accurate superposition this algorithm could be improved.

Acknowledgements The study was supported by the Ministry of science and higher education of the Russian Federation for the Research Station of the Russian Academy of Sciences in Bishkek (AAAA-A19-119020190063-2 and AAAA-A19-119020190064-9). We thank reviewer Smirnova Natalia Alekseevna for comments and our colleague Vitaly Matiukov for geoelectric modelling.

References

1. Typical geological and geophysical models of the environment of seismic and aseismic areas. Ilim, Bishkek (1992). (in Russian)
2. Trapeznikov, Y.A., Andreeva, E.V., Batalev, V.Y., Berdichevsky, M.N., Vanyan, L.L., Volykhin, A.M., Golubtsova, N.S., Rybin, A.K.: Magnetotelluric Soundings in the Kyrgyz Tien Shan. Izvestiya, Phys. Solid Earth **33**(1), 1–17 (1997)
3. Zubovich, A.V., Trapeznikov, Y.A., Bragin, V.D., Mosienko, O.I., Shchelochkov, G.G., Rybin, A.K., Batalev, V.Y.: Deformation field, earth's crust deep structure, and spatial seismicity distribution in the Tien Shan. Russ. Geol. Geophys. **42**(10), 1634–1640 (2001)
4. Bataleva, E.A., Batalev, V.Y., Rybin, A.K.: Interrelation of conductivity, seismic velocities and the seismicity for Central Tien Shan lithosphere. Lithosphere (Russia) **5**, 81–89 (2015). (in Russian)
5. Kissin, I.G., Ruzaikin, A.I.: Relations between seismically active and electrically conductive crustal zones in the Kyrgyz Tien Shan. Izvestiya, Phys. Solid Earth **33**(1), 18–25 (1997)
6. Berdichevsky, M.N., Sokolova, E.Y., Varentsov, I.M., Rybin, A.K., Baglaenko, N.V., Batalev, V.Y., Golubtsova, N.S., Matyukov, V.E., Pushkarev, P.Y.: Geoelectric section of the Central Tien Shan: analysis of magnetotelluric and magnetovariational responses along the Naryn geotraverse. Izvestiya, Phys. Solid Earth **46**(8), 679–697 (2010). https://doi.org/10.1134/S1069351310080057
7. Rybin, A.K., Spichak, V.V., Batalev, V.Y., Bataleva, E.A., Matyukov, V.E.: Array magnetotelluric soundings in the active seismic area of Northern Tien Shan. Russ. Geol. Geophys. **49**(5), 337–349 (2008). https://doi.org/10.1016/j.rgg.2007.09.014

8. Belenovich, T.Y., Sabitova, T.M.: Dynamics of the transorogenic Talas-Fergana fault. In: Modern Geodynamics of the Tien Shan Lithosphere. Nauka, Moscow, pp. 101–105 (1991) (in Russian)
9. Rodi, W.L., Mackie, R.L.: Nonlinear conjugate gradients algorithm for 2-D magnetotelluric inversion. Geophysics **66**(1), 174–187 (2001). https://doi.org/10.1190/1.1444893
10. Bataleva, E.A., Mukhamadeeva, V.A.: Complex electromagnetic monitoring of geodynamic processes in the Northern Tien Shan (Bishkek Geodynamic Test Area). Geodyn. Tectonophys. **9**(2), 461–487 (2018) (in Russian)
11. Lienert, B.R., Havskov, J.: A computer program for locating earthquakes both locally and globally. Seismol. Res. Lett. **66**(5), 26–36 (1995). https://doi.org/10.1785/gssrl.66.5.26

Application of Adaptive Filtering Techniques for Filtering Induced Seismic Noise

V. V. Gravirov and K. V. Kislov

Abstract The complexity and versatility of modern research requires the development and improvement of methods and tools for the operational processing of seismic data. Often, the main task in seismic data processing is to efficiently and reliably extract useful components from the recorded signal for their further use. Obviously, in this case, seismologists need to use the entire arsenal of methods and algorithms for reducing in modern information-measuring systems induced noise caused by various kinds of dynamic processes. An adaptive filter can be proposed as one of the options for solving such problems of filtering a seismic signal in the presence of interference caused by various kinds of dynamic processes. This article is devoted to the problems of using adaptive filters to separate seismic signals with simultaneous digital recording of a seismic signal and atmospheric pressure. The article shows that using adaptive filtering, it is possible to significantly improve the signal-to-noise ratio and successfully to separate seismic signals masked by induced interference. Thus, using the adaptive filtering method, it is possible to solve the problem of improving the data quality of seismic instruments by removing unwanted induced noise from them.

Keywords Adaptive filtration · Digital filtration · Seismic instrument · Noise reducing

1 Introduction

The completeness of knowledge about the processes occurring in the interiors of the Earth largely depends on correct instrumental observations. The variety of phenomena under study involves the use of research seismic waves in the widest

V. V. Gravirov (✉)
Schmidt Institute of Physics of the Earth of the Russian Academy of Sciences (IPE RAS), Moscow, Russia

V. V. Gravirov · K. V. Kislov
Institute of Earthquake Prediction Theory and Mathematical Geophysics of the Russian Academy of Science (IEPT RAS), Moscow, Russia

A. Kosterov et al. (eds.), *Problems of Geocosmos—2020*, Springer Proceedings in Earth and Environmental Sciences, https://doi.org/10.1007/978-3-030-91467-7_23

frequency range. In this regard, today one of the important problems of seismometry is the parametric improvement of seismic equipment, including an increase in the resolution of instruments, an expansion of their frequency and dynamic ranges.

Obtaining high quality seismic data is a major technological challenge. This is primarily because in many cases the instrument resolution approaches the physically achievable limit. In addition, on the way to increasing the sensitivity of seismometers, various types of interference appear, both instrumental and transmitted through the ground, air, temperature and etc. At the same time, a theoretical analysis of the influence of the external environment on instruments is often very difficult, and its nature is ambiguous, difficult to interpret, changing from station to station and from instrument to instrument of the same type. However, if we consider a system, for example, of a seismometer as an additive one, then it will be possible to consider all types of interference separately, and also try to evaluate and reduce their influence.

There are many publications devoted to analytical and experimental studies of the influence of the environment on the level of instrumental noise [1–3] today. As the most complete, which can be used in assessing the noise level of instruments and for new developments, one should definitely mention the works of A. Rykov [4–11], E. Wielandt [12–15], V. Feofilaktov, V. Theophilaktov [16, 17] and E. Linkov [18–21]. However, they usually deal with specific instruments or their elements and individual interference factors. At an expert group workshop in Albuquerque (USA) [22], special attention was paid to the sensitivity of instrument parameters to changes in the environment. The most critical equipment to such disturbances is seismic tiltmeters, since for their correct operation it is necessary to maintain a constant ambient temperature of the order of hundredths of a degree, which is practically difficult to implement. In this case, the use of continuous temperature monitoring [23, 24] with using the adaptive filtration technique can be very relevant. It was supposed to determine the methods of accounting for the effects of temperature, pressure, humidity, groundwater level, rain flows, magnetic fields, electromagnetic interference, shocks and vibrations, acoustic noise, lightning, corrosion and even radiation from a nuclear explosion. This increased attention from leading experts to environmental impacts on seismic instruments was undoubtedly due to the consequences of these impacts. Therefore, for example, a nuisance signal in a seismic record can in some cases exceed the useful signal recorded at the same frequency by orders of magnitude. At the same time, the intrinsic noise of seismic instruments can differ by two or more orders of magnitude with small differences in their designs. Thus, the main obstacle in the use of broadband seismic equipment is various kinds of interference, the power of which and the effect on the device increase as its frequency response expands into the range of long periods. However, by analyzing the complex influence of interference factors and ways of noise propagation into any seismic record, it is possible, by means of certain mathematical operations, to significantly improve the metrological characteristics of the instruments and, accordingly, their recording. In this article, we will not consider the problem of noise control so widely and will limit ourselves to demonstrate the benefit of adaptive filtering for improving the quality of the seismic records by eliminating induced noise caused by changes in atmospheric pressure.

2 Methods to Mitigate Noise in Seismic Records

The temperature and atmospheric pressure noise propagation into the seismic record of any seismological instrument is illustrated in Fig. 1. Since the noise caused by changes in atmospheric pressure is most significantly appeared in the vertical component of the seismometer records, all further arguments and calculations will be given only for the this component. Note, that for the horizontal components, everything can be done similarly. In this diagram, $X(t)$ is a vertical ("clear" or without any noise) component of ground motions, $T(t)$ and $P(t)$ characterize the components of the instrumental noise caused by changes in ambient temperature and atmospheric pressure, $H(f)$ is the transfer function of seismic instrument and, finally, $Y(t)$—obtained data record.

It should be understood that in addition to the "clear" signal, for example, from an earthquake, the movement of soil $X(t)$ may also include components of seismic interference arising from anthropogenic impact, resonance phenomena [25], shrinkage of the concrete of the pedestal and the soil under it [3], as well as induced interference generated by changes in physical fields (changes in humidity, gravity, electromagnetic fields, etc.), but only when they act directly on the elements of sensors and communication channels, including intrinsic noise ("system jitter") [26], analog noise digital transformation [27], etc. The transfer function of the seismic sensor $H(f)$ may not be stationary and can vary over time, sometimes quite significantly, when environmental conditions change [3, 6].

The greatest influence on any seismic equipment is exerted by interference associated with fluctuations in atmospheric pressure (see Fig. 2) and temperature, both transmitted through the pedestal and directly acting on the instruments. Various methods are used to reduce or eliminate such influences. Let's consider one of them further, namely adaptive filtering. In the scientific literature, it has been repeatedly indicated that the wind is one of the most significant noise-generating factors in seismology [28–31]. It was found that the correlation coefficient between the average wind speed and pressure in the time interval 10–50 s is equal to or exceeds 0.8. Such a high correlation has led to the fact that many researchers ignore the effect of wind on the generation of noise, considering only changes in atmospheric pressure, and the wind speed is considered a scalar quantity [32]. Meanwhile, the apparent directionality of both the mean wind and gusts [33] requires considering not the question

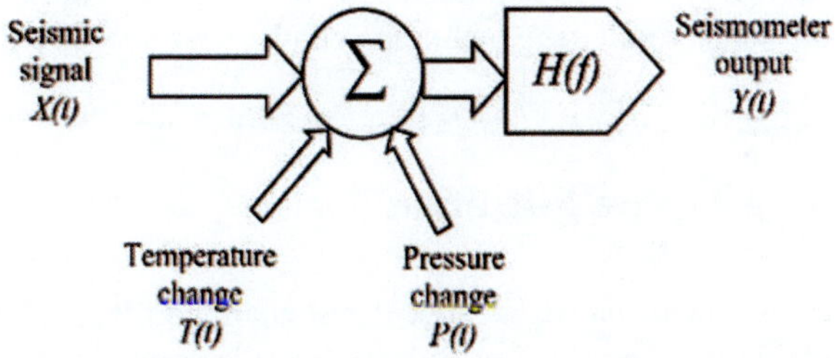

Fig. 1 Simplified scheme of the temperature and atmospheric pressure noise propagation into a seismic record

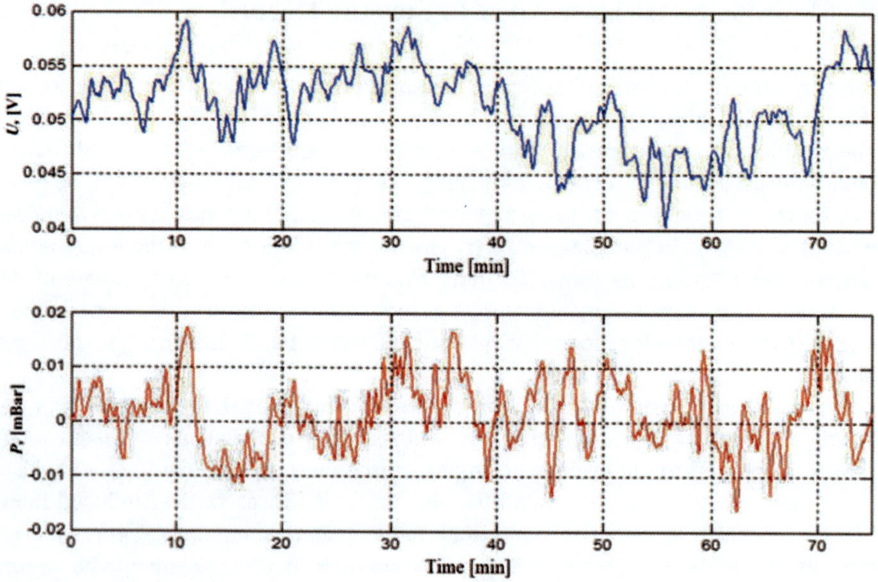

Fig. 2 An example of a record of atmospheric pressure wind fluctuations (lower red graph) and the corresponding vertical component signal from seismometers (upper blue graph)

of the relationship between wind and pressure, but the problem of generating seismic noise by the wind, since the direction and speed of the wind are not uniquely determined only by the pressure gradient [34]. In addition, any obstacle that is in the path of the wind will somehow affect it and disturb the wind field. Such obstacles can be large-scale, such as mountain ranges, and small-scale, such as buildings, trees, forest belts, etc. Accordingly, there are zones of increased wind and zones of wind shadow. In front of an obstacle and behind it, sometimes the so-called windward and leeward eddies can appear. As a result of air friction against the earth's surface and impacts into obstacles, a very complex picture of seismic noise is created, which, naturally, has its own characteristics for each seismic station [35]. The interaction of wind with obstacles, especially buildings and structures, can also cause phenomena such as buffeting, flutter, etc. As a result, it is possible to generate intense noise seismic waves [36]. At a certain strength and direction of the wind, nuisance signal caused by these phenomena appear in the spectrum of the seismic signal and, using the adaptive filtering technique can be very successfully filtered out [37, 38].

3 Adaptive Filtration

Let's now consider in more detail adaptive filtering and methods of its application for the extraction of useful or "clean" seismic signals. In short, adaptive filtering

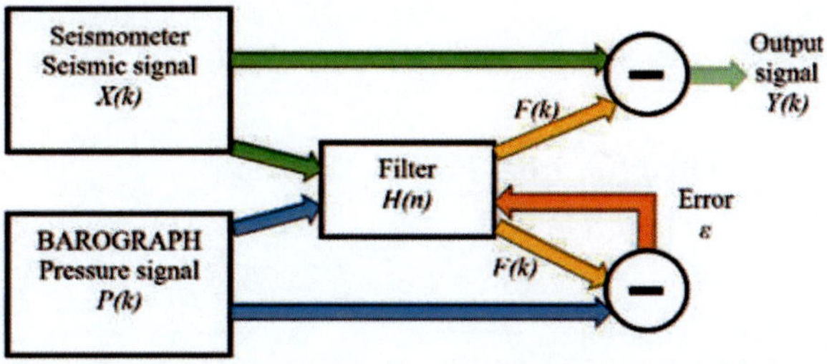

Fig. 3 Simplified block diagram of an adaptive filter for filtering noise generated by atmosphere pressure

is based on the use, in addition to the original noisy seismic signal, one or more interference signals, correlated or weakly correlated with the useful signal. At present, the practical implementation of adaptive filtering can be quite easily implemented due to the appearance of high-performance computers and special programmable microcontrollers. Adaptive filtering differs from the classical digital filter by the presence of a feedback loop and, therefore, the transfer function of the filter is variable and depends directly on all the applied signals. There is typical field of application is usually the solution to the problem of cleaning signals from any unstable interfering signals and noise in the case when their spectral composition almost completely overlaps with the spectrum of the useful signal or when the spectrum of nuisance signal is not known in advance and may change over time. Let us consider the simplest version of an adaptive filter that satisfies the conditions of linear filtering, when the seismic signal $X(k)$ and the interference $P(k)$ are two stationary mutually correlated processes, and the minimum of the root-mean-square error will be considered as the criterion of the filter's optimality. The block diagram of such filter is shown in Fig. 3.

At the inputs of the filter $H(n)$, the recorded time-series of two analog signals come in digital form, $X(k)$ from the seismometer, and $P(k)$ from the barograph. Both signals are normalized and precisely synchronized in time. In this case, the signal from the seismometer can be represented as the sum of the direct "clear" seismic signal $S(k)$, induced interference $Sp(k)$, and seismic noise $N(k)$:

$$X(k) = S(k) + Sp(k) + N(k) = Sn(k) + Sp(k),$$

where $Sn(k)$ is a seismic signal with a noise component. The induced interference $Sp(k)$ is directly related to the signal of the atmosphere pressure interference $P(k)$, that is, their cross-correlation function $Csp(k, k')$ is usually positive:

$$Csp\left(k, k'\right) = Sp(k), \, P\left(k'\right) > 0.$$

As was noted earlier, when the spectral composition of the seismic signal $X(k)$ and the pressure $P(k)$ coincide, the extraction of the useful signal by linear filtering methods usually turns out to be an unsolvable problem [39]. However, using an adaptive filter, one can successfully filter out most of the interference component. In this case, the output signal $F(k)$ can be written as

$$F(k) = \sum_{n} H(n) \cdot X(k-n) \tag{1}$$

and it will differ from $P(k)$ by the value

$$\varepsilon(k) = F(k) - P(k),$$

which characterizes the error signal and is the absolute value of the signal reproduction error at samples k. The Kolmogorov-Wiener filter [40] can be used as one of the possible versions of the software implementation of the filter $H(n)$. In this case, the optimization criterion will correspond to the criterion of the minimum standard deviation

$$\overline{\varepsilon(k)^2} = \overline{[F(k) - P(k)]^2}, \tag{2}$$

which will make it possible to determine the filter coefficients $H(n)$. The bar above means the averaging of the deviation error over all samples. Substituting (1) into Eq. (2), the deviation $\varepsilon(k)^2$ will be equal to

$$\overline{\varepsilon(k)^2} = \overline{\left(\sum_{n} H(n) \cdot X(k-n) - P(k)\right)^2}. \tag{3}$$

Further, from the difference between the original data $X(k)$ and the output of the adaptive filter $F(k)$, we obtain the output seismic signal $Y(k)$:

$$Y(k) = X(k) - F(k) = X(k) - \sum_{n} H(n) \cdot X(k-n).$$

As mentioned before, the minimum of this expression will determine the coefficients of the filter. To find them, we differentiate it by the filter coefficients and equate the resulting equations to zero

$$\frac{d\left(\overline{\varepsilon(k)^2}\right)}{dH(m)} = \overline{\sum_{n} H(n) \cdot X(k-m) \cdot X(k-n)} - \overline{P(k) \cdot X(k-m)} = 0,$$

where

$$\overline{X(k-m) \cdot X(k-n)} = C_{XX}(m-n)$$

is the correlation function of the input signal, and

$$\overline{P(k) \cdot X(k-m)} = C_{XP}(m)$$

is a cross-correlation function of the input and interference signals $P(k)$ and $X(k)$. Then we get

$$\frac{d\left(\overline{\varepsilon(k)^2}\right)}{dH(m)} = \sum_n H(n) \cdot C_{XX}(m-n) - C_{XP}(m) = 0$$

$$\sum_n H(n) \cdot C_{XX}(m-n) = C_{XP}(m),$$

for $n = m = 0, 1, 2, 3, \ldots$ filter order. That is, the criterion (3) will correspond to the equality of the convolution of the response function of the filter with the autocorrelation function of the input signal with the cross-correlation function of output and input signals. This expression can be implemented as a system of linear equations. The solution of such a system of equations will give the desired coefficients that form the transfer characteristic of the filter.

However, it should be emphasized that the described above generalized structure of the adaptive filter in most cases may not lead to a significant improvement in seismic signals, since it does not take into account the possible presence of more complex relationships between the two input signals. In this regard, we use the upgraded 3rd order adaptive filter that has undergone significant changes (see Fig. 4) and based on an adaptive filter described in [41]. It should be noted that if someone will increase in the filter order, the delay introduced by it would increase too, as well as the computational costs including the filter adaptation time constant. This will lead to the fact that any unjustified increase in the order of the adaptive filter is highly undesirable. Therefore, usually, for definition the optimal order of the adaptive filter

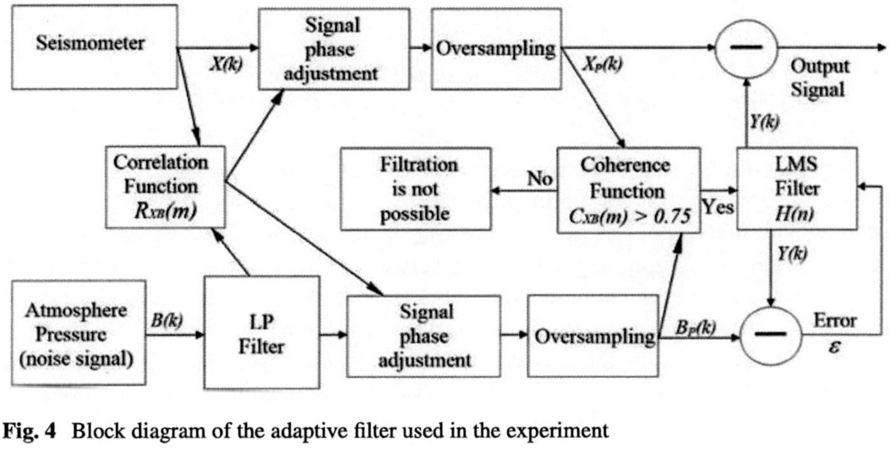

Fig. 4 Block diagram of the adaptive filter used in the experiment

is used a criterion [42] based on the search for such an order at which a minimum signal-to-noise ratio will be observed at its output.

The use of the proposed structure of the adaptive filter is due to the fact that, based on the accumulated experience of our instrumental observations, it was obvious that under the conditions of the described below experiment, the spectrum of atmospheric pressure interference with a high degree of probability will be commensurate with a significant part of the seismic signal spectrum. In this case, a simple adaptive filter (as in Fig. 3) will try to form the required part of the signal frequencies from the insignificant frequencies of the spectrum of the used signal. As a result, huge gain coefficients can be formed at these frequencies, which also affect the frequency components of the noise [43]. It is quite obvious that the result of such filtering will be completely unpredictable.

The adaptive filter used by us is based on the application of the minimum mean square error algorithm [44]. The function of cross-correlation of two input signals R_{XB} was added to the filter structure, which allows making a temporal assessment of the "mutual position" of signals relative to each other, that is, to determine the amount of shift or delay and in-phase of the processes relative to each other. It also allows to consider the delays introduced by the digital low-pass filter. Oversamplers made it possible to significantly reduce the duration of transient processes during the active formation of the coefficients of the adaptive filter. It should be noted that the increase in the sample frequencies of the signals was carried out with the simultaneous interpolation of each of them so that the mean square errors between the interpolated and original samples were minimized. To monitor the quality and the fundamental possibility of carrying out adaptive filtering, the coherence function C_{XB} was used, which is to defined as the ratio of the spectral power density of the input signals to the modulus of the square of their mutual spectral power density. If the coherence function was greater than the established empirical threshold of 0.75, then the decision was made to continue the calculations using the adaptive filtering algorithm. Otherwise, it was decided that further calculations are inexpedient due to the absence of coherent frequencies in the input signals and, therefore, the high probability of the appearance of unreliable data at the filter output, since the proposed filter in this case will try to filter out the frequencies of the interference signal from the insignificant frequencies of the spectrum of the used signal. As a result, most likely, huge gains will be formed at these frequencies, and instead of adaptive filtering, only signal distortions will appear.

As a practical example of using such adaptive filter, let us consider the selection of signals from two distant earthquakes that occurred in the region of the Tonga Islands, with magnitudes of the order of $M_w = 4.2–4.5$, recorded at the Membach seismic station belonging to the Royal Observatory of Belgium. The distance from the registration point to the epicenters of earthquakes was about 16,165 km along the Earth's surface. Figure 5 shows a fragment of the seismic record of the vertical component, including the signals of two earthquakes, from a seismometer installed in a room at the end of the adit. The room was cut in the granite mass at a distance around 40 m from the adit entrance.

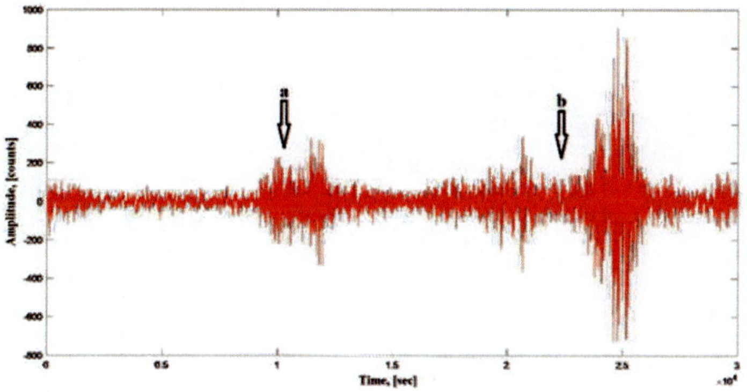

Fig. 5 Fragment of the vertical component record containing two earthquakes, made by a seismometer installed at the end of the adit of the Membach seismic station (Belgium). The first earthquake (**a**) with a magnitude of $M_w = 4.2$ with a source at a depth of 10 km occurred on October 1, 2003 in the Tonga region (15.309° S 173.312° W) at 04:42:18 (UTC) and then after 8 min 5 s approximately in the same area, a stronger earthquake (**b**) also occurred with a source at a depth of 10 km and a magnitude of 4.5

It was so fortunately coincided that at the same time, near the Membach adit, we were working on recording seismic signals using a similar seismometer simply installed on the ground surface. A schematic arrangement of seismometers is shown in Fig. 6, where seismometer No. 1 is installed inside the adit, and seismometer No. 2 is outside. Accordingly, seismometer No. 2 was subject to the additional influence of many noise-generating factors, the main of which was a short-period change in atmospheric pressure due to strong windy weather. A fragment of the record from this seismometer, which also contains a signal of these earthquakes, is represented on Fig. 7. It is clearly seen that the signal from these earthquakes was strongly masked by the signals of atmosphere pressure interference. Simultaneously with

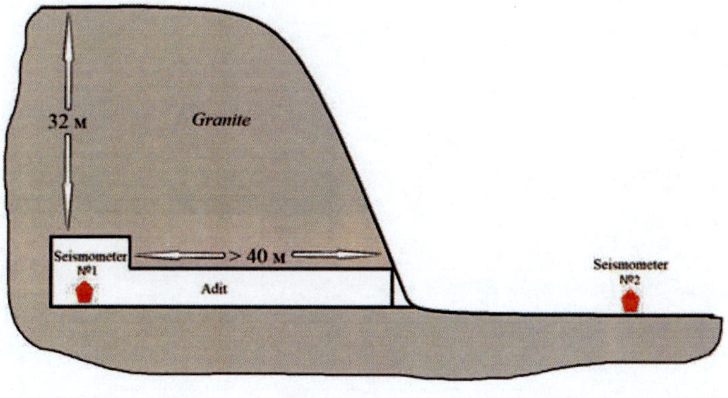

Fig. 6 Layout of seismometers during the experiment

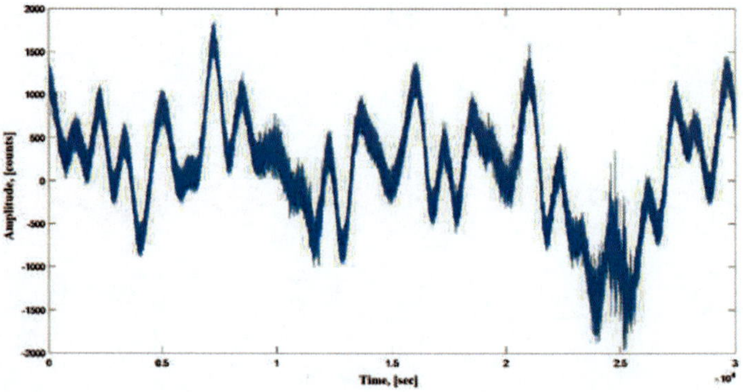

Fig. 7 Fragment of the recording of the seismometer №2 vertical component (corresponding to the data from Fig. 5), installed outside the adit on the ground surface

the recording of seismic signals, changes in atmospheric pressure values from an electronic microbarograph were also recorded.

Thus, we had the opportunity to select the "correct" seismic signal using adaptive filtering and compare it with the record produced inside a closed adit, where temperature and pressure changes were minimized. In order to visually compare the results of adaptive filtering obtained, let us plot on one graph the record from the seismometer from the adit together with the filtering result (see Fig. 8). It clearly shows that, despite the presence of a significant level of interference on the initial signal (see Fig. 7), it is possible with small errors to recover the seismic signal, which is directly related only to seismic events (earthquakes). Of course, a logical question may arise: why use such a complex adaptive filter? It seems quite obvious that since the atmospheric pressure changes caused by gusts of wind are likely to be in the low frequency range, a conventional low-pass filter can be used to filter seismic signals.

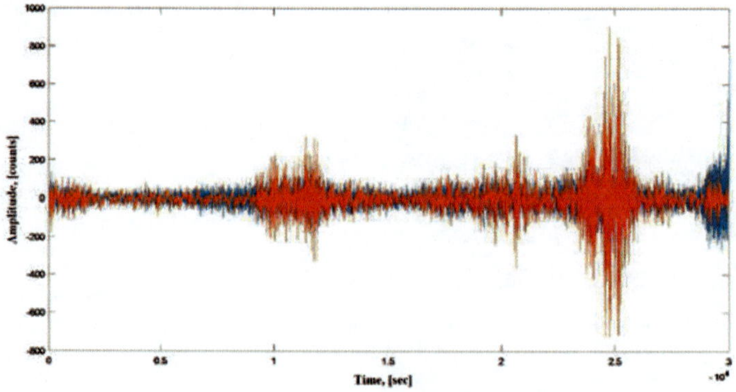

Fig. 8 Comparison of the result of the adaptive filter (blue) with the record of the seismometer №1 installed in the adit (red)

To demonstrate the quality of such filtering let's filter the recorded signal with a low-pass filter (LPF) for comparison. This LPF will help to improve the seismic signal by removing the induced noise. Let us assume that the main interference frequencies lie only in the low-frequency region of the spectrum. This statement is valid for the extraction and further filtration of any slowly varying pressure events (passage over the site of cyclones, etc.).

For the demonstration, we used a LPF described by [45] with a finite impulse response with basic parameters normalized to the signal sample frequency:

$$F_{pass} = 0.1 \cdot F_{sampl},$$
$$D_{pass} = 0.057501127785,$$
$$D_{stop} = 0.00001,$$

where F_{sampl} is the signal sample frequency in Hz; F_{pass}—cutoff frequency of the LPF; D_{pass}—pass band ripple; D_{stop}—stopband attenuation. Exactly the same filter was used in the structure of the proposed adaptive filter (see in Fig. 4), where it was used to filter the signal of interference or baric pressure. Figure 9 represents a fragment of the Euclidean norm time-series, which allows anyone to visually and quantitatively determine the difference between each sample of filter's output signals. In this graph, the Euclidean norm is plotted for comparison between two pairs of signals: between the seismometer's output in the adit and the output of the LPF and between the seismometer's output in the adit and the output of the adaptive filter. As mentioned earlier, all comparisons were made only for the vertical component of the considered signals for each corresponding sample. It is clearly seen that the presented adaptive filter allows for a much better approximation of the seismometer signal that is not directly influenced by atmospheric pressure fluctuations.

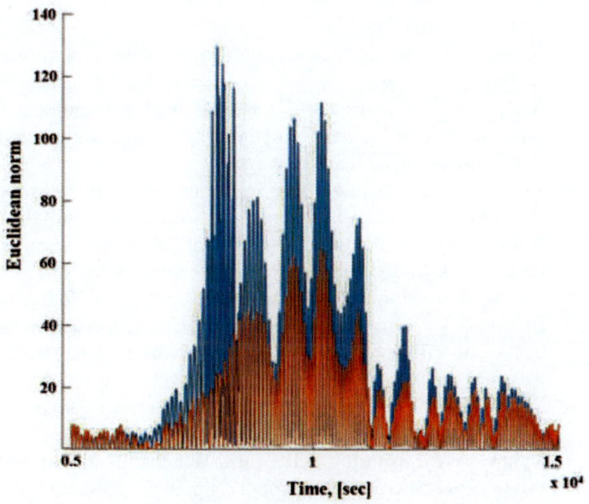

Fig. 9 Euclidean norm time-series that represent difference between two pairs of signals: blue—between seismometer in the audit and LPF output, red—between seismometer in the audit and adaptive filter output

4 Conclusion

As a result, here we practically proved that using the adaptive filtering technique, it is possible to restore a "clean" seismic signal. The applied adaptive filter takes into account the key features of the spectral composition of the signals supplied to its inputs. It is capable of generating transfer functions that provide the extraction of frequencies of "useful" signals from any spectral ranges with maximum suppression of noise corresponding to an additional second input signal [41]. In this case, the filter suppression and gain limits are generated automatically. Similarly, it is possible to filter the original seismic signals taking into account both the external and internal temperatures of the seismic instrument [46, 47]. Thus, adaptive filtering can be the most important reserve for further increasing the resolution of broadband seismometry. Its practical application will make it possible to modernize the technical means of seismic data collection systems, improve the methods and algorithms for the preliminary processing of seismic signals, which together will make it possible to most effectively solve the problems of obtaining high-quality data in specific seismic-geological conditions.

The study was carried out under state tasks of the IEPT RAS and IPE RAS.

Acknowledgements The authors express their gratitude to the staff of the Seismological Department of the Royal Observatory of Belgium for their invaluable assistance in organizing the experiment.

References

1. Bashilov, I.P., Daragan, S.K., Kabychenko, N.V.: Noise parameters of seismic devices. In: Bulletin of the National Nuclear Center of the Republic of Kazakhstan, vol. 2, pp. 19–29 (2002). (in Russian)
2. Kislov, K.V.: Humidity as an interfering factor in seismometry. Nat. Tech. Sci. **6**(38), 161–162 (2008)
3. Kislov, K.V., Gravirov, V.V.: Research of environment influence on self-noise broadband seismic equipments. In: Computational Seismology, vol. 42, 240 p. The Krasand Publishing Group, Moscow (2013). (in Russian)
4. Rykov, A.V.: Astasation of a Seismometer Using Negative Feedback. Seismicheskie pribory (Seismic Devices), vol. 14, pp. 3–5, Nauka, Moscow (1981). (in Russian)
5. Rykov, A.V.: Problems of Metrology of Broadband Seismic Equipment. Seismicheskie pribory (Seismic Devices), vol. 27, pp. 66–76, OIFZ RAS, Moscow (1997). (in Russian)
6. Rykov, A.V.: Modeling a Seismometer, 109 p. OIFZ RAS, Moscow (1996). (in Russian)
7. Rykov, A.V.: Representation of an Electronic Seismometer Using Natural Frequencies. Seismicheskie pribory (Seismic Devices). vol. 27, pp. 39–42. OIFZ RAS, Moscow (1997). (in Russian)
8. Rykov, A.V.: Seismometer and Earth oscillations. Phizika Zemli (Phys. Earth) **2**, 76–80 (1992). (in Russian)
9. Rykov, A.V., Bashilov, I.P.: Ultra-Wideband Digital Set of Seismometers. Seismicheskie pribory (Seismic Devices), vol. 27, pp. 3–6. OIFZ RAS, Moscow (1997). (in Russian)

10. Rykov, A.V., Marchenkov, A.Y.: Modeling a Seismometer in the Area of its Instability. Seismicheskie pribory (Seismic Devices), vol. 24, pp. 90–95. Nauka, Moscow (1993). (in Russian)
11. Rykov, A.V., Marchenkov, A.Y.: Seismometer with Feedback on the Example of Wieland's Seismometer. Seismicheskie pribory (Seismic Devices), vol. 23, pp. 54–58. Nauka, Moscow (1992). (in Russian)
12. Wieland, E., Stein, J.M.: A Digital Very-Broad-Band Seismograph. Ann. Geophys. Ser. B **4**(3), 227–232 (1986)
13. Wieland, E., Streckeisen, G.: The leaf-spring seismometer—design and performance. BSSA **72**, 2349–2367 (1982)
14. Wielandt, E.: Common Seismic Sensors. Datasheet DS 5.1. In: Bormann, P. (ed.) New Manual of Seismological Observatory Practice (NMSOP-2), IASPEI, GFZ German Research Center for Geosciences, Potsdam (2012). https://doi.org/10.2312/GFZ.NMSOP-2
15. Wielandt, E.: Noise in electronic seismograph systems. Z. Geophys. **39**, 597–602 (1973)
16. Feofilaktov, V.D.: Interference in Long-Period Seismometry, 100 p. Nauka, Moscow (1977). (in Russian)
17. Theophilaktov, V.D.: Noises of Vertical Seismometers, 69 p. Nauka, Moscow (1972). (in Russian)
18. Linkov, E.M., Petrova, L.N., Dunaev, A.V.: Observations of Long-Period Earth Oscillations with a Horizontal Seismometer on an Anti-Inclined Platform. Seismicheskie pribory (Seismic Devices), vol. 21, pp. 90–96. Nauka, Moscow (1990). (in Russian)
19. Linkov, E.M.: Seismic Phenomena. 248 p., Publishing house of Leningrad State University, Leningrad (1987). (in Russian)
20. Linkov, E.M.: Superlong-Period Horizontal Seismometric Channel. Seismicheskie pribory (Seismic Devices), vol. 23, pp. 48–53. Nauka, Moscow (1992). (in Russian)
21. Linkov, E.M., Tipisev, S.Y., Butsenko, V.V.: Noise Immunity of a Long-Period Seismograph and Analysis of its Records. Geophysical Apparatus, vol. 75, pp. 78–87. Nedra, Leningrad (1982). (in Russian)
22. Hutt, C.R., Nigbor, R.L., Evans, J.R.: USGS Open-File Report 2009-1055. In: Proceedings of the Guidelines for Seismometer Testing Workshop, 48 p., Albuquerque, New Mexico (2009)
23. Likhodeev, D.V., Gravirov, V.V., Kislov, K.V., Dolov, S.M.: Precision Narrow-Range Differential Temperature Sensors. Nauchnoe Priborostroenie (Scientific Instrumentation) **29**(1), 11–16, Moscow (2019). (in Russian)
24. Likhodeev, D.V., Gravirov, V.V., Kislov, K.V.: Precision differential thermometers for the study of thermal processes based on the North Caucasus geophysical observatory. Sci. Technol. Dev. **97**(1), 15–25 (2018). (in Russian)
25. Chernov, V.V., Gushchin, V.V.: Features of the alignment resonance of seismic instruments. In: Proceedings of the Fifth Scientific Conference on Radiophysics Dedicated to the 100th Anniversary of the Birth of A.A. Andronov, 7 May 2001, pp. 252–253, Nizhny Novgorod (2001)
26. Kolesnikov, Y.A., Matsievsky, S.A.: Noise of Vertical Long-Period Seismometers and Methods for their Reduction. Computational Seismology, vol. 12, pp. 125–144. Nauka, Moscow (1979). (in Russian)
27. Ruscak, S., Singer, L.: Using histogram techniques to measure A/D converter noise. Anal. Dial. **29–2**, 7–8 (1995)
28. Adushkin, V.V., Kharlamov, V.A., Rybnov, Y.S.: Characterization of the Infrasonic Background at Stations Dubna and Zalesovo. In: Physical Processes in Geospheres: Their Manifestation and Interaction (Geophysics of Strong Disturbances): Collection of Scientific Works, pp. 166–175, IDG RAN, Moscow (1999)
29. Trnkoczy, A., Bormann, P., Hanka, W., Holcomb, L.G., Nigbor, R.L.: Site selection, preparation and installation of seismic stations (Chapter 7). In: Bormann, P. (ed.) New Manual of Seismological Observatory Practice (NMSOP-2), IASPEI, GFZ German Research Center for Geosciences, Potsdam, 108 p. (2012). https://doi.org/10.2312/GFZ.NMSOP-2

30. Withers, M.M., Aster, R.C., Young, C.J., Chael, E.P.: High-frequency analysis of seismic background noise as a function of wind speed and shallow depth. BSSA **86**(5), 1507–1515 (1996)
31. Beauduin, P., Lognonne, P., Montagner, J., Cacho, S., Karczewski, J., Morand, M.: The effects of atmospheric pressure changes on seismic signals, or how to improve the quality of a station. BSSA **86**, 1760–1799 (1996)
32. Kolesnikov, Y.A., Toxez, M.N.: Decreasing the Sensitivity of Vertical Seismometers to Fluctuations in Atmospheric Pressure. Computational Seismology, vol. 14, pp. 183–188. Nauka, Moscow (1982). (in Russian)
33. Gravirov, V.V.: Methods for Reducing the Noise of a Seismic Gyrotiltmeter. Thesis ... Cand. Phys.-mat. Sci., 120 p. Moscow (2009). (in Russian)
34. Kislov, K.V.: Theory and Methods of Protecting Broadband Seismometers from Environmental Influences. Thesis ... Cand. Phys.-mat. Sci., 171 p. Moscow (2009). (in Russian)
35. Gravirov, V.V.: Experimental results of registration of slopes of soil of atmospheric origin with a gyrotiltmeter of the KST-1 type. Estestvennie i technicheskie nauki (Nat. Tech. Sci.) **38**(6), 159–160 (2008). (in Russian)
36. Douze, E.J., Sorrells, G.G.: Prediction of pressure-generated Earth motion using optimum filters. BSSA **65**(3), 637–650 (1975)
37. Kislov, K.V., Gravirov, V.V., Labuncov, M.: The Analysis of Wind Seismic Noise and Algorithms of Its Determination. In: Eos Trans. AGU, Fall Meet. Suppl., Abstract S21C-2079, San Francisco, USA (2010)
38. Leontiev, V.I.: Methods and Ways of Increasing the Accuracy of Gravi-Inertial Equipment. Thesis ... Cand. Phys.-mat. n., Moscow (2003). (in Russian)
39. Orfanidis, S.J.: Optimum Signal Processing. An Introduction, 2nd edn. Prentice-Hall, Englewood Cliffs (1996)
40. Ketkov, Y.L., Ketkov, A.Y., Shultz, M.M.: MATLAB 7: Programming, Numerical Methods, 742 p. BHV-Petersburg, Sankt-Peterburg (2005). (in Russian)
41. Gravirov, V.V.: Optimal filters application for allocation of seismic signals in long period channel of the gyrotiltmeter. Seismic Instrum. **46**(1), 13–20 (2010)
42. Haykin, S.: Adaptive filter theory, 936 p. Prentice-Hall, Upper Saddle River (2002)
43. Starovoit, Y.O.: Sensitivity of the Obninsk seismic station to micro-oscillations of atmospheric pressure. Seismic Devices, no. 30, pp. 28–39. Nauka, Moscow (1998). (in Russian)
44. Shynk, J.J.: Frequency-domain and multirate adaptive filtering. IEEE Signal Process. Mag. **9**(1), 14–37 (1992)
45. Yanevich, Y.M., Pavleino, M.A.: Active and digital filters, 280 p., SPb (1999). (in Russian)
46. Gravirov, V.V.: The Program of Adaptive Filtering of the Seismic Signal Taking into Account the Temperature Regime of the Seismometer. Certificate Software Registration Rights, Russian Federation, No. 2019665143, November 20, (2019)
47. Gravirov, V.V., Kislov, K.V.: Adaptive Optimal Filter Program (Slave_Opt_Filter). Certificate Software Registration Rights, Russian Federation, No. 2017661235, October 6, (2017)

On the Question of the Rotational Seismometry Metrology

V. V. Gravirov and K. V. Kislov

Abstract Rotational seismology is a relatively new area of science, which was formed in the past 20 years. During this time, new instruments were created; interesting data was obtained and verified; and main directions for further development of rotational seismology were identified. Now a much greater number of observations are required for further development of rotational seismology. We need to create networks for recording of rotational movements, which should be taken parallel to the traditional seismometrical measurements. This requires development of appropriate metrological standards. It is necessary in order for the seismological data to be useful. More detailed views on this issue are presented in this work.

Keywords Rotational seismometry · Translational seismometry · Rotational ground motions

1 Introduction

Rotational seismology is a relatively new area of science. There are no well-established measurement methods and data quality definitions yet. However, discussions already have been initiated focused on developing data quality metrics [1]. Rotational seismometry differs greatly from translational seismometry. This is due first, to the low rotational movements at teleseismic distances; second, the need to clearly distinguish rotational and linear movements during registration; and third, the low coherence of rotational movements in the near field even at small distances between stations (kilometers and even hundreds of meters, and for engineering seismology—tens of meters). Since rotation does not propagate as a plane-wave, the coherence model based on plane-wave propagation is not applicable to solving the

V. V. Gravirov (✉)
Schmidt Institute of Physics of the Earth of the Russian Academy of Sciences (IPE RAS), Moscow, Russia

K. V. Kislov
Institute of Earthquake Prediction Theory and Mathematical Geophysics of the Russian Academy of Sciences (IEPT RAS), Moscow, Russia

A. Kosterov et al. (eds.), *Problems of Geocosmos—2020*, Springer Proceedings in Earth and Environmental Sciences, https://doi.org/10.1007/978-3-030-91467-7_24

spatial variability of rotational components [2–4]. In so doing, as in translational seismometry, the frequency and dynamic ranges are extremely wide. It was considered that for operation at teleseismic distances, a resolution of 10–10 rad/sec/Hz$^{1/2}$ in the range from 0.1 MHz to 10 Hz is required. For seismological studies of the near field, sensors must have a resolution of 10^{-7} rad/s in the range from 0.01 Hz to 0.1 Hz. Instruments for engineering applications, in addition, must operate in the high-frequency range, up to 100 Hz and have a large dynamic range, recording maximum values of the order of several radians per second. The requirements have a tendency to grow. When theoretical estimates of the capabilities of a large laser gyroscope were published, the resolution requirements for teleseismic studies immediately increased by 2 orders [5–7].

2 Metrology of Translational Seismometry

In translational seismometry, various methods of calibration and validation of instrument are used as metrological procedures. Therefore, the traceability of measurements is violated, as well as the relation between the measurement result and the primary standard [8]. The Global Seismographic Network, which is operated by the Incorporated Research Institutions for Seismology, for the basis of translational seismometry metrology, uses the convergence of measurements. This method for assessing the quality of the data uses the systematic comparison of recorded long-period waveforms with synthetic seismograms calculated for known seismic events. Correlation coefficients and optimal scaling factors between observed and synthetic waveforms are calculated for the three types of data used in the standard centroid moment tensor (CMT) analysis: body waves, with periods in the range 50–150 s, mantle waves, with periods in the range 125–350 s, and surface waves, with periods in the range 50–150 s. This method does not answer a question of a measurement error of the ground motion. However, it is the best approximation of the transfer functions of the channels for seismic events. This is the best of what scientists were able to come up with at present.

Consequently, in translational seismometry, the situation is such that the measuring instrument is not a single sensor, but a seismic network, while the object of measurement is not the ground movement, and the parameters of the source of the earthquake. The serviceability of a separate sensor is judged by its compliance to whole network [9]. As stated in the Federal Law on Ensuring the Uniformity of Measurements, if an international agreement of the Russian Federation establishes rules other than those provided for by the legislation of the Russian Federation on ensuring the uniformity of measurements, the rules of the international agreement shall apply [10]. Thus, this system of metrological support for seismic monitoring is actually the norm in our country. It seems the case with rotational seismometry will stand equal.

3 Metrology of Rotational Seismometry

It is difficult to forecast how will the metrology of rotational seismometry will develop, since rotational seismometers have a completely different principle of operation from classic seismometers and the movements measured by them also differ.

Rotational seismometry and all rotational seismology in general fall into several main areas: teleseismic, near-field, engineering seismology (mainly studies of effects of rotational movements on engineering structures and soils in their foundations and the redistribution of energy in various parts of buildings and structures from linear movements to rotational ones and vice versa) [11]. Separate note should be made on the seismic exploring, where the rotational seismometry can also be used [12].

In translational seismometry, ambient noise is considered the lower sensitivity boundary. Ambient Earth noise is defined as seismic signals in the absence of signals from earthquakes. In 1993, Peterson defined a new low noise model (NLNM) for seismic noise. This envelope reflects the minimum seismic noise level, which is always expected in seismic recordings. The NLNM model is often used for the purpose of site selection or to compare the quality of seismic sites in terms of noise, and as a reference for evaluation of instrument noise. It is proposed to draw up a similar model for rotational movements [13] (see Fig. 1).

Despite some progress, not much is known about the level of rotational Earth noise [14]. Ring laser gyroscopes can obtain the initial approximation. However, there are only a dozen of them available, even when counting those that are not yet in operation. When we record noise by rotational sensors of strong movements, microseismic peaks are not visible in the high-frequency band. This means that the noise of the instruments significantly covers the ambient Earth noise.

Teleseismic rotational movements and movements observed in the near field must be interpreted in different ways. This is caused not only by the difference in signal amplitudes; it also requires application of different metrological schemes.

All characteristics of instruments, such as sensitivity, frequency response, resolution, linearity, self-noise, stability of various specifications over time must be determined. We need to know the clipping level (combined with self-noise, clipping implies the sensor's dynamic range), cross-axis sensitivity, both rotational and translational.

4 The Rotational Teleseismic

Translational seismometry has been very successful in the far field because classical elasticity theory works very well for interpreting records at large distances. The situation is similar with rotational seismology. Large-area ring-laser gyroscopes and seismometers arrays are applicable here. These are very costly methods and it is problematic to create a network from them. A natural reference signal for calibration

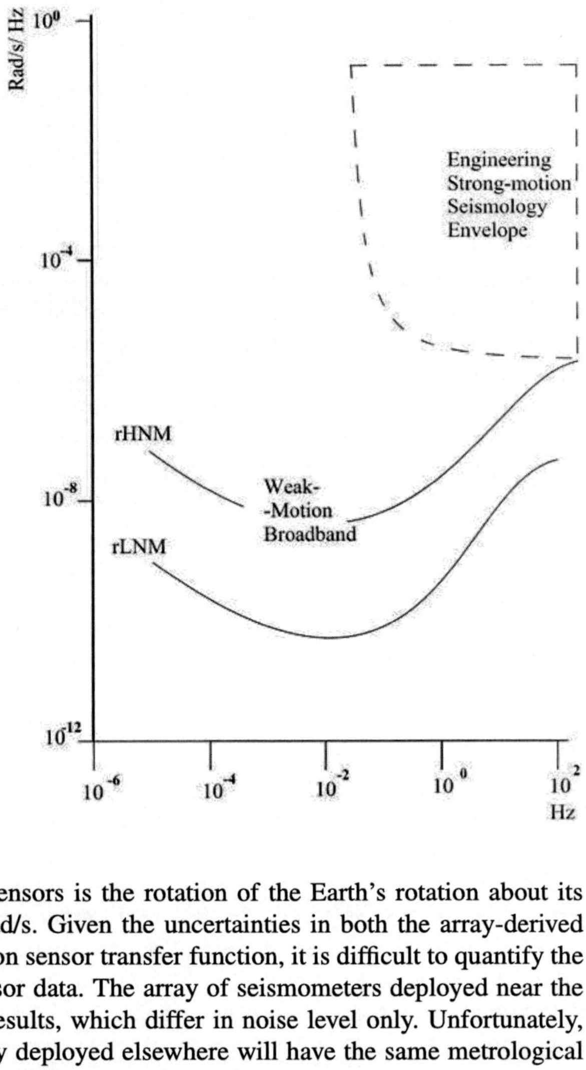

Fig. 1 A sketch of Rotational Low-Noise Model (rLNM). Strong-motion seismometry data is shown in the upper right quadrant

of weak-motion rotational sensors is the rotation of the Earth's rotation about its pole, which is about 73 μrad/s. Given the uncertainties in both the array-derived measurements and the rotation sensor transfer function, it is difficult to quantify the accuracy of the rotation sensor data. The array of seismometers deployed near the laser ring can give similar results, which differ in noise level only. Unfortunately, it is not certain that the array deployed elsewhere will have the same metrological characteristics.

Large-area ring-laser gyroscopes can record short-period variations of the Earth's angular velocity vector. At the same time, the parameters of the Earth's rotation (the duration of the earth day and the coordinates of the pole) are controlled by Very Long Baseline Interferometry (VLBI)—a complex system of radio telescopes distributed on the surface of the planet. This creates additional calibration options.

5 Strong Motion Rotational Seismometry

Near field sensors, engineering seismology sensors and seismic survey devices can be classified as strong motion sensors. The most important parameters for engineering seismology are clip and transverse sensitivity. For seismic surveys, this is the resolution in the appropriate frequency range.

About the near field, any reliable conclusions can be made, having only saved up a large number of data. Increasing efforts toward a sustainable high-quality network for work in the near field is the clear objective. At the same time, the amplitudes of rotational movements are much higher here, and network deployment can based on relatively inexpensive sensors [15, 16].

One of the main indicators is cost. This is not a metrological parameter, but the need to deploy very dense networks in places of possible strong earthquakes forces us to sacrifice even sensitivity for it [17]. Naturally, taking into account the low coherence of rotational movements in the near field, the measuring tool at the given stage is the sensor, and the measured value is the kinematical parameters of rotations recorded by it.

To enable precision observation and to comply with stringent quality requirements, we require well-developed metrology infrastructure, having calibration facilities and that can establish measurement and metrological scheme. The metrological scheme allows constructing models of appearance of error components in order to then carry out the necessary precision calculations. Several important activities are already on the way in this direction [18–22].

At present, strong-motion rotational sensors have self-noise levels well above Earth's ambient rotational noise. Therefore, the Earth's rotation is unavailable for their calibration. Usual method of calibration and test is to rotate the instrument in the Earth's gravitational field. Strong-motion rotational sensors require the use of a precision rotational shake table [5, 18, 23, 24] (see Figs. 2 and 3).

It is essential that each of the strong-motion rotational sensors have to be calibrated individually on a shake table to determine amplitude and phase responses. This is necessary because the sensors for strong-motion work have proved to be less than flat in response and to vary even in mid-band sensitivity.

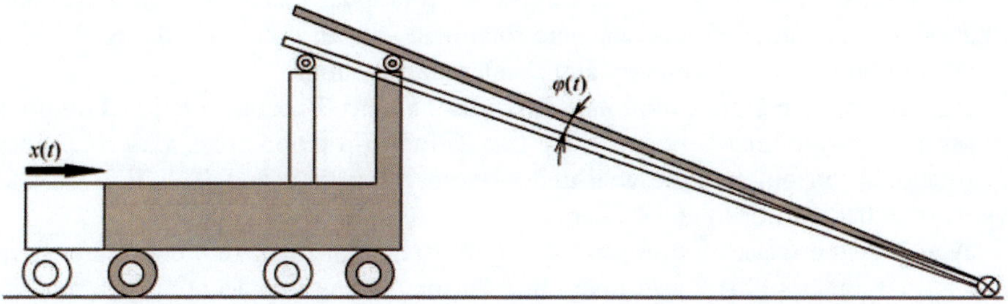

Fig. 2 A scheme of shake table simulating strong tilts

Fig. 3 Electrohydraulic shaking table "Fairy-22" (in the foreground) [23]

When determining the transfer function, it is important to check the linearity, in other words the amplitude and phase response should not depend on the amplitude of the signal.

In order to verify that seismic instruments meet the requirements it is important, to be able to measure the self-noise of seismic sensors.

By recording the sensor output at some quiet location, usually overnight, the Power Spectral Density is computed and all of it is simply attributed to the sensor self-noise.

However, in order to correctly install the sensor, determine the complex of protective measures and the actual level of own noise, it is necessary to determine the environmental sensitivity for such parameters as temperature, pressure, power supply voltage, electromagnetic disturbance, etc.

Measurements of cross-axis sensitivities of rotational sensors to both linear and rotational inputs require both very pure rotational and very pure translational tables capable of both constant-velocity and accelerating motions.

It is difficult to compare motions with differing units. Because rotational motions in earthquakes are relatively small, the translation-to-rotation cross-axis sensitivity of rotational sensors must be well under one part per thousand. This limitation is very strict, but well-designed sensors must adhere to it.

A well-tested weak-motion sensor can be used as primary reference sensors in strong-motion tests [25]. Laser fiber optic sensors, taking into account such modifications, as coils with an increased diameter, modified quadrupole winding, the use of hollow photonic-crystal fiber, are very promising for these purposes.

Note that joint measurements (especially joint operation of sensors of different types) are important for standardization of sensor test methods [26–28].

A separate metrological scheme should be developed for each direction of rotational seismometry. This will allow having a constant confidence in the reliability of seismic data. In any eventual form and presentation of the information, it is important to present to the scientific and user communities the status of data and sensors as clearly and quickly as possible. Instrumentation and methods now available are already being installed and implemented [29]. However, the development of metrological schemes and their implementation is at an initial stage. Terminology and direction of axes and rotations, formats of data recording and archiving are defined [30]. Calibration stands and installations for excitation of waves rotating around the vertical axis and tilts have been designed. Some of the true characteristics of the most commonly used rotation sensors have been determined [18, 31].

6 Conclusion

Observations, analysis, and interpretation of rotational ground motions are beginning to play a significant role in seismology and earthquake engineering. Nevertheless, every manufacturer of devices evaluates the performance in their own way. In addition, the actual performance of the devices turns out to be much lower than the theoretical ones.

There is an urgent need to solve the problem of metrological assurance in rotational seismometry, because the quality of seismic measurements has a decisive influence on a further development of seismology.

The usefulness of rotational seismology will strongly depend on the availability of high-quality data. Confidence in data requires knowledge of their quality.

The study was carried out under state tasks of the IEPT RAS and IPE RAS.

References

1. Kislov, K.V., Gravirov, V.V.: On the question of the rotational seismology. In: Book of Abstracts XIIth International Conference and School Problems of Geocosmos, p. 122. St. Petersburg State University, St. Petersburg, Petrodvorets, Russia (2018)
2. Rodda, G.K., Basu, D.: Coherency model for translational and rotational ground motions. Bull. Earthquake Eng. **16**(7), 2687–2710 (2018). https://doi.org/10.1007/s10518-017-0304-6
3. Kurzych, A., Jaroszewicz, L.R., Dudek, M., Krajewski, Z., Kowalski, J.K., Niespodziany, S., Bernauer, F., Wassermann, J., Igel, H.: Assembly of optical fiber sensors for rotational seismology—data coherence and comparability issues in field application. In: EGU General Assembly 2020, EGU2020-19987 (2020). https://doi.org/10.5194/egusphere-egu2020-19987
4. Stupazzini, M., de la Puente, J., Smerzini, C., Käser, M., Igel, H., Castellani, A.: Study of rotational ground motion in the near-field region. BSSA **99**(2B), 1271–1286 (2009). https://doi.org/10.1785/0120080153

5. Jaroszewicz, L.R., Kurzych, A., Krajewski, Z., Marć, P., Kowalski, J.K., Bobra, P., Zembaty, Z., Sakowicz, B., Jankowski, R.: Review of the usefulness of various rotational seismometers with laboratory results of fibre-optic ones tested for engineering applications. Sensors **16**(12), 2161, 22 (2016). https://doi.org/10.3390/s16122161
6. Takamori, A., Araya, A., Otake, Y., Ishidoshiro, K., Ando, M.: Research and development status of a new rotational seismometer based on the flux pinning effect of a superconductor. BSSA **99**(2B), 1174–1180 (2009). https://doi.org/10.1785/0120080087
7. Velikoseltsev, A.A., Luk'yanov, D.P., Vinogradov, V.I., Shreiber, K.U.: Super-large optical gyroscopes for applications in geodesy and seismology: state-of-the-art and development prospects. Quantum Electron **44**(12), 1151–1156 (2014). https://iopscience.iop.org/article/10.1070/QE2014v044n12ABEH015397. Last accessed 2020/12/21
8. Zakharchenko, N.Z.: Scientific and legal aspects of seismometry metrology. In: Proceedings of I-St Regional Technical Conference on "Problems of Integrated Geophysical Monitoring of the Russian Far East", 11–17 Nov 2007, pp. 122–126, GS RAS, Petropavlovsk-Kamchatsky, Russia (2007). (in Russian)
9. Kislov, K.V., Gravirov, V.V.: On the metrological support of the long-period seismology. In: Proceedings of the 10th International Conference "Problems of Geocosmos", pp. 183–188, St. Petersburg State University, St. Petersburg, Petrodvorets, Russia (2014). http://geo.phys.spbu.ru/materials_of_a_conference_2014/Geocosmos2014proceedings.pdf. Last accessed 2020/12/21
10. Federal Law of Russia No. 102-FZ "On Ensuring the Uniformity of Measurements" as amended in 2020. (in Russian)
11. Guéguen, P., Guattari, F., Aubert, C., Laudat, T.: Comparing direct observation of torsion with array-derived rotation in civil engineering structures. Sensors **21**(1), 142 (2021). https://doi.org/10.3390/s21010142
12. Schmelzbach, C., Van Renterghem, C., Sollberger, D., Häusler, M., Robertsson, J.O.A.: Applications of Seismic Spatial Wavefield Gradient and Rotation Data in Exploration Seismology. In: Book of Abstract: AGU 2017. https://www.researchgate.net/publication/321268366_Applications_of_seismic_spatial_wavefield_gradient_and_rotation_data_in_exploration_seismology. Last accessed 2020/12/21
13. Lee, W.H.K., Celebi, M., Todorovska, M.I., Diggles, M.F.: Rotational seismology and engineering applications. In: Proceedings for the First International Workshop: U.S. Geological Survey Open-File Report 2007-1144, p. 46, Menlo Park, California, U.S.A. (2007). http://pubs.usgs.gov/of/2007/1144/. Last accessed 2020/12/21
14. Kislov, K., Gravirov, V.: Metrology of rotational seismometry. In: Book of Abstracts CTBT Science and Technology Conference 2021 (SnT2021), P3.1–180, pp. 178–179, CTBTO, Vienna, Austria (2021)
15. Yang, Z., Chen, Y.: Current status of rotational seismology studies in PKUCEA Joint Research Center of Modern Seismology of China Theories. In: 2nd International Workshop on Rotation Seismology and Engineering Applications. Programme & Abstracts, p. 24, Masaryk's College, Prague, Czech (2010)
16. Timofeev, V.Y., Timofeev, A.V., Ardyukov, D.G., Boyko, E.V.: Quartz tiltmeters and their use in geophysical studies. Seism. Instr. **56**(2), 134–151 (2020). https://doi.org/10.3103/S0747923920020115
17. Fremd, V.M.: Symmetric triaxial rotational piezoelectric seismometers. Seism. Instr. **56**(4), 373–379 (2020). https://doi.org/10.3103/S0747923920040039
18. Nigbor, R.L., Evans, J.R., Hutt, C.R.: Laboratory and field testing of commercial rotational seismometers. BSSA **99**(2B), 1215–1227 (2009). https://doi.org/10.1785/0120080247
19. Cao, Y., Chen, Y., Zhou, T., Yang, C., Zhu, L., Zhang, D., Cao, Y., Zeng, W., He, D., Li, Z.: The development of a new IFOG-based 3C rotational seismometer. Sensors **21**, 3899 (2021). https://doi.org/10.3390/s21113899
20. Izgi, G., Eibl, E.P.S., Donner, S., Bernauer, F.: Performance test of the rotational sensor BlueSeis-3A in a Huddle Test in Fürstenfeldbruck. Sensors **21**, 3170 (2021). https://doi.org/10.3390/s21093170

21. Jaroszewicz, L.R., Kurzych, A., Krajewski, Z., Dudek, M., Kowalski, J.K., Teisseyre, K.P.: The fiber-optic rotational seismograph-laboratory tests and field application. Sensors **19**, 2699 (2019). https://doi.org/10.3390/s19122699
22. Brokešová, J., Málek, J., Vackář, J., Bernauer, F., Wassermann, J., Ige, H.: Rotaphone-CY: the newest rotaphone model design and preliminary results from performance tests with active seismic sources. Sensors **21**(2), 562 (2021). https://doi.org/10.3390/s21020562
23. Gravirov, V.V., Kislov, K.V.: Development of methods and algorithms for automatic real time identification of waveforms introduction from local earthquakes in increased level of man-induced noises for the purposes of ultra-short-term warning about an occurred earthquake. In: Final Project Technical Report of ISTC #3816. 165 p., IIEPT RAS, Moscow, Russia (2011)
24. Graizer, V.: Tutorial on measuring rotations using multipendulum systems. BSSA **99**(2B), 1064–1072 (2009). https://doi.org/10.1785/0120080145
25. Bernauer, F., Behnen, K., Wassermann, J., Egdorf, S., Igel, H., Donner, S., Stammler, K., Hoffmann, M., Edme, P., Sollberger, D., Schmelzbach, C., Robertsson, J., Paitz, P., Igel, J., Smolinski, K., Fichtner, A., Rossi, Y., Izgi, G., Vollmer, D., Eibl, E.P.S., Buske, S., Veress, C., Guattari, F., Laudat, T., Mattio, L., Sèbe, O., Olivier, S., Lallemand, C., Brunner, B., Kurzych, A.T., Dudek, M., Jaroszewicz, L.R., Kowalski, J.K., Bońkowski, P.A., Bobra, P., Zembaty, Z., Vackář, J., Málek, J., Brokesova, J.: Rotation, strain, and translation sensors performance tests with active seismic sources. Sensors **21**, 264 (2021). https://doi.org/10.3390/s21010264
26. Donner, S., Lin, C.-J., Hadziioannou, C., Gebauer, A., Vernon, F., Agnew, D.C., Igel, H., Schreiber, U., Wassermann, J.: Comparing direct observation of strain, rotation, and translation with array estimates at Pinon flat observatory, California. Seismol. Res. Let. **88**, 1107–1116 (2017). https://doi.org/10.1785/0220160216
27. Brokešová, J., Málek, J.: Comparative measurements of local seismic rotations by three independent methods. Sensors **20**(19), 5679 (2020). https://doi.org/10.3390/s20195679
28. Teisseyre, K.P., Dudek, M., Jaroszewicz, L.R., Kurzych, A.T., Stempowski, L.: Study of rotational motions caused by multiple mining blasts recorded by different types of rotational seismometers. Sensors **21**, 4120 (2021). https://doi.org/10.3390/s21124120
29. Lee, W.H.K., Evans, J.R., Huang, B.-S., Hutt, C.R., Lin, C.-J., Liu, C.-C., Nigbor, R.L.: Measuring rotational ground motions in seismological practice. In: Bormann, P. (ed.) New Manual of Seismological Observatory Practice 2 (NMSOP-2), pp. 1–27, Deutsches GeoForschungsZentrum GFZ, Potsdam (2012). https://doi.org/10.2312/GFZ.NMSOP-2_IS_5.3
30. Evans, J.R., and the International Working Group on Rotational Seismology: Suggested Notation Conventions for Rotational Seismology. BSSA **99**(2B), 1073–1075 (2009). https://doi.org/10.1785/0120080060
31. Kozlov, V.A., Safronov, M.V.: Self-noise of molecular electronic transducers. Tech. Phys. **48**(12), 1579–1582 (2003). https://doi.org/10.1134/1.1634680

Variations of Ambient Temperature and Following Them Instrumental Noise of Seismic Instruments

V. V. Gravirov and K. V. Kislov

Abstract Seismological study requires data of sufficient quality. To this end, it is necessary to understand how to deal with noise caused by non-seismic sources. It is essential to eliminate or reduce the noise occurring in the seismic devices when external temperatures oscillate. At present, high-performance computers allow solving almost any specific problem numerically, and theoretical search for analytical solutions seems unproductive. However, a lack of theoretical knowledge can lead to the emergence of research using vast resources ineffectively. For appropriate evaluation of the cause of noise, it is necessary to understand how external temperature variations penetrate to each element of the device, what the attenuation and phase shift. Theoretical studies are undoubtedly useful in investigating the applicability of sensors, in the development of new sensors, in the development of methods for their installation. Analytic approach in the most general form can be applied to any type of seismic equipment.

Keywords Ambient temperature · Instrumental noise · Seismic instrument

1 Introduction

The problem of increasing the seismic instruments sensitivity and the noise reducing arose almost simultaneously with the emergence of seismology. The improvements of broadband seismic equipment resolution, namely decreasing the amplitude threshold of ground motion, can lead to reduction of the required density of observational network and to distinguishing the poor signals [1, 2].

Interest in the noise studying is primarily related to the fact that the noise mainly defines the fundamental thresholds of measuring. The main obstacles when using

V. V. Gravirov (✉)
Schmidt Institute of Physics of the Earth of the Russian Academy of Sciences (IPE RAS), Moscow, Russia

K. V. Kislov
Institute of Earthquake Prediction Theory and Mathematical Geophysics of the Russian Academy of Sciences (IEPT RAS), Moscow, Russia

A. Kosterov et al. (eds.), *Problems of Geocosmos—2020*, Springer Proceedings in Earth and Environmental Sciences, https://doi.org/10.1007/978-3-030-91467-7_25

broadband equipment are all kinds of noise, the power and influence of which on the device increases with changing in the periods of oscillation towards long ones [3].

Modern broadband seismic instruments can operate in the wide temperature range. Most manufacturers indicate an operating temperature range from −20 °C to +50 °C. However, it only ensures that devices operate in a proper way at constant temperature within this range. Temporal variations of temperature are more importantly, than in what part of the temperature range is the current average temperature.

2 Methods

Firstly, temperature fluctuations can cause ground motion resulting from irregular heating during the day and uneven freezing during the cold season [4]. These phenomena provoke tilts, which have particularly strong detrimental effect on the horizontal pendulums. Secondly, the device is directly affected by temperature fluctuations. Almost all physical and geometric characteristics of devices are sensitive to temperature fluctuations [5, 6].

During the ambient temperature changing the device does not perceives this influence instantaneously. Heat exchange begins as

$$dQ = -hS(t - q)dt,$$

where dQ is the heat [J] that the instrument receives (loses) during $d\tau$ [s]; h—the coefficient of heat transfer, which depends on the device characteristics and the environment [W/m^2 K]; S—the surface area of device through which the heat transfers [m^2]; t—the current temperature of the device [K]; θ—the instantaneous ambient temperature [K].

When the heat exchange is happening, the device temperature is changing as follows

$$dQ = cm\,dt,$$

where c—the thermal capacity of the device material [J/kg K]; m—its mass [kg]; dt—the temperature change of the device [K], caused by the influx of heat dQ.

Thus, we obtain the velocity of perception ambient temperature

$$\frac{dt}{d\tau} = -\frac{1}{\lambda}(t - \theta) \tag{1}$$

the thermal inertia coefficient λ of the instrument in block is given by

$$\lambda = \frac{cm}{hS}.$$

Thus, transforming Eq. (1) we have

$$\frac{dt}{d\tau} = -\frac{1}{\lambda}(t - \theta_0) + \frac{1}{\lambda}\sum_{n=1}^{\infty}(A_n \cos n\omega\tau + B_n \sin n\omega\tau), \tag{2}$$

where θ_0 is the average temperature environment over a certain time interval; A_n и B_n—the maximum amplitude of individual harmonics; ω is the angular frequency.

Introducing a new variable

$$y = t - \theta_0,$$

we will look for a solution of (2) in the general form

$$y = y_1 + y_2,$$

where y_1 is the solution of the homogeneous equation

$$\frac{dt}{d\tau} = -\frac{1}{\lambda}(t - \theta_0)$$

or

$$\frac{dy_1}{d\tau} = -\frac{1}{\lambda}y_1$$

and y_2—partial solution of a differential Eq. (2).

The solution of the homogeneous equation has the following form

$$y_1 = Ce^{-\frac{\tau}{\lambda}},$$

where C is the constant of integration.

The partial solution will be sought in the form of series

$$y_2 = \sum_{n=1}^{\infty} a_n \cos n\omega\tau + b_n \sin n\omega\tau$$

Substituting into Eq. (2) we have

$$\sum_{n=1}^{\infty} -na_n\omega \sin n\omega\tau + nb_n\omega \cos n\omega\tau = -\frac{1}{\lambda}\sum_{n=1}^{\infty} a_n \cos n\omega\tau$$

$$+ b_n \sin n\omega\tau + \frac{1}{\lambda}\sum_{n=1}^{\infty} A_n \cos n\omega\tau + B_n \sin n\omega\tau.$$

where a_n and b_n are amplitudes of the harmonics of temperature fluctuations device. Coefficients comparison of harmonics gives equalities

$$-a_n n\omega = -\frac{b_n}{\lambda} + \frac{B_n}{\lambda},$$
$$b_n n\omega = -\frac{a_n}{\lambda} + \frac{A_n}{\lambda}$$

or

$$a_n = \frac{A_n - B_n k_n}{1 + k_n^2}, \tag{3}$$

$$b_n = \frac{A_n k_n + B_n}{1 + k_n^2}, \tag{4}$$

where $k_n = n\omega\lambda$—the relative coefficient of the thermal inertia.

Appling them to the particular solution we have

$$y_2 = \sum_{n=1}^{\infty} \frac{A_n - B_n k_n}{1 + k_n^2} \cos n\omega\tau + \frac{A_n k_n + B_n}{1 + k_n^2} \sin n\omega\tau.$$

The general solution in the original variables is given by

$$t - \theta_0 = Ce^{-\frac{\tau}{\lambda}} + \sum_{n=1}^{\infty} \frac{A_n - B_n k_n}{1 + k_n^2} \cos n\omega\tau + \frac{A_n k_n + B_n}{1 + k_n^2} \sin n\omega\tau.$$

The integration constant is found from the conditions $\tau = 0$, $t = t_0$:

$$C = t_0 - \theta_0 - \sum_{n=1}^{\infty} \frac{A_n - B_n k_n}{1 + k_n^2}.$$

And now we finally have

$$t - \theta_0 = \left(t_0 - \theta_0 - \sum_{n=1}^{\infty} \frac{A_n - B_n k_n}{1 + k_n^2}\right) \cdot e^{-\frac{\tau}{\lambda}}$$
$$+ \sum_{n=1}^{\infty} \frac{A_n - B_n k_n}{1 + k_n^2} \cos n\omega\tau + \frac{A_n k_n + B_n}{1 + k_n^2} \sin n\omega\tau.$$

Assuming that the instrument for a long time (>>λ) is in contact with the environment, the first term can be neglected; then

$$t - \theta_0 = \sum_{n=1}^{\infty} \frac{A_n - B_n k_n}{1 + k_n^2} \cos n\omega\tau + \frac{A_n - B_n k_n}{1 + k_n^2} \sin n\omega\tau.$$

Having expressed the sum by the cosine, we have

$$t = \theta_0 + \sum_{n=1}^{\infty} \sqrt{\frac{A_n^2 + B_n^2}{1 + k_n^2}} \cos(n\omega\tau - \varphi_n), \tag{5}$$

where

$$\varphi_n = acrtg\left(\frac{A_n k_n + B_n}{A_n - B_n k_n}\right)$$

Similarly, Eq. (2) takes the form

$$\theta = \theta_0 + \sum_{n=1}^{\infty} \sqrt{A_n^2 + B_n^2} \cos(n\omega\tau - \psi_n), \text{ where } \psi_n = arctg \frac{B_n}{A_n}. \tag{6}$$

The harmonics of the device temperature changes are shifted in phase relative to the respective harmonics of environmental temperature at angle

$$\beta_n = \varphi_n - \psi_n.$$

Substituting the values of (5) and (6) we obtain the phase shift

$$\beta_n = \operatorname{arctg} k_n.$$

It is also clear that the ratio of the amplitudes will be

$$I_n = \sqrt{1 + k_n^2}.$$

We remind that k_n is the relative coefficient of thermal inertia, included in these relationships

$$k_n = n\omega\lambda.$$

Thus, the penetration of the temperature fluctuations into the device is determined exclusively by the coefficient of thermal inertia λ.

If the test element is located deep in the device, the influence of low frequencies and temperature will be stronger and the phase shift will be greater. The influence of relatively high-frequency thermal oscillations is also eliminated. In this case, the effect of temperature on separate elements is leveled and this allows reducing the effect of temperature on the drift of the seismometer.

The seismometer is not a monolithic piece of material. It consists of many parts whose temperature changes have different effects on the generation of noise. The coefficients of thermal inertia of the details differ significantly from each other, and there are unequal phase shifts in the generation of noise.

Since each sensor element depends on the temperature in own way, it is necessary to know conditional coefficient of thermal inertia λ_i for each temperature-dependent element (i—number of thermo-dependent element). It is very difficult to calculate λ_i analytically. You can try to obtain these coefficients by finite element method for specific types of sensors. However it is possible to determine λ_i experimentally as it is specified in [7]. If the device is in development, it can be done by its mock-up.

The body with the temperature t_0 put in a room with constant temperature θ_0, then Eq. (1) can be integrated

$$\int \frac{dt}{(t-\theta_0)} = -\frac{1}{\lambda} \int d\tau$$

and thus, we have

$$t - \theta_0 = Ce^{-\frac{\tau}{\lambda}}.$$

The integration constant C can be determined from the boundary conditions $\tau = 0$ and $t = t_0$

$$C = t - \theta_0.$$

Now we have

$$\frac{t-\theta_0}{t_0-\theta_0} = e^{-\frac{\tau}{\lambda}}. \tag{7}$$

If the time equals to the coefficient of thermal inertia λ, Eq. (7) becomes

$$\frac{t-\theta_0}{t_0-\theta_0} = e,$$

i.e., λ can be experimentally determined as the time during which the initial temperature difference between the body and the environment is reduced by a factor e.

The internal temperature of the device is affected by fluctuations of external temperature and also by heating of elements located inside the device [8]. The temperature can vary near heating elements of the electric circuit, during work of the pendulums zero adjustment mechanisms, and when the pressure changes inside the closed volume of the device. In these cases, it is necessary to use the coefficients, which take into account the temperature of the air around the components, instead of the conditional coefficients of thermal inertia, directed only on changes of external temperature.

3 Conclusion

All broadband sensors are sensitive to temperature variations. Firstly, long-period fluctuations of temperature amplitude are significant (especially diurnal variation) and, secondly, and passive methods of device protection must be more complicated for longer periods of temperature fluctuations.

It is essential to understand how to eliminate or reduce the noise occurring in the seismic devices when the outside temperature oscillates.

To determine what part of the total noise is created by this or that element, it is necessary to know how external temperature is transferred to it, what are the amplitudes decay and phase shifts of harmonics. Foreknowing the allowable temperature fluctuations levels and periods in the hardware chambers of a seismic network, we can predict the total temperature noise.

If we are able to determine the conditional thermal inertia coefficient λ_l for each noise generation channel, calculate the phase shifts β_{nl} and the amplitude ratios I_{nl} for the fundamental harmonics and know how each channel generates thermal noise (see, for example [9, 10]), we can determine the most critical elements of the device. If you reduce the noise on these channels, the overall noise level of the device can be reduced efficiently and economically.

The study was carried out for a state assignment of the Institute of Earthquake Prediction Theory and Mathematical Geophysics, Russian Academy of Sciences and Schmidt Institute of Physics of the Earth, Russian Academy of Sciences.

References

1. Fojtíková, L., Kristeková, M., Málek, J., Sokos, E., Csicsay, K., Zahradník, J.: Quantifying capability of a local seismic network in terms of locations and focal mechanism solutions of weak earthquakes. J. Seismolog. **20**(1), 93–106 (2016)
2. Vaezi, Y., Van der Baan, M.: Analysis of instrument self-noise and microseismic event detection using power spectral density estimates. Geophys. J. Int. **197**(2), 1076–1089 (2014). https://doi.org/10.1093/gji/ggu036
3. Rodgers, P.W.: Frequency limits for seismometers as determined from signal-to noise ratios part 2. The feedback seismometer. BSSA **82**(2), 1099–1123 (1992)

4. Bashilov, I.P., Daragan, S.K., Kabychenko, N.V.: Noise parameters of seismic sensors. In: NNC RK Bulletin «Geophysics and Non-Proliferation Problems», vol. 2, pp. 19–29 (2002)
5. Mitronovas, W.: Temperature effects on long-period seismographs: an accurate method to determine the transfer function. BSSA. **66**, 1405–1412 (1976)
6. Gravirov, V.V., Kislov, K.V.: Criticality of seismometer to variations of parameters. Electron. J. Invest. Russ. **26**, 301–312 (2008). (in Russian)
7. Kislov, K.V., Gravirov, V.V.: Research of environment influence on self-noise broadband seismic equipments. In: Computational Seismology, vol. 42, 240 p., The Krasand Publishing Group, Moscow (2013). (in Russian)
8. Gravirov, V., Kislov, K., Likhodeev, D.: A differential highly sensitivity sensor for ac-counting of seismic devices instrumental thermal noise. In: Book of Abstracts: CTBT: Science and Technology Conference SnT-2019, T3.1-P1, Hofburg palace, Vienna, Austria, p. 137 (2019). https://ctnw.ctbto.org/ctnw/abstract/32636. Last accessed 2021/04/21
9. Kislov, K.V., Gravirov, V.V.: Noise of elastic elements seismic instruments. Estestvennye i Tekhnicheskie Nauki (Nat. Tech. Sci.) **5**(37), 142–148 (2008). (in Russian)
10. Kislov, K.V., Gravirov. V.V.: One of the ways generation thermal noise broadband seismometer. Electron. J. Invest. Russ. **27**, 313–321 (2008). (in Russian)

Solar-Terrestrial Physics

Dose Rate Bursts Onboard the ISS and the "Lomonosov" Satellite in the Earth's Outer Radiation Belt

G. I. Antonyuk, V. V. Benghin, T. P. Dachev, O. Yu. Nechaev, I. A. Zolotarev, and V. L. Petrov

Abstract The EXPOSE-R2 experiment was carried out by the European Space Agency on board "Zvezda" module of the ISS. The R3DR2 device was one among the EXPOSE-R2 instruments. The semiconductor detectors were sensitive elements of the R3DR2 device that permit to measure radiation doses behind thin shielding. The results of this experiment are freely available now. We used them to analyse variations in the radiation environment at ISS altitude in the Earth's outer radiation belt. SINP MSU carried out a similar experiment on board The "Lomonosov" satellite. One of its devices DEPRON instrument includes semiconductor detectors. The detectors were similar in design to those in the R3DR2 instrument. The analysis showed the presence of significant enhancements in dose rate. The dose rate of such enhancements reached the level of 28.8 mGy/h according to R3DR2 data and 106.5 mGy/h according to DEPRON data. However, the total dose rate of single enhancements did not exceed 0.8 mGy according to R3DR2 data and 1.6 mGy according to DEPRON data. Comparison with geomagnetic conditions showed that such enhancements are observed mainly during periods of geomagnetic disturbances. The obtained information may be used to estimate the level of radiation impact in the region of the Earth's outer radiation belt.

Keywords Radiation doses · ISS · "Lomonosov" satellite · Outer Van Allen radiation belt

G. I. Antonyuk (✉)
Faculty of Physics, Lomonosov Moscow State University, 1(2), Leninskie gory, Moscow, Russia 119991
e-mail: antonjuk.gi15@physics.msu.ru

G. I. Antonyuk · V. V. Benghin · O. Yu. Nechaev · I. A. Zolotarev · V. L. Petrov
Lomonosov Moscow State University, Skobeltsyn Institute of Nuclear Physics («SINP MSU»), 1(2), Leninskie gory, Moscow, Russia 119991

V. V. Benghin
The Russian Federation State Research Center—Institute of Biomedical Problems of the Russian Academy of Sciences («IMBP RAS»), 76A, Khoroshevskoe Shosse, Moscow, Russia 123007

T. P. Dachev
The Space Research and Technology Institute at the Bulgarian Academy of Sciences («SRTI-BAS»), Bl. 1, Acad. Georgi Bonchev Str., Sofia 1113, Bulgaria

A. Kosterov et al. (eds.), *Problems of Geocosmos—2020*, Springer Proceedings in Earth and Environmental Sciences, https://doi.org/10.1007/978-3-030-91467-7_26

1 Introduction

Radiation from the Earth's radiation belt creates a significant radiation exposure for missions in low-Earth orbits (LEO), in particular on board the International Space Station (ISS) [1]. The inner radiation belt (IRB) protons create an increased radiation level in the South Atlantic Anomaly (SAA) region. There were large fluxes of the outer radiation belt (ORB) electrons observed in the high-latitude area of the Earth's magnetosphere. It is known [2] that ORB electrons fluxes experience significant variations. It may cause malfunctions of spacecraft equipment located in the ORB region.

Large bursts of ORB electrons fluxes were observed at the ISS altitude [3] with high values of radiation absorbed dose rate. The dose rate time plots are shown in Fig. 1 for the measurements on the "Lomonosov" satellite during the passage of the ORB region. Figure 1a presents the plot which corresponds to the passage on 11 May 2016 under calm conditions, while Fig. 1b presents the passage on 6 November 2016 with the presence of an "intensity burst".

The subject of this article focuses on such "bursts", which are of relatively short duration and an intensity noticeably higher than the characteristic values of dose rate when crossing the ORB area. Based on our information, it is not possible to draw a reasonable conclusion whether such "bursts" are caused by a satellite crossing an area of increased flux intensity or by a short-term increase in particle fluxes. To solve the problem of determining the nature of the variations: temporal or spatial, experiments with several satellites flying in close orbits were conducted [4]. We assume that observed "bursts" are caused by the passage of the spacecraft through spatial structures. These structures' typical duration of existence significantly exceeds the

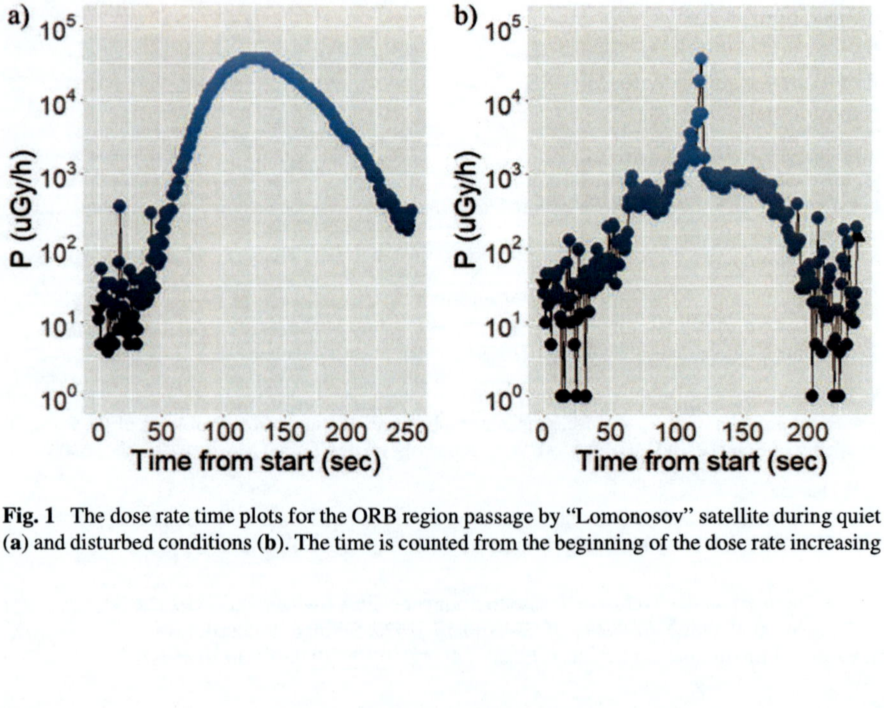

Fig. 1 The dose rate time plots for the ORB region passage by "Lomonosov" satellite during quiet (**a**) and disturbed conditions (**b**). The time is counted from the beginning of the dose rate increasing

time of crossing the high-latitude zone. And these spatial structures are associated with electron precipitations.

The assessment of the ORB electrons' contribution to the radiation dose of astronauts on the ISS is an actual task. It has been found that variations of the geomagnetic field in the form of a storm and sub-storm activity may cause variations of electron fluxes [5]. Further investigation of this relationship may be useful for the impact of electron flux variations on service systems and spacecraft crews in orbit forecasting.

Recently available experimental data obtained with the R3DR2 dosimetry instrument on board the ISS [6] allow us to assess the scale of the radiation exposure of the electrons of the ORB. The observed dose rate variations were examined in detail in [7]. Their contribution to the daily dose values on board the ISS was considered. Dachev (2018) also analyses their correlation with the geomagnetic situation in the near-Earth space. We compared the measurements by the R3DR2 instrument with similar data provided with the DEPRON instrument [8]. The DEPRON performed its measurements on board the "Lomonosov" satellite. It provides a significantly more refined picture of the dosimetry characteristics of high-energy electron flux bursts in the polar oval region.

2 Instruments

The European Space Agency carried out the EXPOSE-R2 experiment on the external surface of the "Zvezda" module. The R3DR2 device was one among the EXPOSE-R2 instruments. The R3DR2 was developed at the Institute of Space Research and Technology of the Bulgarian Academy of Sciences (SRTI-BAS) is one among many dosimeters with a silicon semiconductor detector (SSD) [9–11]. It was distinguished by the small thickness of the material located between the space radiation and the sensitive area of the detector. The R3DR2 instrument's detailed description is provided in [3]. A set of measurements was obtained during the period from 23 October 2014 to 10 January 2016.

Another dosimeter with similar characteristics was the DEPRON instrument [8], developed at the Space Science Department of Skobeltsyn Institute of Nuclear Physics (SINP MSU) [12]. Its detectors are two SSDs and two helium gas-discharge counters of thermal neutrons. The instrument electronics counted the number of pulses per time unit and the sum of amplitudes of these pulses for each SSD, providing the possibility for measurements of the flux and dose rate of cosmic ray charged particles. Gas-discharge counters data is beyond the scope of the present paper. The website [13] provides the DEPRON instrument characteristics. The DEPRON instrument operated on the "Lomonosov" satellite between 5 May 2016 and 10 January 2017. A comparison of the R3DR2 and DEPRON instruments parameters and the measurement conditions is presented in Table 1.

Table 1 Parameters of R3DR2 and DEPRON, and measurement conditions

Parameter	R3DR2 instrument	DEPRON instrument
Instrument specifications		
Radiation detector	Silicon PIN diode	Semiconductor detector
Shielding thickness (g/cm^2)	0.3	0.54 for 1st detector 0.81 for 2nd detector
Thickness of sensors (mm)	0.3	0.3
Proton registration threshold energy (MeV)	19.5	20 for 1st detector 25 for 2nd detector
Electron registration threshold energy (MeV)	0.835	1.25 for 1st detector 1.75 for 2nd detector
Time resolution (s)	10	1
Period of the experiment	From 23 October 2014 Until 11 January 2016	From 5 May 2016 Until 10 January 2017
Spacecraft specifications		
Name of spacecraft	The ISS "Zvezda" module	"Lomonosov" satellite
Flight altitude (km)	From 390 to 430	From 470 to 520
Circulation period (min)	92	93
Orbital inclination (°)	52	97

3 Data Processing Methodology

The data for the R3DR2 instrument was downloaded from the unified network database for Liulin type instruments [6]. The measurement database includes information for the period from 23 October 2014 to 10 January 2016. It is divided into separate text files 10-days long each. These files contain: time of measurement, ISS coordinates in geographical and geomagnetic coordinate systems, flux values, and absorbed dose rates at 10-s resolution.

The DEPRON data contains a time of measurement, coordinates of the spacecraft in both geomagnetic and geographical coordinate systems, dose measurements for each of two SSDs, counts for all four detectors at 1-s resolution. Neutron detectors' data was not used in this analysis.

The "R" programming language [14] and the RStudio shell [15] were used to process the information. The main libraries used were the ggplot2 library [16], designed for data visualization, and the dplyr library [17], designed for data tables processing.

Two data arrays corresponding to the R3DR2 instrument from the EXPOSE-R2 experiment and the DEPRON instrument from the "Lomonosov" satellite were processed. For each of them, databases were compiled containing plots of ORB electron flux enhancements and their properties. Figure 2 shows an example of the enhancement data presentation.

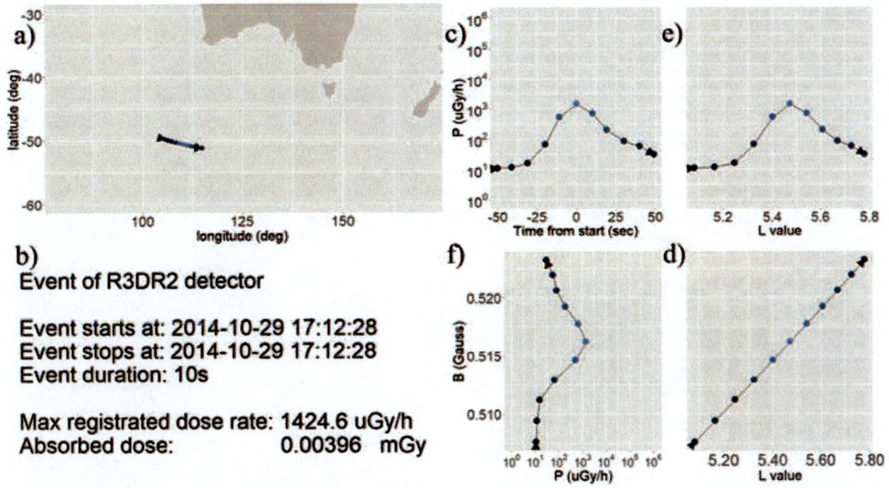

Fig. 2 Dose rate enhancement data set 29 October 2014. The image shows the spacecraft's geographic trajectory (**a**), properties of the enhancement (**b**), dose rate time plot (**c**), LB trajectory (**d**) and dose rate dependence on the L (**e**) and B (**f**) coordinates

The enhancement representation is composed of several plots and parameters of this event. The upper left corner shows the segment of the ISS trajectory corresponding to the period of enhancement detection in the geographical coordinate system (see Fig. 2a). Below this map, the name of the instrument that measured the event and the main properties of the event are listed: the date and time of the first and last detection of the increased electron flux, the calculated duration of the enhancement, the maximum dose rate measured during this event, and the detector absorbed dose during this period (see Fig. 2b). There are four plots on the right side of the image. The upper left panel is the dose rate P (logarithmic scale) time dependence (see Fig. 2c). Time is counted in seconds from the beginning of the enhancement. On the diagonal of this graph, in the lower right panel, the segment of the ISS trajectory in L-B coordinates is shown (see Fig. 2d). There is a graph of dose rate dependence on the L coordinate in the upper right panel (see Fig. 2e). There is a graph of dose rate dependence on the B coordinate in the lower right panel (see Fig. 2f). This graph is as if "laid on a side"—dose rate is plotted along the horizontal axis, while the B coordinate is plotted along the vertical axis.

The chosen form of data presentation allows us to analyse dose rate increases of a more complex shape. Figure 3 shows the data for the event on 26 December 2015 at 16:12:45. The geographical latitude of the event is about 50° S and the longitude is between 110° E and 150° E, corresponding to an area to the south of Australia (see Fig. 3a). The plot of the dose rate (Fig. 3c) shows three spikes in the dose rate value. At the start, the dose rate increases by 2.5 orders of magnitude in about 30 s, from about 0.01 mGy/h to about 3 mGy/h. Then the dose rate falls relatively slowly, but at about 80 s after the start of the event, it rises again rapidly, reaching a maximum value

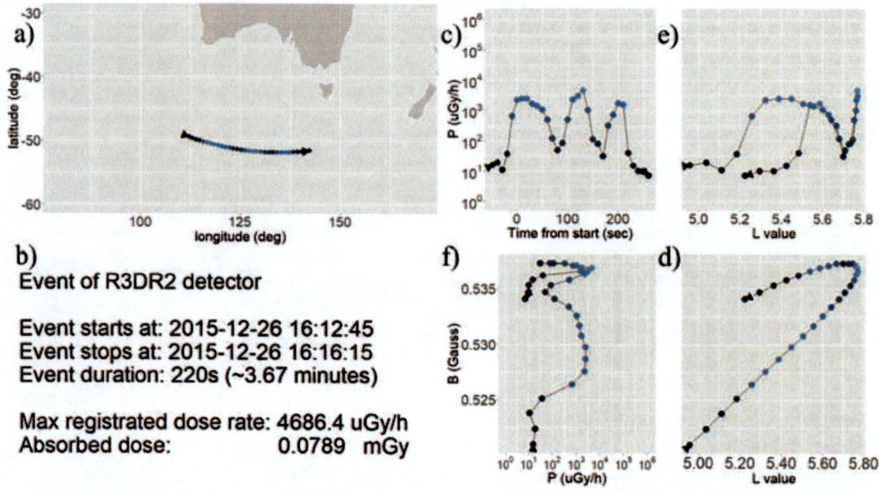

Fig. 3 Dose rate enhancement event on 26 December 2015 started at 16:12:45. The image shows the spacecraft's geographic trajectory (**a**), properties of the enhancement (**b**), dose rate time plot (**c**), LB trajectory (**d**) and dose rate dependence on the L (**e**) and B (**f**) coordinates

of 4.7 mGy/h. This increase is followed by a steep decrease and another steep rise before returning to the baseline dose rate. This event lasted 220 s, with the absorbed dose registered by the instrument reaching 0.08 mGy (see Fig. 3b).

An event with as many as three peaks is quite rare. The ISS trajectory in LB coordinates (Fig. 3d) clearly shows the location of the dose rate peaks, which are represented by the brighter colour of the dots. It can be noticed that the 1st and the 3rd peaks are located at approximately the same L-shell with a value of 5.5, while the 2nd peak corresponds to another L-shell with a value of 5.75. This clearly shows the simultaneous observation of two dose rate enhancements regions on different L-shells during one passage of the high-latitude region. We assume that an additional unsteady flux of electrons was observed in this region of space at the time of the ISS passage.

The "Lomonosov" satellite data provides a more comprehensive picture of the radiation exposure of ORB electrons. Unlike the ISS, the satellite passed through all the L-shells on each rotation in the high-latitude region. The instrument also has a 10 times higher time resolution than the R3DR2. This feature allows investigating faster processes more accurately. However, the higher temporal resolution leads to low statistics in the low dose rate areas resulting in very "jagged" plots.

Figure 4 shows plots characterizing the passage of the "Lomonosov" satellite through the ORB area. It presents a typical event measured by the DEPRON instrument on 5 June 2016 at 14:11:07. The geographical map (Fig. 4a) shows that this event occurred to the south of Australia. Two dose rate spikes are observed at latitudes of 45° S and 50° S.

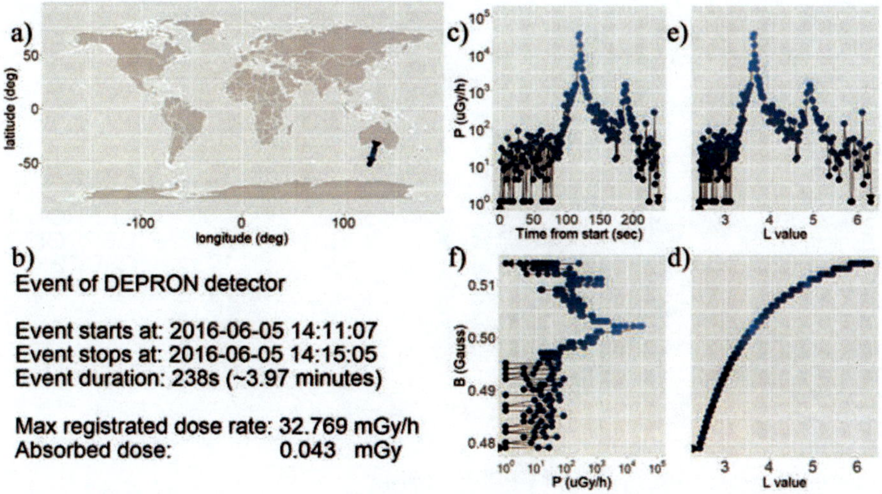

Fig. 4 The "Lomonosov" satellite passage through the ORB area on 5 June 2016. The image shows the spacecraft's geographic trajectory (**a**), properties of the enhancement (**b**), dose rate time plot (**c**), LB trajectory (**d**) and dose rate dependence on the L (**e**) and B (**f**) coordinates

The plots in the right-hand part of the figure show a complex structure of an event (see Fig. 4c). At first, the dose rate increased rapidly by about 3 orders of magnitude from 0.01 mGy/h to 32.8 mGy/h. Then it decreased slowly to about 0.1 mGy/h. At last, it increased again by one order of magnitude before returning to the baseline values. The total duration of the event was 238 s, but the dose rate peaks themselves are much narrower, about 20–30 s each. The absorbed dose during this enhancement was 0.4 mGy (see Fig. 4b).

4 General Patterns of Enhancements

We discovered sixty-three dose rate enhancements in the DEPRON data and 777 ones—in the R3DR2 data during the ORB area passages. Figure 5 presents the maximum dose rates for each event. The R3DR2 measurements are marked with blue dots. The DEPRON data is presented with red and green dots. The first instrument's detector has a shielding of 0.3 g/cm^2. The DEPRON instrument's detectors had shielding of 0.54 g/cm^2 (red) and 0.81 g/cm^2 (green).

The maximum dose rate of 28.8 mGy/h recorded by the R3DR2 instrument occurred on 9 November 2015 at 15:11:38. The first DEPRON detector (red) recorded a maximum dose rate of 106.5 mGy/h. This event occurred on 13 October 2016 between 04:56:39 and 05:00:41.

Noticeably more flux enhancements were recorded during the EXPOSE-R2 experiment, which generally had lower maximum dose rates than similar events recorded

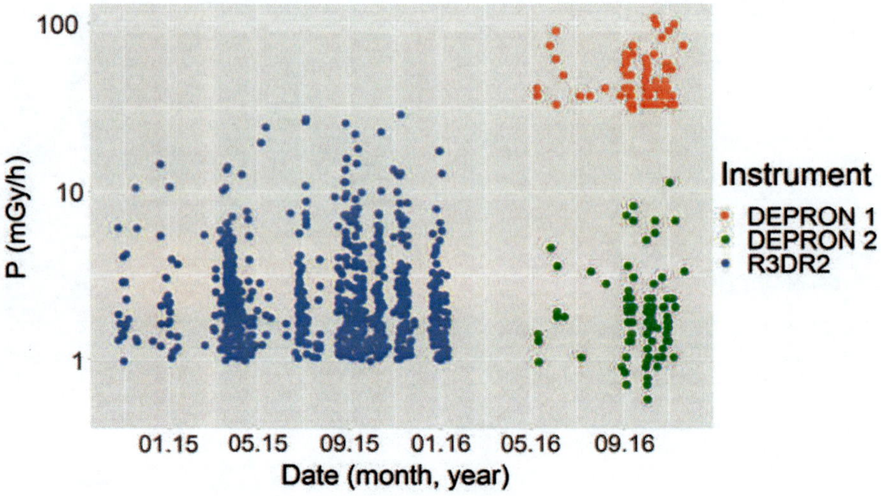

Fig. 5 Time dependence of the maximum dose rate

by the DEPRON instrument. These divergences appear to be caused by differences in spacecraft's inclination, orbital altitude, and the time of measurements. We consider that the higher dose rate measured on the "Lomonosov" satellite is mainly due to the higher orbit.

Figure 6 shows the total doses absorbed by detectors during each electron flux enhancement. The representation form is similar to the one in Fig. 5.

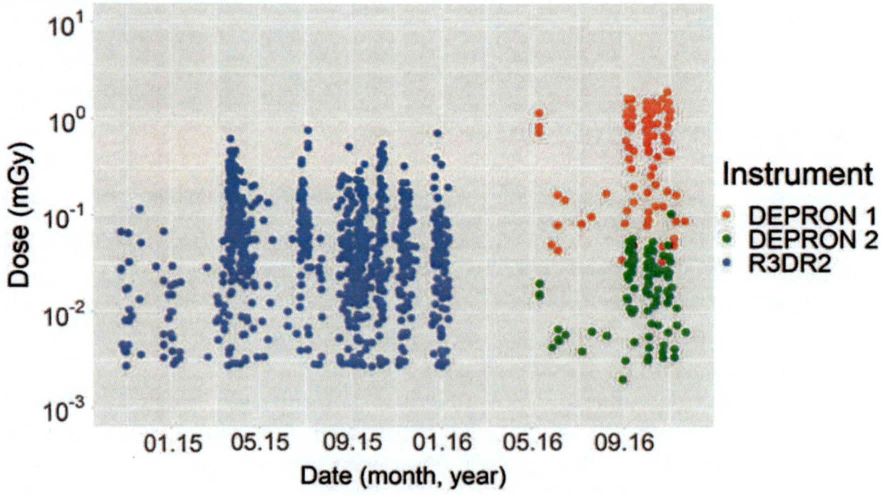

Fig. 6 Absorbed dose for each increase

The event that produced the highest dose rate for the R3DR2 instrument does not match the event with the highest absorbed dose. During this enhancement, there was only a very short increase in dose rate, which did not contribute much to the total value of the absorbed dose, which was about 0.18 mGy. The maximum value of absorbed dose was recorded on 4 July 2015 between 18:57:25 and 19:05:35. During this more long-duration increase in electron flux, the instrument measured the absorbed dose value of 0.77 mGy, and the maximum dose rate value of 27 mGy/h.

A similar picture was observed in the DEPRON instrument data. The event on 13 October 2016 had the highest maximum dose rate. The first detector measured a total absorbed dose of 0.2 mGy during it. The second detector during the same period measured an absorbed dose of only about 0.01 mGy. The maximum absorbed dose measured during an enhancement was 1.6 mGy. It was recorded by the first DEPRON detector on 11 September 2016 between 22:24:06 and 22:29:16. These doses appear to be too small to cause significant damage.

Figure 7 presents a comparison of the enhancements and the widely used Dst and AE geomagnetic activity indices [18].

The upper plot (Fig. 7a) shows events of dose rate enhancements in the ORB area. The lower ones show AE-index (Fig. 7b) and Dst-index (Fig. 7c) values. The red vertical dashed lines mark the time points at which there are significant drops of

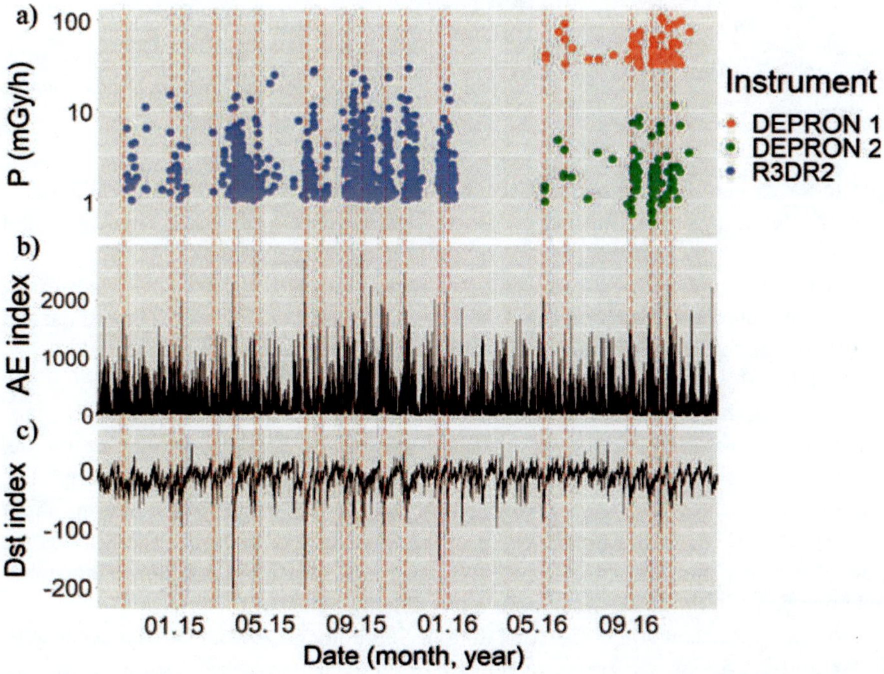

Fig. 7 Comparison of dose rate bursts in the ORB area with geomagnetic activity indices. The image shows enhancements' registration (**a**) along with the AE (**b**) and Dst (**c**) indices fluctuations

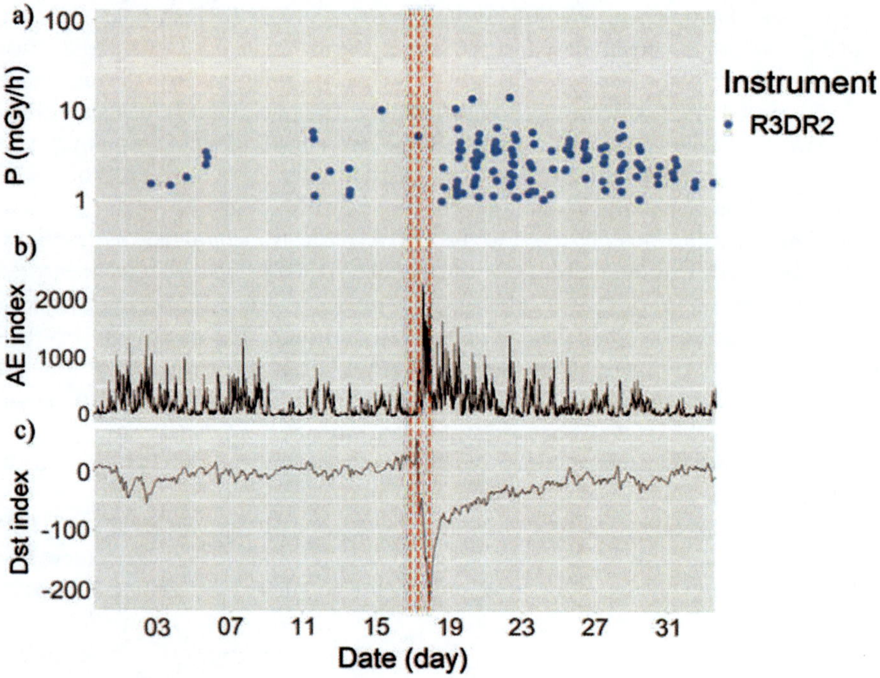

Fig. 8 Comparison between dose rate enhancements and geomagnetic activity during March 2015. The image shows enhancements' registration (**a**) along with the AE (**b**) and Dst (**c**) indices fluctuations

Dst-index and simultaneous increases of AE-index. Most of the "bursts" take place after their geomagnetic disturbances. Several such events are discussed in more detail below.

Figure 8 compares the dose rate enhancements recorded by the R3DR2 instrument with the geomagnetic situation described by the AE and Dst indices. The red dashed lines indicate the phases boundaries of the geomagnetic storm that occurred on 17–18 March 2015. Prior to the first line, the geomagnetic conditions were relatively quiet. During this time, a small number of enhancements after AE index spikes (see Fig. 8b) were observed. There was a steep drop in the Dst index between the 1st and the 2nd lines (see Fig. 8c). This drop indicates the sudden onset of the storm. The main phase of the storm is characterised by the Dst index value decrease to about −200 (see Fig. 8c). It is followed by a recovery phase starting after the 3rd dashed line. The upper plot (Fig. 8a) shows an occurrence of a large number of dose rate enhancements in the ORB area about a day after the start of this phase.

Figure 9 represents the DEPRON data in September–October 2016 and the AE-index and Dst-index variations during this period. There were several periods of geomagnetic disturbance. Sometimes there were observed dose rate enhancements without significant drops in the Dst-index. Let's consider these disturbed periods.

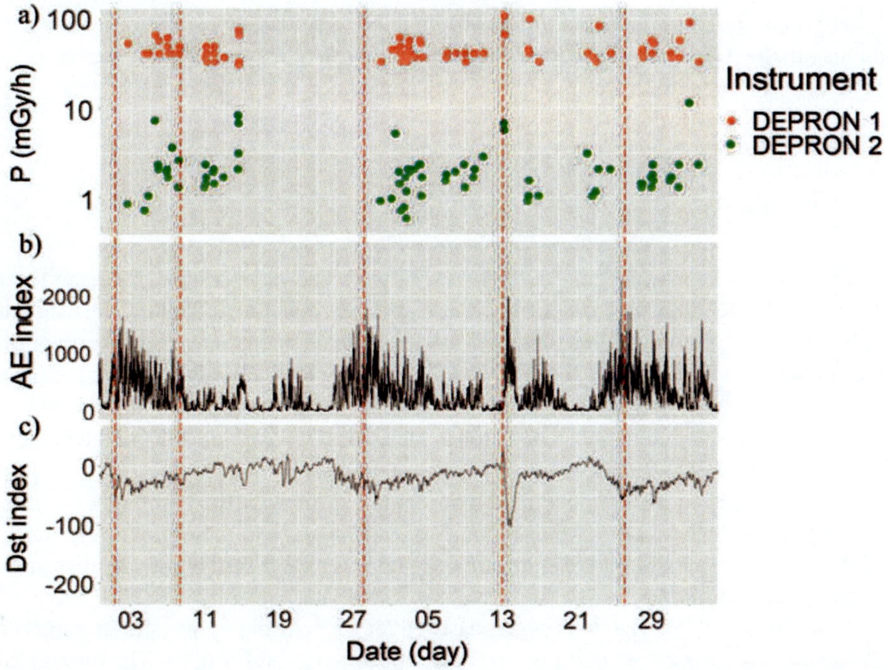

Fig. 9 Comparison between enhancements and geomagnetic activity in September and October 2016. The image shows enhancements' registration (**a**) along with the AE (**b**) and Dst (**c**) indices fluctuations

The first period occurred around 1–2 September. It was a drop in the Dst index to a value of −50 (see Fig. 9c) and an increase in the AE index value to 1.6×10^3 (see Fig. 9b). After these disturbances, there were 12 enhancements (see Fig. 9a).

The second period followed immediately after the first one and took place on 8–9 September. The Dst index did not fall below −40 (see Fig. 9c). However, the AE-index increased to about 1.3×10^3 (see Fig. 9b). The DEPRON instrument recorded 12 more enhancements after these disturbances (see Fig. 9a).

The third disturbance period took place around 28 September. It was characterized by a drop in Dst to −60 (see Fig. 9c) and an increase in AE to 1.7×10^3 (see Fig. 9b). This period was followed by numerous enhancements (see Fig. 9a).

The fourth period may refer to the geomagnetic storm of 13 October, during which the Dst drop reached −100 (see Fig. 9c). Two enhancements occurred during the initial phase of the storm (see Fig. 9a). Then several enhancements occurred about 1.5 days after the beginning of the recovery phase (see Fig. 9a).

The disturbance on 26 October is considered to be the last period. The decrease of the Dst-index to −60 (see Fig. 9c) and growth of the AE-index to 2.5×10^3 (see Fig. 9b) caused a significant number of enhancements. The DEPRON instrument recorded 12 enhancements (see Fig. 9a).

This comparison shows that the recording of a large number of dose rate enhancements in the ORB region occurs approximately 1–2 days after the onset of a geomagnetic storm recovery phase or other geomagnetic disturbances.

5 Conclusion

We performed an analysis of the dosimetry properties of dose rate enhancements based on the R3DR2 instrument on board the ISS and the DEPRON instrument on board the "Lomonosov" satellite data. The scale of the enhancements' radiation hazard behind a thin shielding during the Earth's Outer Radiation Belt region crossing was estimated based on this experimental data.

There were 63 enhancements detected in the DEPRON data and 777 enhancements detected in the R3DR2 data.

A comparison of the enhancement's occurrence time with the geomagnetic environment data in the form of the Dst and AE indices shows a significant increase of the enhancements number approximately one day after the periods of disturbed geomagnetic conditions.

It is confirmed that the dose rate measured on a low-orbiting spacecraft behind a thin shielding during the ORB region crossing can reach high values. The maximum value of 28.8 mGy/h was recorded on the ISS, and 106.5 mGy/h on the "Lomonosov" spacecraft. In our opinion, the higher dose rate observed on the "Lomonosov" satellite is mainly due to the higher orbit.

However, these enhancements are of a short-term type. No potentially hazardous for the crews of manned spacecraft enhancements were found during processing. The maximum value of dose per single enhancement reached 0.77 mGy for the R3DR2 instrument and 1.6 mGy for the DEPRON instrument.

References

1. Petrov, V.M., Tel'tsov, M.V., Mitrikas, V.G., Akatov, Y.A., Benghin, V.V.: Physical conditions in open space: radiation dosimetry in space flight. In: Panasyuk, M.I., Novikov, L.S. (eds.) Model of Cosmos, vol. 1, Chapter 3.10, pp. 642–667. Universitet, Knizhnyi dom, Moscow (2007). ISBN 978-5-98227-419-9
2. Kuznetsov, S.N., Tverskaya, L.V.: Physical conditions in open space: radiation. In: Panasyuk, M.I., Novikov L.S. (eds.) Model of Cosmos, vol. 1, Chapter 3.4, pp. 518–546. Universitet, Knizhnyi dom, Moscow (2007). ISBN 978-5-98227-419-9
3. Dachev, T.P., Bankov, N.G., Tomov, B.T., Matviichuk, Y.N., Dimitrov, P.G., Häder, D.P., Horneck, G.: Overview of the ISS radiation environment observed during the ESA EXPOSE-R2 mission in 2014–2016. Space Weather **15**(11), 1475–1489 (2017). https://doi.org/10.1002/2016SW001580
4. Johnson, A.T., Shumko, M., Griffith, B., Klumpar, D.M., Sample, J., Springer, L., Blake, J.B.: The FIREBIRD-II CubeSat mission: focused investigations of relativistic electron burst

intensity, range, and dynamics. Rev. Sci. Instrum. **91**(3), 034503 (2020). https://doi.org/10.1063/1.5137905
5. Antonova, E.E., Stepanova, M.V.: The problem of the acceleration of electrons of the outer radiation belt and magnetospheric substorms. Earth Planets Space **67**(1), 1–8 (2015). https://doi.org/10.1186/s40623-015-0319-7
6. Semkova, J., Dachev, T., Koleva, R., Tomov, B., Matviichuk, Y., Bankov, N., Krastev, K., Malchev, S., Dimitrov, P., Mitev, V., Chakarov, S.: Unified web-based database with Liulin type instruments' cosmic radiation data. http://esa-pro.space.bas.bg/database
7. Dachev, T.P.: Relativistic electron precipitation bands in the outside radiation environment of the international space station. J. Atmos. Solar Terr. Phys. **177**, 247–256 (2018). https://doi.org/10.1016/j.jastp.2017.11.008
8. Benghin, V.V., Nechaev, O.Y., Zolotarev, I.A., Amelyushkin, A.M., Petrov, V.L., Panasyuk, M.I., Yashin, I.V.: An experiment in radiation measurement using the Depron instrument. Space Sci. Rev. **214**(1), 1–9 (2018). https://doi.org/10.1007/s11214-017-0445-6
9. Dachev, T., Semkova, J., Petrov, V., Redko, V., Bengin, V., Kostereva, T., Zeitlin, C.: Analysis of the pre-flight and post-flight calibration procedures performed on the Liulin space radiation dosimeter. Acta Astronautica **42**(1–8), 375–387 (1998). https://doi.org/10.1016/S0094-5765(98)00132-5
10. Panasyuk, M.I.: Description of the space radiation control system for the Russian segment of ISS ALPHA. In: 12th IAA Man in Space Symposium, Washington, DC (June 1997). https://doi.org/10.1134/S0010952510030020
11. Reitz, G., Beaujean, R., Benton, E., Burmeister, S., Dachev, T., Deme, S., Olko, P.: Space radiation measurements on-board ISS—the DOSMAP experiment. Radiat. Protect. Dosimet. **116**(1–4), 374–379 (2005). https://doi.org/10.1093/rpd/nci262
12. Division of cosmic rays SINP MSU. http://sinp.msu.ru/en/strukted/1650
13. DEPRON scientific equipment of the "Mikhailo Lomonosov" satellite. http://lomonosov.sinp.msu.ru/en/c-2/depron
14. The R Project for Statistical Computing. https://www.r-project.org/
15. RStudio—an integrated development environment (IDE) for R. https://www.rstudio.com/products/rstudio/
16. Wickham, H., Chang, W., Henry, L., Pedersen, T.L., Takahashi, K., Wilke, C., Woo, K., Yutani, H., Dunnington, D.: ggplot2: elegant Graphics for Data Analysis. Springer, New York (2016). ISBN 978-3-319-24277-4. https://ggplot2.tidyverse.org/
17. Wickham, H., François, R., Henry, L., Müller, K.: dplyr: A Grammar of Data Manipulation. https://dplyr.tidyverse.org/
18. The World Data Center for Solar-Terrestrial Physics in Moscow. http://www.wdcb.ru/stp/index.en.html

Kamchatka Meteoroid Effects in the Lithosphere–Atmosphere–Ionosphere–Magnetosphere System

L. F. Chernogor

Abstract A few physics and mathematics models of the processes in all geospheres that were caused by the fall of the meteoroid has been performed. Mechanical, optical, gas-dynamic, thermodynamic, magnetic, electrical, electromagnetic, ionospheric, seismic, plasma, plume, turbulence, acoustic-gravity wave effects that accompanied the fall of the Kamchatka meteoroid have been estimated. The energy release (about 5×10^{14} J) has been shown to take place mainly at altitudes of 26–30 km, where the rate of mass loss reached about 1 kt/s, and deceleration about 6 km/s^2. Near the terminal height, the speed of the meteoroid decreased by about 10%, and its mass by 20%. The main parameters of the shock wave have been considered; the energy and power of the shock wave were estimated to be about 580 TJ and 58 TW; near the epicenter of the meteoroid detonation, the pressure at the shock front was ~1 kPa; the perturbations in atmospheric pressure at ionospheric heights above the epicenter of the detonation attained tens and even hundreds of percent. The energy and power of the fireball light were about 130 TJ and 1 PW respectively. The size of the heated region and its temperature (about 4600 K) have been estimated, and the four stages of its cooling have been considered. The main parameters of the debris trail (dust and plasma densities, as well as electric and magnetic field strengths) and the parameters of electromagnetic and Alfvén pulses have been estimated. The mechanisms for the generation of acoustic, atmospheric gravity, ion-acoustic, and magnetosonic waves, as well as instabilities, have been considered. The parameters of the emerging cloud of dust, as well as the duration and rate of its settling, have been estimated.

Keywords Meteoroid · Physics and mathematics model · Main physical effects · Lithosphere–atmosphere–ionosphere–magnetosphere system · Ionospheric disturbance · Wave disturbance · Atmospheric gravity wave · Electric, magnetic, and electromagnetic effects · Seismic effect

L. F. Chernogor (✉)
V. N. Karazin Kharkiv National University, 4 Svobody Square, Kharkiv 61022, Ukraine

A. Kosterov et al. (eds.), *Problems of Geocosmos—2020*, Springer Proceedings in Earth and Environmental Sciences, https://doi.org/10.1007/978-3-030-91467-7_27

1 Introduction

In view of the uniqueness of the event and the significance of the consequences, an urgent interdisciplinary task is a detailed and comprehensive study of the effects of the fall on the Earth of each sufficiently large celestial body, as well as the consequences of such a fall [1].

A new surge of interest in the effects of meteoroids has been associated with the fall on February 15, 2013 of a meteoroid, the largest in the last 100 years, termed the Chelyabinsk meteoroid [2]. Another example of a similar, but smaller-scale event, is the passage of the Kamchatka meteoroid. The celestial body entered the Earth's atmosphere at 23:48:20 UT on December 18, 2018 (11:48:20 LT on December 19, 2018). The meteoroid moved at an angle of about 80° with respect to the horizon, had an initial velocity $v_0 \approx 32$ km/s, and initial kinetic energy $E_{k0} \approx 173$ kt TNT. Calculations show that the initial body mass $m_0 \approx 1.41$ kt, and the initial diameter $d_0 \approx 9.4$ m. The meteoroid exploded at an altitude of 25.6 km, and its kinematic characteristics indicate that it was a stone. The coordinates of the detonation point were 56.9°N, 172.4°E, and the fireball light energy was $E_r \approx 130$ TJ [3].

The purpose of this study is a comprehensive analysis and modeling of the main physical effects accompanying the fall of the Kamchatka meteoroid in the atmosphere, ionosphere, and in the Earth's magnetic field.

2 Governing Equations

The movement of a meteoroid with a diameter of no more than 20–30 m (for larger sizes, it is necessary to solve the governing equations of gas dynamics) in the Earth's atmosphere is described by the following well-known equations of meteor physics (see, e.g., [4]):

$$m\frac{dv}{dt} = -\frac{C_d}{2}\rho v^2 S + mg\sin\alpha, \tag{1}$$

$$\frac{dm}{dt} = -\frac{C_h}{2Q}\rho v^3 S, \tag{2}$$

$$mv\frac{d\alpha}{dt} = mg\cos\alpha - \frac{mv^2}{R_E}\cos\alpha - \frac{C_L}{2}\rho v^2 S, \tag{3}$$

$$\frac{dz}{dt} = -v\sin\alpha,\ I = -\frac{\tau_I}{2}\frac{d}{dt}mv^2,\ \alpha_i = -\frac{\beta_i}{M_m v}\frac{dm}{dt}, \tag{4}$$

where m is the body mass, v is the body velocity, S is the cross-sectional area (midsection) of the meteoroid, g is the acceleration due to gravity, α is the angle between the tangent to the trajectory and the horizon, $C_d \approx 0.6$ and C_L are the coefficients of the aerodynamic drag and the upward lift forces, $C_h \approx 0.03$ is the heat transfer coefficient, $Q \approx 6.5$ MJ/kg is the heat of sublimation, R_E is the Earth's radius, ρ is the atmospheric density at the altitude level z, τ_I is the luminosity coefficient, α_i is the linear electron density, β_i is the ionization coefficient, $M_m \approx 5 \times 10^{-26}$ kg is the mass of the meteoroid molecule. Equations (1)–(4) describe deceleration, the loss of mass by a celestial body, the change in the trajectory inclination, and the altitude of the body, as well as the luminosity and the effect of ionization.

The kinetic energy of a celestial body is spent on its fragmentation, heating, melting, ablation, ionization, heating and glowing of matter in the trail, on the formation of ballistic and blast shock waves, the generation of plasma, electrical, magnetic, electromagnetic, acoustic-gravity, seismic, and other physical processes.

Let us evaluate some of these energies. Given the specific energy of ablation Q and the mass loss Δm, the energy associated with ablation $E_{Q0} \approx Q\Delta m$. For the values of Q and $\Delta m \approx m_0$ indicated above, we have $E_{Q0} \approx 9.2 \times 10^{12}$ J or 1.3% of E_{k0}. Since Δm is obviously less than m_0, the energy E_Q is in reality always less than E_{Q0}. If the specific energy of fragmentation of a stony celestial body is $\varepsilon_d = 10^5$ J/kg, the total fragmentation energy is $E_d = \varepsilon_d m_0 \approx 1.4 \times 10^{11}$ J or 0.02% of E_{k0}. The number of atoms in the meteoroid is $N_\Sigma = m_0/M_m$. For the ionization coefficient $\beta_i \approx 1.54 \times 10^{-2}$ [4], $M_m \approx 5 \times 10^{-26}$ kg, and for $m_0 \approx 1.41 \times 10^6$ kg, we have $N_\Sigma \approx 2.8 \times 10^{31}$, and the total number of ionized atoms (molecules) $N_{i\Sigma} \approx 4.3 \times 10^{29}$. If on average it takes $\varepsilon_i \approx 50$ eV to produce an electron–ion pair, we have $E_i = N_{i\Sigma}\varepsilon_i \approx 3.5 \times 10^{12}$ J or 0.5% of E_{k0}. From these estimates, it follows that the main part (about 80%) of the initial kinetic energy of the celestial body is spent on the formation of a shock wave and on air heating.

3 Basic Physical Processes in the Geospheres

3.1 Meteoroid Fragmentation

The fragmentation process of each meteoroid is individual. We assume that a quasi-continuous fragmentation took place. The following three mass distributions of fragments are considered: uniform, log–normal, and power–law function with exponents of 0.5, 1, and 1.5. Fragmentation occurred over the 34–30 km altitude range. The initially spherical body gradually turned into a pan-shaped body with the cross-section $S(z)$, and further the fragments moved as a quasi-liquid. With a significant increase in $S(z)$, an explosive release of energy occurred, which was followed by the appearance of the fireball and the emission of electromagnetic energy predominantly in the optical wavelength range.

Celestial body fragmentation occurs under the condition that the dynamic pressure is equal to the strength of the body's material σ_d, i.e., at $0.365\rho v^2 = \sigma_d$, where ρ is the air density at the altitude of fragmentation z_d [1]. Different parts of the meteoroid have different strengths; therefore, the strength of a stony body varies within wide range, $\sigma_d = 10^6$–10^7 N/m^2 [1]. Then, the condition for fragmentation at $v \approx 30$ km/s is satisfied, respectively, at an altitude of $z_d = H \ln\left(0.365\rho(0)v^2/\sigma_d\right) \approx 30$–$34$ km. Here $H \approx 7.5$ km is the atmospheric scale height.

The coefficient [6]

$$k_d = \frac{d}{d_0} = \frac{4H}{d_0 \sin\alpha}\sqrt{\frac{\rho(z_d)}{\rho_b}},$$

describes the degree of broadening of the meteoroid trail during the process of fragmentation. Here d is the diameter of the cloud of fragments near the altitude, z_d, of the terminal detonation point of the fireball, ρ and ρ_b are the densities of air and meteoroid, respectively. Substituting $d_0 = 9.4$ m, $\sin\alpha \approx 0.98$, $\rho(z_d) \approx 2 \times 10^{-2}$ kg/m^3, $\rho_b \approx 3.3$ t/m^3 yields $k_d \approx 8$. Due to sufficiently large values of k_d, an almost instantaneous process of fragmentation takes place over the distance $L_e \approx 4$ km. Near the altitude of fragmentation, the speed of the meteoroid decreased by about 10%, and its mass by 20%. The mass loss rate was about 1 kt/s, and deceleration was approximately 6 km/s^2.

3.2 Optical Effect

About 18% of the initial kinetic energy of the meteoroid was converted into the energy of optical emission, which amounted to 130 TJ [3]. The altitude of the optical flash was 26–30 km, and its transverse size was about 2 km. The temperature of the luminous formation was close to 4600 K, and the wavelength of the maximum radiation was about 0.6 μm.

3.3 Ballistic Shock Wave

The flight of the meteoroid at hypersonic speed was accompanied by the generation of a ballistic shock wave. The meteoroid began to form the shock at a height where the air for the falling body became a continuous medium. The criterion of the continuity is the Knudsen number $\mathrm{Kn} = l/d \ll 1$, where l is the mean free path between collisions of the air molecules, and d is the diameter of the body [4]. A ballistic wave of noticeable intensity arises over the $\mathrm{Kn} \approx (0.7$–$1.2) \times 10^{-2}$ range of values. For $d \approx d_0 \approx 9.4$ m, this took place at an altitude of $z_{\max} \approx 85$–88 km, respectively. Over the $z \leq 30$ km altitude range, the ballistic wave was superimposed on the blast shock wave. The division into ballistic and blast waves is rather arbitrary. The former is less intense and its cone is elongated along the trajectory. The radius of the ballistic wave is given by the relation $R_b = \mathrm{M}d$, where M is the Mach number $\mathrm{M} = v/v_s$, v and v_s are the velocities of the celestial body and sound [5].

Table 1 Dependence on distance (R) of the periods of the acoustic vibrations generated by ballistic (T_b) and blast (T_c) waves

R (km)	50	100	300	500	1000	2000	5000
T_b(s)	25	30	39	44	53	63	79
T_c(s)	72	85	112	128	152	181	228

At $v \approx v_0 \approx 32$ km/s, $v_s \approx 0.3$ km/s, $d \approx d_0 \approx 9.4$ m, we have M ≈ 107 and $R_b \approx 1$ km. The radius R_b corresponds to the period of acoustic vibrations $T_b = 2.81R_b/v_s = 9.4$ s [5]. According to [6], this period, due to dispersion, increases according to the $T_b(R) = T_b(R_b)(R/R_b)^{1/4}$ law (Table 1). Here, R is the distance range that the shock has traveled. The expression for the dependence on the distance R of the relative pressure in the front of the ballistic wave is given in [5]. If $R^2 \gg R_b^2$, then this ratio has the form $\Delta p/p_0 \approx 0.3(R_b/R)^{3/4}e^{I/2}$, where $I = \int_{z_e}^{z} dz/H(z)$ [6].

3.4 *Explosive Shock Wave*

We assume that the explosive release of energy took place over the length of the meteoroid trajectory equal to $L_e \approx 4$ km. Then the linear energy density $E_L \approx 1.45 \times 10^{11}$ J/m. We assume that the main explosive energy release took place near the altitude $z_e \approx 27$ km. Assuming the explosion to be cylindrical, we calculate the radius of the shock wave using the relation $R_c = \sqrt{E_L/\pi p(z_e)} \approx 4.1$ km, where $p(z_e) \approx 2.7 \times 10^3$ Pa is the atmospheric pressure at the altitude of the energy release maximum. The results of calculations of the period $T_c = 2.81R_c/v_s$ are given in (Table 1).

The results of calculations of the altitude dependence $\Delta p(z)$ for $z \leq 13$ and $z \geq 40$ km without taking into account attenuation are given in Tables 2 and 3. In this case, the altitude dependence $H(z)$ was taken into account and numerical integration over z was performed, as it has been done in the technique [9]. Table 3 shows that as the height increases, the excess pressure in the shock front decreases rather quickly, and the relative change in $\Delta p/p_0$ at $z \geq 40$ km increases. At higher ($z \geq 110$ km) altitudes, $\Delta p/p_0 > 1$. In reality, the magnitude of the effect is much smaller, since the calculations have not taken into account the dissipation of the wave energy, which at $\Delta p/p_0 > 0.2$–0.3 is primarily caused by the effect of nonlinearity. Due to self-action, the spectrum of the acoustic signal is enriched with higher harmonics, which dissipate at lower altitudes. For this reason, the shock wave propagating upward is noticeably attenuated over the ~120–180 km altitude range due to self-absorption effect.

The energy of the shock wave in the atmosphere propagates also in the horizontal direction. This is facilitated by stratospheric and thermospheric waveguides. Thus, for a cylindrical wavefront divergence in the waveguide over the $z \approx 100$–150 km altitude range and at $R \approx 1000$ km, $\Delta p/p_0 \approx 0.1$–0.3.

Table 2 Altitude dependence of pressure Δp in the shock wave (p_0 is unperturbed atmospheric pressure)

z (km)	0	5	10	13
p_0 (10 kPa)	100	51	26	17.7
$\Delta p/p_0$	0.008	0.015	0.028	0.04
Δp (Pa)	800	780	730	707

Table 3 Altitude dependences of pressure (p_0) in an undisturbed atmosphere, relative $\Delta p/p_0$ and absolute Δp shock wave pressure a in a loss-free atmosphere

z (km)	p_0 (Pa)	$\Delta p/p_0$	Δp (Pa)
40	2.2×10^2	0.1	22
50	47	0.12	5.6
60	10	0.18	1.8
70	3	0.26	0.8
80	1	0.33	0.33
90	0.3	0.56	0.17
100	0.1	0.84	8.4×10^{-2}
120	10^{-2}	2.07	2.1×10^{-2}
150	10^{-3}	4.95	5×10^{-3}
200	10^{-4}	11.1	1.1×10^{-3}
300	10^{-5}	22.2	2.2×10^{-4}

Assuming that about 80% [6] of the E_{k0} energy was converted into the shock wave energy, the wave energy was close to 5.8×10^{14} J. For the duration of the explosive shock wave $T_{sw} \approx L_e/v_s \approx 10$ s, we have a power of about 58 TW.

3.5 Thermal Effect

Given E_{k0} and E_r, and assuming the form of the bolide (ellipsoid of revolution), the size and the temperature T of the bolide can be estimated. It turns out that the longitudinal and transverse axes are close to 4 and 2 km, respectively, and $T \approx 4600$ K. The meteoroid trail at 4600 K, termed the thermal, is cooled due to thermal radiation, thermal expansion, a rise with the addition of cold air, a hydrostatic expansion as a result of a decrease in the ambient air pressure during the rise of the thermal, and due to an increase in volume as a result of turbulent diffusion [7–9]. We distinguish the following four stages of cooling [9]: (1) over the $t < t_1 \approx 0.1$ s time interval, the density ρ_p of the material in the trail is greater than the density of the unperturbed gas. The value of t_1 is determined from the $\rho_p(t_1) = \rho_0(z_e)$ condition. In this case, ρ_p is determined mainly by the total mass m of the meteoroid fragments. The radius of the cylindrical formation is $r(t) \sim t^{1/2}$; (2) over the $t_2 > t > t_1$ time interval, where $t_2 \approx 10$ s is the time taken by trail expansion from r_e to $r_m \approx R_c \approx 4.1$ km, the density of matter in the thermal is determined mainly by the density of the air heated

$\rho = \rho_e T_0 / T = \rho_e / \theta$, where $\theta = T/T_0$, T and T_0 are the air temperature in the trail and in an undisturbed atmosphere, $\rho_e = \rho_0(z_e) \approx 3.55 \times 10^{-2}$ kg/m^3 is the air density at the explosion altitude $z_e \approx 27$ km. The surface area S of the thermal and its volume V were calculated for an ellipsoid of revolution with semi-axes r, r and $L_e/2$. The ratio $S(t) / V(t) \sim r^{-1}(t) \sim t^{-1/2}$; (3) over the $t_3 > t > t_2$ time interval, where $t_3 \approx 20$ s is the time interval taken for an increase in the upward velocity of the thermal; (4) over the $t > t_3$ time interval, the thermal rises vigorously, takes in large masses of cold air, increases in diameter due to a decrease in the pressure of the ambient air, and its volume increases due to turbulent diffusion. All these factors in a time of ~100 s lead to the accelerated cooling of the thermal, its deceleration and stopping [7–9]. The heated thermal existed for several hours.

3.6 Meteoroid Plume Ascent

As a result of the passage of a meteoroid, a cylinder-shaped trail at about 4600 K is formed in the atmosphere under the action of the shock wave. Its radius is given by the relation $R_s(\tau_e) \approx \sqrt{Ld} = \sqrt{Hd/\sin\alpha}$ [9]. The air having been heated to high temperatures, together with the material evaporated from the meteoroid (the thermal), streams (expanding) along the fireball track and upwards with an initial acceleration of ~140 m/s^2. Due to an almost vertical fall, one can consider only a vertical movement. In this case, the steady-state velocity of the plume is given by the relation $v_{\infty m} \approx \sqrt{4gL_e/3C_d}$ [9]. For $g \approx 9.8$ m/s^2, $L_e \approx 4$ km, and $C_d \approx 0.6$, we have $v_{\infty m} \approx 295$ m/s. The ascent duration is ~100 s, and the maximum ascent altitude is ~15 km.

3.7 Aerosol Descent

In the process of ablation and explosion of the celestial body, its fragments were produced with a radius, r_p, of 10^{-6} to ~0.01 m and a mass, m_p, of 10^{-15} to ~10^{-2} kg [9]. The relatively heavy fragments of the celestial body (with a mass of more than ~ 1–10 g) fell on the Earth's surface in the form of meteorites. The lightest dust grains (aerosols), along with hot explosion products, on the contrary, rose upward by ~15 km, i.e., up to an altitude of $z_p \approx 40$–45 km. Subsequently, the aerosols took part in the following three processes: first, in a slow descent to the surface of the Earth; second, in turbulent mixing with ambient air; and third, in the transfer by the prevailing winds around the Earth, and making a complete revolution in about two days. The dust could exist in the atmosphere for ~10 days.

3.8 Turbulence Effect

The meteoroid trail was highly turbulent over the $z \approx$ 26–80 km altitude range. The outer scale of turbulence is $L_t \approx d(z) \approx$ 10–100 m in the altitude range from 80 to 26 km, respectively. Turbulence scale time is $\tau_t = L_t/v_t = L_t/\alpha_t v \approx 3 \times 10^{-3} - 0.3$ s for the same altitude range. Here $\alpha_t = v_t/v \approx 0.1$, v_t is the characteristic speed of turbulent motion [9]. Turbulent diffusion coefficient $D_t = v_t L_t = v_t^2 \tau_t \approx 3 \times 10^4$ m^2 s^{-1}. The meteoroid trail is filled with a plasma with the plasma conductivity σ. Plasma turbulence is characterized by the magnetic Reynolds number $\mathrm{Re}_m = \mu_0 v_t L_t \sigma$, where μ_0 is the permeability of free space. For the values of v_t, L_t, indicated above, and $\sigma \approx 10^3\ \Omega^{-1}$ m^{-1}, we get $\mathrm{Re}_m \approx 30 < \mathrm{Re}_{m\mathrm{cr}} \approx 40$.

3.9 Plasma Effects

The following main parameters of the plasma in the trail have been estimated: the altitude dependences of the linear and volume densities of the electrons, their relaxation times, particle collision frequencies, plasma conductivity, and relaxation times of the electron temperature. It is shown that the linear and volume electron densities of the electrons in the trail at the initial moment were about (2–40) $\times 10^{23}$ m^{-1} and $(1-4) \times 10^{21}$ m^{-3}, the plasma conductivity was ~$10^3\ \Omega^{-1}$ m^{-1}. The role of the dusty component of the plasma could be significant [9].

3.10 Magnetic Effect

Perturbations of the geomagnetic field can be produced by the dipole moment of celestial bodies and by the diamagnetic disturbance produced by the shock wave of the exploding celestial body [10]. Geomagnetic pulsations (quasiperiodic variations in the geomagnetic field) can be generated by the MHD effect during a periodic movement of the ionospheric plasma in the field of acoustic and atmospheric gravity waves generated during the passage and explosion of the meteoroid. The most significant is the dynamo effect of acoustic and atmospheric gravity waves, and the current density $j = eNv_w$ in the wave field (here v_w is the velocity of neutral particles in the wave). For $N \approx 10^{11}$ m^{-3}, v_w = 4–40 m/s, we have $j \approx 6.4 \times 10^{-8} - 6.4 \times 10^{-7}$ A/m^2. In this case, $\Delta B \approx$ 0.5–5 nT for an acoustic wave. The magnetic effect of the electric current of $I = 10^2 - 10^3$ A in the trail is $\Delta B \approx$ 0.2–2 nT at a distance of R = 100 km.

3.11 Electric Effect

The emergence of an electric current is possible due to a partial separation of the electric charges in the trail [10]. Let us estimate the effect of an external electric field of atmospheric origin with the electric field intensity E_0. For $\sigma \approx 10^3\ \Omega^{-1}\ \mathrm{m}^{-1}$, $E_0(z_e) \approx 2.7 \times 10^{-2}$ V/m, $S \approx 10^3\ \mathrm{m}^2$, $H_E = 3.5$ km (scale of decrease in $E_0(z)$), and $L_e = 4$ km, we have the resistance of the trail $R_\sigma \approx 4 \times 10^{-3}\ \Omega$ and the current $I \approx 2.4 \times 10^4$ A.

Let us estimate the electrostatic effect of a meteoroid [10]. The electric charge of the system composed of the plasma plus ambient air is determined by the balance of the forward and reverse electron fluxes in this system [11]. We assume the surface of the plasma envelope around the meteoroid to be a sphere of radius r_s. From this surface, the electrons stream ahead of the ions, because their thermal speed is greater than that of the ions. The runaway electrons leave the positive charge q, which returns the electrons back. In this case, the rate of the removal of electrons, close to the speed of the meteoroid, becomes equal to the rate of return of electrons with the electron mobility μ_e, i.e., $v = \mu_e E_s$, where E_s is the electrostatic field strength on the surface of the plasma envelope. Let us assume for the Kamchatka meteoroid $r_s = 5$–50 m, $v = 30$–1 km/s, $\mu_e = 3 \times 10^{-3} - 3 \times 10^{-2}\ \mathrm{m}^2\ \mathrm{V}^{-1}\ \mathrm{s}^{-1}$ [11] in the altitude range from 32 to 26 km, respectively. Then, for the same altitude range, we obtain $E_s =$ 10–0.03 MV/m, the potential $\varphi = 50$–1.5 MV, $q \approx 3.1$–0.9 mC, and the energy $E = 155$–1.4 kJ. Lightning discharges could have occurred in the meteoroid trail, where the breakdown field strength was 0.03–0.1 MV/m. During the passage of the Kamchatka meteoroid, a weak lightning discharge could well have occurred, the energy of which was relatively low (~1–100 kJ).

Electromagnetic effects are associated with a sharp change in the magnetic field in the trail, which is accompanied by the generation of an electric field pulse, emissions from the meteoroid trail, and by the propagation of infrasound [10]. The first effect turns out to be insignificant, for the second and the third effects $E \sim 10$–100 V/m.

3.12 Generation of Ion-Acoustic and Magnetosonic Waves by Infrasound

The amplitude of the infrasound in the altitude range greater than 100 km was remaining large enough to create prerequisites for converting the infrasound energy into the energy of the ion-acoustic waves [10]. The generation of the ion-acoustic waves at the infrasound wave frequency is effective at the infrasound wave frequency $\omega \sim \nu_{nn}$, where ν_{nn} is the neutral–neutral collision frequency. For the Kamchatka meteoroid, this frequency is ~1–10 s^{-1}. Over the 130–150 km altitude range, the kinetic pressure of the gas becomes comparable to the magnetic pressure (~10^{-3} Pa). In this case, the infrasonic wave can be transformed into the magnetosonic wave. The latter can reach the magnetosphere and thereby ensure the interaction of subsystems in the Earth–atmosphere–ionosphere–magnetosphere system. Part of the magnetosonic wave energy can be channeled into the ionospheric F region and propagated within this region at speeds of 20–40 km/s. Having passed considerable distances, the magnetosonic wave can be transformed back into infrasound.

3.13 Occurrence of Instabilities

Above 100 km, the infrasonic wave becomes nonlinear due to the self-action effect, since here $\Delta p/p > 0.2$–0.3, and its profile transforms into a sawtooth profile [10]. On the steep front, the gradient-drift instability of the ion sound with the period $T \sim 10$ s can occur. In the presence of an ionospheric electric field with a strength of 0.1–1 mV/m in the direction perpendicular to the direction of the ion-acoustic wave, a short-period (with a period $T \ll 1$–10 s) ion-acoustic wave could be generated. The reason for this generation is the drift-dissipative instability. Its frequency can be found in the frequency range 10–100 MHz.

3.14 Effects of Acoustic and Atmospheric Gravity Waves

The movement of a meteoroid in the atmosphere leads to the generation of density waves in a wide frequency range, from acoustic frequencies ~1 kHz to frequencies $\sim 10^{-4} - 3 \times 10^{-3}$ Hz, corresponding to the internal gravity waves (IGWs) [10]. Even before the explosion of the celestial body, about 1 and 0.2% of the kinetic energy of the meteoroid was converted into the energy of acoustic E_{ac} and internal gravity E_g waves. In this case, we have $E_{ac} \approx 7.2 \times 10^{12}$ J, $E_g \approx 1.4 \times 10^{12}$ J [9].

The IGWs can be generated both directly and indirectly. In the former case, their source is the motion of the celestial body and the associated shock waves (ballistic and explosive). The size of the region of the atmosphere disturbed by shock waves is relatively small and the efficiency of IGW generation is comparatively low. In the latter case, the generation mechanism can be as follows. The relatively excessive pressure in the shock front increases with altitude. At $\Delta p/p > 0.3$, the spectrum of the shock-wave signal is enriched with higher harmonics, which are more strongly absorbed than the fundamental harmonic, the dissipation of the acoustic wave energy and heating of the atmosphere occur. The heated region of the atmosphere serves as a source of secondary IGWs. Taking into account that the IGWs are channeled in a waveguide in the ~100–200 km altitude range, the relative amplitude of the IGWs remain rather large (~3–30%) at distances of about 1000 km.

3.15 Ionospheric Effects

The ionosphere can be influenced through several channels [10]. First, it is affected through the flow of the heated material along the plasma plume. Second, it is affected by means of the shock wave. Third, the ionosphere is influenced by the electromagnetic radiation generated by the trail. Fourth, the influence is produced due to the ionization of NO molecules. The passage of the meteoroid resulted in the formation of a plasma trail, and in the disturbance of the lower and upper atmosphere. Noticeable disturbances from the explosion and the secondary processes spread horizontally over a distance of thousands of kilometers. Theoretical estimates show that the periods of secondary IGWs were 10–30 min, and the relative amplitudes were 1–10%. The relative perturbations in the electron density over the epicenter of the explosion attained tens to hundreds of percent.

3.16 Generation of an Alfvén Pulse and MHD Waves

A sharp change in the plasma conductivity at the boundary of the ionosphere arising as a result of disturbance in the frequency of electron collisions with neutrals led to the generation of the Alfvén pulse [10]. Theoretical estimates show that the pulse amplitude was ~1–10 mV/m during the passage of the Kamchatka body. The Alfvén wave propagated along the magnetic field lines, reaching the magnetosphere and the magnetically conjugate region. The unsteady plasma trail of the meteoroid, interacting with the geomagnetic field, is also a source of the magnetohydrodynamic waves that can propagate in the ionosphere and magnetosphere. As a result, the interaction of these waves with the high-energy electrons in the Earth's radiation belt and the electron precipitation into the atmosphere, where additional ionization is produced, could occur. Thus, the channels of the Kamchatka meteoroid impact on the magnetosphere could have arisen. It should be pointed out that from 11:15 to 11:45 LT on December 19, 2018 the disturbances in the geomagnetic field in the magnetosphere could take place, which were similar to those described in papers [12, 13].

3.17 Photoionization Capability

The NO molecules have a low ionization potential, and their density in the ionospheric E region can reach ~10^{14} m^{-3} [10]. At an altitude of 105 km, an ionized trail with the density ΔN of up to 10^{10} m^{-3} and the length of about 100 km could appear. In the daytime, the background value $N_0 \approx (5–10) \times 10^{10}$ m^{-3}. Then $\Delta N/N_0 \approx 0.2–0.1$. In this case, the effect of photoionization was of practical importance, which means that instabilities associated with it were produced.

3.18 Seismic Effect

Seismic effect was relatively weak due to the explosion being over the sea surface. The earthquake Richter magnitude did not exceed 1.

3.19 The Rate of Falling

The rate of falling of bodies similar to Kamchatka body, on average, is one event per 30 years.

4 Conclusions

A comprehensive analysis of the main processes in the Earth–atmosphere–ionosphere–magnetosphere system that accompanied the fall of the Kamchatka meteoroid has been conducted, and the following has been established.

1. The celestial body was almost completely destroyed in the altitude range of about 30 km. The longitudinal and transverse sizes of the luminous formation were about 4 and 2 km, respectively. Its temperature was about 4600 K. The excess pressure near the Earth's surface under the epicenter of the explosion was about 1 kPa. The relative perturbations of air pressure at ionospheric heights above the epicenter of the explosion reached tens of percent. The energy and power of the blast shock wave were close to 580 TJ and 58 TW. The electric, magnetic, and electromagnetic effects, as well as turbulence and meteoroid plume effects, were significant.
2. The passage and explosion of the Kamchatka meteoroid caused noticeable (or strong) disturbances in all geospheres. The passage of the Kamchatka body was accompanied by mechanical, gas-dynamic, thermodynamic, optical, plasma, magnetic, electrical, electromagnetic, and acoustic effects, by the generation of shock, infrasonic, atmospheric gravity, electromagnetic, magnetohydrodynamic, ion-acoustic, and magnetosonic waves, as well as by instabilities. The seismic effect was insignificant.

Acknowledgements Work was supported by the National Research Foundation of Ukraine for financial support (project 2020.02/0015). Work also was partially supported by Ukraine state research projects #0119U002538, #0121U109881, and #0121U109882.

References

1. Adushkin, V., Nemchinov, I. (Eds.): Catastrophic Events Caused by Cosmic Objects. Springer, Netherlands (2008)
2. Popova, O.P., Jenniskens, P., Emelyanenko, V., et al.: Chelyabinsk airburst, damage assessment, meteorite recovery, and characterization. Science **342**, 1069–1073 (2013)
3. Center for NEO Studies (CNEOS). https://cneos.jpl.nasa.gov/fireballs/. Accessed 22 Mar 2021
4. Bronshten, V.A.: Physics of Meteor Phenomena. Springer, Netherlands (1983)
5. Le Pichon, A., Blanc, E., Hauchecorne, A. (Eds.): Infrasound Monitoring for Atmospheric Studies. Springer, Dordrecht Heidelberg London New York (2019)
6. Chernogor, L.F.: Physical Effects of the Lipetsk Meteoroid: 1. Kinemat. Phys. Celest. Bodies **35**(4), 174–188 (2019)
7. Chernogor, L.F.: Dynamics of the convective rise of thermals in the atmosphere. Izv. Atmos. Ocean Phys. **54**(6), 528–535 (2018)
8. Chernogor, L.F., Mylovanov, Yu.B.: Rise of a meteoroid thermal in the Earth's atmosphere. Kinemat. Phys. Celest. Bodies **34**(4), 198–206 (2018)
9. Chernogor, L.F.: Physical effects of the Lipetsk meteoroid: 2. Kinemat. Phys. Celest. Bodies **35**(5), 217–230 (2019).
10. Chernogor, L.F.: Physical effects of the Lipetsk meteoroid: 3. Kinemat. Phys. Celest. Bodies **35**(6), 271–285 (2019).
11. Raizer, Yu.L.: A debate over the acquisition of an electric potential by a meteoroid. Sol. Syst. Res. **37**(4), 333–335 (2003)
12. Chernogor, L.F.: Magnetospheric effects during the approach of the Chelyabinsk meteoroid. Geomag. Aeron. **58**(2), 252–265 (2018)
13. Chernogor, L.F.: Effects of the Lipetsk meteoroid in the geomagnetic field. Geomag. Aeron. **60**(3), 355–372 (2020)

Features of Ionospheric and Magnetic Effects of August 5–6, 2019 Noticeable Geospace Storm Over China and Ukraine

Leonid F. Chernogor, Yu. Zheng, Q. Guo, Y. Luo, K. P. Garmash, and V. T. Rozumenko

Abstract An increase in the main parameters of the solar wind on August 5, 2019 led to a geospace storm, which was observed mainly on August 5 and 6, 2019. The main phase of the magnetic storm took place on August 5, 2019 from 06:00 UT to 08:30 UT. The recovery phase lasted at least 4 days. The magnetic storm manifested itself in significant variations in all components of the geomagnetic field, an increase by an order of magnitude in the level of oscillations of the geomagnetic field in the range of 400–950 s. During the ionospheric storm, the ionospheric F-region was significantly disturbed. The ionospheric E-region remained weakly perturbed. The ionospheric storm significantly affected the Doppler spectra of radio waves in the 5–10 MHz frequency range. The Doppler spectra exhibited significant broadening, and variations in the Doppler shift of frequency and its quasi-periodic changes occurred with a period of 20–40 min and duration of 120–240 min.

Keywords Geospace storm · Ionospheric storm · Magnetic storm · Oblique sounding of the ionosphere · Variations in the Doppler shift of frequency · Quasi-periodic variations · Storm's main phase

1 Introduction

Solar storms are the cause of the geospace storms. Solar flares, coronal mass ejections, and high-speed streams lead to various effects in the Sun–interplanetary-medium–magnetosphere–ionosphere–atmosphere–Earth (inner shells) system (SIMMIAE).

L. F. Chernogor (✉) · Y. Luo · K. P. Garmash · V. T. Rozumenko
School of Radiophysics, Biomedical Electronics and Computer Systems, V. N. Karazin Kharkiv National University, 4 Svobody Sq., Kharkiv 61022, Ukraine

L. F. Chernogor · Yu. Zheng
College of Electronic and Information Engineering, Qingdao University, 308 Ningxia Road, Qingdao 266071, People's Republic of China

L. F. Chernogor · Q. Guo
College of Information and Communication Engineering, Harbin Engineering University, 145 Nantong Street, Harbin 150001, People's Republic of China

A. Kosterov et al. (eds.), *Problems of Geocosmos—2020*, Springer Proceedings in Earth and Environmental Sciences, https://doi.org/10.1007/978-3-030-91467-7_28

As a result, geospace storms arise, which are comprised of synergistically interacting storms in the magnetic field, the ionosphere, the atmosphere, and in an electric field of magnetospheric, ionospheric, and atmospheric origins [1, 2]. The effects of geospace storms are highly diverse and depend on the characteristics of the solar storm, the solar wind parameters, the state of the geospheres, the position in the solar activity cycle, the time of year and day, the means used to observe, and on the geographic and geomagnetic coordinates of the observation site [1–9]. A statistical approach to the study of ionospheric storms was used in [10, 11]. The generalization of the results of the study of the relationship between solar, magnetic, and ionospheric storms was conducted in [1–9]. The results of recent studies of the effects of ionospheric storms are presented, for example, in papers [12–18]. Despite significant achievements, the study of the effects of each new geospace storm still remains an urgent task.

The aim of the work is to present general information about the geospace storm, and the results of the analysis of the features of August 5–6, 2019 magnetic and ionospheric storms.

2 Instrumentation and Measurement Techniques

Consider the means of observation used in this study.

2.1 The Fluxmeter–Magnetometer

The fluxmeter–magnetometer is situated at the V. N. Karazin Kharkiv National University Magnetometer Observatory (49.64°N, 36.93°E). It makes measurements of fluctuations in the H and D components of the geomagnetic field in the range of $T =$ 1–1000-s periods with 0.5-s temporal resolution. The sensitivity of the magnetometer is of the order of 1 pT–1 nT for $T \approx$ 1–1000 s.

2.2 Magnetometer

The meteomagnetic station LEMI-017 is situated at the RI NASU Low-Frequency Observatory (49.93°N, 36.95°E) [http://geospace.com.ua/data/metmag_mf.php]. It acquires measurements of the H, D, Z components of the magnetic field with 1-s temporal resolution. The magnetometer sensitivity is 10 pT.

2.3 Multi-Frequency Multiple Path HF Doppler Radio System for Oblique-Incidence Sounding the Ionosphere

Multi-Frequency Multiple Path HF Doppler Radio System for Oblique-Incidence Sounding of the Ionosphere is situated at the Harbin Engineering University campus (People's Republic of China, 45.78°N, 126.68°E) [16]. The system conducts continuous monitoring of the state of the ionosphere on 14 radio paths (Fig. 1). The radio transmissions from broadcasting radio stations situated in the People's Republic of China, the Republic of Korea, Japan, the Russian Federation, and Mongolia in the frequency range 5–10 MHz are used. The length of the radio propagation paths is about 1–2 thousand km. Signal reception and processing is performed at the Harbin Engineering University on campus.

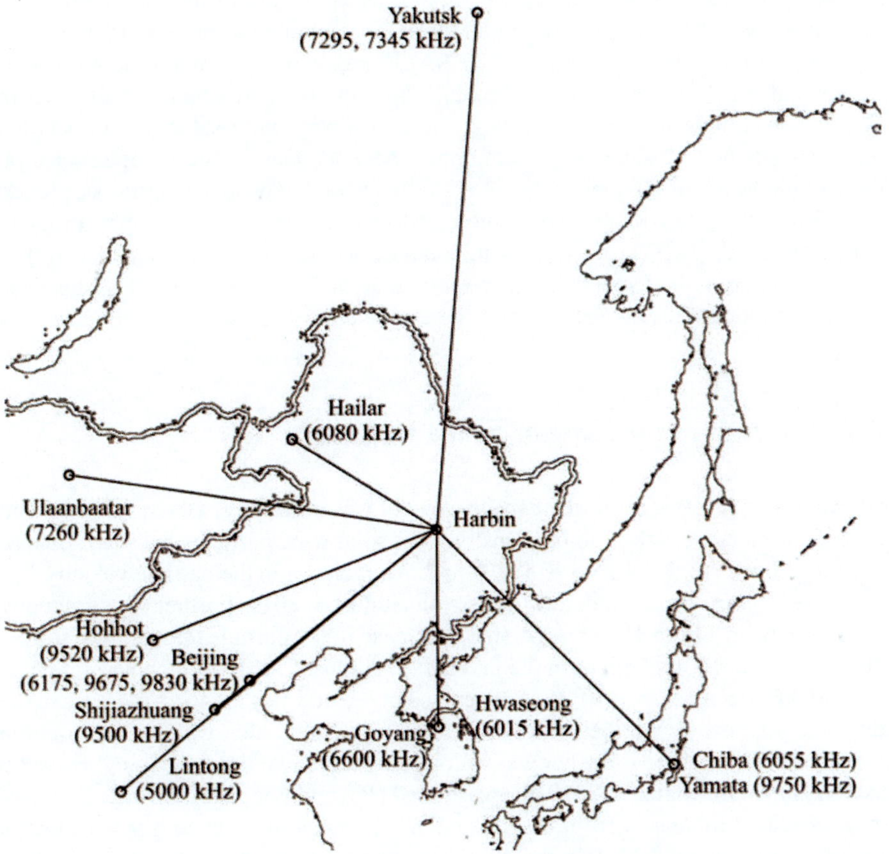

Fig. 1 Layout of the propagation paths used for monitoring dynamic processes acting in the ionosphere

2.4 Ionosonde

Ionosonde is situated in Japan near the city of Wakkanai (URSI code WK546, 45.16°N, 141.75°E) [http://wdc.nict.go.jp/IONO/HP2009/contents/Ionosonde_Map_E.html].

2.5 Analysis Methods

The signal level recorded by the fluxmeter–magnetometer was subjected to systems spectral analysis, which simultaneously uses the short-time Fourier transform, the Fourier transform in a sliding window with a width adjusted to be equal to a fixed number of harmonic periods adaptive, as well as the wavelet transform employing the Morlet wavelet as a basis function. Then, the obtained spectra were analyzed. The three-component magnetometer RI NASU was used for general monitoring of the state of the geomagnetic field. The data from the multi-frequency multiple path system for oblique-incidence sounding the ionosphere were subjected to detailed processing aimed at obtaining the temporal dependences of the Doppler spectra, the Doppler shift of frequency $f_D(t)$ of the main ray, and of the signal amplitude $A(t)$. Further, $f_D(t)$ and $A(t)$ were subjected to secondary processing in order to obtain the trends $\overline{f_D}(t)$ and $\overline{A}(t)$, the fluctuations $\delta f_D(t) = f_D(t) - \overline{f_D}(t)$, $\delta A(t) = A(t) - \overline{A}(t)$, as well as spectra in the period range of $T \approx 1$–60 min and greater. The systems spectral analysis was also used to evaluate the spectra.

2.6 Analysis of the State of Space Weather

On August 2–4, 2019, the state of space weather was quiet (Fig. 2). On August 4 and 5, 2019, an increase in the number density of the solar wind particles n_{sw} was observed to change from ~ 2.5×10^6 to 10×10^6 m^{-3}. An increase in the particle velocity V_{sw} was noted to vary from ~ 350 to 715 km/s on August 5, 2019. Further, V_{sw} gradually decreased from 715 to 450 km/s. A sharp increase in the particle temperature from ~ 10^4 to 5.5×10^5 K took place in the first half of August 5, 2019, which was followed by a gradual decrease to ~ 10^5 K. An increase in n_{sw} and V_{sw} led to an increase in the dynamic pressure p_{sw} in the solar wind from ~ 0.5 to 8.5 nPa, which was observed on August 5, 2019, and it further was followed by a gradual decrease in p_{sw} from 8.5 to (1–2) nPa. On August 5, 2019, the azimuthal (B_y) component of the interplanetary magnetic field increased from ~ 0 to 11.1 nT. Almost simultaneously with this, a decrease from 0 to –7.5 nT in the B_z component was observed to occur. All signs of a magnetic storm have appeared. Powerful bursts of energy per unit time (up to 20 GJ/s), coming from the solar wind into the magnetosphere, were observed. From about 00:00 UT to 04:00 UT on August 5, 2019, the magnitude of D_{st} index values

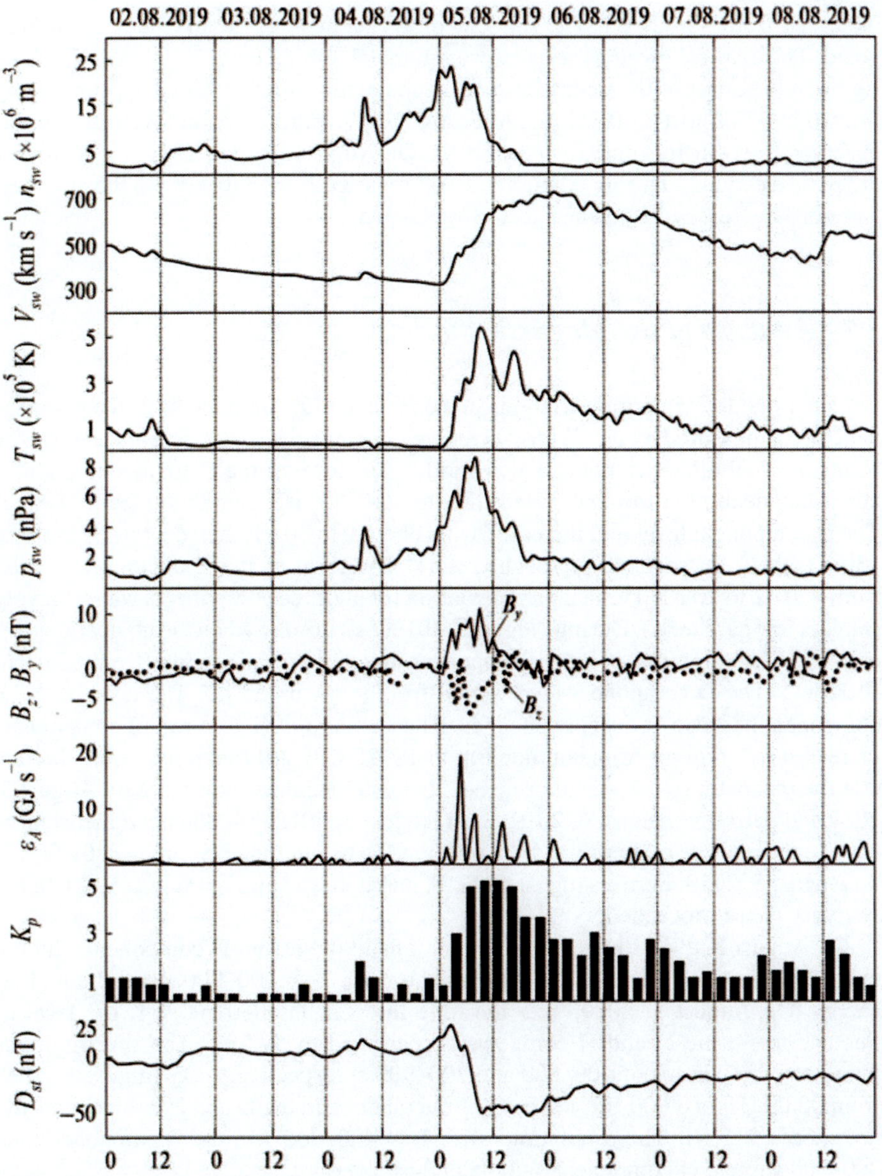

Fig. 2 Universal time dependencies of the solar wind parameters: proton number density n_{sw}, temperature T_{sw}, plasma flow speed V_{sw} (retrieved from https://omniweb.gsfc.nasa.gov/form/dx1.html), calculated dynamic pressure p_{sw}, components B_z and B_y of the interplanetary magnetic field (retrieved from https://omniweb.gsfc.nasa.gov/form/dx1.html), calculated energy input per unit time, ε_A, from the solar wind into the Earth's magnetosphere; K_p- and D_{st}-index (retrieved from https://omniweb.gsfc.nasa.gov/form/dx1.html) for 02–08 August 2019 period. Dates are shown along the upper *abscissa* axis

increased from 0 to 27 nT. The sudden commencement of the magnetic storm was noted. This was followed by a sharp drop in the D_{st} index from 27 to –53 nT, which signified the onset of the main phase of the magnetic storm. It lasted approximately from 05:00 UT to 08:00 UT on August 5, 2019. Further, a recovery phase was observed to occur for more than four days. On August 3 and 4, 2019, the K_p-index did not exceed 1–2. During August 5, 2019, it varied from 3 to 5.3. On the next day, a gradual decrease in K_p values to 1–3 was noted.

2.7 Analysis of the Magnetic Storm

Consider first the temporal variations in the H, D, and Z components of the geomagnetic field, measured by the RI NASU magnetometer (Fig. 3). On reference days, the behavior of all three components was similar. The level of the H component, fluctuating, first decreased from positive values to –(10–12) nT, then increased to 0–8 nT. The D component level first increased from about 0 to 30 nT, then decreased to about –30 nT, then, fluctuating, approached 0 nT. The level of the Z component varied from –20 nT to 20 nT. The standard deviation for all three components was relatively small (a few nanotesla). During August 5, 2019, i.e., the day when the magnetic storm occurred, the variations in the H component increased considerably (from 60 nT to –100 nT). The D component level varied from ~ 40 to –55 nT. The level of the Z component has also noticeably changed. The standard deviation for all components increased by an order of magnitude (up to 10–20 nT). All this represented the fact that the magnetic storm was in progress. Its manifestations took place on August 5, 2019 and partly on August 6, 2019. If on August 5, 2019, a sudden commencement and the main phase of the storm were observed to occur, then beginning at 09:00 UT on August 5, 2019, during August 6, 2019, and until the end of August 9, 2019, the recovery phase proceeded.

On August 2, 2019, the level of fluctuations in the D and H components did not exceed ~ 0.5–1 nT (Fig. 4). The components with $T \approx$ 800–900 s had the highest energy. On August 5, 2019, over the time interval 03:00–07:00 UT, the level of fluctuations in the D and H components increased to 5–7 nT. The spectrum was dominated by periods of 400–600 and 700–900 s, respectively. On August 6, 2019, from 00:00 UT to 07:00 UT, the level of fluctuations in the D and H components did not exceed 2–3 nT. The oscillations with $T \approx$ 400–800 s in the D component and 700–900 s in the H component had the highest energy.

2.8 Analysis of the Ionospheric Conditions

Analysis of the temporal variations in the ionogram parameters have shown that on August 2, 5, and 6, 2019, the $f_{\min}$ frequency fluctuated around 1.5 MHz (Fig. 5). On the same day, the f_{oE} frequency in the daytime reached 3.0–3.3 MHz, and in the

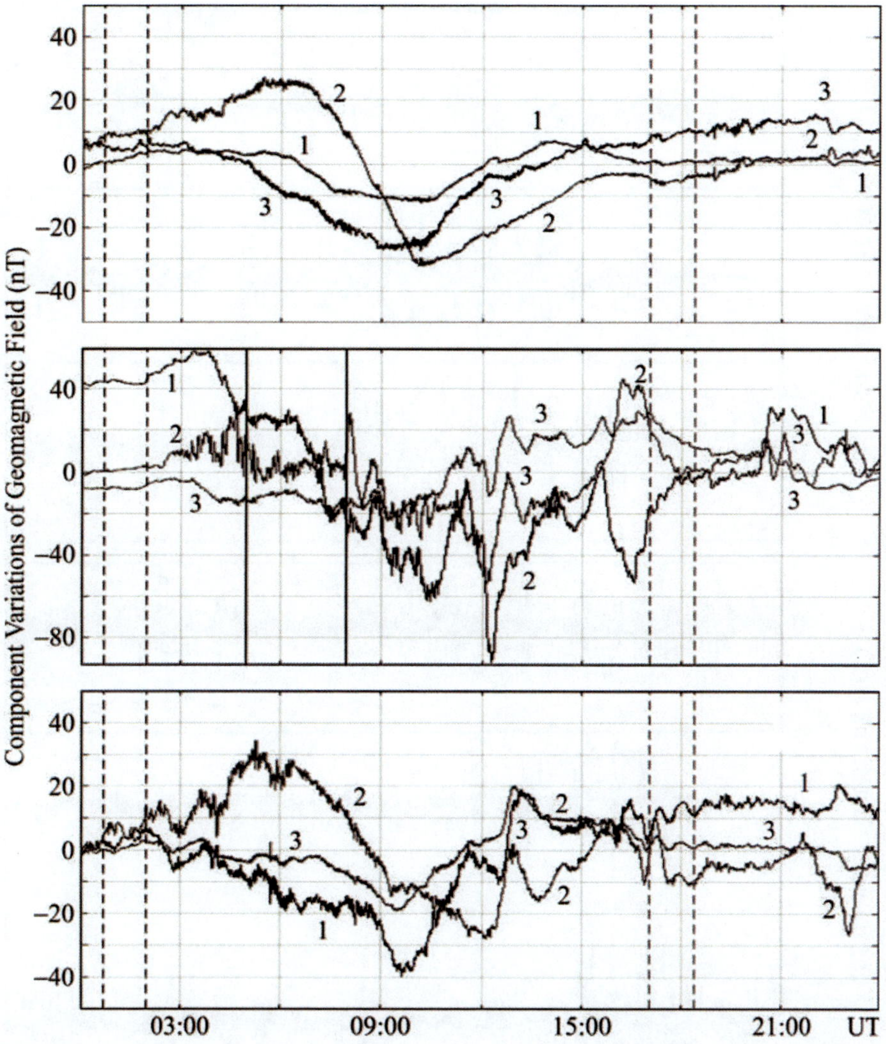

Fig. 3 H (1), D (2), Z (3) components for August 2, 2019, August 5, 2019, and August 6, 2019 (retrieved from http://geospace.com.ua/en/observatory/metmag.html). The solid vertical lines show the time interval corresponding to the main phase of the magnetic storm; dashed lines correspond to the sunrise and sunset moments at the ground and the ionosphere level

morning and in the evening it was observed to be about 2 MHz. The f_{oEs} frequency varied over a wide range, from ~ 3 to 14–16 MHz at some instants of time. During the main phase of the magnetic storm, i.e., approximately over the interval 05:00–09:00 UT on August 5, 2019, f_{oEs} increased from 4 to 8 MHz. The f_{oF2} frequency could be measured very rarely. During the main phase of the magnetic storm, it showed fluctuations from ~ 5 to 10 MHz. During the recovery phase, $f_{oF2} \approx 6$ MHz. After 13:30 UT on August 5, 2019, it dropped to 3 MHz.

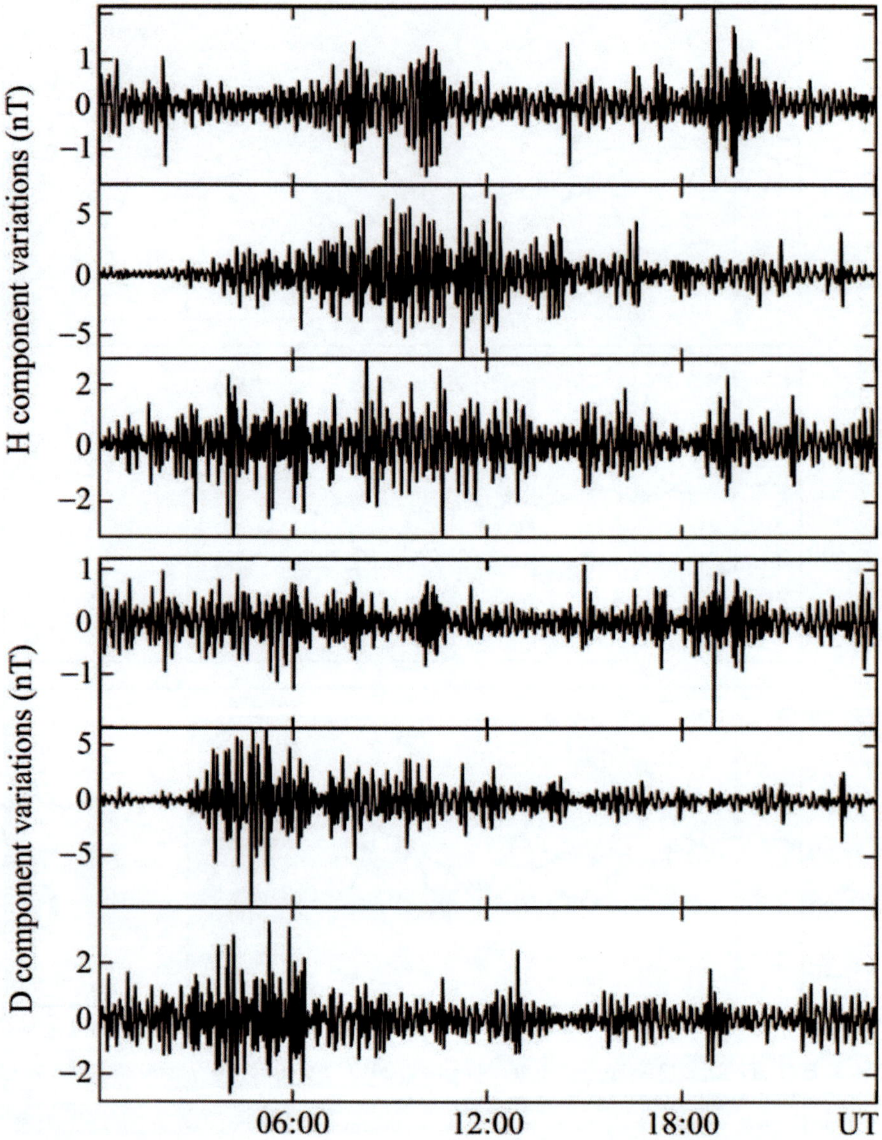

Fig. 4 Magnetic field variations at the V. N. Karazin Kharkiv National University Magnetometer Observatory for August 2, 2019, August 5, 2019, and August 6, 2019 (panels from top to bottom)

The h'_E height varied from 90 to 120–130 km. During the main phase of the magnetic storm, it showed an increase from ~ 95 to 120 km, which continued for about 60 min (Fig. 6). On August 5, 2019, the h'_{Es} height showed fluctuations from 90 to 120–130 km. Measurements of the h'_{F2} height were made very few times. During the main phase of the magnetic storm, the h'_{F2} height showed a sharp increase from

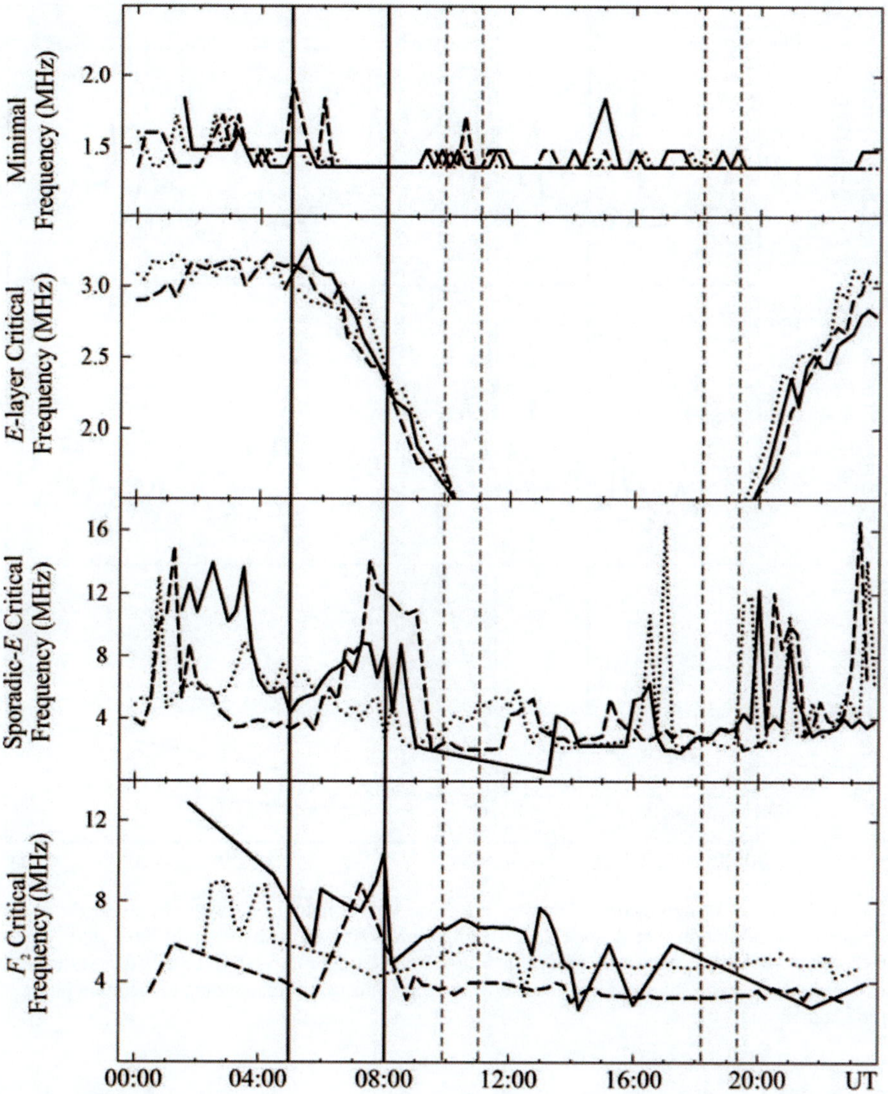

Fig. 5 Temporal variations on ionograms in minimal frequency, E-layer critical frequency, E_S-layer critical frequency, and F_2-layer critical frequency for August 2, 2019 (dotted line), August 5, 2019 (solid line), and August 6, 2019 (dashed line) (retrieved from https://lgdc.uml.edu/common/DIDBYearListForStation?ursiCode=MO155). The solid vertical lines show the time interval corresponding to the main phase of the magnetic storm

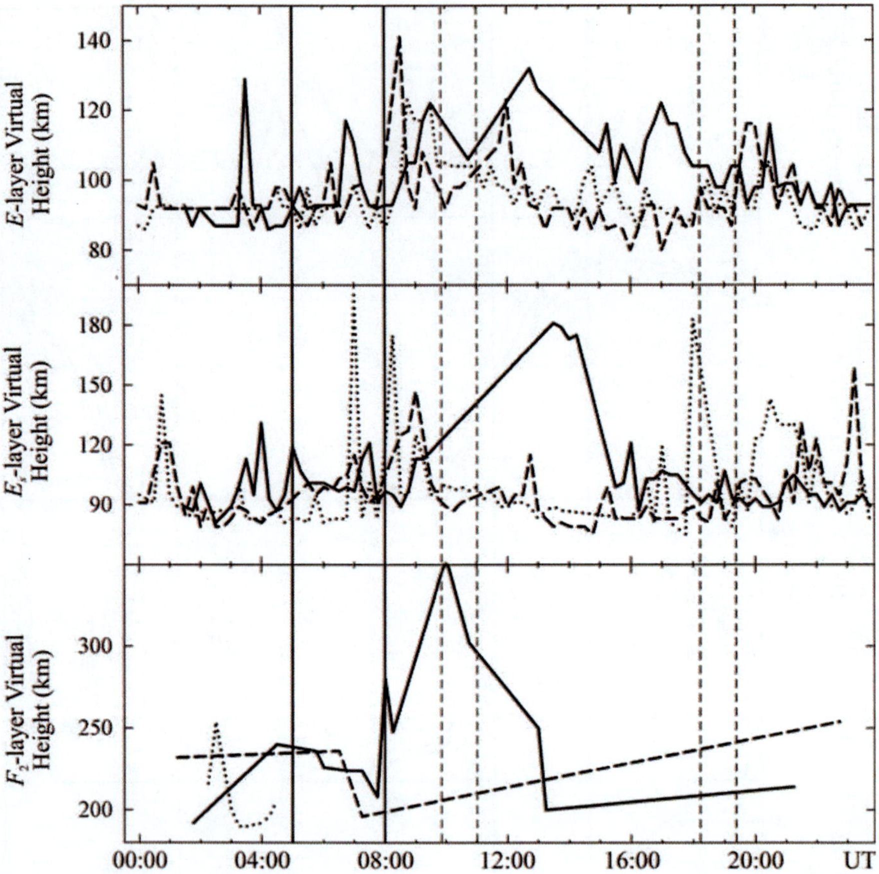

Fig. 6 Temporal variations on ionograms in E-layer virtual height, E_s-layer virtual height and in F_2-layer virtual height for August 2, 2019 (dots line), August 5, 2019 (solid line), and August 6, 2019 (dashed line) (retrieved from https://lgdc.uml.edu/common/DIDBYearListForStation?ursiCode=MO155). The solid vertical lines show the time interval corresponding to the main phase of the magnetic storm

210 to 280 km. Thus, the ionospheric storm relatively little affected the ionospheric E region, but strongly affected the state of the ionospheric F region.

3 Oblique-Incidence Sounding of the Ionosphere

Since radio transmitters do not emit continuously, measurements only from 10 out of the 14 radio paths have been used for further analysis. For technical reasons, as a reference day was selected August 2, 2019 instead of August 4, 2019.

3.1 Lintong/Pucheng to Harbin Radio Propagation Path

The radio station, transmitting at a frequency of 5000 kHz, is situated in the People's Republic of China. The length of the radio path along the Earth's surface is $R \approx 1875$ km. The radio station did not transmit until 06:00 UT. Analysis of the universal time dependences of the Doppler spectra and of the signal amplitude observed on a day with a magnetic storm (August 5, 2019) and two reference days (August 2 and 6, 2019) has shown that the Doppler spectra were predominantly spread on August 2, 5, and 6, 2019. On the day of the storm, the spread increased considerably. Significant changes in the nature of the Doppler spectra were observed in the vicinity of the dusk terminator (from 10:00 UT to 14:00 UT), as well as from 14:00 UT to 20:00 UT. The Doppler spectra covered the range from –2.5 to 2–2.5 Hz. During the storm, the Doppler shift of frequency $f_D(t)$ of the main ray fluctuated mainly in the range from –0.5 to 0.5 Hz. Over the time interval 06:20–09:20 UT on August 5, 2019, considerable (up to 10 dBV) variations in the amplitude $A(t)$ of the signal were observed to occur, which were absent on the reference days.

3.2 Hwaseong to Harbin Radio Propagation Path

The radio station, operating at a frequency of 6015 kHz, is situated in the Republic of Korea at $R \approx 950$ km from the receiver. The station was turned off over the time interval 00:00–03:30 UT. On August 2, 2019, the Doppler spectra were spread mainly near the dawn and dusk terminators (Fig. 7). On August 6, 2019, the Doppler spectra were practically single-ray. On August 5, 2019, the spread increased considerably over the time intervals 07:00–14:00 UT and 17:00–19:30 UT. In addition, from 10:00 UT to 14:00 UT, the quasi-periodic variations of $f_D(t)$ were recorded with a period $T \approx 20$ min and an amplitude $f_{Da} \approx 0.2$ Hz. From 07:00 UT to 11:00 UT on August 5, 2019, the signal amplitude $A(t)$ showed considerable (up to 15 dBV) variations, but on the reference days, such variations were absent.

3.3 Chiba/Nagara to Harbin Radio Propagation Path

The radio transmitter at a frequency of 6055 kHz is situated in Japan at $R \approx 1610$ km from the receiver. The transmitter was turned off from 15:00 UT to 22:00 UT. On August 2, 2019, over the time interval 06:00–15:00 UT, the Doppler spectra were spread. On the day when the ionospheric storm occurred, the spread markedly increased. From about 09:00 UT to 15:00 UT, the Doppler spectra occupied the frequency band from –1.5 to 1.5 Hz. On August 6, 2019, the Doppler spectra were practically single-ray. The Doppler shift of frequency in the vicinity of the dusk terminator varied from –0.5 to 0.5 Hz. On August 5, 2019, from 06:00 UT to 15:00

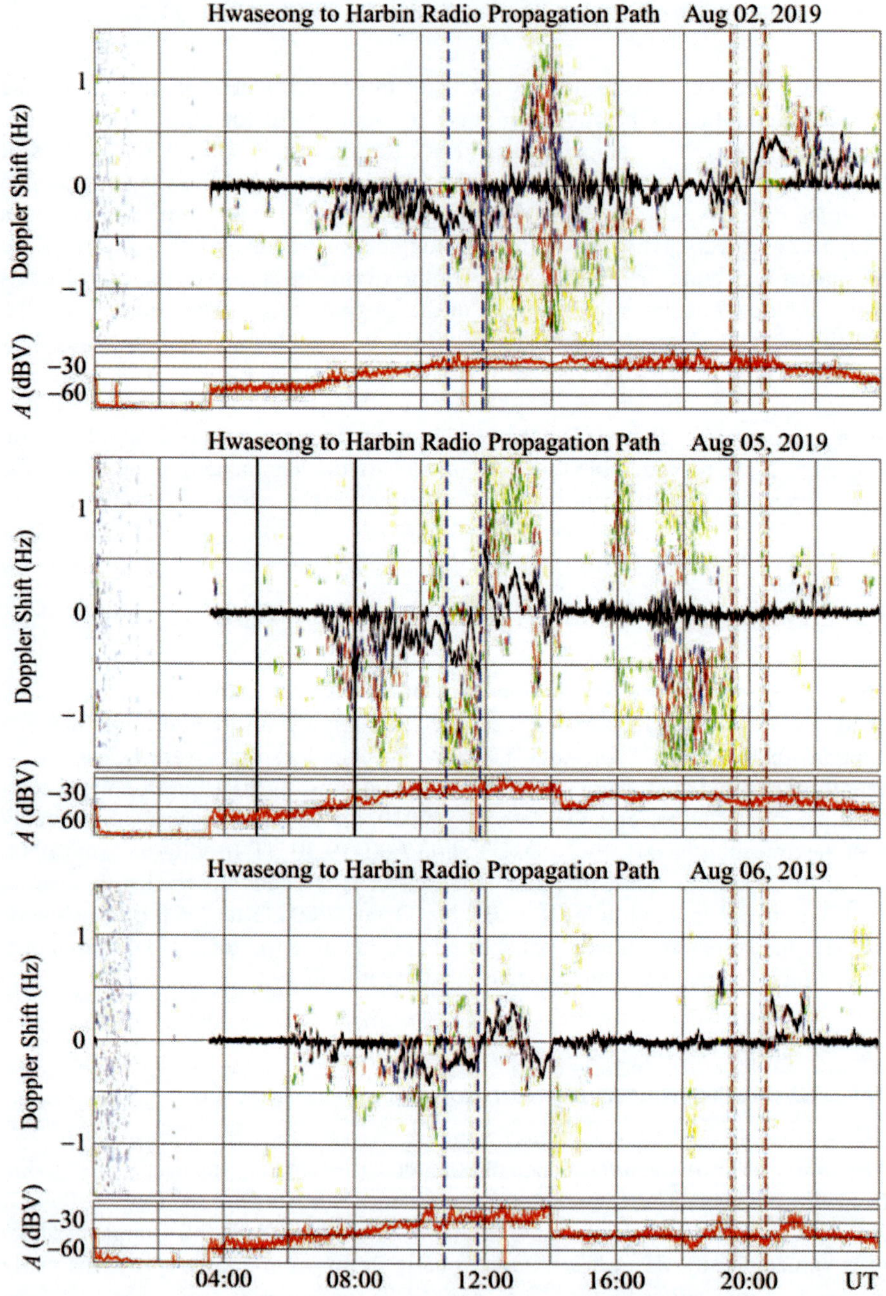

Fig. 7 Example of temporal variations of the Doppler spectra and signal amplitudes at the Hwaseong to Harbin radio path on August 2, 5, and 6, 2019 (panels from *top* to *bottom*). The *solid vertical lines* show the time interval corresponding to the main phase of the magnetic storm, and the *dashed lines* show the times of sunset and sunrise at ground level and at an altitude of 100 km

UT, variations in $A(t)$ occurring within 10 dBV were quasi-periodic. The harmonics with $T \approx 40$ min predominated. This effect was absent on the reference days.

3.4 Beijing to Harbin Radio Propagation Path

The radio station, operating at a frequency of 6175 kHz, is situated in the People's Republic of China at $R \approx 1050$ km from the receiver. The radio station stopped broadcasting from 00:00 UT to 07:00 UT and from 18:05 UT to 20:10 UT. On August 5, 2019, in contrast to the reference days, the Doppler spectra showed a greater broadening. In addition, over the time interval 11:00–14:00 UT, $f_D(t)$ varied according to a quasi-periodic law with $T \approx 24$ min, $f_{Da} \approx 0.20$–0.25 Hz, and duration $\Delta T \approx 180$ min. The variations in $A(t)$ on the day when the ionospheric storm occurred, and on the reference days did not practically differ.

3.5 Goyang to Harbin Radio Propagation Path

The radio station transmitting at a frequency of 6600 kHz is situated in the Republic of Korea at $R \approx 910$ km from the receiver. The signal reception did not occur from 00:00 UT to 09:00 UT. On the day of the ionospheric storm, the Doppler spectra were noticeably different from the spectra on the reference days. First, they were spread more. Second, over the time interval 09:00–14:00 UT, the variations in $f_D(t)$ were quasi-periodic with $T \approx 24$ min and $f_{Da} \approx 0.20$–0.25 Hz. Their duration was at least 240 min. No differences in the character of variations in $A(t)$ have been found.

3.6 Ulaanbaatar to Harbin Radio Propagation Path

The radio transmitter operating at a frequency of 7260 kHz is situated in the area of Mongolia at $R \approx 1496$ km from the receiver. The transmitter was turned off from 05:00 UT to 07:00 UT and from 18:00 UT to 20:25 UT. On August 5, 2019, the behavior of the Doppler spectra was significantly different from the behavior of the spectra on the reference days. The Doppler spectra were spread. They occupied the frequency band from –1.5 to 1.5 Hz. In addition, over the time interval 08:00–10:00 UT, quasi-periodic variations in $f_D(t)$ took place with a period of $T \approx 25$–30 min, $f_{Da} \approx 0.15$–0.20 Hz, and $\Delta T \approx 120$ min. The behavior of $A(t)$ on August 2, 5, and 6, 2019 showed no significant differences.

3.7 Yakutsk to Harbin Radio Propagation Path

The radio station at a frequency of 7345 kHz, is situated on the area of the Russian Federation. In this case, the length of the radio path is close to 1045 km. The radio station transmitted over the time interval 11:00–18:05 UT. On the day of the ionospheric storm, the Doppler spectra were spread more than on the reference days. Variations in $A(t)$ on August 5, 2019 reached 6–12 dBV.

3.8 Shijiazhuang to Harbin Radio Propagation Path

The radio transmitter at a frequency of 9500 kHz is situated in the People's Republic of China at $R \approx 1310$ km from the receiver. On the ionospheric storm day, the turn-offs of the radio transmitter were short-lived. The behavior of the Doppler spectra on August 5, 2019 markedly differed from the behavior of the spectra on the reference days (Fig. 8). Over the time interval 09:00–14:00 UT, quasi-periodic variations in $f_D(t)$ with $T \approx 30$ min, $f_{Da} \approx 0.1$–0.3 Hz, and $\Delta T \approx 240$ min were observed. The variations in $A(t)$ on August 2, 5, and 6, 2019, were not much different.

3.9 Hohhot to Harbin Radio Propagation Path

The radio station at a frequency of 9520 kHz, is situated in the People's Republic of China at $R \approx 1340$ km from the receiver. On the day when the ionospheric storm occurred, the station was shut down from 16:05 UT to 21:40 UT. On August 5, 2019, the variations in the Doppler spectra significantly differed from the variations in the spectra on the reference days. From 10:00 UT to 14:00 UT, anomalously large (from –0.7 to 0.7 Hz) quasi-periodic variations in $f_D(t)$ took place. In this case, $T \approx 40$ min, and $\Delta T \approx 240$ min. The variations of $A(t)$ on August 2, 5, and 6, 2019 showed no significant differences.

3.10 Yamata to Harbin Radio Propagation Path

The radio transmitter at a frequency of 9750 kHz is situated in Japan at $R \approx 1570$ km from the receiver. The radio transmissions were absent for almost half a day. From 08:00 UT to 16:00 UT on the reference day, the Doppler spectra were virtually single-ray. On August 6, 2019, from 10:00 UT to 14:00 UT, a ray from another radio station was observed. On August 5, 2019, the Doppler spectra were spread, and they fluctuated in the frequency range from –1.5 to 1 Hz. The behavior of the amplitudes on August 2, 5, and 6, 2019, differed insignificantly.

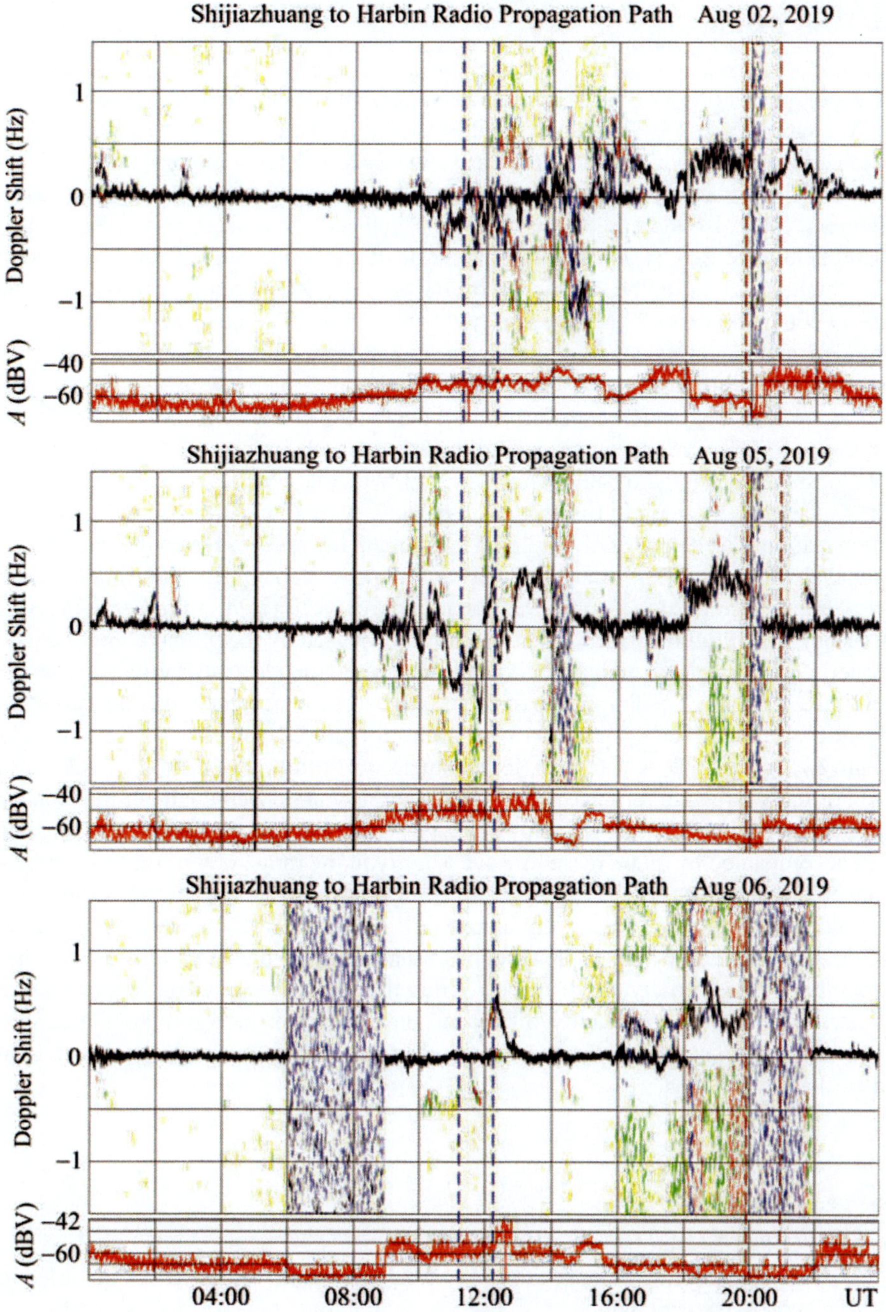

Fig. 8 Same as in Fig. 7, but for the Shijiazhuang to Harbin Radio Propagation Path

4 Discussion

The geospace storm of August 5–6, 2019 was one of the most noticeable in intensity and duration in 2019. The geospace storm led to considerable disturbances in the geomagnetic field, which lasted for at least two days. The level and the standard deviation of all components of the geomagnetic field markedly increased. The fluctuation frequency has also increased considerably. The level of fluctuations in the geomagnetic field increased by an order of magnitude in the period range from 1 to 1000 s. The highest amplitude was observed for harmonics with a period of 400–600 s for the D component and 700–900 s for the H component. The geospace storm caused considerable disturbances in the ionospheric F region, especially during the movement of the solar terminator across the ionosphere, virtually without affecting the ionospheric E region. These disturbances affected the characteristics of radio waves in the 5–10 MHz frequency range. A noticeable broadening of the Doppler spectra was observed on all radio paths. This indicated an increase in plasma turbulence in the ionosphere, an increase in the scattering of radio waves, which led to multiple path propagation. The behavior of the signal in the main ray reflected from the ionosphere noticeably changed. The quasi-periodic variations in the Doppler shift of frequency with periods from 20 to 40 min and durations from 120 to 240 min were observed on a number of radio paths. These variations are caused by the propagation of atmospheric gravity waves, which periodically modulate the electron number density N in the ionosphere. The amplitude of the relative disturbances in N can be estimated from the relation for δ_{Na} [16]. For example, for the Ulaanbaatar to Harbin radio path, $\delta_{Na} \approx$ 10–16%, and for the Shijiazhuang to Harbin radio path, $\delta_{Na} \approx$ 3–10%. Abnormally large amplitudes of the N oscillations were observed along the Hohhot to Harbin radio path, with $\delta_{Na} \approx$ 80–90%. It has been confirmed that geospace storms are accompanied by an increase in wave activity in the range of atmospheric gravity waves. These waves are generated in the auroral oval region and propagated to mid-latitudes. On most radio paths, the ionospheric storm did not considerably affect the signal amplitude. This means that the lower ionosphere, which determines the signal amplitude, was virtually not disturbed during this ionospheric storm. In general, the ionospheric storm significantly influenced the effects of the terminator. It can be argued that a synergistic interaction between two powerful sources of an ionospheric disturbance occurred.

5 Conclusions

1. An increase in the main parameters of the solar wind on August 5, 2019 led to a geospace storm, which was observed mainly on August 5, and 6, 2019.
2. The main phase of the magnetic storm occurred on August 5, 2019, from 05:00 UT to 08:00 UT. The recovery phase lasted at least 4 days.

3. The magnetic storm manifested itself in considerable variations in all components of the geomagnetic field, and in an increase in the level of oscillations of the geomagnetic field by an order of magnitude in the period range of 400–950 s.
4. The ionospheric F region over the People's Republic of China showed a considerable disturbance during the ionospheric storm. The ionospheric E region remained virtually unperturbed.
5. The ionospheric storm significantly affected the Doppler spectra of radio waves in the frequency range 5–10 MHz. The Doppler spectra exhibited considerable broadening. Variations in the Doppler shift of frequency and its quasi-periodic changes occurred with a period of 20–40 min and duration of 120–240 min.
6. Quasi-periodic variations of the Doppler shift of frequency were due to quasi-periodic variations in the electron density, the amplitudes of which varied from 3 to 16%. These variations were caused by the geospace storm.
7. On one of the paths, the amplitude of the Doppler shift of frequency reached 0.7 Hz. In this case, δ_{Na} could reach 80–90%.
8. The ionospheric storm only slightly affected the signal amplitude on most of radio paths.
9. In the course of the terminator impacting the ionosphere, the wave activity was enhanced. The effects from the storm and the terminator were superimposed on each other.

Acknowledgements Work by L. F. Chernogor and Y. Luo was supported by the National Research Foundation of Ukraine for financial support (project 2020.02/0015). Work by K. P. Garmash, V. T. Rozumenko and by L. F. Chernogor was partially supported by Ukraine state research projects #0119U002538, #0121U109881 and #0121U109882. Work by Q. Guo was supported partially by National key R&D plan strategic international science and technology cooperation and innovation (2018YFE0206500).

References

1. Chernogor, L.F., Domnin, I.F.: Physics of Geospace Storms. V. N. Karazin Kharkiv National University Publ, Kharkiv, Ukraine (2014) (in Russian)
2. Chernogor, L.F.: Physics of geospace storms. Space Sci. Tech. **27**(1), 3–77 (2021) (in Ukrainian with English abstract)
3. Prölss, G.W.: Ionospheric F-region Storms. Handbook of Atmospheric Electrodynamics 2. CRC Press, Boca Raton, pp. 195–248 (1995)
4. Buonsanto, M.: Ionospheric storms – A review. Space Sci. Rev. **88**(3–4), 563–601 (1999)
5. Song, P., Singer, H., Siscoe, G. (eds.): Space Weather (Geophysical Monograph). Union, Washington D.C. (2001)
6. Carlowicz, M.J., Lopez, R.E.: Storms from the Sun. Joseph Henry Press, Washington D.C. (2002)
7. Goodman, J.M.: Space Weather and Telecommunications. Springer, Boston, MA (2005)
8. Bothmer, V., Daglis, I.: Space Weather: Physics and Effects. Springer, Berlin (2007)
9. Lilensten, J., Bornarel, J.: Space Weather Environment and Societies. Springer, Dordrecht (2006)

10. Vijaya Lekshmi, D., Balan, N., Tulasi Ram, S., Liu, J.Y.: Statistics of geomagnetic storms and ionospheric storms at low and mid latitudes in two solar cycles. J. Geophys. Res. **116**, A11328 (2011)
11. Yakovchouk, O.S., Mursula, K., Holappa, L., Veselovsky, I.S., Karinen, A.: Average properties of geomagnetic storms in 1932–2009. J. Geophys. Res. **117**, A03201 (2012)
12. Liu, J., Wang, W., Burns, A., Yue, X., Zhang, S., Zhang, Y., Huang, C.: Profiles of ionospheric storm-enhanced density during the 17 March 2015 great storm. J. Geophys. Res. **121**(1), 727–744 (2016)
13. Polekh, N., Zolotukhina, N., Kurkin, V., Zherebtsov, G., Shi, J., Wang, G., Wang, Z.: Dynamics of ionospheric disturbances during the 17–19 March 2015 geomagnetic storm over East Asia. Adv. Space Res. **60**(11), 2464–2476 (2017)
14. Shpynev, B.G., Zolotukhina, N.A., Polekh, N.M., Ratovsky, K.G., Chernigovskaya, M.A., Belinskaya, A.Yu., Stepanov, A.E., Bychkov, V.V., Grigorieva, S.A., Panchenko, V.A., Korenkova, N.A., Mielich, J.: The ionosphere response to severe geomagnetic storm in March 2015 on the base of the data from Eurasian high-middle latitudes ionosonde chain. J. Atmos. Solar-Terr. Phys. **180**, 93–105 (2018)
15. Yamauchi, M., Sergienko, T., Enell, C.-F., Schillings, A., Slapak, R., Johnsen, M.G., Tjulin, A., Nilsson, H.: Ionospheric response observed by EISCAT during the 6–8 September 2017 space weather event: overview. Space Weather **16**(9), 1437–1450 (2018)
16. Chernogor, L.F., Garmash, K.P., Guo, Q., Luo, Y., Rozumenko, V.T., Zheng, Y.: Ionospheric storm effects over the People's Republic of China on 14 May 2019: Results from multipath multi-frequency oblique radio sounding. Adv. Space Res. **66**(2), 226–242 (2020)
17. Luo, Y., Chernogor, L.F., Garmash, K.P., Guo, Q., Rozumenko, V.T., Zheng, Yu.: Dynamic processes in the magnetic field and in the ionosphere during the 30 August–2 September 2019 geospace storm: influence on high frequency radio wave characteristics. Ann. Geophys. **39**(4), 657–685 (2021)
18. Luo, Y., Guo, Q., Zheng, Y., Garmash, K.P., Chernogor, L.F., Shulga, S.N.: Geospace storm effects on August 5–6 2019. Space Sci. Tech. **27**(2), 45–69 (2021)

About Petschek-Type Reconnection Driven by Inhomogeneous Plasma Resistivity

N. V. Erkaev and V. S. Semenov

Abstract Applying a two-dimensional, compressible Hall magnetohydrodynamic (MHD) model, we study reconnection of magnetic fields in various cases: symmetric and non-symmetric antiparallel magnetic fields, and also skewed magnetic fields. The magnetic reconnection process is initiated by switching on a localized resistivity assumed to be a Gaussian function of spatial coordinates. As initial condition we set a one-dimensional steady-state current sheet. The obtained numerical solution indicates that reconnection process evolves asymptotically to the stationary Petschek-type regime in the case of inhomogeneous resistivity. The internal reconnection rate is found to be proportional to square root of the inverse local magnetic Reynolds number. The external reconnection rate is found by matching the external Petschek solution and the internal diffusion region solution for various cases of steady-state two-dimensional reconnection in a compressible plasma. The obtained general formula for the reconnection rate yield those of Sweet–Parker or Petschek in particular cases of pure homogeneous or strongly localized resistivity. In case of skewed reconnecting magnetic field, the reconnection rate is proportional to the sine of angle between magnetic field and reconnection line. The Hall parameter is found to be responsible for generation of the so called Alfvén wings structure. These wings are related to the Hall MHD Alfvén waves propagating faster than the usual Alfvén waves in the ideal MHD model. The obtained Alfvén wings are characterized by intensive field-aligned currents and large variations of the out-of-plane magnetic field and velocity components.

Keywords Magnetic reconnection · Hall MHD simulation · Inhomogeneous plasma resistivity

N. V. Erkaev (✉)
Institute of Computational Modelling of the Siberian Branch of the Russian Academy of Sciences, Krasnoyarsk, Russia
e-mail: erkaev@icm.krasn.ru

V. S. Semenov
Physical Department, S. Petersburg State University, S. Petersburg, Russia

A. Kosterov et al. (eds.), *Problems of Geocosmos—2020*, Springer Proceedings in Earth and Environmental Sciences, https://doi.org/10.1007/978-3-030-91467-7_29

1 Introduction

Magnetic reconnection is considered to be a physical process of fast conversion of magnetic energy into kinetic and thermal energies of plasma. This phenomenon associated with crucial change of magnetic field topology has a wide range of applications in laboratory and space plasma physics discussed in [1] and [2]. In particular, it is of great importance for solar flares and magnetospheric substorms. The resistive magnetohydrodynamic (MHD) model was applied by Sweet [3] and Parker [4] to explain energy release during solar flares. In case of homogeneous plasma resistivity, dissipation of magnetic energy determined by so called reconnection rate is proportional to inverse square root of magnetic Reynolds number. This reconnection rate is rather small in highly conducting plasma. Magnetic reconnection efficiency can be substantially higher in the model proposed by Petschek [5], where dissipation plays important role only inside a small diffusion region, but magnetic energy conversion takes place on the wave fronts or shocks propagating far away from the diffusion region. However that model was not based on a self-consistent solution and provided only upper estimate of the reconnection rate. After that there were many studies of different authors aimed to find necessary conditions for the Petschek-type solution. This question was resolved in [6] on a base of the numerical solution of resistive incompressible MHD equations. The solution demonstrated clearly that strongly localized plasma resistivity can provide magnetic field reconnection with small diffusion region and Petschek-type wave fronts. But in case of sufficiently large resistivity length scale, the magnetic reconnection structure becomes very similar to that of Sweet–Parker. Petschek-type stationary solutions were also obtained in numerical simulations [7, 8] with different model functions (Gaussian and non-Gaussian) describing a non-uniform resistivity. A possible reason for nonhomogeneous anomalous resistivity considered in [9, 10] is related to the ion-acoustic turbulence occurring in thin current sheets when the electric current density exceeds a critical threshold. In the magnetospheric plasma, electric conductivity is typically very large, and thus a very small diffusion length scale is required for efficient magnetic reconnection [11]. When this length scale becomes of the order of the proton inertia length, Hall MHD effects should be taken into account. In particular, they play important role in the magnetospheric conditions [12]. In such case, magnetic reconnection problem has to be studied in a framework of Hall MHD model or **hybrid/kinetic approach**. In this article we address Hall MHD effects on the magnetic reconnection process and analyze a relationship between the internal and external reconnection rates in case of localized resistivity of plasma. In particular, we discuss resemblance between the quasi-stationary magnetic reconnection initiated by a nonhomogeneous localized plasma resistivity and the model of Petschek.

2 Statement of Problem

For magnetic reconnection region, we apply a two-dimensional compressible Hall magnetohydrodynamic (HMHD) equations including nonhomogeneous resistivity:

$$\frac{\partial}{\partial t}\left(\rho \vec{V}\right) + \nabla \cdot \left(\rho \vec{V}\vec{V} - \frac{1}{4\pi}\vec{B}\vec{B}\right) + \nabla\left(P + \frac{1}{8\pi}B^2\right) = 0, \tag{1a}$$

$$\frac{\partial}{\partial t}\left(\frac{\rho V^2}{2} + \frac{B^2}{8\pi} + \frac{P}{(\gamma-1)}\right) + \nabla \cdot \left[\vec{V}\left(\frac{\rho V^2}{2} + \frac{\gamma P}{(\gamma-1)}\right) + \frac{\vec{E}\times\vec{B}}{4\pi}\right] = 0, \tag{1b}$$

$$\frac{1}{c}\frac{\partial \vec{B}}{\partial t} + \nabla \times \vec{E} = 0, \quad \nabla \cdot \vec{B} = 0, \quad \frac{\partial \rho}{\partial t} + \nabla \cdot \left(\rho \vec{V}\right) = 0, \tag{1c}$$

$$\vec{E} + \frac{1}{c}\vec{V}\times\vec{B} = \eta\vec{j} + \frac{1}{nec}\vec{j}\times\vec{B}, \quad \nabla\times\vec{B} = \frac{4\pi}{c}\vec{j}. \tag{1d}$$

For computation convenience we introduce the following normalization suitable for the diffusion region

$$V \to \frac{V}{V_{Ad}}, \quad B \to \frac{B}{B_d}, \quad \rho \to \frac{\rho}{\rho_d}, \quad P \to \frac{P4\pi}{B_d^2}, \quad V_{Ad} = \frac{B_d}{\sqrt{4\pi m_p n_d}}, \tag{2a}$$

$$\lambda = \frac{c^2\eta\sqrt{m_p n_d}}{B_d l_d \sqrt{4\pi}}, \quad \nu_H = \frac{c\sqrt{m_p}}{l_d e\sqrt{4\pi n_d}} = \frac{\delta_p}{l_d}. \tag{2b}$$

Using definitions (2), we rewrite system of equations (1) in the dimensionless form as follows:

$$\frac{\partial}{\partial t}\left(\rho \vec{V}\right) + \nabla \cdot \left(\rho \vec{V}\vec{V} - \vec{B}\vec{B}\right) + \nabla\left(P + \frac{1}{2}B^2\right) = 0, \tag{3a}$$

$$\frac{\partial}{\partial t}\left(\frac{\rho V^2}{2} + \frac{B^2}{2} + \frac{P}{(\gamma-1)}\right) + \nabla \cdot \left[\vec{V}\left(\frac{\rho V^2}{2} + \frac{\gamma P}{(\gamma-1)}\right) + \vec{E}\times\vec{B}\right] = 0, \tag{3b}$$

$$\frac{\partial \vec{B}}{\partial t} - \nabla\times\left(\vec{V}\times\vec{B}\right) + \nabla\times\left(\lambda\nabla\times\vec{B}\right) + \nabla\times\left(\nu_H\frac{1}{n}\left(\nabla\times\vec{B}\right)\times\vec{B}\right) = 0, \tag{3c}$$

$$\nabla \cdot \vec{B} = 0, \quad \frac{\partial \rho}{\partial t} + \nabla \cdot \left(\rho \vec{V}\right) = 0. \tag{3d}$$

We study reconnection of magnetic fields in various cases: symmetric and non-symmetric antiparallel magnetic fields, and also skewed magnetic fields. The magnetic reconnection process is initiated by switching on a localized resistivity assumed to be a Gaussian function of spatial coordinates.

$$\eta = \eta_0 + (\eta_{max} - \eta_0)\exp\left[-\left(\frac{x}{l_d}\right)^2 - \left(\frac{z}{l_d}\right)^2\right]. \tag{4}$$

Simulating magnetic reconnection, we set the initial state as a magnetic configuration with a flat equilibrium current sheet, in which all physical parameters are homogeneous in tangential directions and change only along the normal direction to the plane of the layer. Assuming the plain current sheet, we apply the following initial magnetic field components.

$$B_x = \frac{(B_{x2} - \_B_{x1})}{2}\mathrm{th}(az + b) + \frac{(B_{x1} + \_B_{x2})}{2}, b = 0.5\ln|B_{x1}/B_{x1}|,$$

$$B_y = B_{y0} = \mathrm{const},\ B_z = 0.$$

In simulations, we set the normalized parameters as $a = 20$, $B_{xl} = -B_{x2} = -\!-1$, $\rho_1 = \rho_2 = 1$, $B_{y0} = 0.5$ for symmetric reconnection. In case of non-symmetric reconnection relevant to the magnetopause we set conditions $B_{x2} = 1$, $B_{xl} = -\!-1.8^{1/2}$, $\rho_1 = \rho_2 = 1$, $P_1 = 0.1$, $P_2 = 1$ which correspond to ratios of the plasma and magnetic pressures $\beta_1 = 0.1$ and $\beta_2 = 1$ at the lower and upper boundaries, respectively. The normalized background resistivity used in simulations is $\eta_0 = 0.001$, and the resistivity maximum was varied in a range $\eta_{\mathrm{max}} = 0.005 \div 0.1$. The simulation box along and across the current sheet is 50×4 (in units of l_d) with numerical grid $400 \times\ 200$.

The unsteady MHD equations are solved by the two-step finite difference numerical scheme of Godunov in a uniform right angle grid. For the numerical solution of nonstationary MHD equations, a conservative finite-difference scheme with splitting was used, in which recalculation to the next time step was performed in three stages. At the first stage, with a half time step, the fluxes at the interfaces of the grid cells were determined by the Godunov method based on solving the one-dimensional problem of discontinuity decay in the linearized version. Further, at the second stage, all physical quantities were recalculated for a whole time step using conservative equations in finite difference form and the found fluxes across the cell boundaries without taking into account the diffusion of the magnetic field. At the third stage, the elliptic operator responsible for the diffusion of the magnetic field was switched on, which was approximated by an implicit difference scheme with splitting in directions. In this case, the recalculation to the next time layer was performed using the three-diagonal matrix algorithm.

The reconnection process is characterized by two dimensionless similarity parameters: $\lambda = 1/R_{\mathrm{md}}$ (R_{md} is a local magnetic Reynolds number in the diffusion region) and Hall parameter ν_{H}. The Reynolds number is determined for the maximum value and length scale of the resistivity, and the Hall parameter means a ratio of the proton inertial length to the resistivity length scale.

3 Results of Calculations

The nonstationary dissipative Hall MHD equations integrated numerically yield evolution of the magnetic reconnection and diffusion region structure depending on time.

We obtained results are based on the following runs:

1. Symmetric case: $P_1 = P_2 = 0.1$, $\rho_1 = \rho_2 = 1$, $B_{x1} = -B_{x2} = 1$, $B_{y1} = B_{y2} = 0.5$, $\nu_H = 0.1$, $\eta = 0.001 + 0.05 \exp(-x^2 - z^2)$, simulation time maximum = 30.
2. Non-symmetric case: $P_1 = 0.1$, $P_2 = 0.5$, $\rho_1 = \rho_2 = 1$, $B_{x1} = 1.342$, $B_{x2} = 1$, $B_{y1} = B_{y2} = 0$, $\eta = 0.005 + 0.025 \exp(-x^2 - z^2)$, simulation time maximum = 30.

Figure 1 shows the obtained distribution of the electric current J_y corresponding to dimensionless time $t = 16$. The fronts of intensive electric current resemble crab claws. They correspond to the slow shock fronts generated in the magnetic diffusion region.

Configuration of the magnetic field lines is presented in Fig. 2 at the same time (t = 16) as that for Fig. 1. One can see a strong compression of the field lines before the dipolarization front [13].

Figure 3 shows variations of the convective electric field along x axis for different times. Peaks of the electric field correspond to the dipolarization fronts propagating from the diffusion region in both directions. This figure indicates time when a stationary solution is established. At the sufficiently large time ($t > 30$) the dipolarization fronts reach the boundaries and finally they move away from the calculation domain. After that we have solution corresponding to a quasi-stationary magnetic reconnection.

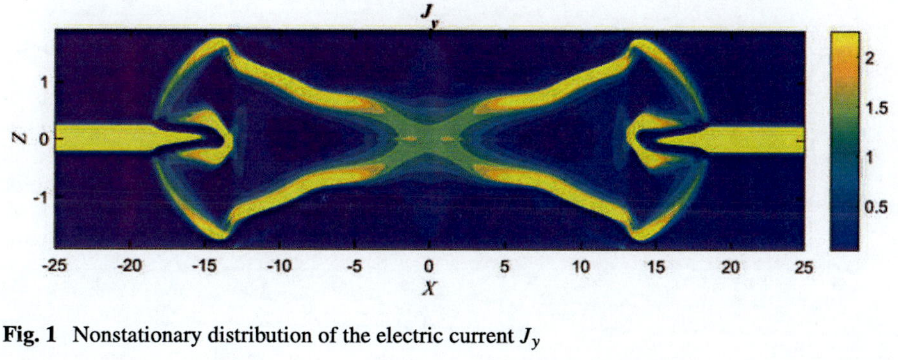

Fig. 1 Nonstationary distribution of the electric current J_y

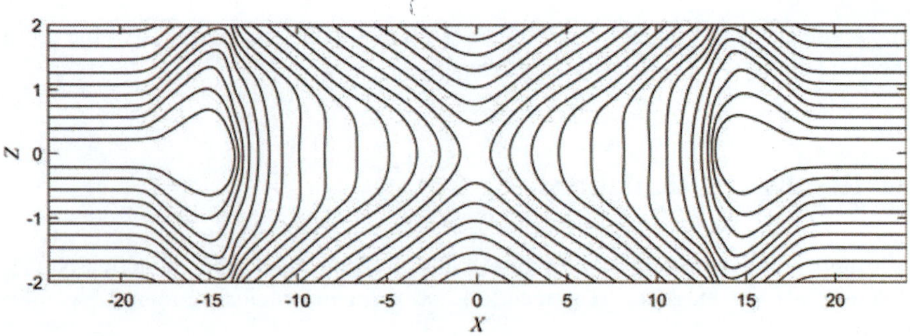

Fig. 2 Magnetic field lines (propagation of the dipolarization front)

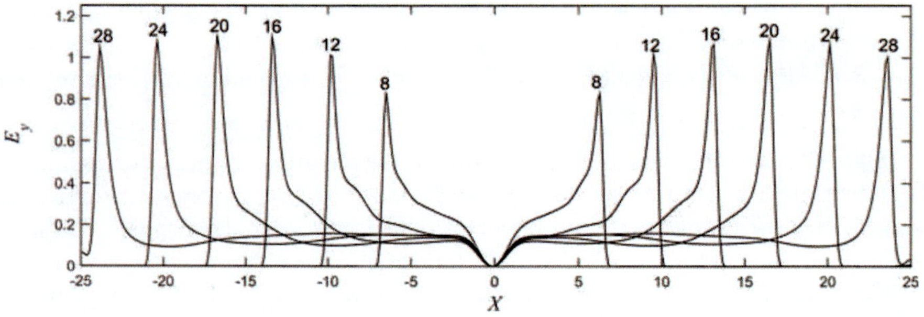

Fig. 3 Convective electric field profiles along the X axis at different times: $t = 8, 12, 16, 20, 24, 28$

Figure 4 shows the calculated configurations of the magnetic field lines corresponding to the established quasi-stationary magnetic reconnection in symmetric cases. Magnetic field lines in the non-symmetric case are shown in Fig. 5. The symmetric case looks rather similar to that of Petschek. The magnetic field lines change their inclinations at the smooth fronts of the slow shocks bounding the accelerated flow. The Hall-related magnetic field and velocity perturbations are determined

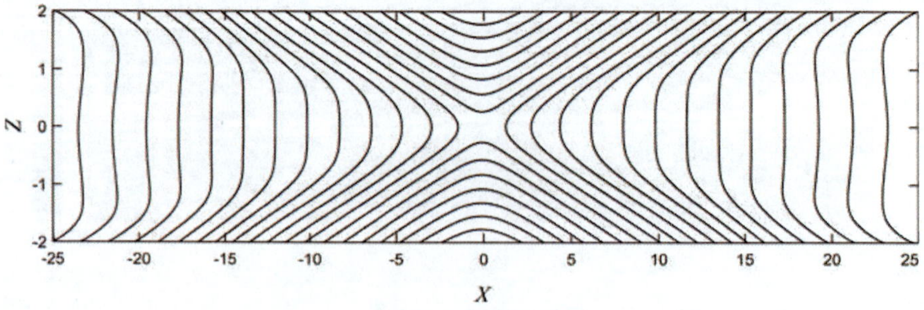

Fig. 4 Quasi-stationary magnetic field lines in symmetric case ($R_{\text{md}} = 20$)

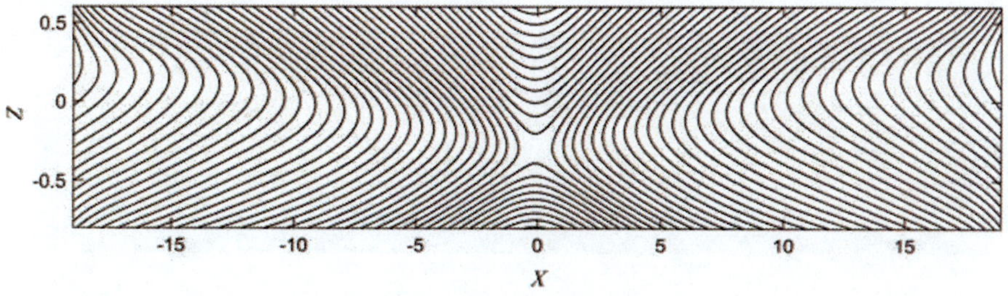

Fig. 5 Magnetic field lines in case of quasi-stationary non-symmetric magnetic reconnection

by the Alfven fronts propagating faster than the slow shock fronts. However, in case of our simulations with Hall parameter about 0.1, the Alfven fronts are rather close to the slow shock fronts, and thus they can hardly be separated from each other.

The non-symmetric case shown if Fig. 5 can be relevant to the dayside magnetopause [14]. Plasma beta is assumed to be 1 and 0.1 at the upper and lower boundaries, respectively. At the lower boundary (z = –-2), the normalized plasma pressure, density and temperature are assumed $P_1 = 0.1$, $\rho_1 = 1$, $T_1 = 0.1$. And at the upper upstream boundary, the parameters are assumed $P_2 = 0.5$, $\rho_2 = 1$, $T_2 = 0.5$. The initial magnetic field profile is given by analytical formula:

$$B_x = 1.171\text{th}(20z + 0.14708) - 0.171.$$

At the initial moment the x-line was at $z = 0$, and finally it was shifted to the negative z coordinate during reconnection process. This is corresponding to the erosion of the subsolar magnetopause for southward interplanetary magnetic field.

Figure 6 shows the normal profiles of the tangential velocity component normalized to the Alfvén speed related to the inflow region. One can see strong acceleration of flow along the current sheet. Here z and x are the normal and tangential coordinates normalized to the length scale l_d given in Eq. (4). The maximum tangential velocity is reached at distance $x = 5$.

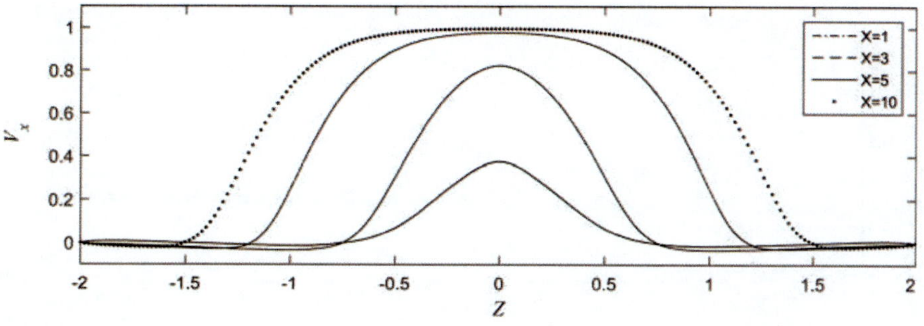

Fig. 6 Tangential velocity profiles along the normal direction for different distances from the stagnation point: $x = 1, 3, 5, 10$

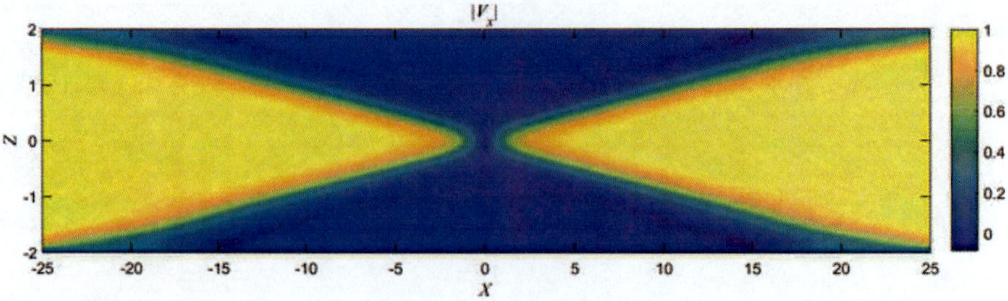

Fig. 7 Outflow velocity distribution in case of symmetric quasi-stationary reconnection

Quasi-stationary distribution of the tangential velocity component normalized to the Alfvén speed is presented in Fig. 7 in case of symmetric magnetic reconnection ($R_{\mathrm{md}} = 20$). The accelerated flow with Alfvén velocity is constrained clearly between the slow shock fronts.

Figure 8 shows electric current profiles along z direction for the different distances from the x-point. At all profiles besides the central one can see two peaks of the current density corresponding to the slow shock fronts. The peaks become to be separated from each other at the distances exceeding three from the stagnation point.

Figure 9 shows the quasi-stationary electric current density distributions in the case of symmetric reconnection. The straight dashed lines mark symmetric stationary slow shock fronts, where the current density has large intensity. These shock fronts has clear resemblance with the model of Petscheck. The case of non-symmetric reconnection is presented in Fig. 10, where exists only the upper slow shock front separating the upper inflow region from the outflow one.

The Hall parameter is found to be responsible for generation of the so called Alfven wings structure [15]. These wings are related to the Hall MHD Alfven waves propagating faster than the usual Alfven waves in ideal MHD model. The obtained Alfven wings are characterized by intensive field-aligned currents and large variations of the out-of-plane magnetic field and velocity components. Figure 11 shows distribution of the out-of-plane B_y magnetic field component which is related to the Hall effects. The Hall related magnetic perturbations are carried to large distances by

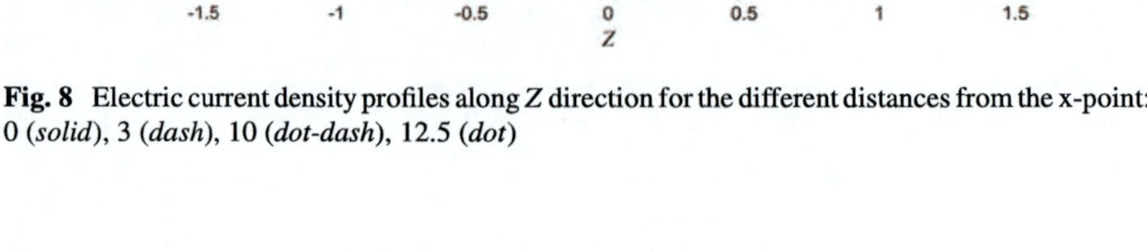

Fig. 8 Electric current density profiles along Z direction for the different distances from the x-point: 0 (*solid*), 3 (*dash*), 10 (*dot-dash*), 12.5 (*dot*)

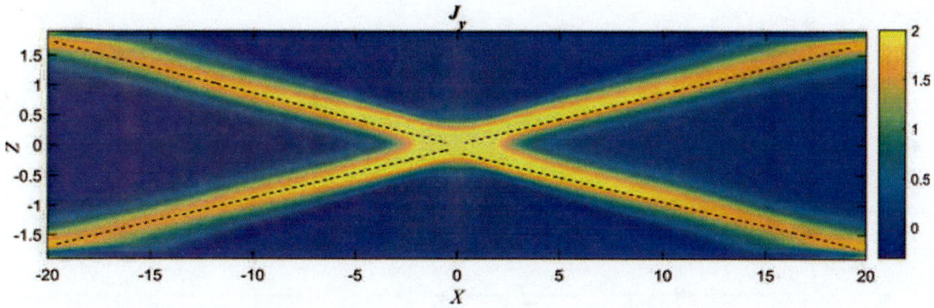

Fig. 9 Quasi-stationary distribution of the electric current J_y in case of symmetric reconnection

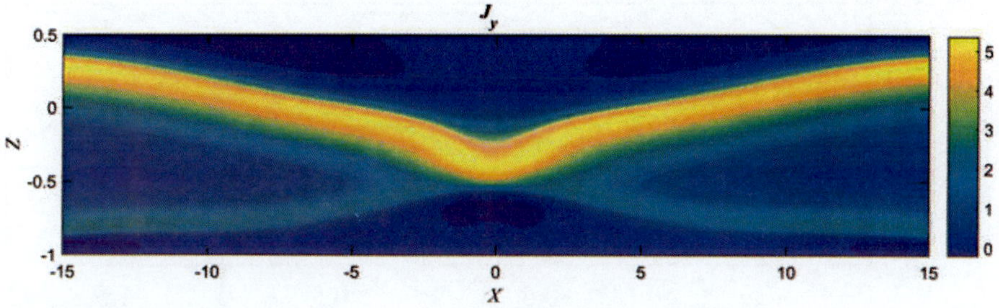

Fig. 10 Electric current distribution in case of non-symmetric magnetic reconnection

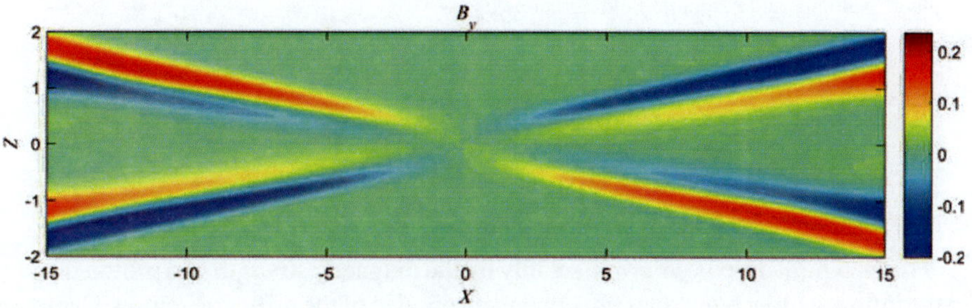

Fig. 11 Distribution of the B_y component related to the Hall effect ($\nu_{\rm H} = 0.1$)

the Alfven waves originated from the diffusion region of the magnetic reconnection. The thickness of these wave fronts is of the order of the diffusion region length scale (l_d).

Variations of the out-of-plane velocity and magnetic field components across the reconnection layer at distance $x = 12.5$ are given in Fig. 12a. The amplitudes of these components are proportional to the Hall parameter (ratio of the proton inertia and diffusion region length scales).

Behavior of the Hall electric field and corresponding electric potential are presented in Fig. 12b. In dimensional units, the maximum potential drop at the wave

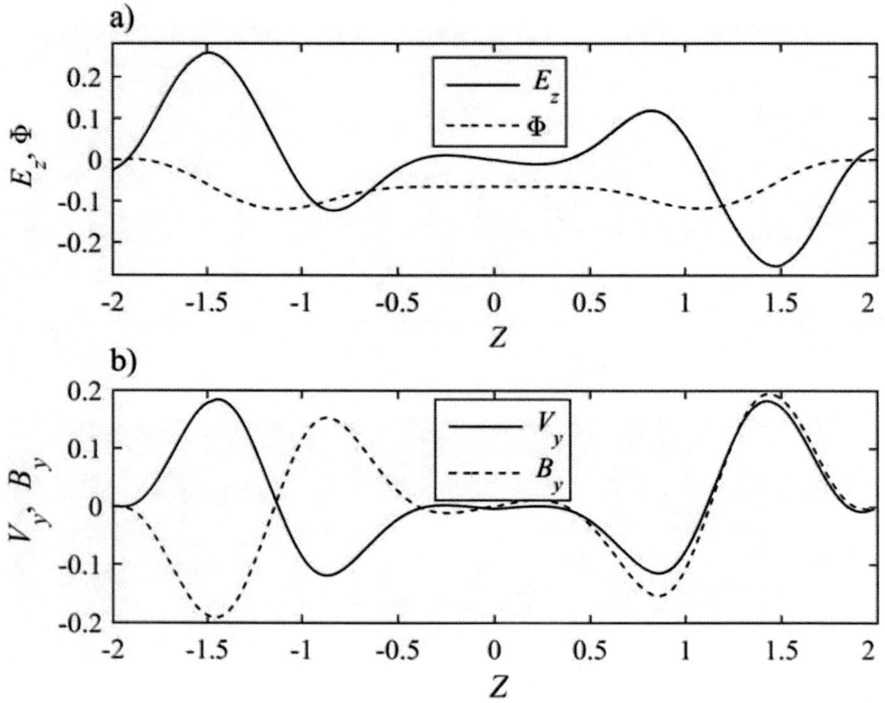

Fig. 12 Profiles of the Hall related quantities at $x = -2.5$: **a** Electric field and potential; **b** "Out-of-plane" velocity and magnetic field components

front is about

$$\Delta\Phi_{\max} \approx \nu_H V_{Ad} B_d \frac{l_d}{c} = B_d^2 \frac{l_p}{c\sqrt{4\pi n_d m_p}} = \frac{B_d^2}{4\pi n_d e}.$$

This potential drop is determined only by the magnetic strength and proton density in the inflow region, but it does not depend on the size of the diffusion region. Decrease of the diffusion length scale just leads to increase of the electric field component across the wave front.

A peculiarity of Alfven wave propagation in Hall magnetohydrodynamics (HMHD) was studied comprehensively in [15]. It was shown that in the reference frame of plasma the HMHD Alfven waves propagate along magnetic field faster than the usual MHD Alfven waves, and their group velocity is an increasing function of the Hall parameter,

$$V_{AH} = V_A \left[k l_p \frac{1}{2} + \sqrt{1 + \frac{1}{4} k^2 l_p^2} \right].$$

Applying this formula to the Hall Alfven wave induced by the magnetic reconnection, we estimate $k \sim 1/l_d$, and thus $V_{AH} = V_A\left(\frac{1}{2}\nu_H + \sqrt{1 + \frac{1}{4}\nu_H^2}\right)$.

We can estimate Hall parameter, taking possible length scales l_d of the diffusion region. If the diffusion length scale is of the order of the proton inertial length: $\lambda \sim 1$ then the speed of the Hall Alfvén wave would exceed the Alfvén speed by 60%. The HMHD Alfven speed can be even substantially greater if the length of the diffusion region is less than the proton inertial length, and the Hall parameter, accordingly, much exceeds unity. In the particular case $k\, l_p \gg$ formula (7) yields the dispersion equation for whistler mode. This is relevant to the kinetic PIC simulations of the reconnection diffusion region in collisionless plasmas [16, 17]. The quadrupole structure of the Hall magnetic field, which appears in kinetic models of the diffusion region, is the initial stage of Hall Alfvén waves, which propagate over long distances without damping and carry information about reconnection faster than other waves.

4 External Reconnection Rate in General Case of Skewed Magnetic Fields

Numerical MHD simulation yields the internal reconnection rate ε_d near the diffusion region:

$$\varepsilon_d = E_y \frac{c}{V_{Ad} B_d} = \frac{\tilde{\varepsilon}}{\sqrt{R_{md}}}, \quad R_{md} = \frac{4\pi V_{Ad} l_d}{c\eta}, \quad V_{Ad} = \frac{B_d}{\sqrt{4\pi n_d m_p}}.$$

Here the index "d" denotes the parameters at the upper boundary of the diffusion region, is a constant dimensionless parameter obtained from the numerical solution (about 0.7). Magnetic Reynolds number R_{md} can be expressed through the ratio of the Bohm's resistivity η_B to the plasma resistivity maximum η_m and Hall parameter

$$R_{md} = \frac{\eta_B}{\eta \nu_H}, \quad \eta_B = \frac{B_d}{n_d e c}.$$

To argue importance of Bohm's resistivity we refer to results of two-dimensional hybrid simulations of tangential discontinuity at the magnetopause [18]. There was obtained cross-field diffusion with the Bohm's rate.

The external reconnection rate can be found by matching the external and internal solutions for magnetic field for various cases of steady-state two-dimensional reconnection in compressible plasma. In case of quasi-stationary magnetic field in the inflow region, the out-of-plane electric field component E_y has to be constant

$$E_y = -\frac{1}{c} V_z B_x = \text{const.}$$

Applying equation for E_y at the lower and upper boundaries of the inflow region and using definition for the reconnection rate, we get

$$\varepsilon_{0x} = \varepsilon_d \frac{B_d^2}{B_{0x}^2}. \tag{5}$$

Here index "0" corresponds to the upper boundary of the inflow region.

Figure 13 shows the calculated profiles of magnetic field and current density along the z axis. One can see that the electric current density is a strongly decreasing function of z coordinate normalized to the length scale l_d, and the current density becomes rather small above the diffusion region. Therefore the magnetic field can be considered as a potential one in the inflow region, and thus it can be described by harmonic functions, which obeys the Laplace's equation with Neumann boundary condition at the lower boundary of the inflow region. Hence, for B_x component we use the Poisson integral [19]

$$B_x = B_d + \frac{1}{\pi} \int_{-\infty}^{+\infty} \left[\frac{\partial B_z}{\partial x} \ln\left(\frac{\sqrt{(x-s)^2 + z^2}}{|x-s|} \right) \right] ds$$

Variation of B_z component along the reconnection layer can be described by function

$$B_z(x) = \kappa B_{0x} \varepsilon G(x),$$

where coefficient κ is between 1 and 2 (in symmetric incompressible case) and function $G(x)$ has asymptotic limit ~ 1 for large x/l_d. For large z Eq. (13) can be

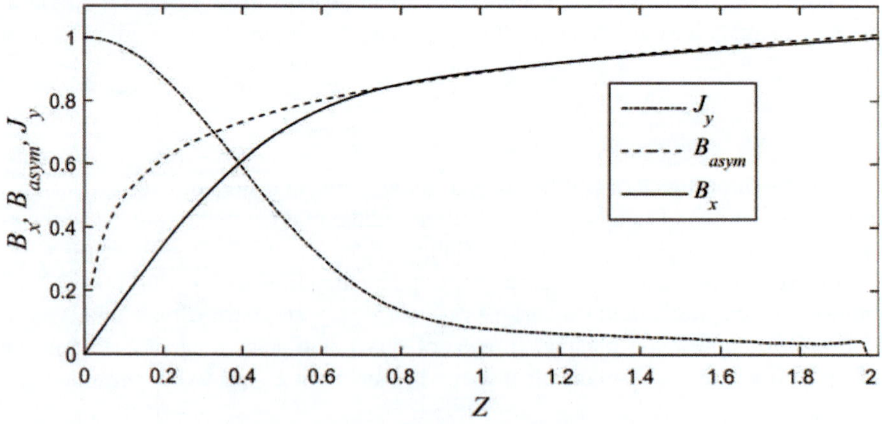

Fig. 13 B_x component (*solid*), logarithmic asymptotic B_{asym} (*dash*), current density J_y (*dot-dash*) as functions of z ($x = 0$)

simplified to asymptotic formula

$$B_x = B_d + B_{0x}\frac{2}{\pi}\varepsilon_{0x}\kappa\ln(z/l_d).$$

where $B_{0x} = B_0 \sin(\theta)$. Figure 13 demonstrates rather clearly that this logarithmic asymptotic function approximates quite well the B_x variation in the inflow region within a wide interval between the diffusion region and the external boundary: $0.5\, l_d < z < L$.

Assuming $B_x = B_{0x}$ in the external region for $z = L$, we get equation to determine ratio B_d/B_{0x},

$$\frac{B_d}{B_{0x}} = 1 - \frac{2}{\pi}\varepsilon_{0x}\kappa\ln\left(\frac{L}{l_d}\right). \tag{6}$$

Using Eqs. (5) and (6), we derive equation for the external reconnection rate

$$\varepsilon_{0x} = \varepsilon_d\left[1 - \frac{2}{\pi}\varepsilon_{0x}\kappa\ln\left(\frac{L}{l_d}\right)\right]^2.$$

Introducing definitions

$$f = \sqrt{\varepsilon_{0x}/\varepsilon_d},\ A = \frac{2}{\pi}\varepsilon_d\kappa\ln\left(\frac{L}{l_d}\right),$$

we get quadratic equation with respect to f

$$Af^2 + f - 1 = 0.$$

Taking positive solution of this equation, we find

$$\varepsilon_{0x} = \frac{4\varepsilon_d}{(\sqrt{1+4A}+1)^2}.$$

Therefore, the external reconnection rate is found as

$$\varepsilon_{0x} = \frac{4\,\tilde{\varepsilon}}{\left(\sqrt{R_{md}^{1/2} + \frac{8}{\pi}\kappa\,\tilde{\varepsilon}\,\ln\left(\frac{L}{l_d}\right)} + R_{md}^{1/4}\right)^2} < \frac{\pi}{2\kappa\ln\left(\frac{L}{l_d}\right)}.$$

For low values of parameter A the external reconnection rate is similar to that of Sweet–Parker. But in case of sufficiently large parameter A, the reconnection rate is similar to that of Petschek. Using the reconnection rate expression, we can find the electric field in the reconnection region

$$E_y = -\frac{1}{c} V_z B_x = \frac{\varepsilon_{0x}}{c\sqrt{4\pi n_0 m_p}} B_0^2 \sin(\theta)^2.$$

However, in case of the magnetospheric conditions, $\ln(L/l_d)$ has rather moderate values. In particular, we can take reasonable scales $L \sim 10^{10}$ and $l_d \sim \delta_p \sim 10^8$ cm for the magnetotail plasma sheet which yields estimate $\ln(L/\delta_p) \sim 4.6$.

5 Discussion and Conclusion

We use the numerical dissipative Hall MHD model taking into account finite nonhomogeneous anomalous resistivity of plasma in the diffusion region. Inside the diffusion region, we assumed plasma resistivity scaled to that of the Bohm's diffusion. We obtained quasi-stationary reconnection rate in various cases of antiparallel and skewed magnetic fields, symmetric and non-symmetric plasma sheet configurations. The obtained numerical solution indicates that reconnection process evolves asymptotically to the stationary Petschek type regime in case of non-uniform resistivity.

In case of uniform resistivity, two possible solutions reported in [20] were obtained in the framework of two-fluid MHD model with finite ratio of the proton and electron masses assumed ~25. There was proposed the critical resistivity threshold above which only Sweet–Parker solution was possible. And the Petschek-type solution was shown to be possible only for the sufficiently low resistivity, below the critical value.

Our results do not contradict to those obtained in [20] because we considered non-uniform resistivity for which we obtained the Petschek-type reconnection structure even for the resistivity exceeding the critical one estimated in [20].

The internal reconnection rate is found to be proportional to square root of the inverse local Reynolds number determined for the maximum resistivity and its length scale. The external reconnection rate is found by matching the external magnetic perturbations with the internal diffusion region solution for various cases of steady-state two-dimensional reconnection in a compressible plasma. The obtained general formula for the reconnection rate reduces to those of Sweet–Parker and Petschek in particular cases of pure homogeneous or strongly localized resistivity. In case of skewed reconnecting magnetic field, the reconnection rate is proportional to the sine of angle between magnetic field and reconnection line. The influence of the Hall effects results in the generation of the special Alfvén waves, which are absent in the usual ideal MHD model. These waves arise in the diffusion region have the group velocity greater than the usual Alfvén speed, and propagate far away, keeping their amplitude. The group speed of the Hall Alfven waves increases with the Hall parameter. At the front of such waves, there are intense field-aligned currents, as well as variations in the corresponding Hall components of the magnetic field and velocity proportional to the Hall parameter. Hall Alfvén waves propagating faster than conventional MHD waves can serve as precursors of impulse reconnection. It is important to note that the solution obtained within the framework of the Hall MHD reconnection model depends on only two parameters—the diffusion magnetic Reynolds number and the Hall parameter. The first parameter determines the speed of reconnection, and the second is the intensity of the Hall Alfvén waves. These parameters can be used to match the Hall MHD reconnection model with the existing kinetic models of the diffusion region. In the PIC simulations, magnetic reconnection occurs in a very thin electron diffusion region, where electron inertia and nongyrotropic pressure plays a dominant role. However, it is possible to determine the effective plasma conductivity as a ratio of the local current density to the electric field component along the reconnection line. This conductivity is related to some effective magnetic Reynolds number. For example, comparing the internal reconnection rate in our model ($\varepsilon_d = 0.6/R_{md}^{1/2}$) with a typical estimate of the reconnection rate in kinetic models ($\varepsilon_d \sim 0.1$), we obtain an estimate of the effective Reynolds number for the diffusion region as $R_{md} \sim 40$. We can also estimate Hall parameter, taking into account the characteristic dimensions of the diffusion region in kinetic models of the order of the proton inertial length: $\nu_H \sim 1$. With this parameter, the speed of the Hall Alfvén wave can exceed the Alfvén speed by 60%. This velocity can be even greater if the length of the diffusion region is less than the proton inertial length, and the Hall parameter, accordingly, exceeds unity. The quadrupole structure of the Hall magnetic field, which appears in kinetic models of the diffusion region, evolves further to the Hall Alfvén fronts, which propagate over long distances without damping and carry information from the diffusion region faster than other waves.

References

1. Yamada, M., Kulsrud, R., Hantao, J.i.: Magnetic reconnection. Rev. Modern Phys. **82**, 603–662 (2010)
2. Uzdensky, D.A.: Magnetic reconnection in extreme astrophysical environments. Space Sci. Rev. **160**, 45–71 (2011)
3. Sweet, P.A.: The neutral point theory of solar flares. In: Lehnert, B. (ed.) Electromagnetic Phenomena in Cosmic Physics, pp. 123–134. Cambridge University Press, London (1958)
4. Parker, E.N.: Sweet's mechanism for merging magnetic fields in conducting fluids. J. Geophys. Res. **62**(4), 509–520 (1957)
5. Petschek, H.E.: Magnetic field annihilation. In: Hess, W.N. AAS-NASA Symposium on the Physics of Solar Flares NASA SP-50, pp. 425–439. National Aeronautics and Space Administration, Washington, DC (1964)
6. Erkaev, N.V., Semenov, V.S., Jamitzky, F.: Reconnection rate for the inhomogeneous resistivity Petschek model. Phys. Rev. Lett. **84**(7), 1455–1458 (2000)
7. Baty, H., Forbes, T.G., Priest, E.R.: Petschek reconnection with a non-localized resistivity. Phys. Plasmas **16**, 012102 (2009)
8. Baty, H., Forbes, T.G., Priest, E.R.: The formation and stability of Petschek reconnection. Phys. Plasmas **21**, 112111 (2014)
9. Kulsrud, R.M.: Magnetic reconnection. Sweet-Parker versus Petschek. arXiv:astro-ph/0007075v1
10. Uzdensky, D.A.: Petschek-like reconnection with current driven anomalous resistivity and its application to solar flares. Astrophys. J. **587**, 450–457 (2003)

11. Birn, J., Drake, J.F., Shay, M.A., Rogers, B.N., Denton, R.E., Hesse, M., Kuznetsova, M., Ma, Z.W., Bhattacharje, A., Otto, A., Pritchett, P.L.: Geospace environmental modeling (GEM). Magnetic reconnection challenge. J. Geophys. Res. **106**(A3), 3715–3719 (2001)
12. Treumann, R.A., Jaroschek, C.H., Nakamura, R., Runov, A., Scholer, M.: The role of the Hall effect in collisionless magnetic reconnection. Adv. Space Res. **38**, 101–111 (2006)
13. Wu, P., Shay, M.A.: Magnetotail dipolarization front and associated ion reflection: particle-in-cell simulations. Geophys. Res. Lett. **39**, L08107 (2012)
14. Scholer, M.: Asymmetric time-dependent and stationary magnetic reconnection. J. Geophys. Res. **94**(A11), 15099–15111 (1989)
15. Sallago, P.A., Platzeck, A.M.: Alfven waves and wings in Hall magnetohydrodynamics. J. Geophys. Res. **109**, A04218 (2004)
16. Divin, A., Lapenta, G., Markidis, S., Semenov, V.S., Erkaev, N.V., Korovinskiy, D.B., Biernat, H.K.: Scaling of the inner electron diffusion region in collisionless magnetic reconnection. J. Geophys. Res. **117**, A06217 (2012)
17. Divin, A., Semenov, V., Korovinskiy, D., Markidis, S., Deca, J., Olshevsky, V., Lapenta, G.: A new model for the electron pressure nongyrotropy in the outer electron diffusion region. Geophys. Res. Lett. **43**, 10565–10573 (2016)
18. Winske, D., Omidi, N.: Diffusion at the magnetopause. Hybrid simulations. J. Geophys. Res. **100**(A7), 11923–11933 (1995)
19. Vladimirov, V.S.: Equations of Mathematical Physics. Publisher "Mir", Moscow (1984)
20. Cassak, P. A., Shay, M. A. and Drake, J. F.: Catastrophe model for fast magnetic reconnection onset. Phys. Rev. Lett. 95, 235002 (2005)

Sensitivity of Surface Meteorology to Changes in Cloud Microphysics Associated with IMF B_y

Arseniy Karagodin, Irina Mironova, and Eugene Rozanov

Abstract In recent decades, the response of surface meteorology to the variation of the B_y component of the interplanetary magnetic field (IMF) at high latitudes, known as the Mansurov effect, has been extensively studied. However, the role of the global electric circuit (GEC) in these processes is still unclear. Recent studies suggested that the response of cloud microphysics to IMF B_y-induced changes in cross-polar cap potential and fair weather downward current density J_z may be one of the possible ways for solar wind-surface meteorology coupling. Such a mechanism requires confirmation, but numerical studies that would show the Mansurov effect discovered in observations have not been carried out earlier. In this work, we evaluated a sensitivity of surface meteorological parameters to a certain change in the stratiform-cloud autoconversion rate imposed by the IMF B_y-related changes in the J_z using the chemistry-climate model SOCOLv3. It should be noted that in this first study, we use the non-realistic approach where the IMF B_y-induced anomalies staying constant for the whole period of simulation. Nevertheless, we found that even small changes of $\pm 12\%$ in the J_z and corresponding changes in the autoconversion rate in regions where IMF B_y-induced anomalies in the cross-polar cap potential occur, can lead to large-scale anomalies in surface pressure (up to 2 hPa) and air temperature (up to 1.5 °K) at high and middle latitudes that are in good agreement in magnitude with those detected in observations. This study shows the importance of further investigating solar-terrestrial coupling through the GEC using global climate models.

A. Karagodin (✉) · I. Mironova · E. Rozanov
St. Petersburg State University, Saint Petersburg, Russia
e-mail: st058394@student.spbu.ru

I. Mironova
e-mail: i.a.mironova@spbu.ru

E. Rozanov
e-mail: eugene.rozanov@pmodwrc.ch

I. Mironova
National Research Nuclear University MEPhI, Moscow, Russia

E. Rozanov
The Physikalisch-Meteorologisches Observatorium Davos/World Radiation Center (PMOD/WRC), Davos, Switzerland

A. Kosterov et al. (eds.), *Problems of Geocosmos—2020*, Springer Proceedings in Earth and Environmental Sciences, https://doi.org/10.1007/978-3-030-91467-7_30

Keywords Atmospheric electricity · Climate · Cloud microphysics · Mansurov effect · Surface pressure and air temperature · Numerical modeling · Solar-terrestrial coupling

1 Introduction

The relationship between changes in the solar wind and electric current in the lower atmosphere, as well as associated anomalies in surface meteorological parameters, have been poorly studied yet compared to studies of the relationship between the magnetosphere and ionosphere with the solar wind. Currently, there are studies showing a relationship between the variation in the dawn-to-dusk component (B_y) of the interplanetary magnetic field (IMF) in Geocentric Solar Magnetospheric (GSM) coordinate system and anomalies in surface meteorological parameters, known as the Mansurov effect [2, 13, 13, 17] showed the hypothesized chain of connection between the solar wind and surface meteorology. It begins with the influence of IMF B_y on the cross-polar cap potential, which was studied earlier and presented in [19]. Investigation of the relationship between the cross-polar cap potential and the ionospheric potential of the lower atmosphere showed that a contribution to the ionospheric potential is present at high latitudes, where anomalies in the cross-polar cap potential, caused by IMF B_y, are formed [4, 12, 16].

Despite that the relationship between the solar wind and anomalies in surface meteorological parameters has been established, the role of atmospheric electricity in this connection is still a matter of debate [12, 14]. Several studies have suggested that the fair-weather downward current density of GEC (J_z) which is determined by ionospheric potential may affect cloud microphysics, thereby altering the lifetime of stratiform clouds in fair-weather regions [9, 10, 22]. It was assumed that the current can lead to the ion-induced formation of cloud condensation nuclei and, also, can affect the rate of cloud droplet coalescence (autoconversion rate). It has also been suggested that the formation of cloud condensation nuclei under the influence of J_z is a weak process compared to the change in the autoconversion rate [3, 20]. In turn, the current that flows through the cloud layer affects the horizontal cloud-air boundary layer where there is a conductivity reduction and where droplets form or evaporate thereby charging droplets. Charges can substantially enhance droplet-droplet collisions (for smaller droplets (less than 10 μm) due to the image force attraction), therefore, modify the evolution of cloud droplets and their size distribution that in turn affects the radiative properties of clouds [10]. This process leads, as expected, to the formation of large-scale anomalies in near-surface meteorology. It should be said that despite that the effect of atmospheric electricity on cloud microphysics has been discussed in many studies, there are no quantitative estimates showing the level of change in the microphysical parameters of clouds under the influence of a current of a certain magnitude [9]. This work aims at the estimation of the sensitivity of surface pressure and air temperature to a certain change in the stratiform cloud autoconversion rate in regions where IMF B_y-induced anomalies in the cross-polar cap potential appear. In this study, the chemistry-climate model SOCOLv3 (Solar Climate Ozone Links) is used. Section 2 introduces the SOCOLv3 model and the performed numerical experiments. The results of this study are presented in Sect. 3. Discussion and summary are provided in Sect. 4.

2 The SOCOL Chemistry-Climate Model and Conducted Experiments

The chemistry-climate model SOCOLv3 [21] consists of a dynamic core, which is represented by the spectral general atmospheric circulation model MA-ECHAM5.4 (the Middle Atmosphere version of the European Center/Hamburg Model version 5.4) [18], which interactively coupled with the MEZON chemistry-transport module (Model for the Evaluation of oZONe Trends) [6]. The interaction carried out by radiative forcing caused by O_3, H_2O, N_2O, CH_4, and chlorofluorocarbons. The SOCOLv3 uses a horizontal Gaussian grid with triangular truncation T42 (64 latitudes × 128 longitudes), dividing the model space into 2.5 × 2.5-degree grid cells. The model grid in the vertical direction consists of 39 levels in a hybrid sigma-pressure coordinate system, covering heights from the Earth's surface up to 0.01 hPa (about 80 km). The time step of the model is 15 min for dynamic and physical processes, while for calculations of total radiation and to treat atmospheric chemistry it is 2 h. MEZON shares horizontal and vertical spatial resolution with MA-ECHAM5.4 and includes 95 chemical compounds and 215 gas phase, 16 heterogeneous and 75 photolysis chemical reactions. The autoconversion rate in the SOCOL model is calculated using the parameterization presented in [15].

The SOCOL model included the GEC scheme, which has, in particular, the ionospheric potential and the downward current density J_z, which has already been used to calculate the GEC parameters [8, 11]. The parameterization of air conductivity that used in our study to calculate J_z is based on the ion pair concentration produced by galactic and solar cosmic rays as well as by radon emissions and ion mobility in the atmosphere (for more details see [1, 16]).

In this work, a numerical experiment (EXP) was developed, consisting of three ensemble members, which differed in a slight perturbation in CO_2 initial concentration during the 1st month of the calculation period. We also designed a reference experiment (REF) that also consists of 3 ensembles. For both experiments, the period of the simulation was four years, with all conditions corresponding to the period from 1999 to 2002. The assessment of the statistical significance of the results obtained was carried out using the Student's test applied to the members of the ensemble of experiments.

Lam et al. [12] showed that the anomalies in the cross-polar cap potential are about 30 kV at high IMF B_y ($\geq |3$ nT$|$). Anomalies form over Greenland and northern Canada in the Northern Hemisphere, where the anomaly has a negative sign, and over Antarctica over the Southern Hemisphere, where a positive anomaly is formed

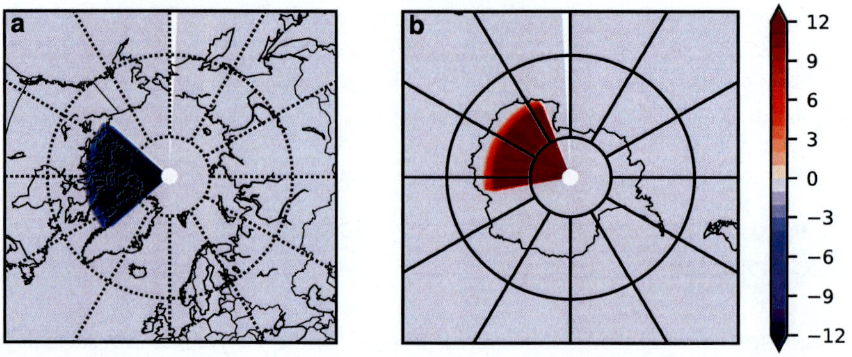

Fig. 1 Percentage difference in downward current density J_z between EXP and REF experiments simulated with SOCOLv3, calculated as: ((EXP-REF)/REF) * 100%. **a** Anomaly in the northern hemisphere; **b** anomaly in the southern hemisphere

(anomalies are calculated as the absolute difference between the magnitude of potential on days with a strongly positive IMF B_y ($\geq 3\,\text{nT}$) and potential on days with strongly negative IMF B_y ($\leq -3\,\text{nT}$) (see [12]).

In this work, for simplicity, a constant anomaly in the ionospheric potential was set for the entire calculation period (1999–2002), which in the Northern Hemisphere is $-30\,\text{kV}$ and $+30\,\text{kV}$ in the Southern hemisphere, respectively. Next, the downward current density J_z was calculated using an updated ionospheric potential, and after this J_z was compared with the reference J_z calculated without taking into account the $|30\,\text{kV}|$ anomaly in the ionospheric potential. Figure 1 shows the percentage difference of the obtained J_z current density against the one from the reference experiment.

Figure 1 shows that the anomaly in the J_z imposed by IMF B_y-induced changes in cross-polar cap potential corresponds to $\pm 12\%$ of the total current density. The quantitative assessment of the relationship between current and autoconversion rate is missing so in this study we assume a linear relationship between these two quantities. This means that increases/decreases of the autoconversion rate is by the same amount by which increases/decreases the J_z current density, i.e. $\pm 12\%$ in our case depending on the hemisphere. Thus, a constant effect on the autoconversion rate in two regions was made, equal to $\pm 12\%$ of its full value, and then the obtained fields of surface pressure and air temperature were compared with the fields from the reference experiment.

3 Results

3.1 Surface Pressure Anomalies

Anomalies were calculated as the absolute difference in surface pressure and air temperature from the experiment, where the anomaly in the autoconversion rate is

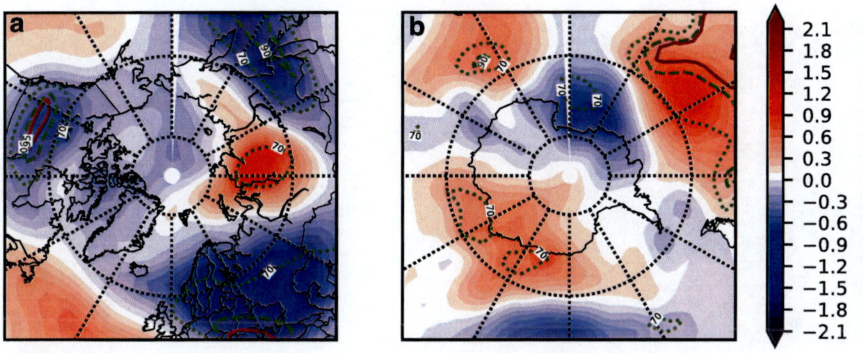

Fig. 2 Surface pressure anomalies caused by changes in autoconversion rate averaged for the 1999–2002 period. **a** Anomalies in the northern hemisphere; **b** anomalies in the southern hemisphere. The green line indicates a statistical significance of more than 90%; the purple line indicates a statistical significance of more than 95%

applied, and those from the reference experiment (EXP-REF). Anomalies presented here are the average of three ensemble members. Figure 2 shows surface pressure anomalies in both hemispheres of the Earth.

In the Northern hemisphere, at high latitudes (above 60 N), positive surface pressure anomalies are found with a maximum of about 1.5 hPa over the northern regions of Russia. In the middle latitudes, there are several widespread zonal negative anomalies with a value of up to 2 hPa. Over the Southern hemisphere, the picture is the opposite, at high latitudes, there is a negative pressure anomaly up to 1.5 hPa and several positive anomalies at middle latitudes, with some statistically significant regions. Figure 2 indicates that the anomalies in the northern hemisphere are deeper in the upper latitudes than in the southern one.

It can be also featured that a negative anomaly in autoconversion rate leads in the Northern hemisphere to several negative pressure anomalies in middle latitudes whereas a positive anomaly Southern hemisphere leads to positive pressure response in the mid-latitudes. What the pressure response would be if the anomalies are swapped should be verified.

3.2 Surface Air Temperature Anomalies

Figure 3 shows surface air temperature anomalies in both hemispheres of the Earth.

Figure 3 shows that in the Northern hemisphere at high latitudes, there are two positive anomalies in the surface air temperature and two negative anomalies up to 1 °K. In the Southern hemisphere, more pronounced and positive anomalies up to 1.5 °K and a few negative anomalies were found. According to the temperature response depicted in Fig. 3, it is difficult to say about the dependence of temperature anomalies on the sign of J_z-imposed changes in autoconversion rate.

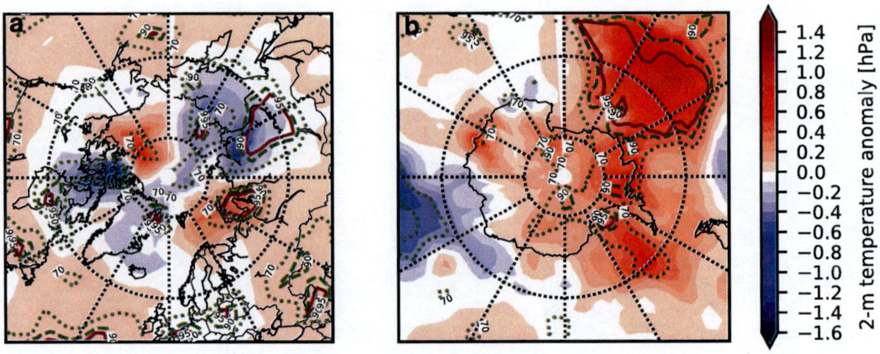

Fig. 3 Surface air temperature anomalies caused by changes in autoconversion rate averaged for the 1999–2002 period. **a** Anomalies in the northern hemisphere; **b** anomalies in the southern hemisphere. The green line indicates a statistical significance of more than 90%; the purple line indicates a statistical significance of more than 95%

It is also worth mentioning that, despite the autoconversion rate is influenced by IMF B_y-related changes in J_z in only two local regions, anomalies in air pressure and temperature are formed even in mid-latitudes. This is due to the atmospheric transport associated, in particular, with large-scale planetary (Rossby) waves, which transfer clouds from the region of influence on it, which leads to extensive anomalies, even at mid-latitudes [7, 14].

4 Discussion and Summary

In this work, a sensitivity of the surface pressure and air temperature to certain change in the stratiform-cloud autoconversion rate imposed by the variation of the downward current J_z due to the IMF B_y was studied. It was found that even local and small changes in the autoconversion rate of $\pm 12\%$ lead to the appearance of large-scale anomalies in surface meteorology at high and middle latitudes. We revealed the anomalies in the surface pressure of about 1 hPa at the maximum and in the surface air temperature that not exceed 1.5 °K.

In our work we assumed a linear dependence of changes in J_z and changes in autoconversion rate. However, the major question remains as to what is the quantitative J_z-related change in the autoconversion rate, due to the complexity of microphysical processes in clouds. It should be noted that modern global climate models still have a rather low resolution to accurately reproduce the microphysical processes occurring in the clouds. Even the clouds themselves in the models are presented in a simplified form since the model's grid is many times greater than the diameter of the cloud. Such a study can be carried out, for example, using local meteorological cloud models.

Currently, anomalies in surface meteorology found in observations associated with the Mansurov effect are not statistically confirmed and have a statistical significance below 95% [5]. In our work, several statistically significant anomalies were found, however, more ensemble members are needed to make an accurate conclusion on significance. Also, the dependence of anomalies in air temperature and pressure on the sign of changes in autoconversion rate due to J_z should be studied further as based on our results it is still hard to bring out the conclusion.

It is should be mentioned that in this first study we used non-realistic case where we analyze the sensitivity of surface meteorology to the constant IMF B_y-related anomalies in J_z current density applied for the whole period of simulation. Nevertheless, even in this case, the obtained anomalies in surface meteorology are in good agreement in magnitude with those detected in observations [7, 14]. This study will be continued with the use of more ensemble members and with the inclusion of IMF B_y in the model to link the anomalies in J_z to variation of IMF B_y and investigate a cross-correlation between IMF B_y and anomalies the surface meteorological parameters, since these anomalies strongly depend on the IMF B_y state (positive or negative). Nevertheless, this work showed the importance of further studying the IMF-induced atmospheric electricity effects on weather and climate.

Acknowledgements We are grateful to the Russian Foundation for Basic Research (RFFI) (project No. 19-35-90134) for the financial support of this study. The work was done in the SPbSU "Ozone Layer and Upper Atmosphere Research Laboratory" supported by the Ministry of Science and Higher Education of the Russian Federation under agreement 075-15-2021-583. ER was partially supported by the Russian Science Foundation (RNF) (grant No. 21-17-00208). The authors also thank ETH's High-Performance Computing Center (ID SIS) for the possibility to use the Euler Linux cluster to conduct our numerical experiments and the Center for Climate Systems Modeling (C2SM) for their support. IM was partially supported by a grant from the Government of the Russian Federation (contract no. 075-15-2019-1892).

References

1. Baumgaertner, A.J.G., Thayer, J.P., Neely, R.R., Lucas, G.: Toward a comprehensive global electric circuit model: atmospheric conductivity and its variability in CESM1(WACCM) model simulations. J. Geophys. Res. (Atmos.) **118**(16), 9221–9232 (2013). https://doi.org/10.1002/jgrd.50725
2. Burns, G.B., Tinsley, B.A., French, W.J.R., Troshichev, O.A., Frank-Kamenetsky, A.V.: Atmospheric circuit influences on ground-level pressure in the Antarctic and Arctic. J. Geophys. Res. (Atmos.) **113**(D15), D15112 (2008). https://doi.org/10.1029/2007JD009618
3. Carslaw, K.S., Harrison, R.G., Kirkby, J.: Cosmic rays, clouds, and climate. Science **298**(5599), 1732–1737 (2002). https://doi.org/10.1126/science.1076964
4. Cowley, S.W.H., Lockwood, M.: Excitation and decay of solar wind-driven flows in the magnetosphere-ionosphere system. Ann. Geophys. **10**(1–2), 103–115 (1992)
5. Edvartsen, J., Maliniemi, V., Tyssøy, H.N., Asikainen, T., Hatch, S.M.: The Mansurov effect: real or a statistical artefact? Earth Space Sci. Open Arch. **14** (2021). https://doi.org/10.1002/essoar.10505897.1
6. Egorova, T., Rozanov, E., Zubov, V., Karol, I.: Model for investigating ozone trends (Mezon). Izv. Atmos. Ocean Phys. **39**, 277–292 (2003)

7. Freeman, M.P., Lam, M.M.: Regional, seasonal, and inter-annual variations of Antarctic and sub-Antarctic temperature anomalies related to the Mansurov effect. Environ. Res. Commun. **1**(11), 111007 (2019). https://doi.org/10.1088/2515-7620/ab4a84
8. Golubenko, K., Rozanov, E., Mironova, I., Karagodin, A., Usoskin, I.: Natural sources of ionization and their impact on atmospheric electricity. **47**(12), e88619 (2020). https://doi.org/10.1029/2020GL088619
9. Harrison, R.G., Lockwood, M.: Rapid indirect solar responses observed in the lower atmosphere. Proc. Roy. Soc. Lond. Ser. A **476**(2241), 20200164 (2020). https://doi.org/10.1098/rspa.2020.0164
10. Harrison, R.G., Nicoll, K.A., Ambaum, M.H.P.: On the microphysical effects of observed cloud edge charging. Q. J. Roy. Meteorol. Soc. **141**(692), 2690–2699 (2015). https://doi.org/10.1002/qj.2554
11. Karagodin, A., Rozanov, E., Mareev, E., Mironova, I., Volodin, E., Golubenko, K.: The representation of ionospheric potential in the global chemistry-climate model SOCOL. Sci. Total Environ. **697**, 134–172 (2019). https://doi.org/10.1016/j.scitotenv.2019.134172
12. Lam, M.M., Chisham, G., Freeman, M.P.: The interplanetary magnetic field influences mid-latitude surface atmospheric pressure. Environ. Res. Lett. **8**(4), 045001 (2013). https://doi.org/10.1088/1748-9326/8/4/045001
13. Lam, M.M., Chisham, G., Freeman, M.P.: Solar wind-driven geopotential height anomalies originate in the Antarctic lower troposphere. **41**(18), 6509–6514 (2014). https://doi.org/10.1002/2014GL061421
14. Lam, M.M., Tinsley, B.A.: Solar wind-atmospheric electricity-cloud microphysics connections to weather and climate. J. Atmos. Solar-Terr. Phys. **149**, 277–290 (2016). https://doi.org/10.1016/j.jastp.2015.10.019
15. Lohmann, U., Roeckner, E.: Design and performance of a new cloud microphysics scheme developed for the ECHAM general circulation model. Clim. Dyn. **12**(8), 557–572 (1996). https://doi.org/10.1007/BF00207939
16. Lucas, G.M., Baumgaertner, A.J.G., Thayer, J.P.: A global electric circuit model within a community climate model. J. Geophys. Res. (Atmos.) **120**(23), 12054–12066 (2015). https://doi.org/10.1002/2015JD023562
17. Mansurov, S.M., Mansurova, L.G., Mansurov, G.S., Mikhnevich, V.V., Visotskii, A.M.: North-south asymmetry of geomagnetic and tropospheric events. J. Atmos. Terr. Phys. **36**, 1957–1962 (1974). https://doi.org/10.1016/0021-9169(74)90182-2
18. Manzini, E., Giorgetta, M.A., Esch, M., Kornblueh, L., Roeckner, E.: The influence of sea surface temperatures on the northern winter stratosphere: ensemble simulations with the MAECHAM5 model. J. Clim. **19**(16), 3863 (2006). https://doi.org/10.1175/JCLI3826.1
19. Pettigrew, E.D., Shepherd, S.G., Ruohoniemi, J.M.: Climatological patterns of high-latitude convection in the northern and southern hemispheres: dipole tilt dependencies and interhemispheric comparisons. J. Geophys. Res. (Space Phys.) **115**(A7), A07305 (2010). https://doi.org/10.1029/2009JA014956
20. Pierce, J.R.: Cosmic rays, aerosols, clouds, and climate: recent findings from the CLOUD experiment. J. Geophys. Res. (Atmos.) **122**(15), 8051–8055 (2017). https://doi.org/10.1002/2017JD027475
21. Stenke, A., Schraner, M., Rozanov, E., Egorova, T., Luo, B., Peter, T.: The SOCOL version 3.0 chemistry-climate model: description, evaluation, and implications from an advanced transport algorithm. Geosci. Model Dev. **6**(5), 1407–1427 (2013). https://doi.org/10.5194/gmd-6-1407-2013
22. Tinsley, B.A.: The global atmospheric electric circuit and its effects on cloud microphysics. Rep. Prog. Phys. **71**(6), 066801 (2008). https://doi.org/10.1088/0034-4885/71/6/066801

UV Pulsations in the Auroral Region According to Measurements on the Lomonosov Satellite

P. A. Klimov and K. F. Sigaeva

Abstract The Lomonosov satellite was launched into a 97.3° polar sun-synchronous orbit on April 26, 2016, and operated in orbit until December 2017. The TUS detector is a highly sensitive telescope, designed for both detection of ultra-high-energy cosmic rays and the Earth's atmosphere UV-radiation in the 240–400 nm range. To measure variety of atmospheric phenomena the telescope has four modes of operation with different temporal resolutions (from 0.8 μs to 6.6 ms). Among the events recorded above 50° N in the mode with a temporal resolution of 6.6 ms, 66 were identified with an unusual spatiotemporal structure, representing local pulsations of glow in the field of view of the instrument. Geographical distribution was analyzed and it was shown that events are mainly located in the area of equatorial boundary of auroral oval (more than half of events were registered on L-shells 4–6). Characteristic frequencies of pulsations of glow intensity are of the order of 1–10 Hz, and areas of pulsating glow are localized in the space with characteristic horizontal size of 10–15 km. A comparison with the THEMIS ground-based all-sky cameras was made. 11 joint observations were found and the presence of aurora lights at the time of TUS events registration was shown by ground-based cameras, although no obvious coincidences of pulsations were found.

Keywords Flickering aurora · Pulsating aurora · Orbital telescope

1 Introduction

Numerous ground-based observations demonstrate the existence of a fine spatial and temporal structure of auroras, also mentioned as a flickering or pulsating aurora. Pulsating auroras are quasi-periodic variations of intensity of aurora luminescence

P. A. Klimov (✉) · K. F. Sigaeva
Skobeltsyn Institute of Nuclear Physics Lomonosov Moscow State University, 1(2), Leninskie gory, Moscow 119234, Russia

K. F. Sigaeva
Faculty of Physics, Lomonosov Moscow State University, 1(2), Leninskie gory, Moscow 119234, Russia

A. Kosterov et al. (eds.), *Problems of Geocosmos—2020*, Springer Proceedings in Earth and Environmental Sciences, https://doi.org/10.1007/978-3-030-91467-7_31

caused by precipitation of energetic electrons. They occur mainly in the MLT sector from midnight to morning after expansion of the auroral oval and during the substorm recovery phase. They appear as irregular patches of luminosity with quasi-periodic (2–20 s or more) temporal fluctuations, which are often accompanied by rapid complex movements of their bright part, synchronized with changes in luminosity [1]. The characteristic position for this type of auroras is on the equatorial edge of the auroral oval, shorter periods of pulsations appear at lower latitudes [2].

Analysis of ground-based optical observations of the flickering aurora showed the following features of the spatiotemporal structure. The width of the flickering column is in the range of 1–12 km [3], the lifetime of the flickering column is 1–2 s [4]. The luminosity modulation amplitude is less than 10–20% of the background luminosity, and typical frequencies are 3–15 Hz [5].

Various types of pulsating aurora are distinguished by the spatial structure and intensity modulation: pure pulsations (fixed size and shape), expanding pulsations, streaming aurora, etc. [6]. Three categories of pulsating aurora were proposed based on the stability of patch shape and spatial extent: Amorphous Pulsating Aurora (APA), Patchy Pulsating Aurora (PPA) and Patchy Aurora (PA) [7].

New measurements of fast pulsations in the near-ultraviolet (NUV) wavelength range in the auroral and subauroral regions were conducted by the TUS detector onboard the Lomonosov satellite with a high sensitivity and temporal resolution (6.6 ms). In what follows we present some results of these new data.

2 The TUS Detector and Data Selection

The Lomonosov satellite was launched on April 26, 2016 into a polar sun-synchronous orbit with an altitude of 470–500 km and an inclination of 97.3°, which allowed observations in the polar region. The TUS detector consists of two main parts: a mirror-concentrator and a photodetector in its focal plane. The photodetector pixels are 256 Hamamatsu R1463 photomultiplier tubes (PMTs). Each pixel has a black hood that protects it from side illumination, as well as a UV filter (300–400 nm) with a diameter of 13 mm and a thickness of 2.5 mm. PMTs are arranged in the following way—256 pixels are grouped into 16 identical modules, having a common system of high-voltage power supply, system of data acquisition and primary processing. More details about the TUS telescope can be found in [8].

The electronics of the photo detector operate in four modes, designed to register transient atmospheric events with different temporal resolution. The main mode has a resolution of 0.8 μs. It is designed to measure extensive air showers as well as fastest transient atmospheric events, such as elves. Other modes of the TUS detector have lower temporal resolution: 25.6 μs and 0.4 ms for recording slower events, such as sprites and various jets, and 6.6 ms for recording the slowest events: micrometeors, space debris, and auroras, the latter are analyzed in this paper.

In this mode, the TUS detector operated for three small intervals of time: 26.12.2016–10.01.2017, 28.02.2017–21.03.2017, 08.11.2017–15.11.2017,

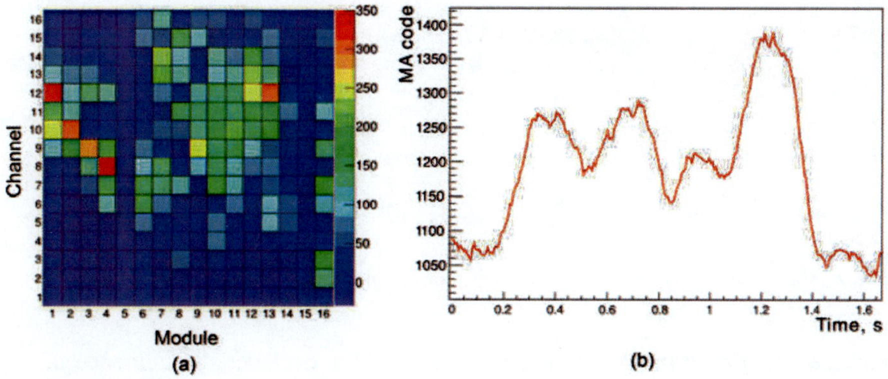

Fig. 1 Example of the event with NUV pulsations measured by the TUS detector November, 10, 2017 at 13:31 UTC: Pixel map (**a**), waveform of one pixel: module 7, channel 12 (**b**)

and during this time about ten thousand events were registered, 2575 of them at high latitudes (more than 50°) in the Northern Hemisphere. In the Southern Hemisphere at that time at the same latitudes was polar day, and registration of events was not conducted there.

Among all data several types of events were identified: (1) Stationary noise (majority of events where signal intensity fluctuates around a certain value). (2) Stationary source on the surface of the Earth or in the atmosphere: cities, clouds (maximum intensity of signal moves along the detector matrix at the speed of a satellite). (3) Thunderstorms and lightning discharges (events with sharp peaks of high intensity, which corresponds to lightning return strokes). (4) Events with unusual pulsations of the NUV signal at high latitudes.

The first three types of events were excluded from the further analysis. And only 66 events of the last type were considered.

Figure 1a shows pixel map of one such pulsating event and demonstrates the spatial structure of the event. Figure 1b shows the waveform of a single pixel (module 7, channel 12). The signal growth is observed for 1.2 s with a pulsation. The pulsation period is about 0.3 s, and the amplitude is 10% of the background level. These temporal features are close to ones described in [5].

3 Results

3.1 Geographical Distribution and Geomagnetic Activity Analyses

The geographical location of the events is shown in Fig. 2. As can be seen, all events are located along the aurora region at latitudes from 52 to 71° N. The events were

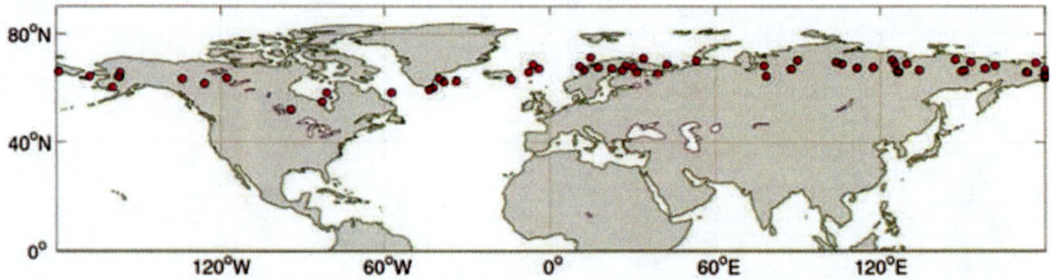

Fig. 2 Map of selected events with NUV pulsations (violet dots)

recorded both over the ocean and over land, which excludes the influence of the anthropogenic factor. The distribution of these events over the L-shell was studied. It appears that most of the events were measured when the satellite was between 4 and 6 L-shell, i.e. close to the equatorial edge of the auroral oval and even further south.

In this work, the AE-index was used to estimate the level of geomagnetic activity at the time of measuring the events. The AE-index data were provided by the World Data Center for Solar-Terrestrial Physics in Moscow (http://www.wdcb.ru). The majority of events is located during high geomagnetic activity during the substorm recovery phase. The example of AE-index behavior during the third period of measurements when a half of events were registered is shown at Fig. 3. All events are recorded at the local maxima of the AE index or at the stage of its decrease after the peak. Distribution of the number of NUV pulsation events by the AE-index is shown in Fig. 4a. It can be seen that the maximum number of events (33) was measured in the 200–400 nT interval. Comparison with the distribution of all events registered by the detector was also carried out to obtain the real frequency of occurrence of events in different periods of geomagnetic activity. The black histogram in the Fig. 4b shows the distribution of observation times by AE-index. And the green histogram shows

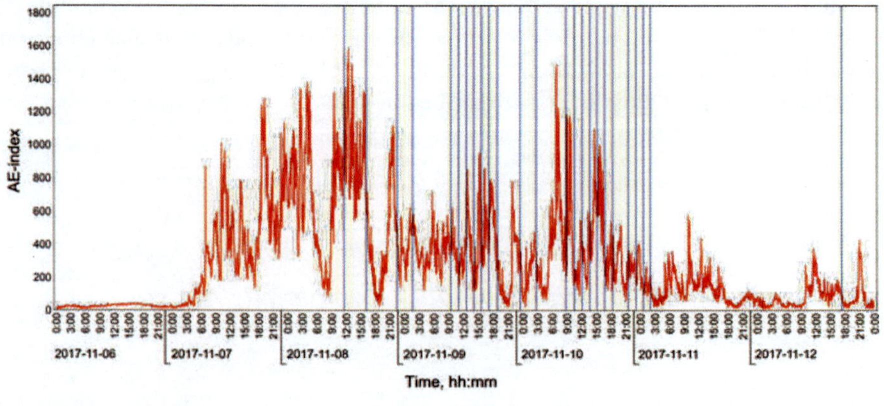

Fig. 3 AE-index variations during the period of measurements 06.11.2017–12.11.2017. Blue lines indicate the TUS events

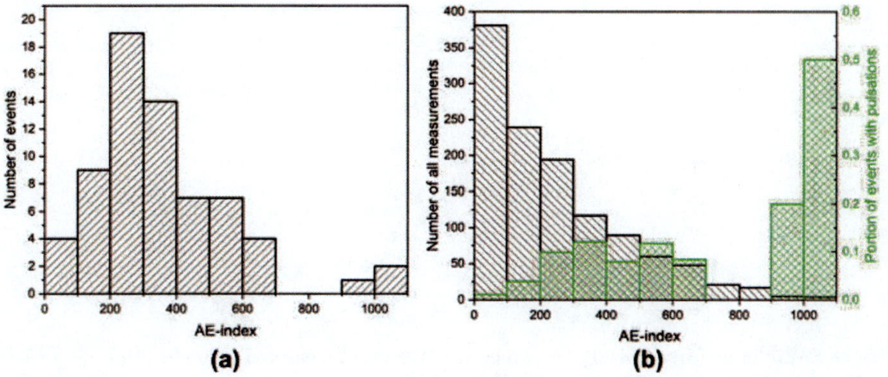

Fig. 4 Distribution of events with NUV pulsations by AE-index (**a**). Distribution of all measurements at high latitudes by AE-index (black) and portion of events with pulsations (green) (**b**)

events frequency normalized by the observation time for a given AE-index. Thus, it is clearly seen that the effective fraction of events with pulsations during strong geomagnetic perturbations increases.

The distribution of events by MLT is shown in Fig. 5. It can be clearly seen that the most of NUV pulsation events were observed around MLT midnight, but is it mostly due to the satellite orbit parameters. The Lomonosov satellite had a sun-synchronous polar orbit and passed over aurora latitudes equator at a fixed local time around midnight.

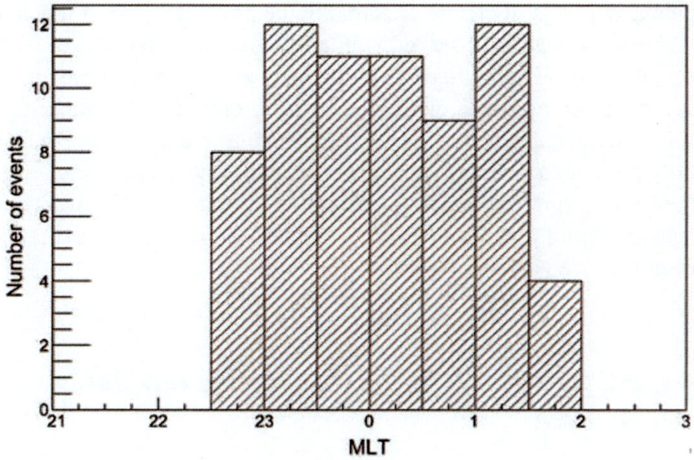

Fig. 5 Distribution of events with NUV pulsations by MLT

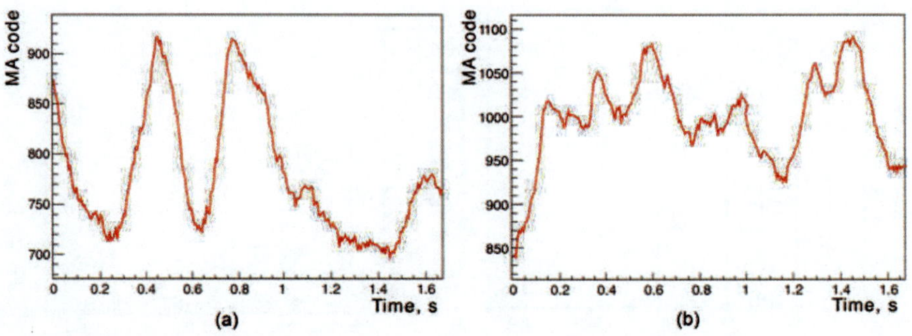

Fig. 6 Waveforms of two pixels for two events. The event measured on 06.01.2017 at 05:18:26 UTC, demonstrates low frequency (2 Hz) repeating structures (**a**). The event measured on 06.01.2017 at 05:18:26 UTC with higher frequency pulsations (> 10 Hz) (**b**)

3.2 Temporal and Spatial Structure of Events

The time structure of the registered events is quite diverse: different intensity, number of pulsations and their period. Examples of such events are shown in Fig. 1 and in Fig. 6. It is well seen from the figures that the frequencies of pulsations differ clearly. The study of frequency distribution demonstrates that most frequently measured pulsations lie in the region of 3 and 5 Hz, but there are events with frequencies up to 20 Hz.

The recorded events are characterized by the different temporal structures in different parts of the matrix. An example of the distribution of temporal structures over a part of the detector matrix is shown in Fig. 7. It demonstrates the oscillograms of 20 neighboring pixels. It can be seen that in the down left part of this matrix fragment only one slow peak is observed, while in the central part there are pulsations with 3 Hz modulation. There are three pixels with zero signal. These channels were broken. So, in the photo detector matrix various temporal structures can be seen simultaneously which means that spatially they occur in a part of the detectors FOV.

The FOV of one instrument pixel is 10 mrad, which corresponds to an area of 5×5 km on the Earth's surface (and 4×4 km at an altitude of 100 km). The observed structure changes from pixel to pixel in the recorded events suggest that the typical horizontal source size is about 10 km.

3.3 Comparison with THEMIS All-Sky Ground-Based Cameras Data

The data recorded by the TUS detector was compared with data from the THEMIS all-sky ground-based cameras [9]. This system of cameras is located so that it covers the entire sky over North America in the area of aurora observations. Temporal

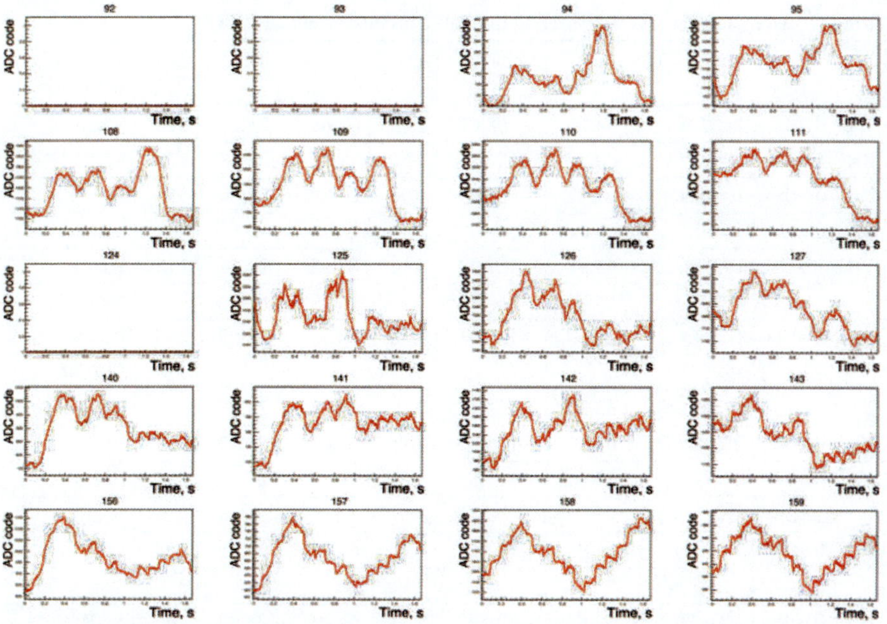

Fig. 7 Waveforms of part of photodetector. Event measured on 10.11.2017 at 13:31 UTC above Chukotka. Three pixels with zero signal are inoperative

resolution is only 3 s, and thus it cannot measure fast pulsations, but its data allows one to understand the general activity and global structure of aurora emission.

A total of 11 data intersections were found (see Table 1) in measurements of the TUS detector and THEMIS camera array. Analysis of the joint observations of events showed that during the registration of events by the TUS detector active aurora luminescence was observed over North America.

One example of joint observations carried out on 06 January, 2017 at 05:17:33 UTC is presented in Fig. 8. On the panel **a** of the figure there is a map of images from all-sky THEMIS cameras. The red dot marks the place of event registration by the TUS detector. According to THEMIS cameras data the presence of auroras at the time of event registration is clearly visible (bright arc over the north of Canada). In the camera which corresponds to the TUS event only dim diffuse light is seen. It might be closed by clouds at the time of measurements or its sensitivity is not enough to register the same pulsations as TUS. On the panel **b** of the figure one of the oscillograms of the detector matrix at the moment of pulsations measurements is shown. Variations of luminescence at this moment are clearly seen on the oscillogram. The same situation is seen in Fig. 9: bright and variable aurora luminescence in the majority of THEMIS cameras. TUS events are on the edge of westernmost camera, but above the cloud coverage.

So, the TUS detector can provide an additional information of fine temporal structure of faint NUV emission in aurora due to its high aperture and sensitivity.

Table 1 List of TUS events measured in the THEMIS all-sky cameras FOV

Date	Time UTC	Lat	Long	L	MLAT	MLT	Ae	THEMIS station code
2017-01-03	11:19:57	64.23	−168.06	4.46	61.68	23.17	614	KIAN
2017-11-10	10:24:54	60.17	−159.75	3.69	59.15	22.88	175	MCGR
2017-01-08	10:38:19	64.22	−157.69	4.90	63.30	23.09	555	KIAN
								MCGR
2017-11-10	10:23:52	63.97	−157.23	4.84	63.15	22.83	175	KIAN
								MCGR
2017-01-01	10:39:41	65.67	−156.91	5.50	64.78	23.09	286	KIAN
								MCGR
2017-11-10	08:49:54	63.30	−134.19	6.19	66.32	22.80	462	WHIT
2017-01-06	08:24:59	61.62	−126.00	6.15	66.06	23.12	220	FSIM
2017-01-07	07:57:16	63.64	−117.82	8.25	69.22	23.14	287	FSIM
								EKAT
2017-01-01	6:00:49	51.95	−94.16	4.40	60.52	23.45	285	GILL
								TPAS
2017-01-06	5:18:26	54.82	−82.57	5.32	64.05	23.63	266	SNKQ
2017-01-06	5:17:33	58.11	−80.95	6.73	67.40	23.72	265	SNKQ

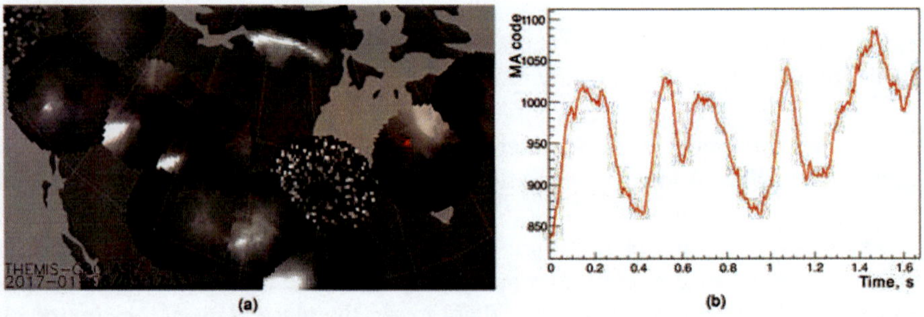

Fig. 8 Map of images from THEMIS all-sky ground cameras, red dote shows the place of TUS detector at that moment (**a**). Waveform of one pixel for the event measured on 06.01.2017 at 05:17:33 UTC (**b**)

4 Conclusions

During the operation of the TUS detector on orbit NUV pulsations of atmospheric glow at high latitudes were registered. It became possible due to high sensitivity of the telescope, and the polar orbit of the spacecraft.

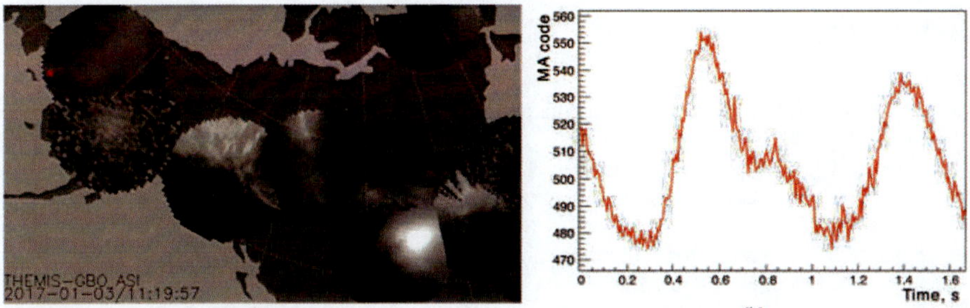

Fig. 9 Map of images from THEMIS all-sky ground cameras, red dote shows the place of TUS detector at that moment (**a**). Waveform of one pixel for the event measured on 03.01.2017 at 11:19:57 UTC (**b**)

The structure of the recorded signals is quite diverse, but all the described events have the following characteristic features in common: frequencies of the order of 1–10 Hz and the spatial size of the pulsations about 10 km. In the FOV of the telescope we can simultaneously observe several different regions of pulsations with different temporal structures.

Analysis of geographical distribution and geomagnetic conditions showed that events were recorded at latitudes from 52 to 71° N. The location of these events does not depend on the level of geomagnetic activity. Most pulsations were registered at L-shells from 4 to 6, which corresponds to the equatorial edge of the auroral zone. However, the event's appearance frequency correlates with the geomagnetic activity.

A comparison was made between the TUS detector measurements and THEMIS all-sky ground-based camera data. The analysis showed the presence of bright polar lights at the time of registration of events in most cases. However, the pulsations themselves are not visible on the cameras, possibly because the TUS detector is much more sensitive than the cameras and recorded more faint emissions. Due to the TUS detector design, it can record events with even faster pulsations (such as those recorded in [10]), if sufficient exposure is achieved.

The spatial and temporal structure of the registered events is similar to the previously observed flickering auroras [5] and to the internal modulations of pulsating auroras associated with the high-energy part of the precipitating electrons caused by LBC waves. However, their nature and mechanism of occurrence are not yet clear. To study and clarify the nature of this phenomenon, further experiments on high-sensitivity orbital detectors, as well as the comparison of data obtained on satellites with data from ground-based observatories, are needed.

Acknowledgements The authors are grateful to the Lomonosov collaboration and the TUS detector development team. The work was done with partial financial support from the State Space Corporation ROSCOSMOS and the Interdisciplinary Scientific and Educational School of Moscow University "Fundamental and Applied Space Research".

References

1. Yamamoto, T.: On the temporal fluctuations of pulsating auroral luminosity. J. Geophys. Res.: Space Phys. **93**, 897–911 (1988)
2. Duncan, C., Creutzberg, F., Gattinger, R., Harris, F., Jones, A.: Latitudinal and temporal characteristics of pulsating auroras. Can. J. Phys. **59**, 1063–1069 (2011). https://doi.org/10.1139/p81-140
3. Sakanoi, K., Fukunishi, H.: Temporal and spatial structures of flickering aurora derived from high-speed imaging photometer observations at Syowa station in the Antarctic. J. Geophys. Res. Space Phys. **109**, A01221 (2004). https://doi.org/10.1029/2003JA010081
4. Kunitake, M., Oguti, T.: Spatial-temporal characteristics of flickering spots in flickering auroras. J. Geomagn. Geoelectr. **36**, 121–138 (1984). https://doi.org/10.5636/jgg.36.121
5. Sakanoi, K., Fukunishi, H., Kasahara, Y.: A possible generation mechanism of temporal and spatial structures of flickering aurora. J. Geophys. Res. **110**, A03206 (2005). https://doi.org/10.1029/2004JA010549
6. Nishimura, Y., Lessard, M.R., Katoh, Y., Miyoshi, Y., Grono, E., Partamies, N., Sivadas, N., Hosokawa, K., Fukizawa, M., Samara, M., Michell, R.G., Kataoka, R., Sakanoi, T., Whiter, D.K., Oyama, S.I., Ogawa, Y., Kurita, S.: Diffuse and pulsating aurora. Space Sci. Rev. **216**, 4 (2020). https://doi.org/10.1007/s11214-019-0629-3
7. Grono, E., Donovan, E.: Differentiating diffuse auroras based on phenomenology. Ann. Geophys. **36**, 891–898 (2018). https://doi.org/10.5194/angeo-36-891-2018
8. Klimov, P.A., Panasyuk, M.I., Khrenov, B.A., Garipov, G.K., Kalmykov, N.N., Petrov, V.L., Sharakin, S.A., Shirokov, A.V., Yashin, I.V., Zotov, M.Y., Biktemerova, S.V., Grinyuk, A.A., Grebenyuk, V.M., Lavrova, M.V., Tkachev, L.G., Tkachenko, A.V., Park, I.H., Lee, J., Jeong, S., Martinez, O., Salazar, H., Ponce, E., Saprykin, O.A., Botvinko, A.A., Senkovsky, A.N., Puchkov, A.E.: The TUS detector of extreme energy cosmic rays on board the Lomonsov satellite. Space Sci. Rev. **212**, 1687–1703 (2017). https://doi.org/10.1007/s11214-017-0403-3
9. Donvan, E., Mende, S., Jackel, B., Frey, H., Syrjäsuo, M., Voronkov, I., Trondsen, T., Peticolas, L., Angelopoulos, V., Harris, S., Greffen, M., Connors, M.: The THEMIS all-sky imaging array-system design and initial results from the prototype imager. JASTP **68**(13), 1472–1487 (2006). https://doi.org/10.1016/j.jastp.2005.03.027
10. Kataoka, R., Miyoshi, Y., Hampton, D., Ishii, T., Kozako, H.: Pulsating aurora beyond the ultra-low-frequency range. J. Geophys. Res. (Space Phys.) **117**, A08336 (2012). https://doi.org/10.1029/2012JA017987

Polar Cap ULF Pulsations: Coordinated Radar-Magnetometer Observations

O. V. Kozyreva, V. A. Pilipenko, X. Shi, E. C. Bland, and L. Baddeley

Abstract This study uses multi-instrument geomagnetic and ionospheric observations in Antarctica to examine the causative mechanisms of the ultra-low-frequency (ULF) variations of the electrodynamics of the polar cap—the least explored part of the coupled solar wind-magnetosphere-ionosphere system. The ionospheric oscillations corresponding to Pc5–6 pulsation band (quasi-periods about 3–15 min) were detected by SuperDARN Mac-Murdo radar. The ground geomagnetic response *B* was examined using magnetometers at polar latitudes in Antarctica. We have selected events with quasi-periodic variations of the ionospheric Doppler velocity *V* and simultaneously observed by ground magnetometers. Supposedly, this class of events is produced by magnetospheric MHD waves. We have determined the effective wave impedance, that is *V/B* ratio, which enabled us to identify the physical nature of specific polar cap pulsations. This type of polar Pc5–6 pulsations has been associated with Alfven mode, though the occurrence of periodic disturbances in the region with open or strongly extended field lines seems puzzling.

Keywords ULF pulsations · Polar cap · SuperDARN radar · Magnetometers

1 Introduction: Specific Polar Cap ULF Pulsations

The electrodynamic coupling between the solar wind energy input and magnetospheric dynamics mostly occurs at the boundary layers and in the magnetotail, which

O. V. Kozyreva (✉) · V. A. Pilipenko
Institute of Physics of the Earth, Moscow, Russia
e-mail: kozyreva@ifz.ru

V. A. Pilipenko
Space Research Institute, Moscow, Russia

X. Shi
Virginia Tech, Blacksburg, USA

High Altitude Observatory, NCAR, Boulder, USA

E. C. Bland · L. Baddeley
University Centre in Svalbard, Longyearbyen, Norway

A. Kosterov et al. (eds.), *Problems of Geocosmos—2020*, Springer Proceedings in Earth and Environmental Sciences, https://doi.org/10.1007/978-3-030-91467-7_32

are geomagnetically conjugate to the cusp/cleft and polar cap. The polar cap (the region near the magnetic pole bounded by the auroral oval) often has been considered as a quiet region surrounded by the intense auroral activity along which all interesting processes occur. However, such a view has to be radically changed, as polar disturbances have been established to have a fundamental connection with magnetospheric dynamics. Thus, the polar region can be considered as the stage of a variety of fascinating phenomena which are vital in understanding the solar wind—magnetosphere interaction.

Studies of electromagnetic phenomena in the polar cap have been hampered so far by the lack of continuous observations due to difficulties with operating extensive permanent arrays at very high latitudes. The area of our particular interest is the Antarctic continent—the only landmass at polar latitudes, which is a uniquely suited location, though not easily accessible, for stable arrays of instruments. Antarctica offers great geophysical advantages such as a vast territory of uniform surface, free from man-made electromagnetic pollution, no interference from lightning, and a specific configuration of geomagnetic field.

The terrestrial magnetosphere is a turbulent system and cannot be fully characterized without accounting for turbulence into its energy balance. The occurrence of natural MHD resonators and waveguides results in a quasi-periodic response in the Ultra Low Frequency (ULF) frequency range (time scales from seconds to tens of minutes) to turbulent driving. The characteristics of ULF activity in the polar cap are much less known because of limited station coverage at very high latitudes. The magnetic activity in the polar caps was found to be richer and more important than has been generally thought [1]. The occurrence of specific long-period ULF signatures (periods ~4–20 min), which might be classified as Pi3 pulsations inside the polar cap, and different from common Pc5 waves, was found by *Yagova* et al. [2, 3]. The physical mechanism of these polar cap signals has not been established yet.

The intense wave activity in the ULF frequency range is also observed in the region of the ionospheric projection of the magnetospheric boundary layers. At dayside, irregular long-period magnetic pulsations are observed almost every day over a wide period range (3–20 min) when a magnetic observatory approaches the projection of the dayside cusp/cleft. These cusp-associated broad-band disturbances were denoted IPCL (Irregular Pulsations at Cusp Latitudes) by *Bolshakova* et al. [4]. Though, later the epicenter of IPCL power was found to be ~2–3° equatorward from the cusp equatorward boundary [5]. At nightside, persistent geomagnetic and auroral poleward boundary intensifications (PBI) are observed near the poleward boundary of the auroral oval [6]. The polar cap ULF disturbances may be traces of the dayside cusp or nightside auroral poleward boundary activity. The drivers and mechanisms of IPCL and PBI have not been firmly established, though in some events an association of these pulsations with transient phenomena in the magnetosheath [7] or with magnetotail plasma flow bursts [8] was found.

Magnetospheric ULF waves can be detected not only by ground magnetometers, but by ionospheric radars as well. The ULF waves can modulate the ionospheric velocity and plasma density, and thus can be recorded by the Doppler sounders [9], SuperDARN coherent scatter radars [10–12] and EISCAT incoherent scatter radars [13, 14], and even by the global navigation satellite system GPS [15]. On the other hand, in the frequency range overlapping with Pc5–6 ULF band the terrestrial ionosphere is filled with a broad spectrum of acoustic-gravity waves (AGWs), which may be generated by various atmospheric sources. Variations of the ionospheric plasma and currents in the E-layer are produced by dragging of the ionized component by oscillatory movement of the neutral particles. As a result, AGWs in the ionosphere may be accompanied by modulation of geomagnetic field [16].

Thus, even coordinated ionospheric and geomagnetic observations face ambiguity in interpretation of the mechanism of periodic oscillations, especially striking at very high latitudes. An investigation using data from all Northern hemisphere SuperDARN radars indicated that oscillations of the ionospheric plasma velocity in the Pc5 band (with a mean frequency ~2 MHz) occurred often on the night-side [17]. An automated algorithm detected a significant number of ionospheric oscillations above 80° magnetic latitude. Ionospheric signatures of periodic disturbances in the Pc5 band were surveyed using high-resolution data from SuperDARN radars [18]. Ionospheric wave events in this band were found to dominate at high and polar latitudes in the dusk sector. However, the association of the ionospheric quasi-periodic variations reported in the studies mentioned above with the magnetospheric Pc5 waves seems questionable, because morphological properties of the ionospheric variations (e.g., latitudinal and LT distributions) differ from known ones for ground geomagnetic activity. Moreover, no comparison with simultaneous geomagnetic data has been made. Therefore, mechanisms of the ionospheric fluctuations in the Pc5–6 band and their association with geomagnetic pulsations at very high latitudes still have not been established.

Probably, specific long-period polar cap waves and transients may be driven by oscillatory modes of the magnetotail or tail lobes, theoretically predicted in [19, 20]. The region of the nightside magnetosphere near the inner plasmasheet boundary forms a cavity for compressional mode [21]. Compressional modes of this cavity also may be responsible of some ULF high-latitude nightside phenomena. Thus, more work is necessary to understand physics of quasi-periodic ionospheric pulsations and their possible interrelations with geomagnetic pulsations. Combined magnetometer-radar measurements are very promising because they enable one to determine the *E/B* ratio, that is a proxy of the wave impedance [22]. This characteristic may help to discriminate different types of MHD waves at very high latitudes.

The main focus of this study is to determine the causes of electromagnetic fluctuations with time scales about minutes—ten minutes in the Southern polar cap. In order to obtain additional clues about the mechanisms of high-latitude ULF waves, we compare data from the Antarctic magnetometers with simultaneous monitoring of the ionospheric plasma velocity variations derived from SuperDARN radar observations. Some implications of the observational results are discussed.

2 Observational Facilities

The virtue of the Antarctic array is the opportunity to monitor ULF wave activity deep in the polar cap. The geographic and geomagnetic coordinates of the Antarctic stations, noon UT, are given in Table 1. Each site is equipped with flux-gate magnetometer with flat frequency response and provides 1 min measurements of local geomagnetic H, D, and Z components.

An essential issue of the studies at very high latitudes is that the standard geomagnetic coordinates calculated from internal sources of geomagnetic field only, do not characterize properly a site location in respect to magnetospheric domains and orientations of the coordinate axes. Moreover, magnetic local time (MLT) may differ from ordinary local time (LT) by up to 8 h. To characterize the actual position of stations in respect to the ionospheric projections of main magnetospheric domains we have used the OVATION model, which provides the poleward and equatorward boundaries of the auroral oval [23]. This auroral precipitation model parameterized by the interplanetary parameters was developed using energetic particle measurements from the Defense Meteorological Satellite Program (DMSP) satellite. Distinguishing feature of the model includes an optimized solar wind-magnetosphere coupling function which predicts auroral power significantly better than any interplanetary parameter or geomagnetic index.

We present an analysis of ULF oscillations observed in the Southern polar ionosphere by the SuperDARN high-frequency (3–30 MHz) coherent scatter radar at McMurdo (MCM) which provides measurements of the ionospheric flow in the F-layer. The MCM field of view (FOV) covers the polar cap. The magnetic stations DMC, VOS, CSY are within the radar FOV. The DMC magnetometer is located directly under the beam 8 of the MCM radar. The radar FOV is divided into 16 azimuthal beams spaced ~3.2° apart, with each beam divided into 75 range gates of 45 km length. The total field of view thus covers ~52° azimuthal sector extending from 180 km to over 3500 km in range. The observations were obtained during the SuperDARN "common mode" operation when beams are sampled sequentially with an integration time of ~3 s per beam. The final time resolution of the data stream along a particular beam direction is 1 min. However, one can observe velocity

Table 1 List of Antarctica stations used in the study. Corrected geomagnetic (CGM) coordinates and UT of local magnetic noon have been computed for an altitude of 100 km using the web facility (nssdc.gsfc.nasa.gov/space/cgm/cgm.html)

Station	Code	Geo. lat	Geo. long	CGM lat	CGM long	Noon (UT)
McMurdo	MCM	−77.85	166.67	−79.9	327.5	18:56
Casey	CSY	−66.17	110.32	−80.7	155.2	06:53
Vostok	VOS	−78.46	106.82	−83.3	54.4	13:12
Terra Nova Bay	TNB	−74.69	164.12	−80.0	306.9	20:05
Dome C	DMC	−75.10	123.38	−88.9	54.7	13:00
ZhongShan	ZHS	−69.37	76.37	−74.5	97.1	10:12

oscillations with radar only if a suitable backscattering from field-aligned inhomogeneity presents. Signatures of ULF waves are identified in the radar data as periodic fluctuations in the Doppler velocity **V** arising due to mostly horizontal plasma motion.

The Doppler broadening of the spectral width of the radar backscattered signal due to turbulence inside the plasma scattering volume, as measured in SuperDARN radar returns, is frequently considered as a benchmark of the ionospheric projections of magnetospheric boundary domains [24]. The accuracy of such qualitative determination of ionospheric projections of the magnetospheric domains is about 1°. The spectral width boundary is shown to represent a good proxy for the equatorward cusp boundary and the open/closed field line boundary (OCB) close to midnight and noon, but is located ~2–4° equatorward of the OCB across much of the morning sector [25].

A direct penetration of long period variations from the solar wind and boundary regions along reconnected field lines, convected with the solar wind flow into the distant magnetotail, into the polar cap may occur. To examine this possibility, variations of solar wind and interplanetary magnetic field (IMF) parameters are analyzed using the 1 min OMNI data base. To characterize the energy flow from the solar wind into the high-latitude ionosphere we have used PC-index (https://pcindex.org/archive), which measures the intensity of the trans-polar current across the polar cap.

3 Observational Results

3.1 Typical Events

Quasi-periodic variations of the ionospheric plasma velocity in the Pc5–6 band that we have observed by the MCM radar during summer time can be roughly classified into 3 classes:

(1) simultaneously observed by ground magnetometers (DMC, VOS stations);
(2) geomagnetic pulsations without ionospheric signatures;
(3) ionospheric pulsations without accompanying geomagnetic fluctuations.

Supposedly, the first and second classes are produced by magnetospheric MHD waves. We examine the ionosphere-ground relationship of the first class of polar ULF activity (~20 events). The occurrence of 3rd class of ULF phenomena may have different non-magnetospheric origin (e.g., atmospheric AGWs) and they will be considered elsewhere.

3.2 *Examples of Simultaneous Ionospheric-Magnetic Pulsations in the Polar Cap*

Dayside Pulsations (2013/09/27)

Transient quasi-periodic (~12 min) variations are observed on 1250–1400 UT by Antarctic magnetometers in the dayside sector. According to the OVATION model, the stations VOS and DMC are deep inside the statistical polar cap, whereas the poleward auroral boundary is around ~80° S (Fig. 1).

Peak-to-peak amplitude of these pulsations is ~10–15 nT (Fig. 2h). Even trans-polar current is modulated by these pulsations, as evident from PC index variations (Fig. 2f). No evident trigger/driver either in the solar wind or in IMF data can be seen (Fig. 2a, b, c, d and e). The solar wind parameters Np ~5 cm^{-3}, V ~300 km/s are steady and demonstrate just small irregular fluctuations. The IMF magnitude is about 1 nT, whereas IMF Bz > 0.

MCM radar observes a similar response at the same time, with peak-to-peak velocity variations V ~400 m/s (Fig. 2g). However, the disturbance waveform in the

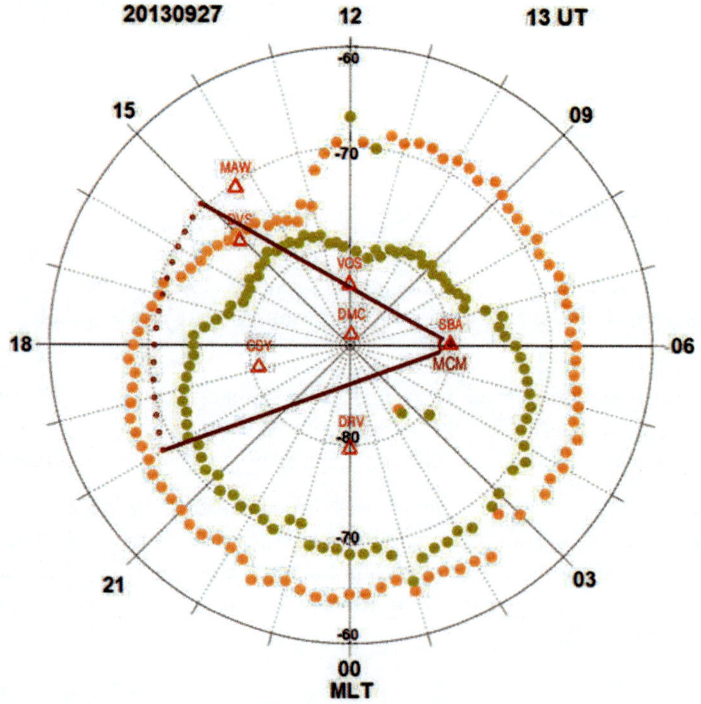

Fig. 1 The position of Antarctic magnetic stations (open triangles) in respect to the ionospheric projections of the poleward (army green) and equatorward (orange) boundaries of the auroral oval given by the OVATION model during the event 2013/09/27, 13 UT. FOV of the MCM radar is shown with straight lines

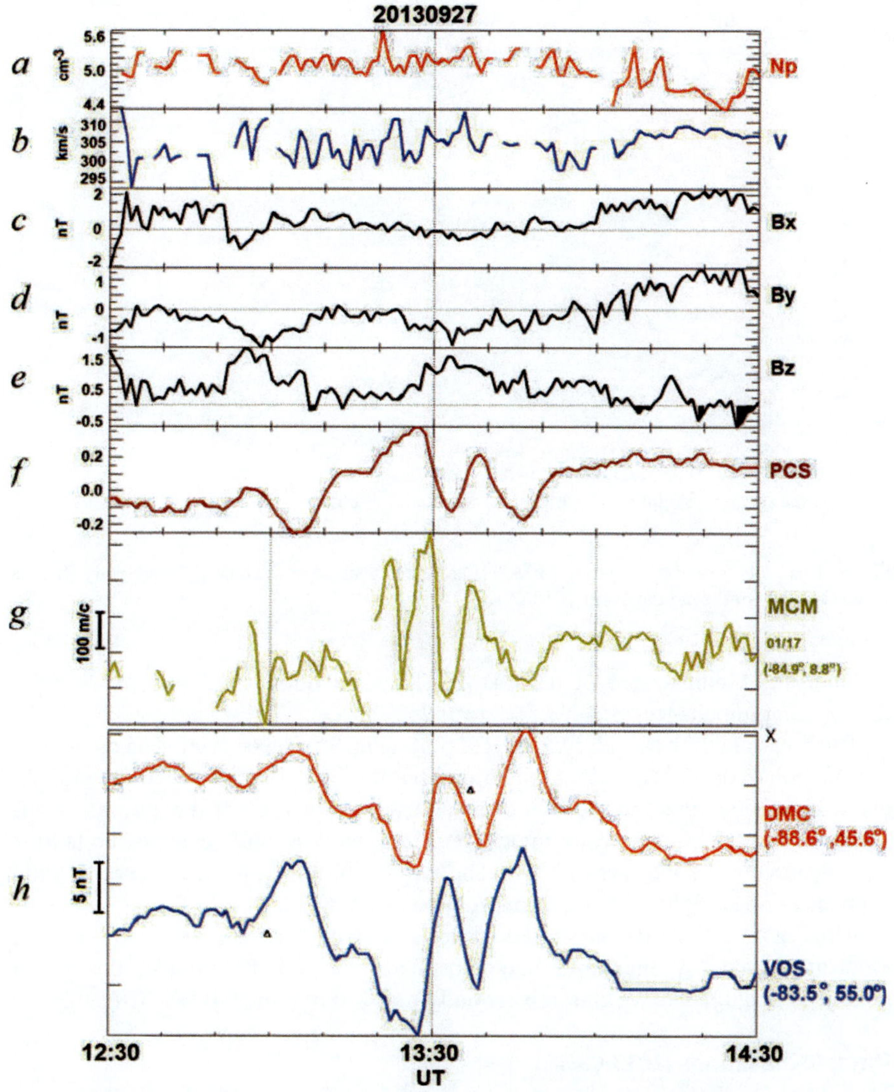

Fig. 2 Variations of the solar wind density *N*p (**a**) and velocity *V* (**b**), IMF Bx (**c**), By (**d**), and Bz (**e**) components, PC-index for Southern hemisphere (**f**); Doppler velocity variations (beam/gate and coordinates are indicated near right-hand Y-axis) (**g**); Magnetic variations (X-component) at stations DMC (red) and VOS (blue) (**h**) during the event 2013/09/27

ionosphere is distorted as compared with geomagnetic response, probably owing to the superposition of second harmonic.

Comparison of the resulting velocity time series for the MCM radar along with the magnetic field signatures recorded by Antarctic magnetometers provides an important estimate for the theoretical interpretation. Peak-to-peak amplitudes are $V \sim$

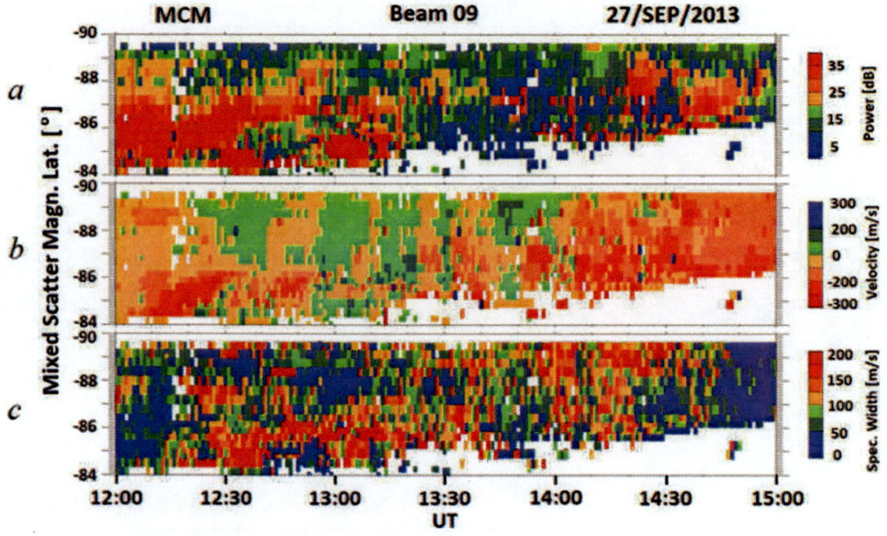

Fig. 3 RTI plots of variations along MCM radar beam 09 of power (**a**), Doppler velocity (**b**), and spectral width (**c**) during the event 2013/09/27

420 m/s, and *X* component ~ 8 nT at DMC. Thus, the normalized ionospheric effect (or apparent impedance) is *V/X* ~52.5 (m/s)/nT.

The spectral width boundary from RTI plots (Fig. 3c) can be used as an estimate of the location of dayside equatorward auroral boundary. Using this spectral width plot one can conclude that the location of equatorward boundary of the auroral oval (> 50 m/s) during the time interval under consideration is around geomagnetic latitude ~84° S, around the latitude of VOS. The OCB inferred from radar spectral width happens to be at higher latitude than predicted by OVATION model.

Although VOS station is shifted towards dayside from the geomagnetic pole, according to the Tsyganenko-01 magnetospheric magnetic field model, the apex of field lines conjugate to VOS is substantially dragged into the tail lobe (Fig. 10).

Dayside Pulsations (2013/08/12)

This event on 0850–1010 UT was found and described in [26]. During this event the Antarctic stations are both deep inside the polar cap, according to the OVATION model (Fig. 4).

Pc6 pulsations can be seen at VOS and DMC (Fig. 5h). Fluctuations are noticeable in PC index (Fig. 5f). No evidence of the polar pulsation triggering or driving by the solar wind or IMF variations can be seen (Fig. 5a, b, c, d and e). The average amplitude of velocity variation for the MCM radar is *V* ~220 m/s, whereas amplitude of *X* component of the magnetic field signature recorded at DMC is X ~10 nT. Thus, the normalized ionospheric effect is *V/X* ~22 (m/s)/nT.

Nearly synchronous ULF oscillations of the Doppler velocity were extended over 10–15 range gates (450–675 km) for the MCM radar within the Southern polar cap,

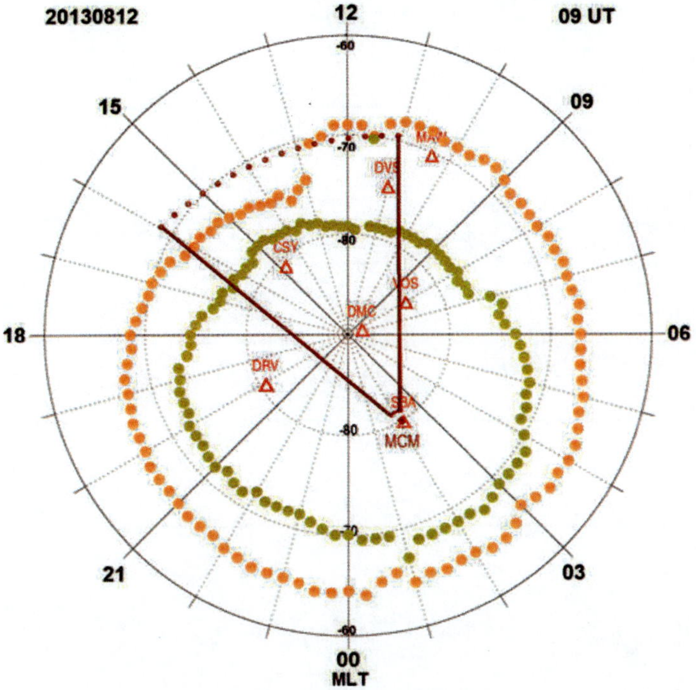

Fig. 4 The position of Antarctic magnetometers (triangles) in respect to the ionospheric projections of the poleward (army green) and equatorward (orange) boundaries of the auroral oval given by the OVATION model during the event on 2013/08/12, 09 UT. FOV of the MCM radar is shown with straight lines

covering magnetic latitudes of 80–90° S. This wave feature gives a possibility to average over several ranges the velocity data along a selected beam [26]. The polar cap oscillations, detected by the MCM radar along beam 8, exhibit frequencies in the range 1.1–1.3 MHz over the entire FOV (Fig. 5g).

As follows from RTI plots (Fig. 6b), there is no clear sense of phase propagation in latitudinal direction, as typical for resonant Pc5 pulsations. The spectral width boundary at RTI indicates that the location of dayside equatorward auroral boundary is around 84.5° S (Fig. 6c).

Nightside Pulsations (2013/07/03)

According to OVATION model, the FOV of the MCM radar and Antarctic stations are inside the polar cap (Fig. 7), and shifted towards nighttime from geomagnetic pole.

Series of geomagnetic pulsations in the nominal Pc6 band are observed inside the polar cap at stations DMC and VOS, during the interval 0150–0430 UT (Fig. 8h). No evident trigger can be seen in the solar wind or IMF data (Fig. 8a, b, c, d and e). IMF Bz is weakly positive. Pulsations reveal themselves even in PC index (Fig. 8f). No pulsations with the same period or waveform can be seen at lower auroral latitudes.

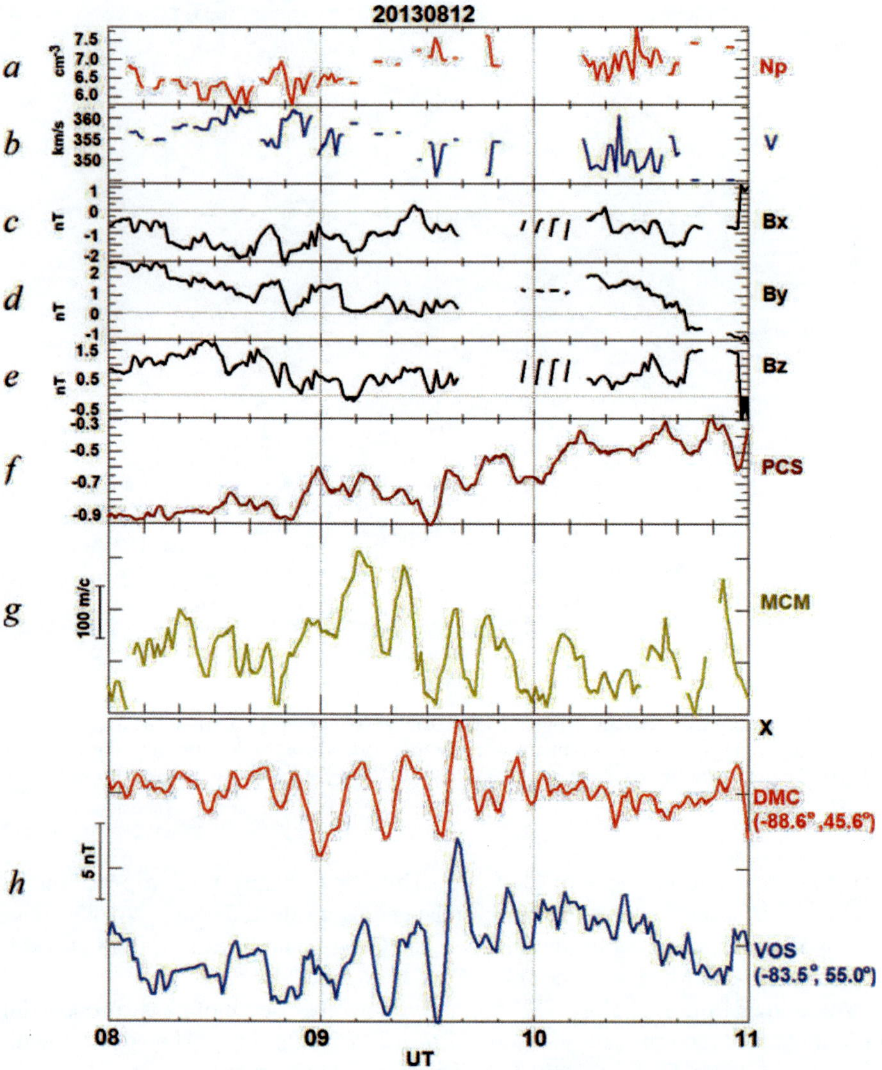

Fig. 5 Variations of the solar wind density Np (**a**) and velocity *V* (**b**), IMF Bx (**c**), By (**d**), Bz (**e**) components, PC-index (**f**), Doppler velocity gate-averaged variations along beam 08 (**g**), magnetic variations (*X*-component) at stations DMC and VOS (**h**) during the event 2013/08/12

The MCM range-averaged data from beam 9 show similar pulsations in the Doppler velocity (Fig. 8g). The average amplitude of velocity variation is V ~250 m/s, whereas amplitude of X component of the magnetic field signature recorded at DMC is X ~15 nT. Thus, the normalized ionospheric effect is V/X ~16.7 (m/s)/nT.

RTI plots do not show any latitudinal propagation, pulsations are synchronous at all gates (Fig. 9b). The spectral width at RTI plot indicates the poleward boundary

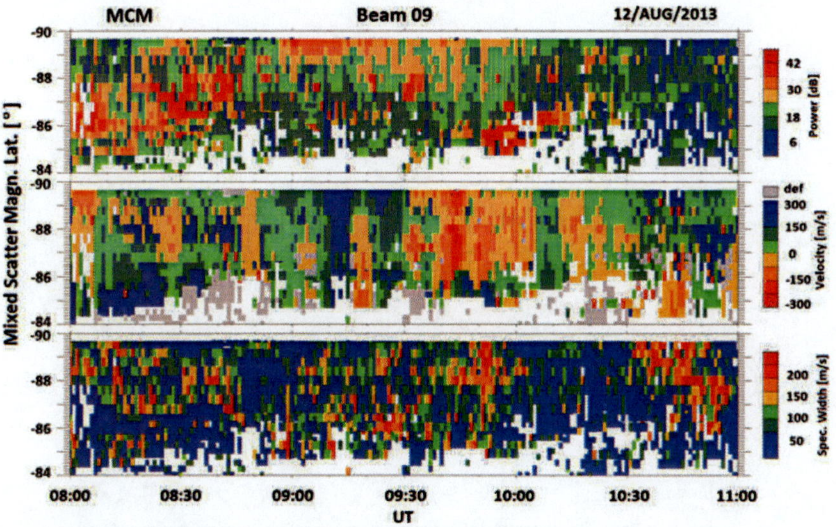

Fig. 6 RTI plots along MCM radar beam 09 showing variations of power (**a**), Doppler velocity (**b**), and spectral width (**c**) during the event 2013/08/12

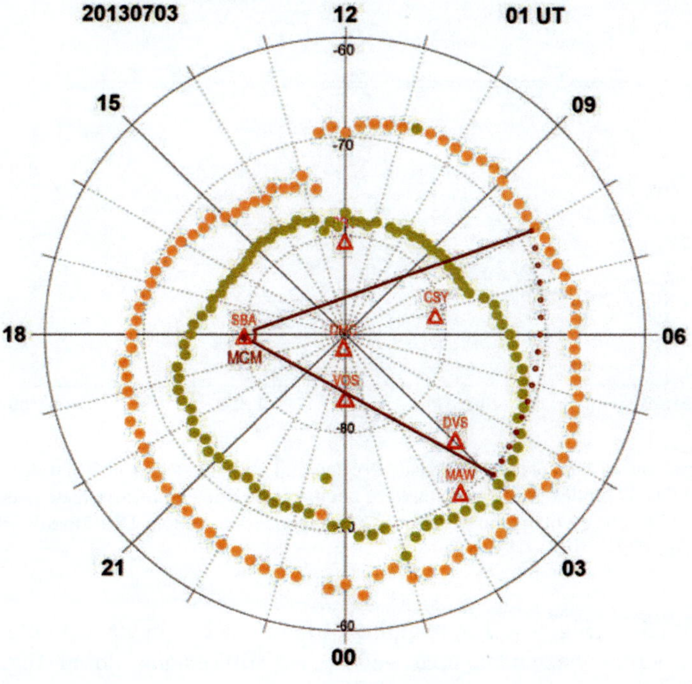

Fig. 7 The position of Antarctic magnetic stations (triangles) in respect to the ionospheric projections of the poleward (army green) and equatorward (orange) boundaries of the auroral oval given by the OVATION model during the event on 2013/07/03, 01 UT

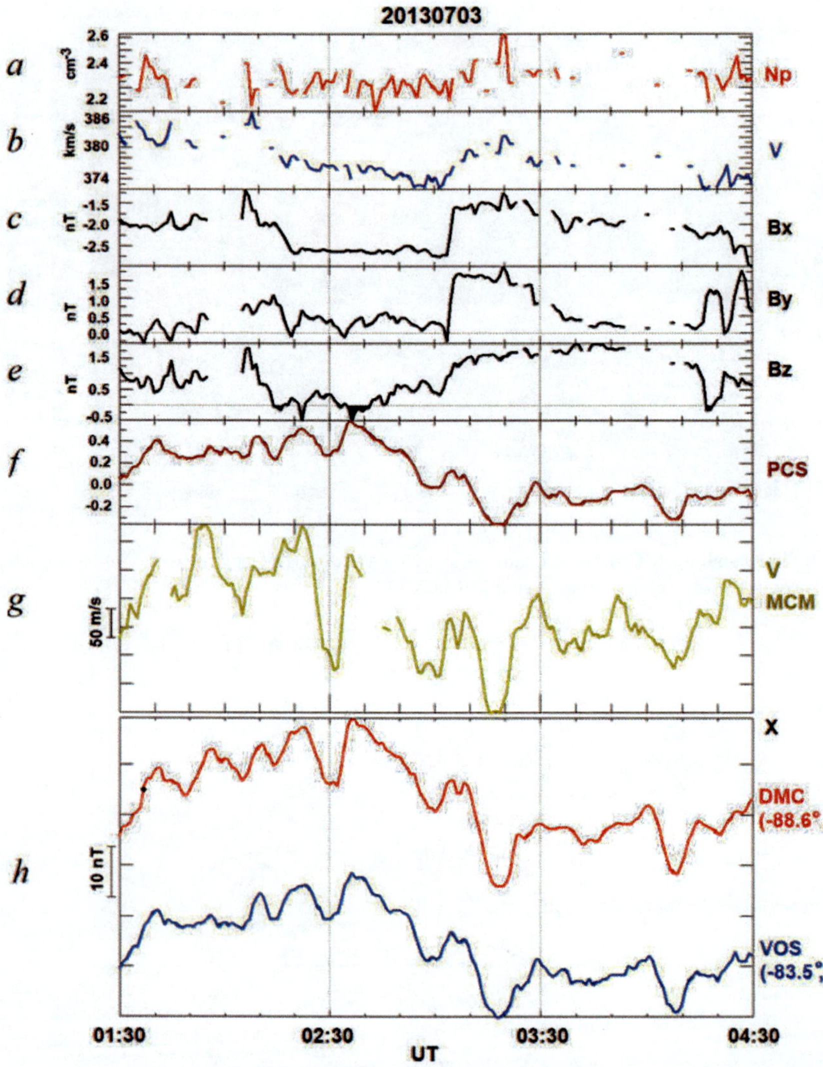

Fig. 8 Variations of the solar wind density *N*p (**a**) and velocity *V* (**b**), IMF Bx (**c**), By (**d**), Bz (**e**) components, PC-index (**f**) for the Southern hemisphere, Doppler velocity gate-averaged variations along 09 beam (**g**), magnetic variations (*X*-component) at stations DMC and VOS (**h**) during the event 2013/07/03, 0130–0430 UT

location at 85° S (Fig. 9c). The mapping of field line conjugate to VOS shows that field line is strongly extended, up to ~60 R_E, but still remains closed (Fig. 10).

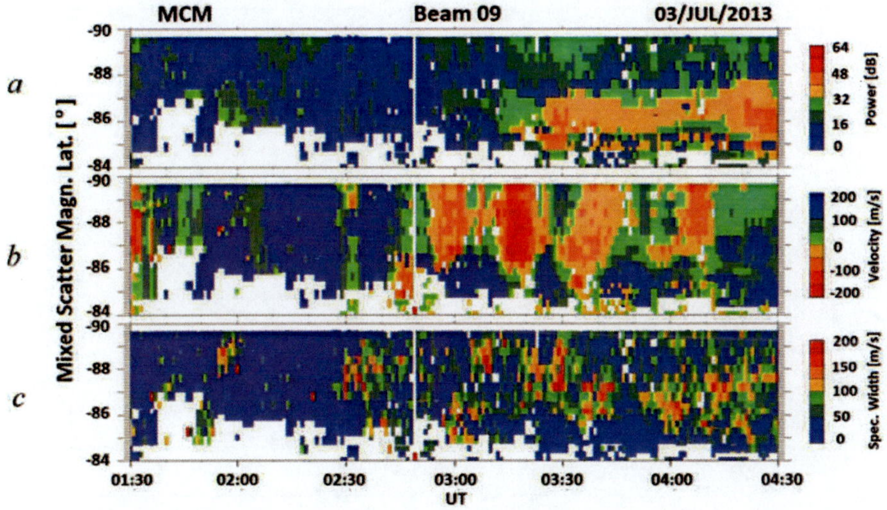

Fig. 9 RTI plots of Doppler velocity variations along MCM radar beam 9 of power (**a**), Doppler velocity (**b**), and spectral width (**c**) during the event 2013/07/03

4 Mechanisms of the Ionospheric Velocity Modulation by MHD Waves

To understand the possible generation mechanisms, the distinctive features of ULF signatures, including their morphological characteristics, amplitude-phase spatial structures, and solar wind/IMF correlation, should be examined in greater detail. Here we examine one aspect of the problem—the ionosphere-ground relationship, which nonetheless provides an important information about ULF wave properties if compared with some theoretical estimate.

We use a coordinate system where the *X*-coordinate corresponds to the geomagnetic North–South direction, and *Y* corresponds to the East–West direction. The modulation of the ionospheric plasma velocity may be caused by the electric field **E** of an incident MHD wave. The geomagnetic field $\mathbf{B}_0$ is inclined to the ionosphere by the inclination angle *I*. The radar observations of ULF waves in the upper ionosphere (F-layer) can detect oscillations of the ionospheric plasma with velocity $\mathbf{V} = [\mathbf{E} \times \mathbf{B}_0]/B_0^2$. The corresponding **E**-field can be obtained as $\mathbf{E} = -[\mathbf{V} \times \mathbf{B}_0]$ from radar **V** data (the order of magnitude estimate is $V[\text{m/s}] = 20 \cdot E\ [\text{mV/m}]$). At high latitudes the radar backscatter is sensitive to the near-horizontal plasma motion $\mathbf{V} = \{V_x, V_y\}$. The horizontal components of the plasma velocity are related to the horizontal components of the wave electric field as

$$\begin{pmatrix} V_x \\ V_y \end{pmatrix} = \frac{1}{B_o} \begin{pmatrix} E_y \sin I \\ E_x / \sin I \end{pmatrix} \tag{1}$$

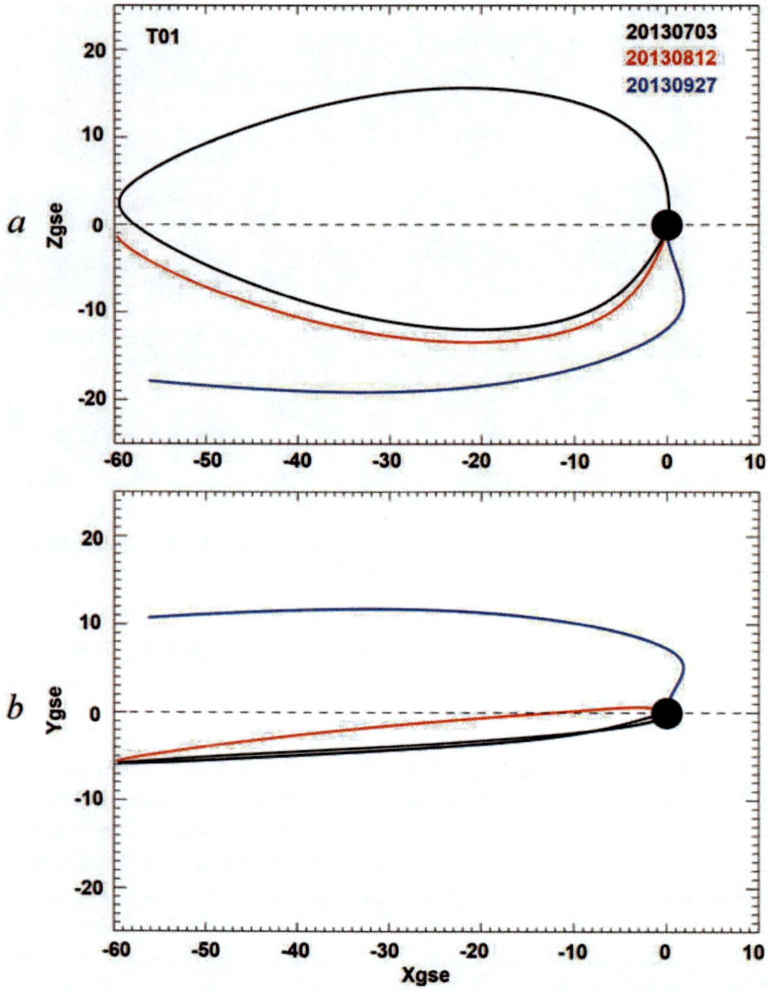

Fig. 10 The mapping of field lines conjugate to VOS station into the magnetotail made with the Tsyganenko-01 magnetospheric magnetic field model for events under study: X–Z panel (**a**), X–Y plane (**b**)

A contribution from V_z, which can be obtained from a ground scatter, for the event under consideration is weak.

The interaction of long-wavelength MHD waves with the coupled magnetosphere—ionosphere—atmosphere—ground system can be considered under the "thin ionosphere" approximation. In this approximation the E-layer of the ionosphere is approximated as an infinitely thin film with the height-integrated Hall and Pedersen conductances, located at altitude h above the ground.

4.1 *Alfven Mode Incidence on the Ionosphere*

Here we examine the possibility that polar cap geomagnetic pulsations are associated with the Alfven mode oscillations. In realistic observations it is hard to discriminate poloidal and toroidal electric or magnetic field components, because the actual structure of the Alfven velocity distribution and wave front is not known. In a toroidal Alfven mode (the azimuthal scale is much larger than radial scale, $\lambda_y > > \lambda_x$) the dominant components in the ionosphere are E_x, V_y, and no magnetic field compression *B*. Though, in realistic situations (especially in the polar cap) both spatial scales in the azimuthal and radial directions may be comparable, $\lambda_y \geq \lambda_x$, so both horizontal components V_y, V_x and E_x, E_y are present in an incident wave with comparable amplitudes.

Upon the transmission of an Alfvenic mode through the ionosphere the $\pi/2$ wave ellipse rotation should occur $B_y \rightarrow B_x^{(g)}$(*X*-component), $B_x \rightarrow B_y^{(g)}$(*Y*-component) [27]. The relationship between the ionospheric plasma velocity V_y and the ground magnetic disturbance $B_x^{(g)}$, for toroidal Alfven mode is to be as follows [22].

$$\frac{V_y}{B_x^{(g)}} = \frac{V_H}{B_0 \sin^2 I} \tag{2}$$

The characteristic ionospheric velocities $V_{P,H,C} = \left(\mu_o \Sigma_{P,H,C}\right)^{-1}$ are determined by the height-integrated Pedersen Σ_P, Hall Σ_H, and Cowling Σ_C conductances. For the order of magnitude estimate the following relationship can be used $V[\text{km/s}] \approx 800\Sigma^{-1}[\text{S}]$. Assuming that for the polar ionosphere Σ_H ~1 S, V_H ~800 km/s, $B_0 =$ 30 000 nT, $\sin^2 I = 0.95$, the theoretically predicted ratio $V/B_x^{(g)} \sim 28$ (m/s)/nT. This estimate is about the measured ratio *V/X* ~ 20–50 (m/s)/nT.

4.2 *Fast Mode Incidence on the Ionosphere*

The ground magnetic disturbance can be a response to magnetospheric cavity or waveguide modes. These modes belong to a fast magnetosonic (compressional) wave. Azimuthally large-scale fast mode has dominant components B_x, E_y, V_x with vanishing field-aligned current *j* = 0. Upon the transmission through the ionosphere no ellipse rotation occurs. The ratio of the ground magnetic response to the incident compressional wave is determined by the transmission properties of the whole ionosphere-atmosphere-ground system. For the highly conductive ground, the penetration of the fast mode through the ionosphere is controlled by the parameter $p = \omega h / V_C$. Typically, $|p| < < 1$, so the ionosphere may be considered to be transparent for the fast mode. Thus, the incident magnetospheric wave is partially screened by the ionosphere from the ground and reflected mainly by Earth's surface. Above the nearly transparent ionosphere, the dominant electric component of the fast mode V_x at altitude z is related to the magnetic ground response $B_x^{(g)}$ by the following relationship [22]:

$$\frac{V_x(z)}{B_x^{(g)}} = -i\frac{\omega(z+h)\sin I}{B_o} \tag{3}$$

For a compressional wave with $\omega = 0.01\ s^{-1}$, $z + h = 300$ km, $B_0 = 3{\cdot}10^4$ nT, $\sin I = 0.95$, the relationship (3) shows that the "apparent impedance" is *V/B* ~0.1 (m/s)/nT, that is about 2 orders of magnitude less than that of Alfven mode. Thus, the electric field and velocity induced in the ionosphere by a fast compressional mode is much weaker than that by an Alfven mode for the same magnitude of the ground magnetic signal.

5 Discussion

SuperDARN HF radars recorded oscillations of the Doppler velocity in the Pc5–6 frequency band at F region altitudes that visually resemble ULF wave signatures specific for the polar cap, named as $Pi_{CAP}3$ pulsations in [3]. In many events a ground magnetic response to ionospheric pulsations can be revealed, though after detrending of raw data. In considered events the recorded polar cap pulsations modulated PC index, or trans-polar current. This observation is rather unexpected because the algorithm of PC index calculation assumes that this integrated characteristic of the energy transfer from the solar wind into the high-latitude magnetosphere is independent on interplanetary or geomagnetic fluctuations. However, in fact the PC index has turned out to be vulnerable to polar cap wave processes.

The mapping of field lines conjugate to VOS station into the magnetotail made with T-01 magnetospheric magnetic field model is shown in Fig. 10. During events under study the field lines are strongly extended into magnetotail, up to ~ 60 R_E and even beyond. The modeling predicts that field lines may be closed, though their geometry is far from quasi-dipole field lines in the inner magnetosphere. The period of oscillations is hard to estimate because a distribution of plasma density along the field lines in the deep magnetotail is unknown. Anyway, because field lines are so extended their Alfven period is to be larger than the Alfven period at auroral latitudes (Pc5 band).

An adequate interpretation of ULF observations in the polar regions inevitably demands a revision of many concepts developed earlier for lower latitudes. The occurrence of quasi-periodic signatures in the polar cap region poses a problem: can Alfvenic quasi-resonators exist in a region with open or very extended field lines?

In the high-latitude magnetosphere the propagation of Alfven waves can be significantly affected by local variations in the geometry of the geomagnetic field lines. These variations may take place in the tail current sheet or in the regions adjacent to the magnetopause. A quasi-resonator may occur due to partial Alfven wave reflection from regions with a rapid change of magnetic field geometry. Quasi-periodic oscillations can be excited whose period is determined by the dimensions of Alfven quasi-resonators between the polar ionosphere and strong kinking of field lines. The feasibility of partial Alfven wave reflection from a magnetic field kink for a simplified geometry was demonstrated analytically by *Pilipenko* et al. [28]. However, predictions of this model should be further verified, both experimentally and numerically with more realistic numerical models.

The sources of ULF waves in the Pc5–6 band (1–6 MHz) in the polar cap are probably different from those in the auroral oval. Indeed, studies of *Yagova* et al. [2] of Pc5–6 variations at the array of Antarctica magnetometers showed that ULF activity inside the polar cap was decoupled from that at auroral latitudes. Information about plasma processes can be transported to the high latitude ionosphere along extended magnetic field lines by Alfven waves. During this transport, a ground signal could be enhanced as compared with that in the tail lobe by a geometric factor, which is about the ratio between the scales of the lobe and polar cap. Rare satellite observations of ULF waves in the lobes [29, 30] have shown low wave amplitudes in the lobes compared with that in the plasma sheet [31, 32]. At steep interface between different magnetotail domains a conversion of compressional mode into guided Alfven oscillations may occur [33, 34].

Cavity or waveguide oscillations in the nighttime magnetosphere associated with fast compressional mode can illuminate an essential part of the nightside magnetosphere. Just ground magnetometer observations cannot resolve the mechanism of polar ULF disturbances. The wave impedance, that is *V/B* ratio, estimated from the combined data of ionospheric radar and ground magnetometer has indicated that ionospheric pulsations are produced by magnetospheric Alfven waves. The impedance method proves that pure compressional oscillations (e.g., cavity mode or lobe waveguide mode) cannot be responsible for the radar detected Pc5–6 pulsations, but they may be responsible for magnetic pulsations absent in the ionospheric data. However, in the real magnetosphere, the Alfven mode waves are always coupled with magnetosonic (fast compressional) mode. Hopefully, the impedance method may help to discriminate the ionospheric pulsations produced by AGWs.

The epicenter of power of geomagnetic ULF pulsations under study was inside the polar cap, and similar pulsations were not observed at lower latitudes. The correspondence the polar cap and auroral pulsation activity needs a special consideration. Also, a possible occurrence of similar polar cap pulsations simultaneously in the Northern and Southern hemispheres will be examined elsewhere.

We often met a mismatch between the locations of the auroral oval boundary predicted by the OVATION model and resulting from the MCM radar spectral width plots. This issue should be more deeply examined using other Antarctic SuperDARN radars, e.g. HLB and SYS.

Satellite observations have revealed a realm of dynamical phenomena of different spatial–temporal scales in the geomagnetic tail, such as bursty earthward flows, impulsive acceleration processes in the plasmasheet, multiple sporadic ion beams, flapping of the neutral sheet, etc. Ground images of many magnetotail dynamical phenomena have not been identified yet. Some of them may be responsible for polar cap Pc6/Pi3 pulsations, but their magnetospheric counterpart has not been identified so far.

6 Conclusion

SuperDARN observations at polar latitudes in Antarctica have revealed the persistent occurrence of quasi-periodic variations of the Doppler velocity in the Pc5–6 band (periodicity ~3–15 min) in the polar cap. The combined usage of ionospheric radar and ground magnetometer gives us a possibility to estimate an apparent impedance, that is *V/B* ratio. The application of this criterium indicated that ionospheric pulsations of the ionospheric plasma velocity detected by the MCM radar and magnetometers in the polar cap are produced by magnetospheric Alfven waves, though the occurrence of periodic disturbances in the region with open field lines seems puzzling. For the class of ULF geomagnetic variations without an ionospheric response the magnetospheric compressional cavity modes may be responsible.

Acknowledgements This work was supported by grant 20-05-00787 from Russian Science Foundation (OVK), grant ES647692 from the program INTPART of Research Council of Norway (EMB, LJB, VAP), and grant AGS-1341918 from U.S. National Science Foundation (XS). The SuperDARN data are available at http://vt.superdarn.org. The authors acknowledge the use of SuperDARN data - a collection of radars funded by agencies of Australia, Canada, China, France, Italy, Japan, Norway, South Africa, UK and USA. Magnetometer observations at stations VOS and DMC are maintained by the Arctic and Antarctic Research Institute (Russia), Ecole et Observatoire des Sciences de la Terre (France), and Istituto Nazionale di Geofisica e Vulcanologia (Italy). We acknowledge the usage of PC-index (https://pcindex.org). The Tsyganenko model was downloaded from the website (https://ccmc.gsfc.nasa.gov/requests/instant/tsyganenko). The OVATION-Prime product can be downloaded from the website (https://www.ngdc.noaa.gov/stp/ovation_prime/data).

References

1. Engebretson, M.J., Posch, J.L., Pilipenko, V.A., Chugunova, O.M.: Ground based observations of ULF waves at high latitudes. In: Takahashi, K., Chi, P.J., Denton, R.E., Lysak R.L. (eds.) Magnetospheric ULF Waves: Synthesis and New Directions, Geophysics Monograph, vol. 169, pp. 137–156. AGU, Washington DC (2006)
2. Yagova, N., Lanzerotti, L., Villante, U., Pilipenko, V., et al.: Magnetic activity in the ULF Pc5–6 band at very high latitudes in Antarctica. J. Geophys. Res. **107**(N8), 1195 (2002)
3. Yagova, N.V., Pilipenko, V.A., Lanzerotti, L.J., et al.: Two-dimensional structure of long-period pulsations at polar latitudes in Antarctica. J. Geophys. Res. **109**, A03222 (2004)
4. Bolshakova, O.V., Kleimenova, N.G., Kurazkovskaya, N.A.: Dynamics of polar cap according to the observations of long period geomagnetic pulsations. Geomag. Aeron. **28**(4), 661–665 (1988)
5. Pilipenko, V.A., Kozyreva, O.V., Lorentzen, D.A., Baddeley, L.J.: The correspondence between dayside long-period geomagnetic pulsations and the open-closed field line boundary. J. Atmos. Sol. Terr. Phys. **170**, 64–74 (2018)

6. Lyons, R.L., Nagai, T., Blanchard, G.T., et al.: Association between Geotail plasma flows and auroral poleward boundary intensifications observed by CANOPUS photometers. J. Geophys. Res. **104**, 4485–4500 (1999)
7. Kozyreva, O., Pilipenko, V., Lorentzen, D., Baddeley, L., Hartinger, M.: Transient oscillations near the dayside open-closed boundary: evidence of magnetopause surface mode? J. Geophys. Res. **124**, 9058–9074 (2019)
8. Keiling, A.: Alfvén waves and their roles in the dynamics of the Earth's magnetotail. A review. Space Sci. Rev. **142**, 73–156 (2009)
9. Pilipenko, V.A., Fedorov, E.N., Teramoto, M., Yumoto, K.: The mechanism of mid-latitude Pi2 waves in the upper ionosphere as revealed by combined Doppler and magnetometer observations. Ann. Geophys. **31**, 689–695 (2013)
10. Sakaguchi, K., Nagatsuma, T., Ogawa, T., Obara, T., Troshichev, O.A.: Ionospheric Pc5 plasma oscillations observed by the King Salmon HF radar and their comparison with geomagnetic pulsations on the ground and in geostationary orbit. J. Geophys. Res. **117**, A03218 (2012)
11. Baddeley, L.J., Yeoman, T.K., McWilliams, K.A., Wright, D.M.: Global Pc5 wave activity observed using SuperDARN radars and ground magnetometers during an extended period of northward IMF. Planet. Space Sci. **55**, 792–808 (2007)
12. Norouzi-Sedeh, L., Waters, C.L., Menk, F.W.: Survey of ULF wave signatures seen in the Tasman international Geospace environment radars data. J. Geophys. Res. **120**, 949–963 (2015)
13. Fenrich, F.R., Waters, C.L., Connors, M., Bredeson, C.: Ionospheric signatures of ULF waves: passive radar techniques. In: Takahashi, K., Chi, P.J., Denton, R.E., Lysak, R.L. (eds.) Magnetospheric ULF Waves: Synthesis and New Directions, AGU Geophysical Monograph 169, Washington DC (2006)
14. Belakhovsky, V., Pilipenko, V., Murr, D., Fedorov, E., Kozlovsky, A.: Modulation of the ionosphere by Pc5 waves observed simultaneously by GPS/TEC and EISCAT. Earth Planets Space **68**, 102 (2016)
15. Kozyreva, O.V., Pilipenko, V.A., Bland, E.C., Baddeley, L.J., Zakharov, V.I.: Periodic modulation of the upper ionosphere by ULF waves as observed simultaneously by SuperDARN radars and GPS/TEC technique. J. Geophys. Res. **125**, e2020JA028032 (2020). https://doi.org/10.1029/2020JA028032
16. Shalimov, S., Kozlovsky, A.: High-latitude E and F region coupling signature: a case study results from rapid-run ionosonde. J. Geophys. Res. **120**, 3033–3041 (2015)
17. Bland, E.C., McDonald, A.J., Menk, F.W., Devlin, J.C.: Multipoint visualization of ULF oscillations using the SuperDARN. Geophys. Res. Lett. **41**, 6314–6320 (2014)
18. Shi, X., Ruohoniemi, J.M., Baker, J.B.H., et al.: Survey of ionospheric Pc3–5 ULF wave signatures in SuperDARN high time resolution data. J. Geophys. Res. **123** (2018). https://doi.org/10.1029/2017JA025033
19. McKenzie, J.F.: Hydromagnetic oscillations of the geomagnetic tail and plasma sheet. J. Geophys. Res. **75**, 5331–5339 (1970)
20. Ershkovich, A.I., Nusinov, A.A.: Geomagnetic tail oscillations. Cosmic Electrodyn. **2**, 471–480 (1972)
21. Leonovich, A.S., Mazur, V.A.: Why do ultra-low-frequency MHD oscillations with a discrete spectrum exist in the magnetosphere? Ann. Geophys. **23**, 1075–1079 (2005)
22. Pilipenko, V., Belakhovsky, V., Kozlovsky, A., Fedorov, E., Kauristie, K.: Determination of the wave mode contribution into the ULF pulsations from combined radar and magnetometer data: method of apparent impedance. J. Atmos. Sol. Terr. Phys. **77**, 85–95 (2012)
23. Newell, P.T., Liou, K., Zhang, Y., et al.: OVATION prime-2013: extension of auroral precipitation model to higher disturbance levels. Space Weather **12**, 368–379 (2014)
24. Baker, K.B., Dudeney, J.R., Greenwald, R.A., et al.: HF radar signatures of the cusp and low-latitude boundary layer. J. Geophys. Res. **100**, 7671–7695 (1995)
25. Baker, K.B., Greenwald, R.A., Ruohoniemi, J.M., et al.: Simultaneous HF-radar and DMSP observations of the cusp. Geophys. Res. Lett. **17**, 1869–1872 (1990)
26. Bland, E.C., McDonald A.J.: High spatial resolution radar observations of ultralow frequency waves in the southern polar cap. J. Geophys. Res. **121**, 4005–4016 (2016)

27. Alperovich, L.S., Fedorov, E.N.: Hydromagnetic Waves in the Magnetosphere and the Ionosphere, Astrophysics and Space Science Library, vol. 353, p. 456, Springer Netherlands (2007)
28. Pilipenko, V., Mazur, N., Fedorov, E., Engebretson, M.J., Murr, D.L.: Alfven wave reflection in a curvilinear magnetic field and formation of Alfvenic resonators on open field lines. J. Geophys. Res. **110**, A10S05 (2005)
29. Siscoe, G.L.: Resonant compressional waves in the geomagnetic tail. J. Geophys. Res. **74**, 6482–6486 (1969)
30. Sarafopoulos, D.V.: Long duration Pc5 compressional pulsations inside the Earth's magnetotail lobes. Ann. Geophys. **13**, 926–937 (1995)
31. Bauer, T.M., Baumjohann, W., Treumann, R.A., Sckopke, N., Luhr, H.: Low-frequency waves in the near-Earth plasma sheet. J. Geophys. Res. **100**, 9605–9617 (1995)
32. Chen, S.H., Kivelson, M.G.: On ultralow frequency waves in the lobes of the Earth's magnetotail. J. Geophys. Res. **96**, 15711–15723 (1991)
33. Allan, W., Wright, A.N.: Hydromagnetic wave propagation and coupling in a magnetotail waveguide. J. Geophys. Res. **103**, 2359–2368 (1998)
34. De Keyser, J.: Magnetohydrodynamic wave mode conversion in the Earth's magnetotail. J. Geophys. Res. **105**, 13009–13016 (2000)

The Asymmetry of Magnetospheric Configuration and Substorms Occurrence Rate Within a Solar Activity Cycle

Marina Kubyshkina, Vladimir Semenov, Nikolai Erkaev, Evgeniy Gordeev, and Igor Kubyshkin

Abstract The aim of this work is to test the hypothesis by Kivelson and Hughes (Kivelson and Hughes in Planet. Space Sci. 38:211–220, 1990), according to which, for asymmetric configurations with a curved current sheet and large curvature of magnetic field lines in the tail of the magnetosphere, the substorm breakdown threshold decreases. The work used two databases on the substorms onset: (1) (Frey et al. in J. Geophys. Res. 109:A10304, 2004)—4700 events for 2000–2005, and (2) SUPERMAG—18,800 events for 2000–2010 (Gjerloev in J. Geophys. Res. 117:A09213, 2012). OMNI data was used for the solar wind parameters. We try to answer the question what solar wind parameters are responsible for the bending of the current sheet in the tail of the magnetosphere, how these parameters affect the number of emerging substorms, and what structures of the solar wind are capable of carrying them. Among these possible factors of symmetry break, we explore the alfvénic-type excitations, the dipole tilt, and the direction of solar wind flow, which were traced within the solar activity cycle. As a result, we found a notable correlation (R = 0.65) of yearly substorm occurrence rate and the yearly amount of earthward propagating alfvénic disturbances, while no correlation (R = 0.01) was observed for substorms and sunward propagating alfvénic fluctuations. We also found that the yearly averaged absolute value of dipole tilt at a substorm explosion is always larger (by 0.5°–2.0°) than the mean value of absolute dipole tilt and the difference is larger during the solar cycle maximum when substorm occurrence rate also maximizes. The above difference becomes about 0.5° larger if we take into account the effective tilt angle and consider the input of solar wind flow direction to the actual dipole tilt. We also show that the averaged AL-index for a portion of substorms, which occurred under the dipole tilt less than 8° was −244 nT, and the AL-index for substorms with larger tilt (20°–30°) was −192 nT, which gives more than 20% difference.

Keywords Magnetospheric plasma sheet · Substorms · Asymmetric configurations

M. Kubyshkina (✉) · V. Semenov · E. Gordeev · I. Kubyshkin
Saint Petersburg University, Saint Petersburg, Russia
e-mail: m.kubyshkina@spbu.ru

N. Erkaev
Siberian Federal University, 660041 Krasnoyarsk, Russia

A. Kosterov et al. (eds.), *Problems of Geocosmos—2020*, Springer Proceedings in Earth and Environmental Sciences, https://doi.org/10.1007/978-3-030-91467-7_33

1 Introduction

The aim of our study is to determine the role of symmetry breaking in the dynamics of the magnetosphere, namely, to assess the influence of asymmetry on the development of a substorm. This problem is especially interesting, since asymmetric configurations of the magnetosphere (determined by the tilt of the Earth's dipole and variations in the direction of the solar wind) are normal and ubiquitous, while symmetric configurations, on which the vast majority of existing studies are concentrated, are extremely rare. In addition, earlier Kivelson and Hughes [1] suggested that for asymmetric configurations (with a curved current sheet and large curvature of magnetic field lines), the substorm breakdown threshold decreases due to local increase of magnetic gradients as well as electric current density. However, Korovinskiy et al. [4] studied the generalization of well-known Harris-Faddeev-Kan-Manankova equilibrium [5] for a tilted two-dimensional current sheet and showed that with respect to all basic parameters (current density, magnetic field gradient, specific volume and entropy of a flux tube) symmetric and asymmetric current sheets are very similar. The effect of asymmetry turned out to be very delicate and it needs substantial experimental study. The role of fast variations of solar wind direction (first of all Vz component) in substorm dynamics was investigated in [6–9]. The Bx component of IMF as an additional factor of asymmetry was discussed in [10–12]. Within the framework of this work, we consider several specific problems, such as:

(1) a statistical study of substorm activity with respect to changes in solar wind parameters;
(2) a study of the influence of the dipole tilt on substorm activity;
(3) an estimation of the accumulated magnetic flux in the tail for symmetric and asymmetric configurations of the magnetosphere.

2 Theoretical Background and Datasets

The asymmetry of the current sheet in the tail of the magnetosphere is associated with three main factors.

(1) The Earth's magnetic dipole tilt. Figure 1 shows the changes in the dipole tilt as a function of time. One can see that only 4 months a year near the positions of the equinoxes the dipole can be vertical twice a day. In the remaining 8 months, the dipole is inclined, and the tail current sheet is always asymmetric (bent and curved), and the near-Earth part of the magnetospheric tail current sheet is displaced from the ecliptic plane. The diurnal variation of the dipole tilt angle is approximately 20°.
(2) The presence of the Z-component of the solar wind flow velocity—Vz. The presence of a non-radial component of the solar wind velocity leads to a deviation of the far tail from the Sun-Earth direction and, accordingly, the remote part of the current sheet from the ecliptic plane. The deflection angle is usually

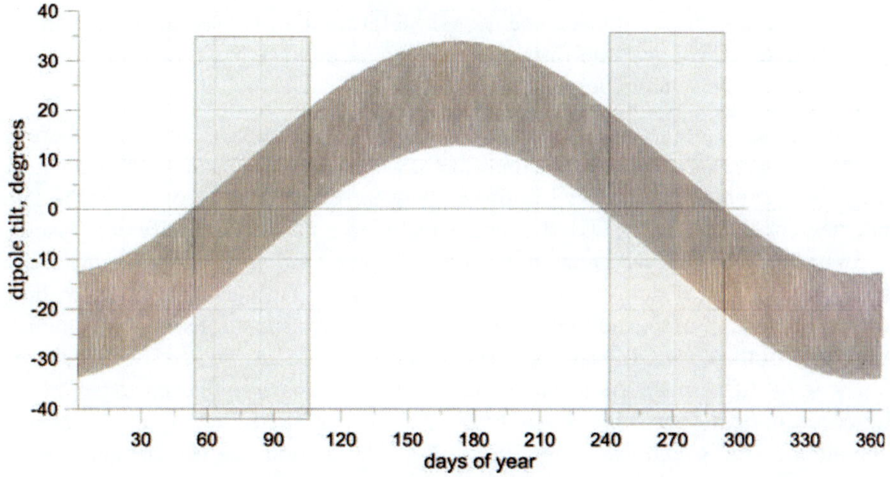

Fig. 1 The value of the Earth's dipole tilt depending on the day of year. The periods when the dipole tilt angle may equal to 0 twice a day are highlighted

small, within ± 6–8°, but it can change rapidly, leading to abrupt (within 10–15 min) changes in the position of the current sheet

(3) The influence of the Bx component of the Interplanetary Magnetic Field (IMF), which, in the southward direction of the IMF, causes a shift in the reconnection line at the magnetopause and different loading of the tail lobes. Its effect is comparable to that of the Vz-component of the solar wind velocity, but a particularly effective displacement occurs at negative values of the Bz component of the IMF [10, 12]. Figure 2 illustrates schematically this mechanism. Under the action of the IMF Bx component, the current sheet can shift by 1–2 Re. To

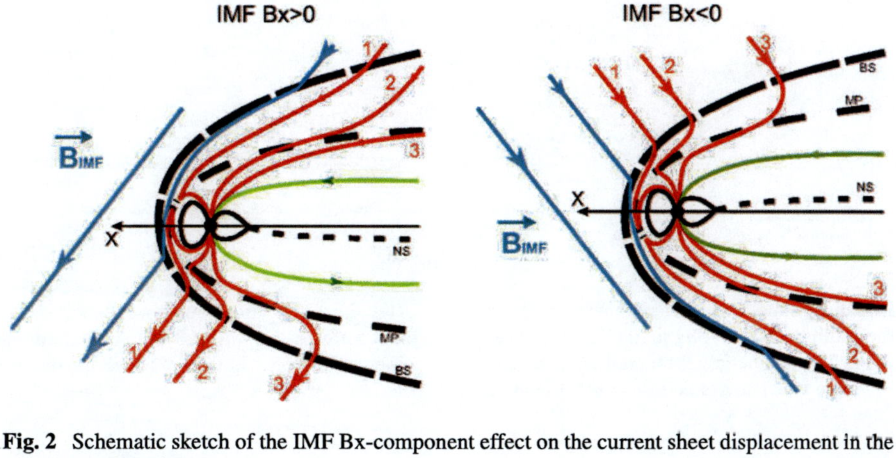

Fig. 2 Schematic sketch of the IMF Bx-component effect on the current sheet displacement in the tail of magnetosphere

check this effect, a three-dimensional MHD simulation was carried out on the NASA CCMC website (https://ccmc.gsfc.nasa.gov/). Numerical simulations confirm the scheme proposed in Fig. 2 [12].

Due to the combined influence of all three factors, the Earth's magnetosphere is in an asymmetric state most of the time, as shown in the bottom panel of Fig. 3 where the field lines start from magnetic latitudes between 60 and 70° according to the T96 magnetic field model [13] with the same solar wind parameters in different seasons.

In order to find out which parameters of the solar wind or their combinations are responsible for the bending of the current sheet in the tail of the magnetosphere, how these parameters affect the number of magnetospheric substorms, and also which structures of the solar wind are capable of transporting them, we used two databases obtained by different instruments. The first one is a database on the substorms onsets based on optical data from the IMAGE satellite compiled by Frey et al. [2] and containing 4700 events for years 2000–2005. The second one is a database taken from the SUPERMAG resource [3], where the onset of substorms were determined

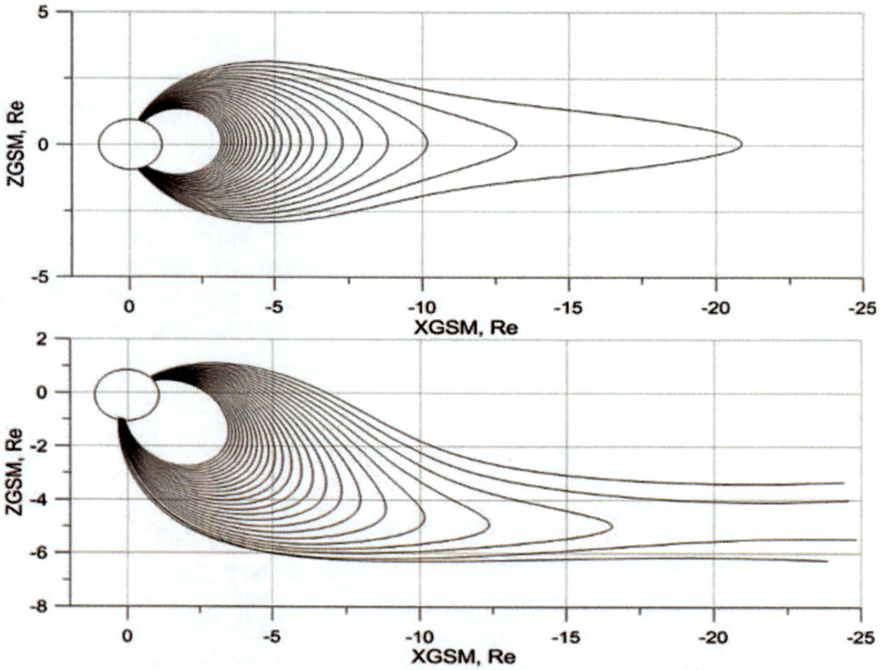

Fig. 3 Symmetrical and asymmetric magnetosphere of the Earth. The upper and lower panels show the results of calculating field lines starting from magnetic latitudes between 60 and 70° according to the T96 magnetic field model with the same solar wind parameters. On the top panel, the dipole tilt angle is 0°, on the bottom panel it is minus 33°

by the SML index (analog to the AL-index), and containing 18,800 events for 2000–2018. The solar wind data used in this work were taken from the OMNI database (http://omniweb.nasa.gov).

3 Observations and Results

3.1 Correlation Between IMF Bz and Bx and Solar Wind Velocity Vz

Taking into account how the position of the current sheet changes under the influence of a non-radial solar wind flow and an interplanetary field with a non-zero Bx component (see Fig. 2), the main attention was paid to the combined effect of the Vz component of the solar wind velocity and the Bx component of the IMF. We assumed that the cumulative effect of these two factors (that is, the product (Vz·Bx) will lead to an increase in the asymmetry of the overall configuration of the magnetosphere. More precisely: the coincidence of the signs of the product (Vz·Bx) and Bz increases the asymmetry of the magnetosphere (bending and displacement of the current sheet in the tail of the magnetosphere), and different signs of the product (Vz·Bx) and Bz make the magnetosphere more symmetric (see Fig. 2). Thus, we examined the 5 min averaged values of Vz, Bx and Bz taken 5 min prior to the substorm onset. The results of the study for two significantly different databases [2] and [3] showed that asymmetric substorms occur approximately twice as often as symmetrical ones, see Fig. 4. This seems to be in full agreement with the hypothesis of Kivelson and Hughes [1], but it soon became clear that this ratio is a reflection of the characteristics of the solar wind. For verification, five-minute solar wind data for years 2000–2010 were ordered according to the signs of (Vz·Bx) product and Bz, and the results are shown in Fig. 5.

In fact, the coincidence of the signs of the product (Vz·Bx) and Bz is a property of Alfvén waves propagating from the Sun; if the signs of (Vz·Bx) and Bz are different, the Alfvén waves propagate towards the Sun. The relationship between the product signs (Vz·Bx) and Bz in the Alfvén wave is illustrated in Fig. 6, although this can be shown rigorously (see [14]).

Thus, it turns out that Alfvén waves carrying geoeffective Bz < 0 from the Sun also automatically transfer the combination (Vz·Bx) < 0, which contributes to the maximum bending of the current sheet in the tail, and thereby makes the configuration of the magnetosphere as asymmetric as possible. Since, on average, there are two times more Alfvén waves coming from the Sun than outgoing ones, the number of substorms also corresponds to this proportion. The effect of Kivelson and Hughes [1] turns out to be more subtle and requires an additional analysis.

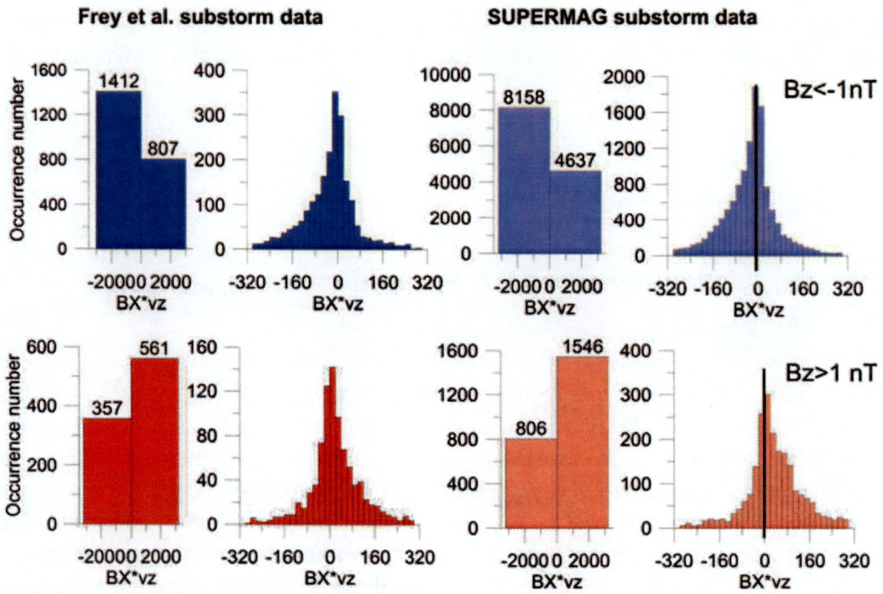

Fig. 4 Distribution of the number of substorms depending on the product value (Vz·Bx) for negative IMF values (Bz < −1) (blue) and for positive Bz > + 1 (red) for two databases. Two variants of histograms are presented—a two column histogram with a division only into positive and negative values (Vz·Bx), and a detailed 20-column histogram which shows that with an increase in the value (Vz·Bx), the prevalence of the number of substorms, where (Vz·Bx) and Bz have the same signs are preserved

3.2 Superimposed Epoch Analysis of Substorm Onset Data

We carried out a statistical study to check whether the distribution of (Vz·Bx) changed during the substorm loading one hour before a break. A superimposed epoch analysis showed a maximal asymmetry roughly at the time of the beginning of the substorm explosion. If Bz has a negative value, then substorms occur almost twice (more precisely, 1.75 times) more often at (Vz·Bx) < 0 than at (Vz·Bx) > 0. Similar results are also true for Bz > 0. Moreover, according to the super epoch analysis, this difference increases before the onset of the substorm, slowly over 40 min, and very sharply in the last 10–15 min, which is illustrated in Fig. 7.

The key question of the hypothesis of Kivelson and Hughes [1] is whether the substorm initiating instability threshold for an asymmetric current sheet decreases. To solve it, the method of superimposed epochs was applied, where the values of the IMF Bz component were investigated, and the moment of the beginning of the explosive phase of the substorm was taken as the zero point (Fig. 7). As can be seen, the dependence Bz (t) for both databases [2, 3] has a pronounced minimum near the beginning of the explosive phase. It is important to note that for asymmetric substorms (where the signs of Bz and (Vz·Bx) coincide) the minimum Bz is −3.4 nT,

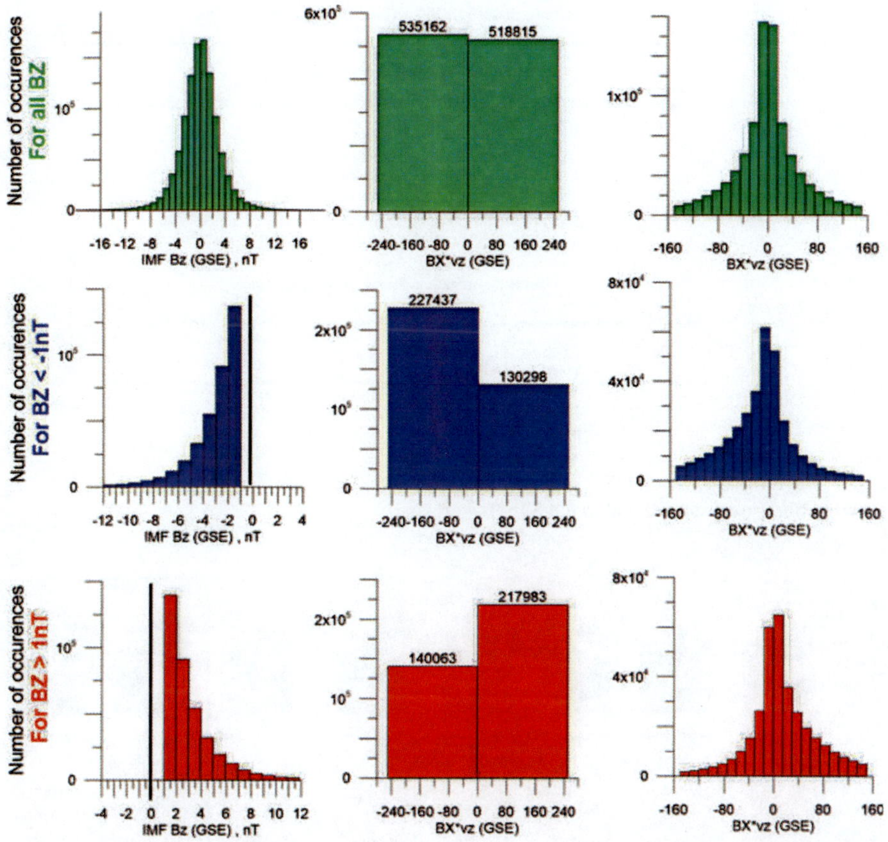

Fig. 5 Histograms for the distributions of the product value (Vz·Bx) at negative (Bz < −1, blue, middle row) and positive (Bz > + 1, red, bottom row) values of the IMF Bz component in the solar wind. The top panel shows the total distributions (Vz·Bx), without taking into account the sign of Bz, from which it can be seen that both Bz itself and (Vz·Bx) have a distribution symmetric relative to 0

and for symmetric substorms it is −4.0 nT according to the SUPERMAG base [3]. Frey et al. [2] data gives for asymmetric substorms Bz = −2.2 nT, for symmetric substorms −2.5 nT (Fig. 8).

If we consider this minimum Bz value as some estimate of the substorm breakdown threshold, then it turns out that, in accordance with the Kivelson and Hughes hypothesis, this threshold actually decreases: by 18% according to the SUPERMAG base, and slightly less by 14% according to the Frey et al. base. Moreover, we can see that during the entire preliminary phase, the value of Bz continues to remain lower in value in the case of an asymmetric configuration, which indicates a lower loading (lower reconnected magnetic flux).

The magnitude of the asymmetry effect turned out to be not very large, but it should be borne in mind that most substorms develop on X lines at a distance greater

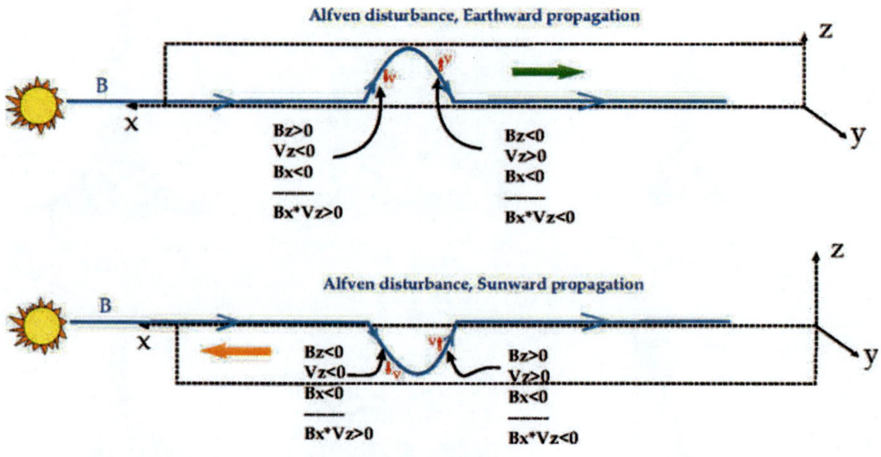

Fig. 6 Relationship between the product signs (Vz·Bx) and Bz in the Alfvén wave propagating from the Sun (upper panel) and to the Sun (lower panel)

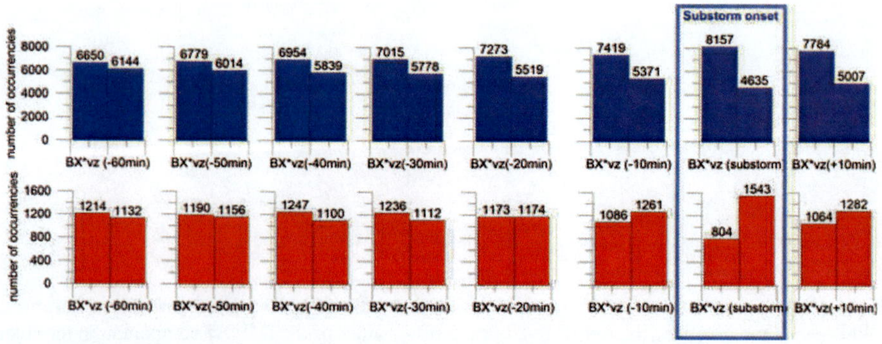

Fig. 7 The method of superimposed epochs to check the change in the ratio of the number of positive and negative realizations of the product (Vz·Bx) before the onset of a substorm for Bz < 0 (blue, top panel) and Bz > 0 (red, bottom panel)

than 20 Re in the tail, when the current sheet becomes already symmetric, and for the first arrivals the asymmetry effect should be stronger.

3.3 Alfvén Waves and Substorm Occurrence in a Solar Cycle

Another significant factor indicating the important role of earthward moving Alfvén waves on the frequency of substorms occurrence is the dependence of the number of these waves on the phase of the solar cycle. We know that the maximum number of substorms is observed at the phase of decline in solar activity, shortly after its

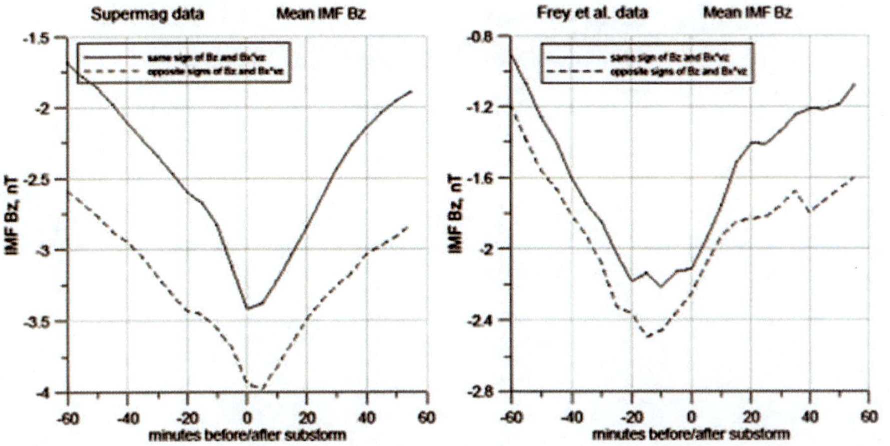

Fig. 8 Superimposed epochs method for the IMF Bz component during a substorm for two databases. The solid line corresponds to substorms for which the values Bz and (Vz·Bx) had the same sign, the dashed line—opposite signs. Zero corresponds to the beginning of the explosive phase. Five-minute averages of Bz and (Vz·Bx) were used)

maximum. Similar behavior was observed for a number of earthward propagating Alfvén waves, this fact is illustrated in Fig. 9. Figure 9 shows the yearly number of periods when the signs of Bz and (Vz·Bx) coincide (i.e., the number of Alfvén waves coming from the Sun)—the red curve, and when the signs do not coincide (i.e., the number of Alfvén waves directed towards the Sun)—blue curve. The black line is the averaged values of the Wolf numbers over two solar cycles. One can see a good correspondence in a shape of these curves, also during periods of greater solar activity, the number of incoming Alfvén waves increases strongly, and this number reaches maximum at the decay phase of the solar cycle. In the years of minimal solar activity, the difference in the number of incoming and outgoing waves is also minimal.

The next obvious step was to compare the annual number of substorms with the number of incoming and outgoing Alfvén waves and Fig. 10 demonstrates the comparison. Here, the brown curve gives the annual number of substorms in the longest list (SUPERMAG [3]), and the blue curve shows the percentage difference between the number of incoming and outgoing Alfvén waves. It is clearly seen that an increase in the number of substorms in most cases coincides with an increase in the number of incoming Alfvén perturbations, which fits well into the hypothesis about the influence of an increase in the asymmetry of the magnetospheric configuration on the frequency of occurrence of substorms.

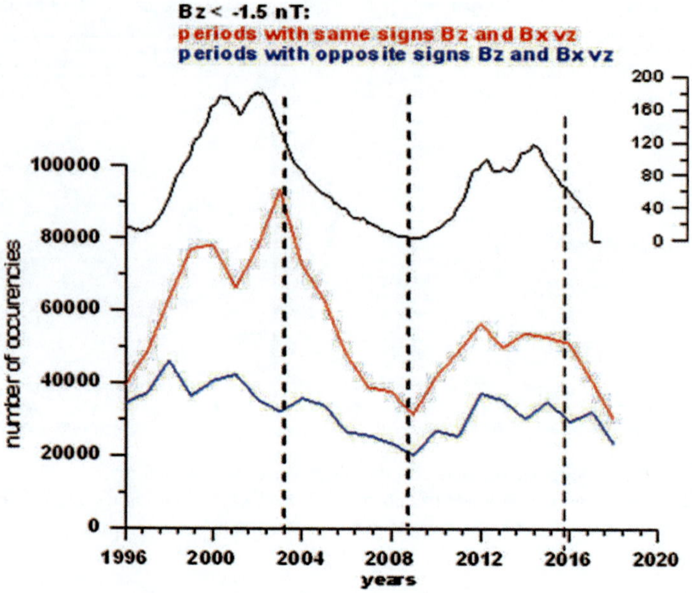

Fig. 9 Red curve—the yearly number of periods when the signs of Bz and (Vz·Bx) coincide (i.e., the number of Alfvén waves coming from the Sun), blue curve—the yearly number of periods when the signs do not coincide (i.e., the number of Alfvén waves directed towards the Sun). The black line gives the averaged values of the Wolf numbers over two solar cycles

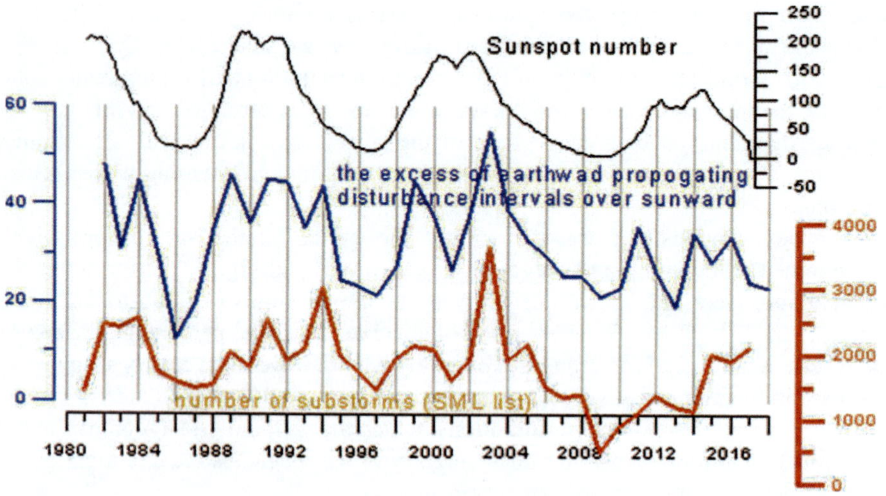

Fig. 10 Averaged Wolf numbers (black line), percentage difference between the annual number of incoming and outgoing Alfvén waves (blue line) and the annual number of substorms (brown line)

3.4 *Dipole Tilt Importance*

We discussed above the influence of solar wind on magnetospheric configuration and substorms probability, but did not mention the importance of dipole tilt value and its slow but considerable variations. Previously the importance of dipole tilt in substorm probability was established statistically in [8]; here we confirm these results on larger data and with different approach.

First, we estimated the yearly averaged dipole tilt absolute value for all substorm onsets during a given year as:

$$< \Psi \text{ subst} > = \Big(\sum_{substorms} |\Psi i|\Big)/N_{substorms}$$

For these estimations we used the mentioned above substorm onset datasets by Frey et al. [2] and SUPERMAG [3], additionally we used the database of geosynchronous injections compiled by Borovsky and Yakimenko [15]. All three databases showed that the dipole tilt at the moments of substorms onsets exceeded the yearly mean value of dipole tilt by 5–20%, with maximal difference close to maximum of solar activity (see Fig. 11, left panel). It is important that if we consider the additional rotation of plasma sheet caused by non-radial solar wind flow and calculate the total tilt as:

$$< \Psi \text{ tot} > = \Big(\sum_{substorms} \left| \Psi i + \operatorname{atan}\left(\frac{Vz}{Vx}\right)\right| \Big)/N_{substorms},$$

where Vx and Vz are the components of solar wind velocity, the resulting tilt will practically always be larger than the pure dipole tilt (Fig. 11, middle and right panels). This fact reflects the situation where the substorm begins at a time when the solar

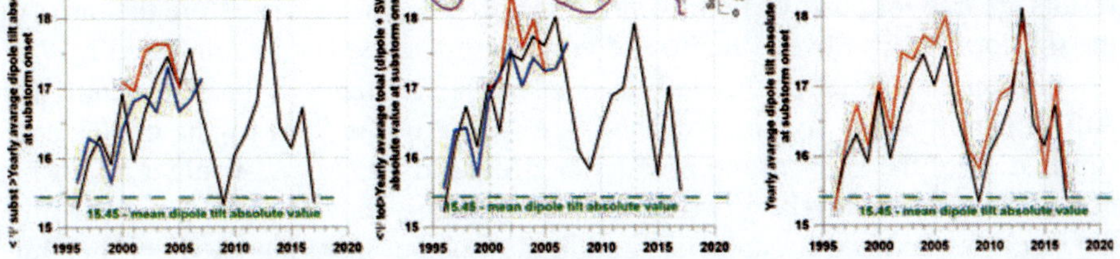

Fig. 11 Yearly averaged dipole tilt absolute value for three data sets (left panel), yearly averaged plasma sheet total tilt (dipole + solar wind additional rotation) for three data sets with yearly averaged Wolf numbers on the top (middle plot), yearly averaged dipole tilt (black) and total tilt (red) for the longest data set (SUPERMAG [3]). Green line on each plot gives the yearly mean dipole tilt absolute value for comparison

wind flow shifts the plasma sheet in a direction that would increase the plasma sheet total tilt.

Figure 11 proves the hypothesis that substorms (unlike magnetic storms) more often start at times with larger dipole tilts, and especially if solar wind flow increases the plasma sheet tilt.

The decrease of substorm instability threshold and lower intensity of substorms starting under larger tilts may be obtained from statistical comparison of substorm properties in two groups of substorms: those which started under dipole tilts close to zero ($\Psi = [-7;0]$ degrees) and those, which started under largest tilts ($\Psi = [-30;-20]$ degrees). For this study we used the largest data set based on SML index (SUPERMAG [3]) which covers two solar cycles. The intervals of tilts were chosen so that the number of substorms in both sets would be comparable. Histograms in Fig. 12 show the results of comparison: upper panel gives the tilt intervals and the number of substorms in both sets, middle panel shows the distribution of IMF Bz near substorm onset for substorms in two sets, and lower panel shows the peak AL index distribution for substorms in both sets. One can see that the mean value of Bz is larger for substorms starting with more symmetric configuration, which again confirms the suggestion of lower loaded magnetic flux in tilted configuration, and, hence, lower instability threshold. Accordingly, the mean AL index value is also larger for symmetric substorms; the substorms which start at larger tilts produce less wide AL distribution centered at smaller values.

4 Discussion and Conclusions

The statistical analysis of substorm onsets in a vast period covering two solar cycles enabled us to verify the Kivelson and Hughes [1] hypothesis of lower substorm instability threshold for a curved and bent current sheet.

We also found that the distribution of the product (Vz·Bx) in the solar wind has a very noticeable asymmetry at the start of the substorm. If IMF Bz is negative, then substorms occur 1.75 times more often for $(Vz \cdot Bx) < 0$ than for $(Vz \cdot Bx) > 0$. Similar results are true for positive Bz as well. This difference increases before the onset of the substorm, slowly over 40 min, and very sharply in the last 15 min (see Fig. 7).

An important result of the study was a simple new method for determining the Alfvén perturbations coming from the Sun (going to the Sun) by the coincidence (mismatch) of the signs of the product (Vz·Bx) and Bz. A statistical study for the period 1996–2018, covering two 11-year solar cycles, showed that 66% of the time the signs (Vz·Bx) and Bz coincide, and in the years of maximum activity this time increases to 75%, and in the years of minimum decreases to 60% (Fig. 9). This difference in the number of earthward and sunward propagating Alfvén perturbations observed near the Earth determines the number of fluctuations born near the solar surface, where the Alfvén velocity is greater than the solar wind flow velocity, and,

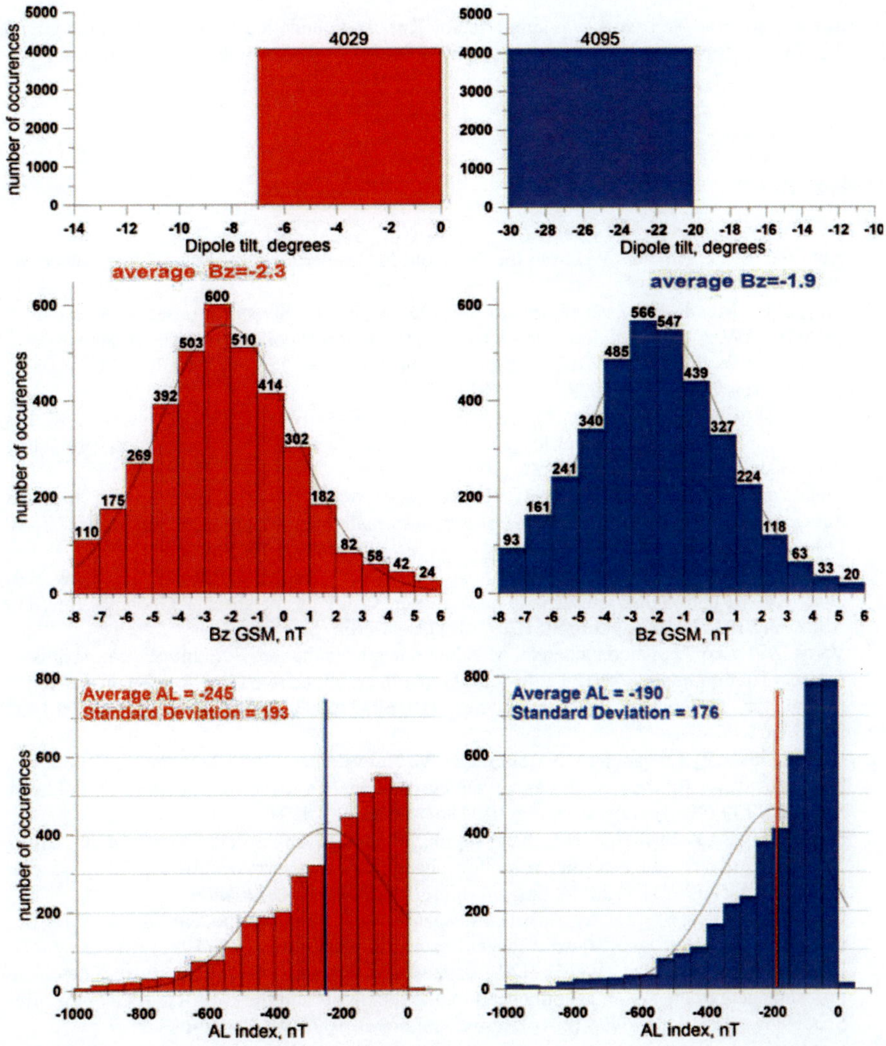

Fig. 12 Upper panel—tilt range and number of substorms in the subsets with small and large tail current sheet bending; middle panel—the distribution of IMF Bz-component near the onset for substorms in two subsets, lower panel—distribution of peak AL index for substorms in both sets

hence sunward perturbations do not reach the Earth. This explanation makes reasonable much larger difference between earthward and sunward propagating fluctuations during maximum solar activity periods.

Finally, an important conclusion follows that the geoeffective Bz component of the IMF, together with the combination of parameters (Vz·Bx) that determine the asymmetry of the magnetosphere, is carried almost exclusively by Alfvén waves propagating from the Sun.

Acknowledgements This work was supported by Russian Foundation for Basic Research grant # 20-55-53011\20 "Artificial and real magnetosphere: methods of presentation and diagnostics".

References

1. Kivelson, M.G., Hughes, W.J.: On the threshold for triggering substorms. Planet. Space Sci. **38**(2), 211–220 (1990)
2. Frey, H.U., Mende, S.B., Angelopoulos, V., Donovan, E.F.: Substorm onset observations by IMAGE-FUV. J. Geophys. Res. **109**, A10304 (2004). https://doi.org/10.1029/2004JA010607
3. Gjerloev, J.W.: The superMAG data processing technique. J. Geophys. Res. **117**, A09213 (2012). https://doi.org/10.1029/2012JA017683
4. Korovinskiy, D.B., Kubyshkina, D.I., Semenov, V.S., Kubyshkina, M.V., Erkaev, N.V., Kiehas, S.A.: On application of asymmetric Kan-like exact equilibria to the Earth magnetotail modeling. Ann. Geophys. **36**, 641–653 (2018)
5. Manankova, A.V., Pudovkin, M.I., Runov, A.V.: Stationary configurations of the two-dimensional current-carrying plasma sheet: exact solutions. Geomag. Aeron. **40**, 430–438 (2000)
6. Panov, E.V., Nakamura, R., Baumjohann, W., Kubyshkina, M., Artemyev, A.V., Sergeev, V.A., et al.: Kinetic ballooning/interchange instability in a bent plasma sheet. J. Geophys. Res. **117**, A06228 (2012). https://doi.org/10.1029/2011JA017496
7. Voros, Z., Fasko, G., Khodanchenko, M., Honkonen, I., Janhunen, P., Palmroth, M.: Windsock memory conditioned RAM (CO-RAM) pressure effect: forced re-313 connection in the Earth's magnetotail. J. Geophys. Res.: Space Phys. **119**, 6273–6293 (2014). https://doi.org/10.1002/2014JA019857
8. Kubyshkina, M., Tsyganenko, N., Semenov, V., Kubyshkina, D., Partamies, N., Gordeev E.: Further evidence for the role of magnetotail current shape in substorm initiation. Earth Planets Space **67**, 139 (2015). https://doi.org/10.1186/s40623-015-0304-1
9. Semenov, V.S., Kubyshkina, D.I., Kubyshkina, M.V., Kubyshkin, I.V., Partamies, N.: On the correlation between the fast solar wind flow changes and substorm occurrence. Geophys. Res. Lett. **42**(13), 5117–5124 (2015). https://doi.org/10.1002/2015GL064806
10. Cowley, S.W.H.: Asymmetry effects associated with the x-component of the IMF in a magnetically open magnetosphere. Planet. Space Sci. **29**(8), 809–818 (1981)
11. Hoilijoki, S., Souza, V.M., Walsh, B.M., Janhunen, P., Palmroth, M.: Magnetopause reconnection and energy conversion as influenced by the dipole tilt and the IMF Bx. J. Geophys. Res. Space Phys. **119**, 4484–4494 (2014). https://doi.org/10.1002/2013JA019693
12. Gordeev, E.I., Sergeev, V.A., Amosova, M.V., Andreeva V.A.: Influence of interplanetary magnetic field radial component on the position of the Earth's magnetotail neutral sheet. In: Geophysical Methods of Survey the Earth and its Subsoil, pp. 11–21 (2016). https://doi.org/10.13140/RG.2.1.4821.6565
13. Tsyganenko, N.A.: Modeling the Earth's magnetospheric magnetic field confined within a realistic magnetopause. J. Geophys. Res. **100**, 5599 (1996)
14. Kubyshkina, M., Semenov, V., Erkaev, N., Gordeev, E., Dubyagin, S., Ganushkina, N., Shukhtina, M.: Relations between vz and Bx components in solar wind and their effect on substorm onset. Geophys. Res. Lett. **45** (2018). https://doi.org/10.1002/2017GL076268
15. Borovsky, J.E., Yakymenko, K.: Substorm occurrence rates, substorm recurrence times, and solar wind structure. J. Geophys. Res. Space Phys. **122**, 2973–2998 (2017). https://doi.org/10.1002/2016JA023625

Lightning Activity of Eruptive Clouds from Shiveluch Volcano (Kamchatka, Russia)

Evgeniy Malkin, Pavel Firstov, Nina Cherneva, and Gennadiy Druzhin

Abstract One of the local features of the Kamchatka Peninsula is volcanic lightning and aeroelectric formations resulting from volcanic eruptions. When an eruptive cloud (EC) arises and is formed, it becomes electrified, which leads to the occurrence of numerous lightning strikes. This indicates a significant role of electrostatics in the formation of EC even for weak eruptions and demonstrates the informative value of volcanic lightning for estimating the EC size. The article presents cases of recording of thunderstorm activity using a VLF direction finder (30 Hz–40 kHz) located 390 km from Shiveluch volcano. Based on its data, electromagnetic pulses (EMP) from volcanic lightning for 14 explosions were recorded for the period 2017–2019. The number of pulses per minute, EMP counting rate (**v**), was taken as the parameter characterizing the process intensity. Two phases are distinguished in **v** dynamics depending on eruption character and intensity. Phase **I** characterizes the eruption beginning character. A peak-like phase **I**, lasting up to 5 min, is observed for the eruptions, which begin with magma fragmentation according to an explosion scenario. For the eruptions, which begin according to a «smooth» scenario, only phase **II** can be observed. It characterizes lightning activity of a developed eruptive cloud.

Keywords Volcano · Eruption · Lightning discharges

Pavel Firstov: deceased

E. Malkin · N. Cherneva (✉) · G. Druzhin
Institute of Cosmophysical Research and Radio Wave Propagation FEB RAS, Paratunka, Kamchatka, Russia
e-mail: nina@ikir.ru

E. Malkin
e-mail: malkin@ikir.ru

P. Firstov
Kamchatka Branch of Federal Research Center, Geophysical Survey RAS, Petropavlovsk-Kamchatsky, Russia

A. Kosterov et al. (eds.), *Problems of Geocosmos—2020*, Springer Proceedings in Earth and Environmental Sciences, https://doi.org/10.1007/978-3-030-91467-7_34

1 Introduction

Volcanic explosive eruptions are accompanied by complicated geophysical processes occurring during the ejections of a great amount of hot ash and volcanic gases, containing 90% of water steam, into the atmosphere. The water, dissolved in magma and constituting 7% from the weight composition, causes its fragmentation (breaking) when ground pressure is released. Moreover, a large amount of dispersed rock (pyroclastics) is produced. It forms EC together with volcanic gases. Owing to different physical processes EC are electrified that is confirmed by the observed lightning strokes [9].

Volcanic ash is the greatest hazard for the population and urban territories in a far zone. It is also dangerous for aircraft engines. EC propagation monitoring carried out by different methods allows one to announce ash hazard that gives the possibility to take preventive actions in order to reduce the consequences. One of the methods of EC motion tracing is recording of EMP from accompanying lightning discharges. That was shown in a number of papers [1, 2].

Shiveluch volcano on Kamchatka peninsula is one of the most active volcanoes (Fig. 1, open source [12]). Its latest catastrophic eruption took place on November 12, 1964. After a 16-year pause of volcano activity, an extrusive dome is growing within the newly formed crater from August 1980 up to the present time with short breaks. It is accompanied by explosive eruptions. During powerful explosions, EC rises up to the tropopause height (10–12 km) and propagates for hundreds of kilometers. In this case, significant lightning activity is observed [7]. In some cases, during the eruptions

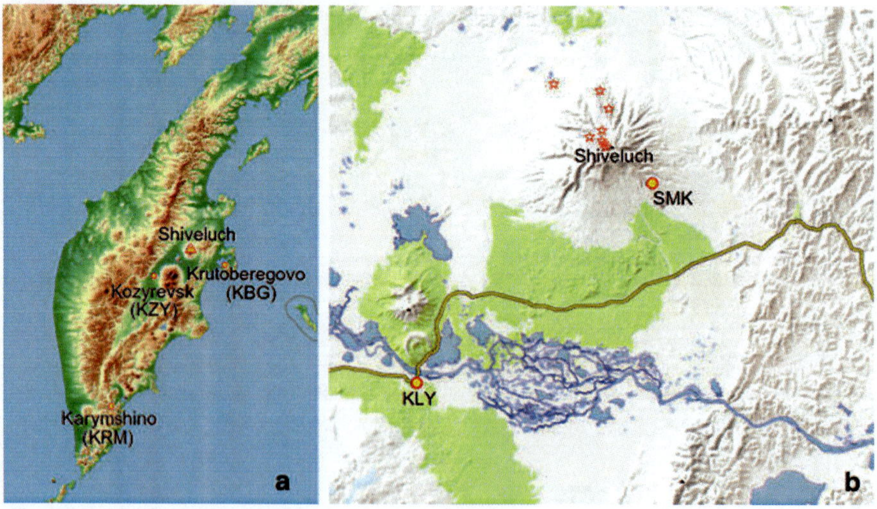

Fig. 1 Locations of Shiveluch volcano and VLF direction finder on Kamchatka peninsula (**a**), lightning strokes recorded by WWLLN during Shiveluch volcano eruption on March 09, 2019 at 05:49 (**b**). Open source [12]

Fig. 2 Explosive eruption of Shiveluch volcano and formation of a pyroclastic flow on December 03, 2012 (photo by A.V. Shevchenko)

we observed pyroclastic flows, which are a mixture of hot gas (up to 800 °C), ash and stones moving along the volcano slopes with the velocity up to 700 km/h (Fig. 2). This paper presents the observation results for EMP accompanying separate volcanic explosions during the period 2017–2019.

2 Existing Methods

The main method to detect high-current electric discharges of atmospheric origin is the differential rangerfinder one. It consists in accurate determination of the delay time between the same pulse records at spaced basic stations followed by hyperbolas set reconstruction and search for their intersection. Basic observation sites form a lightning location network. The most known representative of such a network is the NLDN (National Lightning Detection Network) located in the USA. Lightning discharges are located with TOA (Time of Arrival) algorithm [5].

Within the framework of this algorithm, time is fixed when the maximum of signal «ground» component first half-wave is achieved. The signal is recorded in the frequency band of 1–400 kHz. Within the inner zone of this network, the accuracy of lightning discharge location is ~500 m. The World Wide Lightning Location Network (WWLLN) is similar in method but is different in realization. This network also refers to differential rangerfinder locating systems (DRLS) but the method to determine the time delay obeys the TOGA algorithm (Time of Group Arrival). The essence of this method is measurement of signal group delay recorded in the frequency band of 6–22 kHz. Figure 1b shows lightning discharges recorded by WWLLN network during Shiveluch volcano eruption on June 16, 2017 [6].

If we compare the systems mentioned above, we should note that lightning discharge detection by WWLLN network relatively the NLDN network is ~10% at discharge currents of > 25 kA and ~35% and more at 130 kA. In volcano explosive eruptions, the discharges with the currents of < 75 kA form the majority. Thus, it is not possible to rely on the data of WWLLN global network to monitor EC propagation, but it makes sense to apply the network data to check the presence of a lightning in the region [8, 10, 11].

For the DRLS to function properly, high accuracy of timing at all basic station and dense station network are required. Based on that, the method is not suitable for constructing a local network.

Single-site direction finding method was chosen as the basic one to construct a local network. This method consists in the determination of an azimuth of TEM (Transverse Electromagnetic) wave arrival based on separately recorded field components. Basic station is an antenna system composed of two mutually perpendicular frame magnetic and one rod electric antennas, signal transfer and processing systems. For the network to function at the location accuracy of ~500 m in the radius of 1000 km from the stations, three observation sites with the base up to 300 km are required. In this case the time convergence accuracy should not be more than 100 μs [4].

3 Data and Analysis

Based on the data of Kamchatka Branch of Federal Research Center, Geophysical Survey RAS (http://www.emsd.ru/~ssl/monitoring/main.htm), a catalogue of the strongest events on Shiveluch volcano was composed estimating EC heights for the period 2017–2019. On the first stage, EMP azimuthal distribution in the sector of 0–80° was under the consideration for the selected events. As it is clear from Fig. 3a, lightning activity is distinctly observed for half an hour during the eruption on May 11, 2017 from the azimuth to Shiveluch volcano (25.6° $\pm$ 10°).

During the second stage, graphs of counting rate dynamics of EMP from the direction to the volcano were plotted (Fig. 3b). Simultaneously with that, EMPs with the coordinates near the volcano were selected for that time from the WWLLN catalogues.

4 Lightning Activity of Eruptive Clouds

The eruption on May 11, 2017 began with a burst after which ash-gas mixture began to flow out. EC rose to the tropopause height (10 km). We can distinguish two phases

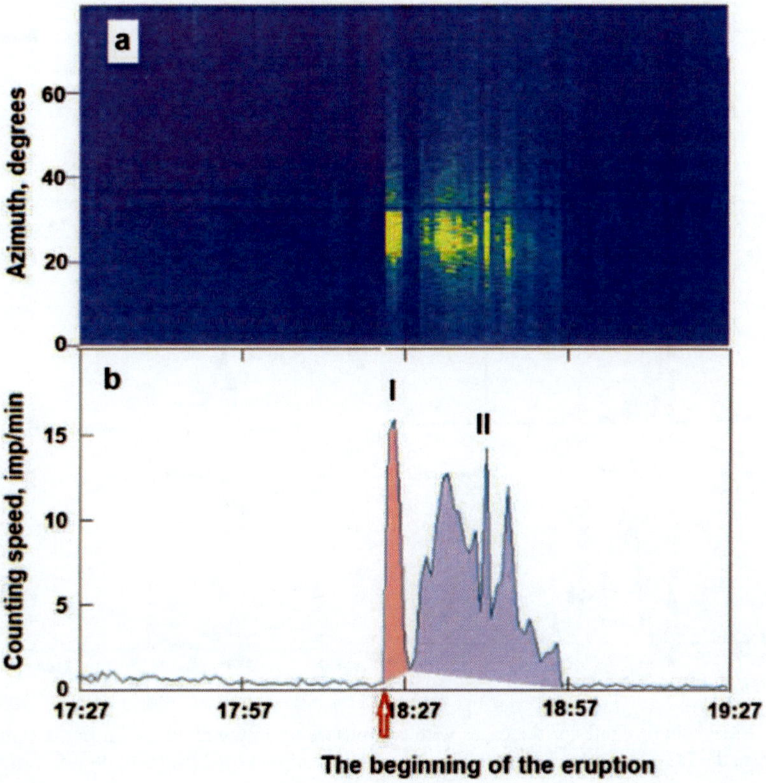

Fig. 3 EMP azimuthal distribution in the sector of 0–80° (**a**), counting rate of EMP from the direction to Shiveluch volcano (**b**) during the eruption on May 11, 2017

in the dynamics of the counting rate (**v**) for this eruption. Phase **I** lasted for ~3 min and was characterized by a sharp increase followed by a sharp decrease. Slower decrease of **v** and long duration of ~30 min are characteristic for the next phase **II** (Fig. 3b).

Dynamics of **v** with two phases was observed for separate explosions on Alaska volcanoes, Augustine in 2006 and Redoubt in 2009. Phase **I** occurred at the first stage of the eruption in the region of thermal formation and rise that determined its longest duration. Phase **II** characterized EC electrification on the stage of floating when it was washed away and moved by wind. In this case, EC contacted the clouds of meteorological origin that caused water vapor crystallization and additional electrification resulting in powerful lightning activity [3].

Figures 4 and 5 show **v** dynamics for 14 explosions of Shiveluch volcano. They also illustrate EC height and EMP number detected by WWLLN. We can note for the whole data set that phase **I** lasted for several minutes (< 8) and phase **II** lasted for more than 15 min.

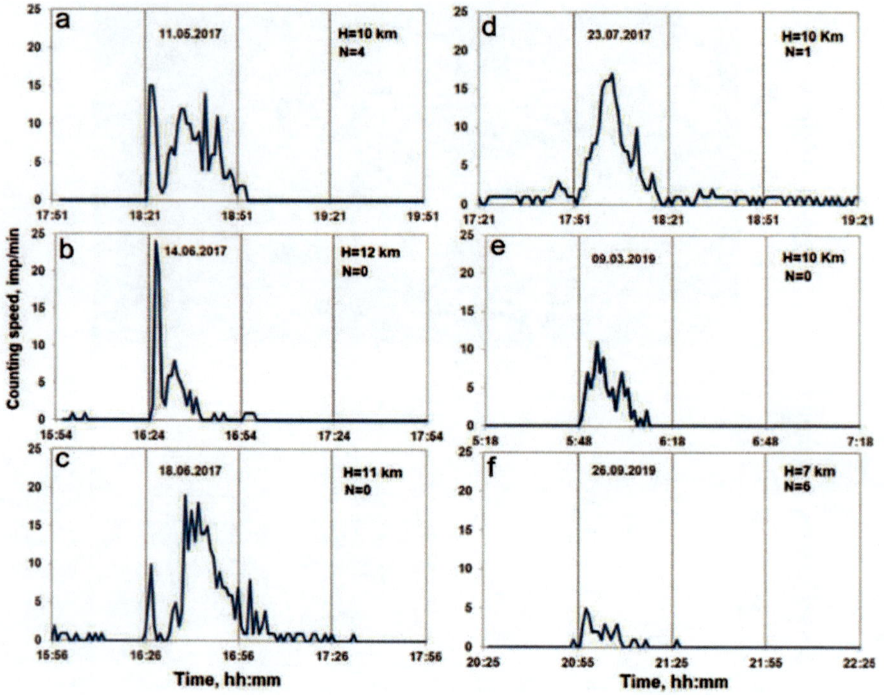

Fig. 4 EMP counting rate for the cases with two phases or only with phase **II**. In the right upper corner are the EC upper edge height (H) and the number of recorded EMPs by WWLLN (N)

Figure 4 illustrates the graphs of EMP counting rate and shows the cases with the second phase or two phases of volcanic lightning. The events presented in Fig. 4a, b and c were accomponied by shock infrasonic waves recorded at Kluchi site. When recording the events, illustrated in Fig. 4e, d and f, there were no shock waves. For all the cases from Fig. 4, we should note the duration of the explosive earthquake with the duration of more than 15 min. In the events, shown in Fig. 5, all the eruptions began with a burst. The duration of the explosive earthquakes was about 15–20 min.

5 Conclusions

In recent years, much attention has been paid to the atmospheric-electrical effect accompanying explosive eruptions. This is due to the increasing density of the population on Earth and the ash hazard for the population, primarily for aviation. The paper compares the dynamics of lightning activity accompanying explosions. The features of the development of two stages of volcanic thunderstorms associated with the formation of an eruptive column at the first stage and the development of EC at the second are highlighted.

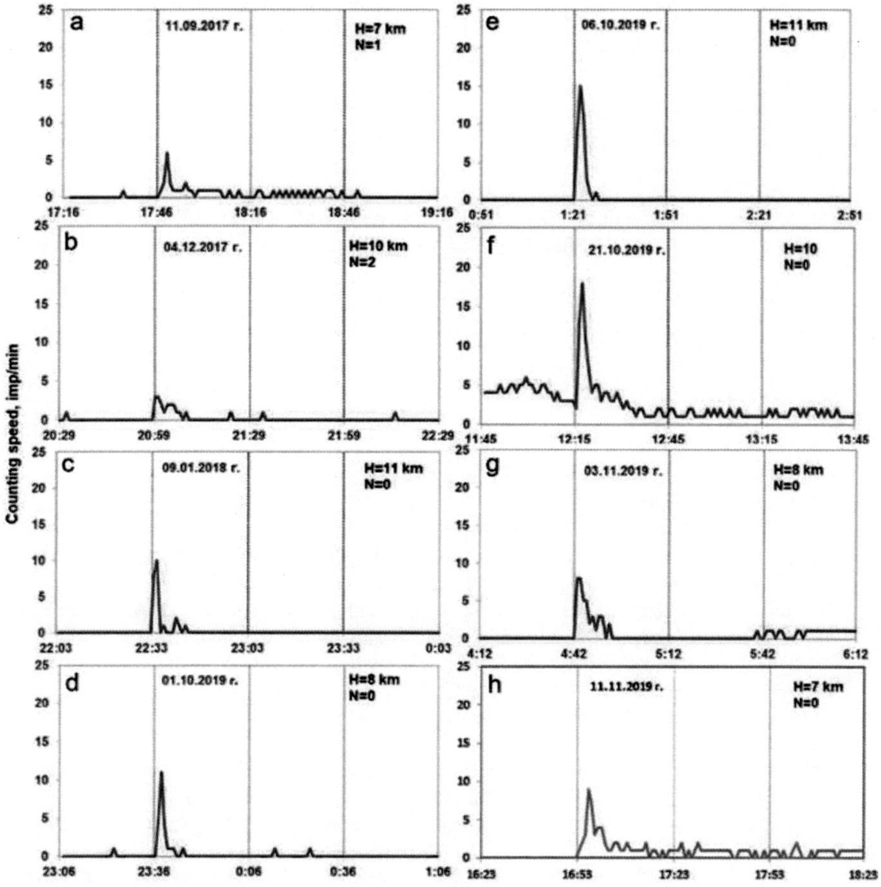

Fig. 5 EMP counting rate for the cases only with phase **I**

Two phases are distinguished in the **v** dynamics depending on eruption character and intensity. Phase **I** reflects the character of earthquake beginning. For the eruptions, which begin with magma fragmentation according to an explosion scenario, a peak-like phase **I** is observed. It lasts up to 5 min. For the explosions, which begin according to a «smooth» scenario, only phase **II** is observed. It characterizes lightning activity in a developed eruptive cloud. This phase intensity depends not only on the eruption mechanism but also on the meteorological conditions in the region. That explains the fact that, on the whole, there is no relation between **v** dynamics and EC height and EMP number based on WWLLN data.

Monitoring of electromagnetic pulses from discharges of "volcanic lightning" expands the informative value of the eruption process and the characteristics of eruptive cloud propagation depending on meteorological state. There is no doubt that with the modern development of technology and computational methods, monitoring of VLF radiation can be included in the aviation systems of warning about the ash hazard for numerous flights on the routes along the eastern coast of Kamchatka.

Acknowledgements The work was carried out within the framework of State Task Theme (2021–2023) «Physical Processes in the System of Near Space and Geospheres under Solar and Lithospheric Impacts» (AAAA-A21-121011290003-0) and using the data of large-scale research facilities «Seismic infrasound array for monitoring Arctic cryolitozone and continuous seismic monitoring of the Russian Federation, neighboring territories and the world» (https://ckp-rf.ru/usu/507436/). The authors are grateful to A.V. Shevchenko for the photos illustrating the event described in the paper.

References

1. Behnke, S.A., McNutt, S.R.: Bull. Volcanol. **76**, 1–12 (2014)
2. Behnke, S.A., Thomas, R.J., Edens, H.E., Krehbiel, P.R., Rison, W.: J. Geophys. Res. Atmos. **119**, 833–859 (2014). https://doi.org/10.1002/2013JD020781
3. Behnke, S.A., Thomas, R.J., McNutt, S.R., Schneider, D.J., Krehbiel, P.R., Rison, W., Edens, H.E.: J. Volcanol. Geotherm. Res. **259**, 214–234 (2013)
4. Druzhin, G.I., Pukhov, V.M., Sannikov, D.V., Malkin E.I.: Vestnik KRAUNTs. Fiziko-matematicheskie nauki—Bull. KRASEC Phys. Math. Sci. **27**(2), 95–104 (2019)
5. Dowden, R.L., Brundell, J.B., Rodger, C.J.: VLF lightning location by time of group arrival (TOGA) at multiple sites. J. Atmos. Solar Terr. Phys. **64**(7), 817-830 (2002)
6. Firstov, P.P., Akbashev, R.R., Holzworth, R., Cherneva, N.V., Shevtsov, B.M.: Izvestiya Atmos. Oceanic Phys. **53**(1), 24–31 (2017)
7. Firstov, P.P., Malkin, E.I., Akbashev, R.R., Druzhin, G.I., Cherneva, N.V., Holzworth, R.H., Uvarov, V.N., Stasiy, I.E.: Registration of atmospheric-electric effects from volcanic clouds on the Kamchatka Peninsula (Russia). Atmosphere **11**, 634 (2020). https://doi.org/10.3390/atmos11060634
8. Hutchins, M.L., Holzworth, R.H., Rodger, C.J., Brundell, J.B.: Far field power of lightning strokes as measured by the world wide lightning location network. J. Atmos. Oceanic Technol. **29**, 1102–1110 (2012)
9. McNutt, S.R., Williams, E.R.: Volcanic lightning: global observations and constraints on source mechanisms. Bull. Volcanol. **72**(10), 1153–1167 (2010)
10. Malkin, E., Druzhin, G., Firstov, P., Cherneva, N., Uvarov, V., Sannikov, D., Stasiy, I.: E3S Web Conf. **127**, 02021 (2019)
11. Smith, C., Said, R., Eaton, A.V., Holzworth, R.H.: Volcanic lightning as a monitoring tool during the 2016–2017 eruption of Bogoslof Volcano, AK. In: 25th International Lighting Detection Conference. March 12–15, pp. P. 1–7. Ft. Lauderdate, Florida, USA (2018)
12. http://www.emsd.ru

Dynamic Characteristics of Field-Aligned Ionospheric Irregularities Under the Conditions of Ionosphere Modification

Vladimir Sivokon' and **Nina Cherneva**

Abstract Telecommunication systems of decameter range are applied to transfer information for long distances. However, electromagnetic radiation suffers significant attenuation determined by re-reflection from the ground surface. These losses can be considerably decreased if the propagation would take place in an interlayer ionospheric waveguide. The idea to apply the interlayer ionospheric waveguide for data transfer was formulated quite a long time ago but its practical implementation is hampered by the problem of effective excitation of the interlayer ionospheric waveguide and radiation coupling. One of the possible ways to solve this problem is the application of scattering on magnetically oriented irregularities of the ionosphere. It is known that scattering on irregularities is accompanied by fading. In the works, carried out in this area of investigation by the specialists of Kharkov Institute of Radio Astronomy, scintillation index was used to estimate fading. The index allows one to calculate the fading level quantitatively. In addition, it is considered that the best approximation of signal intensity distribution is Nakagami distribution. Taking into account irregularity motions, intensity distributions, which are significantly different from those used for the scintillation index calculation, are possible. Thus, it is topical to investigate intensity distributions, for example, from the power of impact on the ionosphere. The attempt to carry out such a study is made in this paper.

Keywords Ionospheric waveguide · Field-aligned irregularities · Scintillation index

1 Introduction

Fading, i.e., random change of signal level at a receiving site, is typical for telecommunication systems of decameter range. The communication channel using the interlayer ionospheric waveguide is not an exception.

Fading may be determined by different reasons, the main of which are:

V. Sivokon' (✉) · N. Cherneva
Institute of Cosmophysical Research and Radio Wave Propagation FEB RAS, Paratunka, Kamchatka, Russia

A. Kosterov et al. (eds.), *Problems of Geocosmos—2020*, Springer Proceedings in Earth and Environmental Sciences, https://doi.org/10.1007/978-3-030-91467-7_35

1. Interference of several beams propagating by different ways in the ionosphere.
2. Scattering on ionospheric inhomogeneities.
3. Magneto-ionic splitting in the ionosphere, in the result of which linear polarization waves, falling on the ionosphere, are transformed into two waves of elliptic polarization.
4. Presence of travelling ionospheric inhomogeneities on the propagating path.

The indicated causes may exist simultaneously that results in the degradation of the communication channel capacity and increase of error probability. When exciting the inter-ionospheric waveguide applying artificial inhomogeneities of the ionosphere, all these causes are a priori realized. The principle of energy pumping into the inter-layer waveguide is based on signal energy scattering on artificial inhomogeneities. As the result, many waves, interfering at the observation site, are formed. In order to increase the impact efficiency, selective polarization excitation is used, i.e., in the general case the transmitting antenna of the heating facility forms a rotating polarization wave. Inhomogeneities move that is similar to travelling ionospheric inhomogeneities.

The law of signal amplitude distribution is a significant characteristic of fading. Fast fading is described well by the Rayleigh law. During slow fading, amplitude distribution agrees well with the normally logarithmic law. These laws have one peak and refer to one-modal ones. However, in a number of cases, for example when travelling ionospheric inhomogeneities are presented, distributions have two (bimodal) or several (polymodal) peaks.

To estimate fading, according to the recommendations of the International Advisory Committee on Radio Communication [1], scintillation index is used.

$$S_4 = \sqrt{\frac{\langle I^2 \rangle - \langle I \rangle^2}{\langle I \rangle^2}}$$

where I is the signal intensity (proportional to squared signal amplitude). It denotes the ensemble average. The scintillation index S_4 is associated with signal intensity fluctuation peak-to-peak and depend on amplitude distribution character. There are three kinds of fading, they may be of low intensity, average intensity and high intensity. Low intensity corresponds to $S_4 < 0.3$. In this case, amplitude distribution corresponds to logarithmically normal distribution. For the fading of average intensity, the scintillation index is from 0.3 to 0.6 and it corresponds to Nakagami distribution. In the case of high intensity, $S_4 > 0.6$ and signal amplitude behavior is better described by the Rayleigh distribution. In the theory of signal transition, anomalous distribution of amplitudes is also considered. It occurs quite rarely and is explained by the presence of moving ionospheric disturbances on the propagation path. In the active impact region, ionospheric plasma turbulence is formed. It is accompanied by plasma motion that may cause amplitude anomalous distribution.

The paper [2] describes experiments on electromagnetic wave propagation in ionospheric waveguide on Alaska-Antarctic Path. The scintillation index values mentioned in the paper correspond to the fading of average intensity, though at some frequencies they are quite close to the fading of high intensity. We should note that one of the components of artificial radio-frequency radiation of the ionosphere was used as a signal analogue during the experiments. It is obvious that its fading is determined by the scattering on artificial irregularities, because that is just the way how ionospheric waveguide is excited, and by the propagation effects. In this case there is no possibility to determine the primary reason for fading occurrences. The SDR receivers in Alaska do not allow one to solve the problem of fading selection. Thus, we organized observations of the experiments carried out on EISCAT facilities, which are surrounded by SDR receivers [3, 4]. That makes it possible to make observations at the line-of-sight distances of the impact area and to observe fading occurrences determined mainly by the scattering effect.

Software Defined Radio (SDR) is the radio equipment in which the greater part of physical level capabilities is realized via a computing machine with special software under the condition of operative modification of the rest of hardware part according to the requirements of communication operational standard. This technology allows one to set or to changes operational radio-frequency parameters such as frequency range, modulation type, excluding, however, the operational parameters, determined by radio device settings (according to the specification of system).

Software realization of the majority of functions on the processing of high-frequency signals and operative programmable control of the equipment provide cardinal increase of the functional capabilities of the radio technical device by supporting the operation in different services, wide frequency band and in different standards. Thus, demodulation task is realized in digital form that allows a receiver to operate with different types of signals having the same hardware part.

The following parameters were set on the receivers:

- demodulation band—20 kHz;
- USB demodulation mode—one-band demodulation of the upper side frequency;
- amplification—0 dB;
- automatic gain control—turned off.

2 Data and Analysis

In October 2019, some experiments were carried out on the EISCAT facilities. During them step change of power was observed. Processing of the data recorded at Arctic and Alta (north of Norway) and Mala (Sweden) is shown in Fig. 1. The HAARP and EISCAT facilities have different parameters, thus it does not make sense to make direct comparison of the results, however the following features attract our attention: scintillation index values, illustrated in the paper [2], and those, which we observed during the experiments, are comparable that may show that the majority of the fading occurrences are caused by scattering; the scintillation index for Arctic site correlates well with the impact power whereas for Alta and Mala sites this relation is weak; scintillation index maximum for Arctic site corresponds to 70% of power,

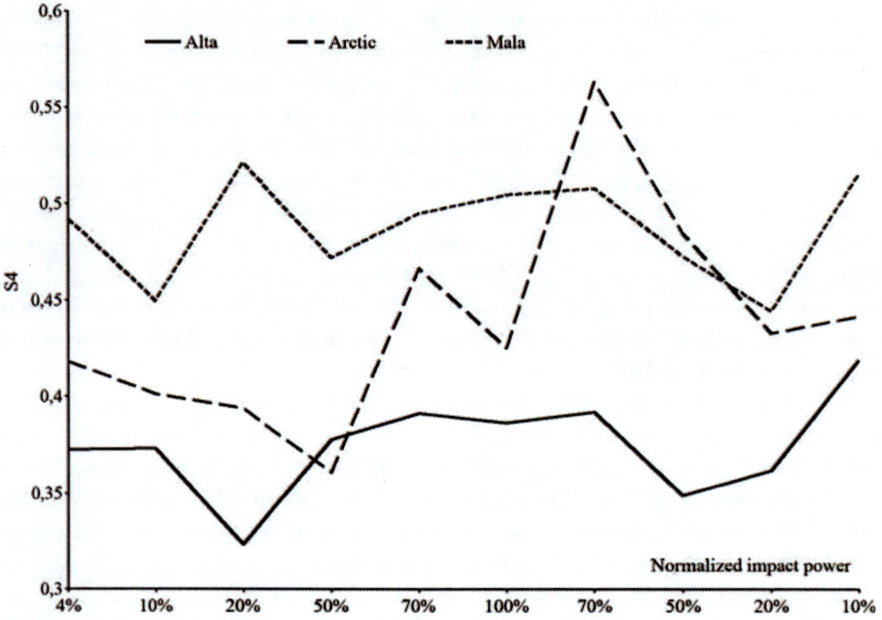

Fig. 1 Scintillation index (S4) variations during impact power increase

not to 100%, that may indicate inertial phenomena in the impact region; scintillation index variations at different sites are not almost connected that may indicate structure inhomogeneity in the impact region.

Scintillation index describes signal level fluctuations and helps to solve telecommunication tasks. As our experience shows [5–8], sometimes it makes more sense to apply the ratio of dispersion to the average, shown in Fig. 2, to solve geophysical problems.

The upper left part of the figure shows azimuth distribution of the directions to observation sites from the impact region. The dependence of the ratio of dispersion to scattered wave average on impact power for Arctic site is traced very clearly in the graph. It is not almost traced for Alta site. The disturbance peak corresponds to the impact power of 70% from the maximum level. This effect is also observed in velocity distribution. That can be explained by inertial processes in plasma.

We can state several reasons for bimodality in amplitude distribution.

1. Interference of two beams, which have the frequencies that slightly differ owing to the Doppler effect. Such a relation between signal level fluctuations and inhomogeneity velocities may indicate one of the reasons of fading, the Doppler effect.

   The authors of the paper [9] obtained an approximated expression for probability density, process envelope for two-beam propagation and the presence of Doppler effect

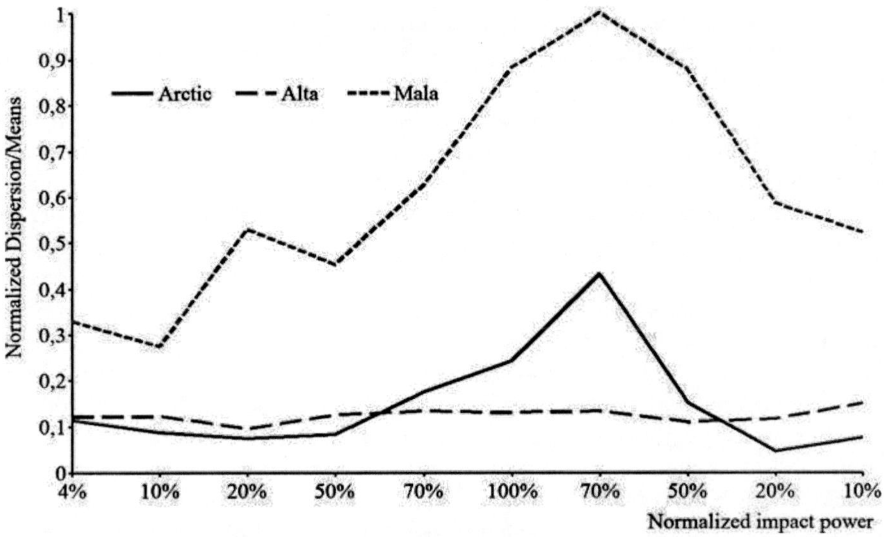

Fig. 2 Variations of the ratio of dispersion to the average

$$W(R) \cong \frac{R}{12\sigma^2} exp\left[-\frac{1}{12\sigma^2}(R^2 + E_1^2 + E_2^2\right]$$
$$\times \left\{ P I_0\left[\frac{E_2}{\sigma^2}(R - E_1)\right] + \frac{1}{P} I_0\left[\frac{E_2}{\sigma^2}(R + E_1)\right]\right.$$
$$\left. +2\left[P I_0(\frac{E_2}{\sigma^2}\sqrt{R^2 - RE_1 + E_1^2}) + \frac{1}{P} I_0(\frac{E_2}{\sigma^2}\sqrt{R^2 + RE_1 + E_1^2})\right]\right\}$$

where $P = exp\left(\frac{-RE_1}{\sigma^2}\right)$ and it was shown that the distribution is bimodal.

Depending on the values in the formula, amplitude distribution may have two peaks. In this case, interfering waves should not be correlated. Frequency shift, determined by Doppler effect, is enough.

2. Interference of two waves, correlated between each other and having, for example, normal distribution. In this case, frequency shift is not required.
3. The resulting signal is the sum of amplitude-modulated signal and Gaussian noise.
4. The resulting signal is the result of interference of three and more beams correlated between each other.

To our point of view, the most likely mechanism of formation of amplitude anomalous distribution is the first cause since, according to the dimensions of the impact region which much exceed ten wave lengths, the supposition on scattered radiation correlation is highly improbable. The supposition on the signal amplitude modulation is realizable by focusing and defocusing of scattered radiation on artificial ionospheric inhomogeneities but it is not likely.

We decided to check how the impact power affects the amplitude distribution character. It turned out that during impact power increase, amplitude distribution can be bimodal and polymodal Fig. 3.

In order to understand the reasons of distribution character change, we considered the change of magnetically oriented irregularity topology in the impact region using the multiposition diagnostics method. We should note that we observe the magnetically oriented irregularities, the dimensions of which are resonant for a concrete frequency of a heating wave. In our case it is 4540 kHz.

It turned out that longitudinal dimensions of irregularities and the distance between them along and across the Earth magnetic field change during impact intensity increase. This dependence manifests, in a greater degree, for the distance between irregularities along the Earth magnetic field [10], however such structuring may increase scattered wave intensity but not the increase in fading occurrences during impact power growth. We should note that in Fig. 2 high variations are observed at Arctic and Mala sites for which quasi-longitudinal orientation of propagation path relative to the Earth's magnetic field lines is typical. Irregularity motion is mainly observed along the Earth magnetic field line. Then the most likely reasons of amplitude distribution modality change are the variations of irregularity motion velocity and direction.

The difference in the changes of fading index S4 and amplitude distribution at different observation sites may indicate the extremely inhomogeneous structure of

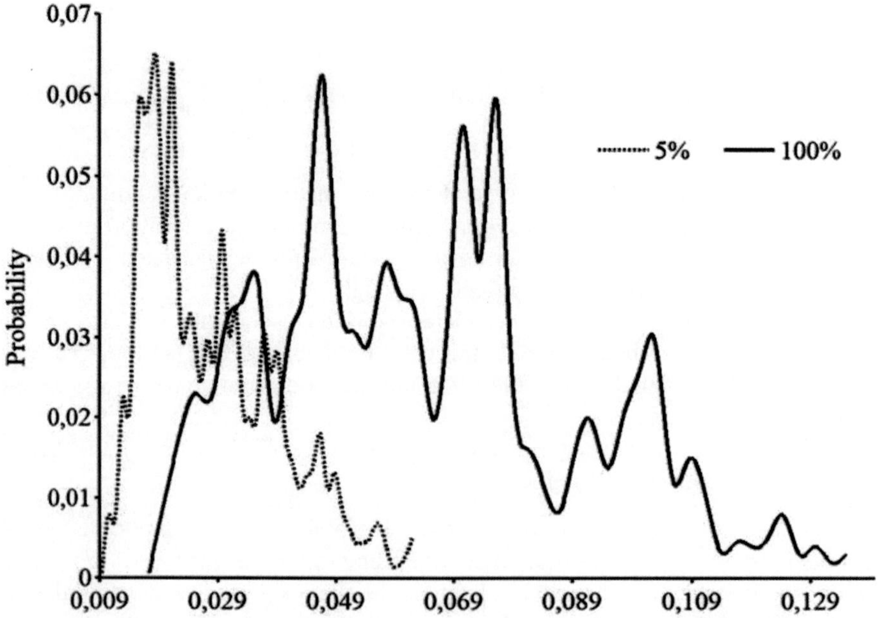

Fig. 3 Amplitude distribution according to impact intensity

the scattering region. Thus, the model of scattering on magnetically oriented inhomogeneities of the ionosphere, which was suggested in the paper [2], is simplified. It is more likely that formation of the second field by scattering on artificial inhomogeneities has local character that is confirmed by the observations at the sites slightly different in azimuth and remoteness (Arctic-Alta).

3 Conclusions

It has been determined that scintillation index and amplitude distribution of the radiation scattered on ionospheric magnetically oriented irregularities depend on impact power.

This dependence is explained, to some degree, by the change of irregularity topology, and it is likely to be explained, to a great degree, by irregularity motion.

In order to create a transionospheric channel for data transition, it makes sense to apply the approach developed by the researchers from USA and Kharkov. In this case data transfer should be organized at the frequencies adapted to the parameters of ionospheric magnetically oriented irregularities and heating should be made by the radiation with the parameters minimizing turbulent phenomenon occurrences in the impact region.

Acknowledgements The work was carried out within the framework of State Task Theme (2021–2023) «Physical Processes in the System of Near Space and Geospheres under Solar and Lithospheric Impacts» (AAAA-A21-121011290003-0).

References

1. https://www.itu.int/dms_pubrec/itu-r/rec/p/R-REC-P.531-12-201309-S!!PDF-R.pdf
2. Yampolski, Y., Milikh, G., Zalizovski, A., Koloskov, A., Reznichenko, A., Nossa, E., Bernhardt, P.A., Briczinski, S., Grach, S.M., Shindin, A., Sergeev, E.: Front. Astron. Space Sci. (2019). https://doi.org/10.3389/fspas.2019.00012
3. http://www.ve3sun.com/KiwiSDR/
4. http://rx.linkfanel.net
5. Sivokon', B.P., Druzhin, G.I.: Geomag. Aeron. **46**(4), 521–524 (2006) (in Russ.)
6. Sivokon', B.P.: Izvestiya vysshikh uchebnykh zavedeniy. Fizika—Russ. Phys. J. **59**(12–13), 56–60 (2016) (in Russ.)
7. Sivokon', B.P.: Vestnik Kamchatskogo gosudarstvennogo tekhnicheskogo universiteta—Bull. Kamchatka State Tech. Univ. (40), 30–36 (2017) (in Russ.)
8. Sivokon', B.P.: Geomagnetizm i aeromomiya—Geomag. Aeron. **57**(3), 343–352 (2017) (in Russ.)
9. Dorofeev, V.M.: Radiotekhnika **28**(8), 39–42 (1973) (in Russ.)
10. Bakhmet'eva, N.V., Belikovich, V.V., Kagan, L.M., Ponyatov, A.A., Tolmacheva, A.V.: Vestnik rossiyskogo fonda fundamental'nykh issledovaniy (3), 9–36 (2007) (in Russ.)

Estimation of the Parameters of Field-Aligned Ionospheric Irregularities in the Area of Active Impact

Vladimir Sivokon' and **Nina Cherneva**

Abstract Ionospheric irregularities significantly affect radio-technical and telecommunication systems. The degree of this influence depends on the electromagnetic radiation range of these systems and irregularity properties. In recent times the possibilities of generation of artificial ionospheric irregularities by heating facilities are under intensive investigation. The facilities include a powerful decameter-range transmitter and a directed antenna. In the result of such impact on the ionosphere, electron concentration irregularities, including magnetically oriented ones, are formed. The irregularities are the structures aligned along the Earth's magnetic field lines. Their longitudinal dimension is many times as large as their lateral dimension. A heating wave is intensively scattered on such irregularities. It is known as a self-scattering effect (Galushko et al. in Radiofizika i radioastronomiya—Radiophys. Radioastron. 17(2):112–124, 2012 (in Russ.) [1]). In this case, the scattering character is determined by irregularity parameters and topology (Sivokon' et al. in Elektromagnitnye volny i elektronnye sistemy—Electromag. Waves Electron. Syst. 24(8):23–28, 2019 (in Russ.) [2]). On the basis of these features of magnetically oriented irregularities and applying the network of SDR receivers, located within the line-of-sight of the impact area, we investigate the dependence of irregularity motion velocity variations and direction on the intensity of impact on the ionosphere.

Keywords Field-aligned irregularities · Soft defined radio

1 Introduction

One of the examples of possible application of artificial magnetically-oriented irregularities is the USA Patent 9,423,495. It involves a ship, generating electromagnetic lightning field, and a ship group, receiving radio impulses. It is assumed that in order to obtain current information on ionospheric parameters, a vertical sounding station will be installed on the lightning ship. As an alternative, they consider the variant to

V. Sivokon' (✉) · N. Cherneva
Institute of Cosmophysical Research and Radio Wave Propagation FEB RAS, Paratunka, Kamchatka, Russia

A. Kosterov et al. (eds.), *Problems of Geocosmos—2020*, Springer Proceedings in Earth and Environmental Sciences, https://doi.org/10.1007/978-3-030-91467-7_36

obtain these data based on the results of ionosphere sounding from a satellite. In any case, the key element of this technique is the ionosphere. If they include a mobile heating facility, generating a region of artificial magnetically oriented irregularities, into the ship group, the efficiency of this technique will be improved due to the more effective lightning signal re-emission and will not depend on diurnal and seasonal changes of ionosphere properties.

Magnetically oriented inhomogeneities of the ionosphere can be also used in telecommunication technologies. It is known that during transmission of signals for long distances applying the wave length decameter range, it undergoes strong absorption. To decrease the losses, an interlayer ionospheric waveguide can be used. That allows one to provide information transmission in global scale. In order to excite the interlayer waveguide, heating facilities are used. They form artificial magnetically oriented inhomogeneities in the ionosphere. During artificial inhomogeneity motion, one should take into account the Doppler Effect, which causes the change in fading distribution law. As a rule, in decameter connection channels, such distributions have one maximum, for example normal distribution law. It is also known that in the case when there are travelling ionospheric disturbances on signal propagation path, the amplitude distribution law becomes anomalous. Scintillation index increases that affects negatively the communication channel capacity and its resistance to noise. For example, when scintillation index increases from 0.1 to 0.25 (average intensity fading), the error probability P_{er} may increase by 3 orders. It is obvious that complete realization of information transmission in the interlayer ionospheric waveguide is impossible without investigation of dynamic characteristics of ionospheric artificial inhomogeneities. Thus, to realize these techniques, we need to know artificial magnetically oriented inhomogeneity properties.

2 Data and Analysis

In order to study these properties, we apply the heating facility experiments during which artificial magnetically oriented irregularities of the ionosphere are formed. When Soft Defined Radio (SDR) technologies appeared and developed, multipoint observations based on computer-controlled radio receiver networks were organized [3]. Software Defined Radio (SDR) is the radio equipment in which the greater part of physical level capabilities is realized via a computing machine with special software under the condition of operative modification of the rest of hardware part according to the requirements of communication operational standard. This technology allows one to set or to changes operational radio-frequency parameters such as frequency range, modulation type, excluding, however, the operational parameters, determined by radio device settings (according to the specification of system).

Software realization of the majority of functions on the processing of high-frequency signals and operative programmable control of the equipment provide cardinal increase of the functional capabilities of the radio technical device by supporting the operation in different services, wide frequency band and in different standards. Thus, demodulation task is realized in digital form that allows a receiver to operate with different types of signals having the same hardware part.

The following parameters were set on the receivers:

- demodulation band—20 kHz;
- USB demodulation mode—one-band demodulation of the upper side frequency;
- amplification—0 dB;
- automatic gain control—turned off.

Receivers located within the line-of-sight of the active impact area are used for the observations. Thus, possible ionospheric effects on electromagnetic radiation propagation from the impact region to the observation site are minimized. Initially, we used university network Twenty but the capability of frequency tuning is limited in it. Thus, at present time we apply the network Kiwi receivers. In most cases this network is represented by radio fans, thus preliminary selection of receivers is required. The receivers are selected on the basis of two parameters: frequency stability and sensitivity. For this purpose, we use Moscow frequency signal transmitter at the frequency of 4996 kHz and with the power of 5 kW. As a result, three points were selected, two of them are in the north of Norway, Arctic and Alta, and one in Sweden, Mala.

In October 2019, some experiments, under the direction of N.F. Blagoveshchenskaya, were carried out on EISCAT facilities. As it was stated on the site [4], the aim was to study the development and the features of artificial turbulences and to generate artificial magnetically oriented irregularities in the ionosphere. On one of the days, in particular on October 18 step variation of heating wave power was observed at the frequency of 4540 kHz. Heat time was set to 2 min and the pause was 1 min. At the start, the impact power increased and then decreased. The parameters of impact intensity increase and decrease were symmetric. We proceeded from the assumption that impact intensity variation will probably cause the changes in irregularity motion velocity and direction.

On the example of the data from Arctic site, we shall show the record processing results. Velocity values changed quite rapidly, thus we needed to construct velocity distribution histograms. It is difficult to plot velocity variations during one heating cycle on one graph, so we divided the cycle into two stages, power increase and its decrease. Figure 1 shows irregularity velocity distribution variations at the stage of power increase.

It is clear from the graph that during low impact power, irregularities moved mainly from the observation site. As the impact power increased, both irregularity motion velocity and direction changed.

As the impact power decreased (Fig. 2), we could expect synchronous changes in irregularity motion velocity and direction. However, as we can see from the figure, when the power was 75%, the tendency of velocity growth to the observation point maintained that may indicate inertial processes in the region of active impact. We should note that irregularity velocity variations differ during power increase and its decrease.

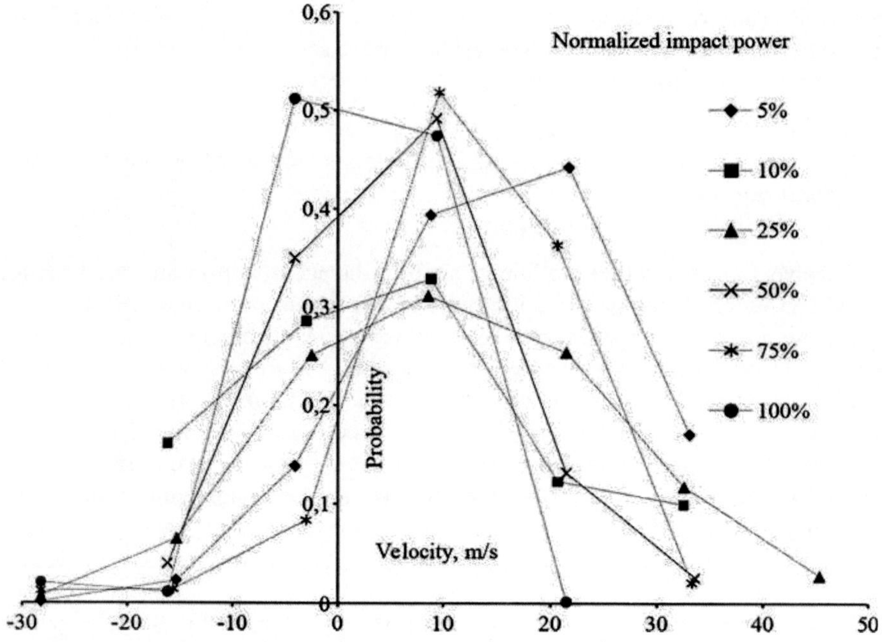

Fig. 1 Velocity variations during impact power increase

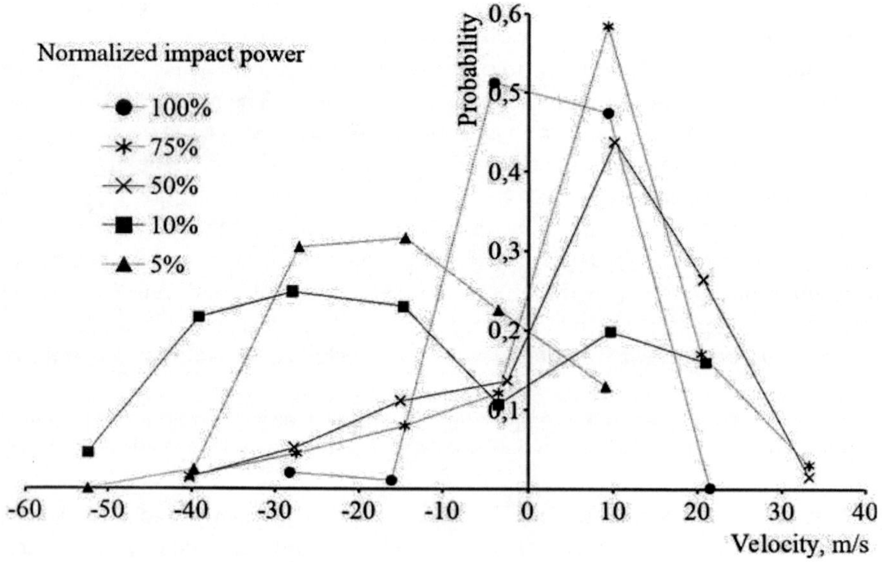

Fig. 2 Velocity variations during impact power decrease

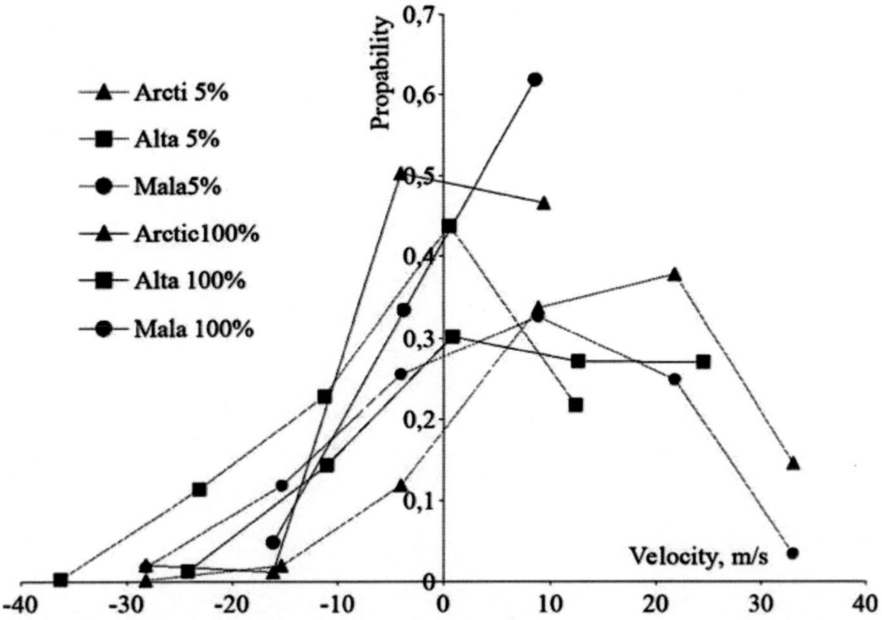

Fig. 3 Comparison the velocities at three observation sites

We use multipoint observations; thus, we have the opportunity to estimate velocity variations in different directions (Fig. 3). The comparison is made for the impact power minimum and maximum. For the reference, the left part of the figure shows the relative location and the distance between the observation sites. Arctic and Alta are in the North and Mala is in the South.

It is clear from the figure that velocity variation character for the directions, which differ by 6° (Arctic and Alta), is not the same both for the impact power minimum and maximum. This tendency is also typical for the opposite directions of Alta and Mala that is likely to be explained by impact area inhomogeneity. Moreover, during the impact maximum, decrease in the number of expected velocities, i.e., some kind of ordering, is observed for at least two observation sites. We should note that quasi-longitudinal, relatively the Earth magnetic field lines, irregularity motion is characteristic for these sites.

The heating area is observed at different angles at receiver location points; thus, we can determine vertical and horizontal components of the velocity and the direction of irregularity motion. In 48% of cases, irregularities move south-westward. Irregularities move south-eastward and north-westward with almost the same probability, 25% and 22%, respectively. The probability of moving north–eastward is very low, 5% of cases.

In the course of observations, such phenomena as spectrum broadening and splitting are recorded [5]. In a number of cases, we observe the effects which we consider to be associated with vertex processes in the area of active impact (Fig. 4). There

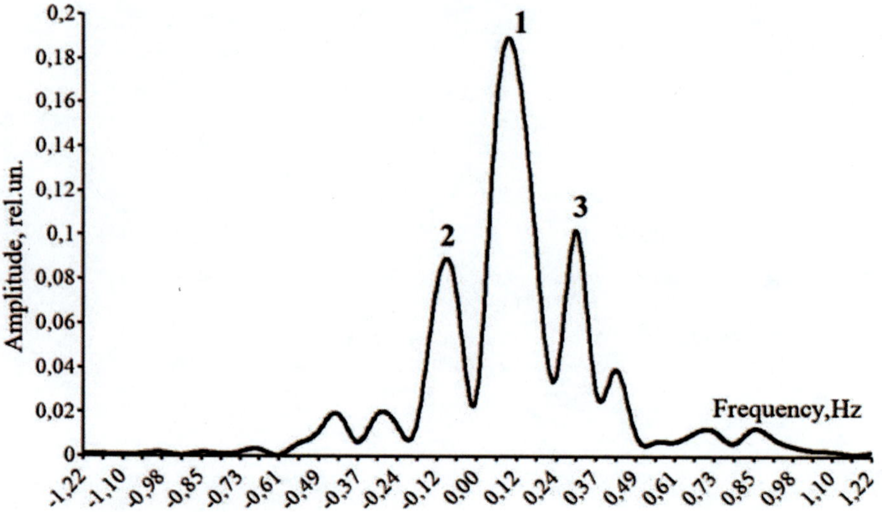

Fig. 4 Spectrum splitting

are three components in the scattered radiation spectrum whereas 2 and 3 are almost symmetric to the first one. If we assume that the first component is determined by the radial drift [6, 7], then the side components and their symmetry may indicate vertex motion of irregularities. The paper [8] shows that vertex structure provides additional stability to plasma irregularities aligned along magnetic field lines, i.e., magnetically oriented ionospheric irregularities in our case.

Based on the frequency shift, radial drift velocity is 12.1 m/sec. The velocities corresponding to the components 2 and 3 relatively the first one are 26 and 24 m/sec, respectively. Radial drift velocity is comparable with the results described in the papers [9, 10].

In publications of Chernogor et al. [11], there is always information about the state of space weather at the time of the experiments. We appreciate this approach, and we also use it. From publication [11] it follows that the phenomena described in this work were observed at $Dst_{min} \approx -32$ nT, $Kp_{max} = 7$ °C from 13:00 to 14:00 UT on November 18, 2019, when our observations were carried out, the state of the magnetic field was different $Dst_{min} \approx -2$ nT, $Kp_{max} = 1$ [12]. Consequently, there is reason to believe that the velocity variations are caused precisely by the active influence on the ionosphere, and not by geomagnetic pulsations.

It was determined as the result of observations that a number of effects arise during artificial magnetically oriented inhomogeneity formation. They are determined by inhomogeneity motion in the region of active impact in the ionosphere. They are the frequency Doppler shift, spectrum widening and splitting. Together with scattering, i.e., the possibility of interference of several waves arrived by different ways at the observation point, there is quite a difficult but not unpromising situation. To our point of view, when creating artificial ionospheric inhomogeneities which are supposed to be used to increase radiotechnical and telecommunication tasks, one should adapt the impact mode to geophysical conditions and not to try to generate higher turbulence in active impact region. For example, a hybrid frequency was used in the experiment which we observe. That frequency is equal to the square root from plasma and hyro-magnetic frequency square. We know heating wave polarization. If we assume that an ordinary wave was excited, these two factors could lead to such velocity distribution, spectrum widening and splitting, formation of vertex processes. When applying hyro-magnetic frequency harmonics and the polarization coordinated with unusual component, we can obtain artificial ionospheric inhomogeneity properties acceptable for radiotechnical systems.

3 Conclusions

We have tested the possibility to investigate the velocities and directions of ionospheric magnetically oriented irregularity motion by SDR technology.

The observations showed the dependence of motion velocity and direction on heating wave power.

Irregularity motion determined by the active impact on the ionosphere is mainly southward.

Acknowledgements The work was carried out within the framework of State Task Theme (2021–2023) «Physical Processes in the System of Near Space and Geospheres under Solar and Lithospheric Impacts» (AAAA-A21-121011290003-0).

References

1. Galushko, V.G., Bezrodnyy, V.G., Koloskov, A.V., Zalizovskiy, A.V.: Radiofizika i radioastronomiya—Radiophys. Radioastron. **17**(2), 112–124 (2012) (in Russ.)
2. Sivokon', V.P., Kolesnikov, R.V., Demichev, I.V.: Metod otsenki parametrov iskusstvennykh magnitoorientirovannykh neodnorodnostey ionosfery [Methods for estimating of artificial magnetically oriented irregularities of the ionosphere]. Elektromagnitnye volny i elektronnye sistemy—Electromagn. Waves Electron. Syst. **24**(8), 23–28 (2019) (in Russ.)
3. http://www.ve3sun.com/KiwiSDR/
4. https://portal.eiscat.se/schedule/schedule.cgi?year=2019&month=10&S=on&HEA=on
5. Ponomarenko, P.V.: J. Geophys. Res. **104**(A5), 10081–10087 (1999)
6. Avdeev, V.B., Beley, V.S., Belenov, A.F., Galushko, V.G., Erukhimov, L.M., Myasnikov, E.N., Ponomarenko, P.V., Sergeev, E.N., Sinitsyn, V.G., Yampol'skiy, Y.M., Yarygin, A.P.: Izvestiya VUZov. Radiofizika—Radiophys. Quantum Electron. **37**(4), 479–492 (1994) (in Russ.)
7. Koloskov, A.V., Beley, V.S., Leyzer, T.B., Yampol'skiy, Y.M.: Radiofizika i radioastronomiya—Radiophys. Radioastron. **4**(3), 247–260 (1999) (in Russ.)
8. Izhovkina, N.I.: Geomag. Aeron. **53**(3), 345–353 (2013)
9. Uryadov, V.P., Vertogradov, G.G., Ponyatov, A.A.: Izvestiya VUZov. Radiofizika—Radiophys. Quantum Electron. **51**(12), 1011–1025 (2008) (in Russ.)

10. Bakhmet'eva, N.V., Vyakhirev, V.D., Grigor'ev, G.I., Egerev, M.N., Kalinina, E.E., Tolmacheva, A.V., Zhemyakov, I.N., Vinogradov, G.R., Yusupov, K.M.: Geomag. Aeron. **60**(1), 99–115 (2020)
11. Chernogor, L.F., Vertogradov, G.G., Uryadov, V.P., Vertogradova, E.G., Shamota, M.A.: Izvestiya VUZov. Radiofizika **53**(12), 766–785 (2010). (in Russ.)
12. http://wdc.kugi.kyoto-u.ac.jp/dst_realtime/201910/index.html

Uncertainty of Sunspot Parameters Reconstructed from Early Telescopic Sunspot Observations

Nadezhda Zolotova and Mikhail Vokhmyanin

Abstract We present the results of the reconstruction of the parameters of solar activity from astronomical observations in the seventeenth century. Drawings of the solar disk made by various observers differ in the observing methodology and accuracy. The distribution of the sunspot group areas from historical archives was compared with that of the modern epoch. A lack of small groups of sunspots was found in early observations. This, in turn, leads to an uncertainty in the number of sunspot groups in tens of percent. Additional uncertainty of group numbers is introduced by sorting the sunspots into groups. The latitude-time distribution of sunspots is also discussed.

Keywords Sunspots · Solar cycle · Observations

1 Introduction

Counting sunspots and their mapping on the solar disk apparently might be viewed as the longest-lasting astrophysical measurements. Here, the primary is the so-called long-time series on solar activity. However, these indexes, i.e. the Sunspot Number and the Group Sunspot Number [1], gradually diverge [2] before the Greenwich epoch (RGO/USAF/NOAA database, hereafter RGO). This suggests that at least one of these measures incorrectly represents sunspot activity in the past. A large-scale revision of sunspot indexes began in 2011 on the Sunspot Number Workshops [3]. Revised and uncovered new sunspot data [4–9] gradually improve our knowledge of solar activity through the ages [10, 11].

Figure 1 schematically shows the history of the butterfly diagram. The period of the Maunder Minimum is known as a period of weak solar activity and dormant

N. Zolotova (✉) · M. Vokhmyanin
Saint-Petersburg State University, Saint-Petersburg, Russia
e-mail: ned@geo.phys.spbu.ru

M. Vokhmyanin
ReSoLVE Centre of Excellence, Space Climate Research Unit, University of Oulu, Oulu, Finland

A. Kosterov et al. (eds.), *Problems of Geocosmos—2020*, Springer Proceedings in Earth and Environmental Sciences, https://doi.org/10.1007/978-3-030-91467-7_37

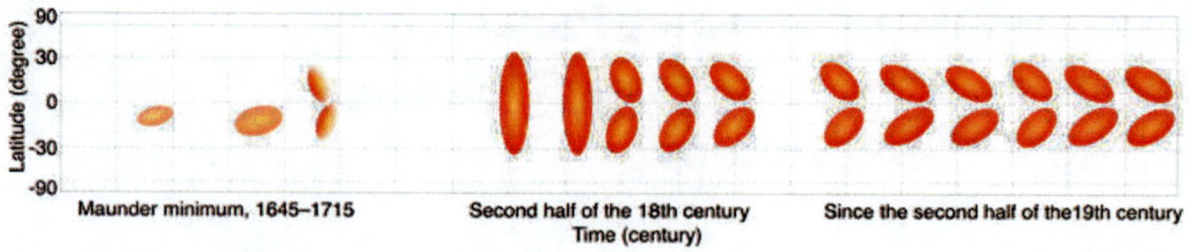

Fig. 1 Schematic butterfly diagram through the centuries

northern hemisphere [12]. Observations by Johann Staudacher [13] of the second half of the eighteenth century demonstrate the remarkable deviation of the sunspot distribution from the modern butterfly [14]. During two cycles, the solar equator was populated by sunspots transforming the butterfly into a kind of paraglider. Here, we would like to extend the revision of the activity indices to the revision of the latitude-time distribution of sunspots.

In this study, we present the results of sunspot parameters reconstruction of the seventeenth century. Section 2 discusses the uncertainties of counting the sunspot group numbers. Section 3 deals with the latitude-time distribution of sunspot groups. Finally, Sect. 4 summarises the results.

2 Uncertainty of Sunspot Groups

The first issue of processing sunspot drawings is sorting spots into groups. Commonly, we rely on the classification by McIntosh [15] and assign that the linear size of a sunspot group is on average limited to 7° of latitude and 15° of longitude. However, in complicated cases, groups are assigned arbitrarily.

Figure 2 demonstrates the uncertainty introduced by sorting spots into groups performed by various researchers. On the *Top*, drawings of 12–15 September 1613 made by Fabio Colonna with the underlaid heliographic grid are shown. Arrow defines the direction of rotation, i.e. Colonna drew an inverted image of the Sun from right to left. On 12th September, both Hoyt and Schatten [1] (hereafter, H&S) and we assigned 3 groups (grey ovals). On the next day, activity complex A apparently was counted by H&S as one group (dashed oval), while we assigned two groups (solid ovals). Here, we say "apparently", since we only know from the H&S database how many groups they assigned per day. Note that such a discrepancy in sorting spots into groups also occurs in databases of modern observatories. On 14th September, on contrary, two activity complexes near the eastern limb were sorted into four groups by H&S and into two groups by us. Finally, on the next day, both H&S and we assigned 7 groups, while the sorting seems to have been done differently.

Figure 2 (*Bottom*) compares the number of groups defined by H&S (grey dots) and by us (open circles). On average, H&S and we assigned nearly the same number of groups. However, in the second half of September, either H&S assigned 1–2 more

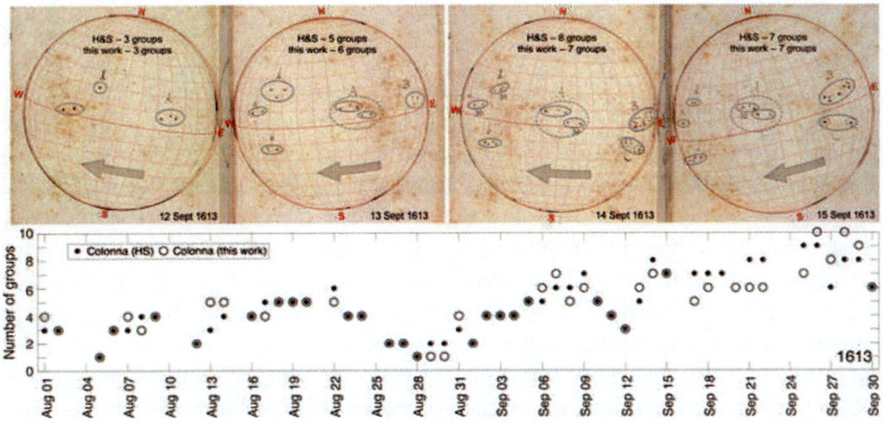

Fig. 2 Top: Sorting sunspots into groups in drawing by Colonna. Bottom: Number of groups defined by Hoyt and Schatten [1] and in this work

groups, or we assigned 1–2 more groups. If the average activity level in the second half of September is about 7 groups per day, then such a discrepancy introduces an error of about 15–25%. Hence, reconstructed parameters of solar activity in the past not only depend on accuracy, aim, and method of historical observations [16], but are also affected by the processing of these historical reports by modern researchers.

To evaluate the uncertainty of sunspot group numbers reported by historical observers themselves, we compare the fractional distribution of sunspot group areas of historical and modern observations. In other words, we would like to estimate how many small sunspot groups might be missed on historical drawings. Here, we hypothesize that the abundance of sunspot groups of the first half of the seventeenth century and during the twentieth century was approximately the same. In colored stairs, Fig. 3 shows distributions of historical observations. The filled colored bars correspond to the RGO dataset. The period of the ascending phase of cycle 17 has one

Fig. 3 Fractional distribution of sunspot group areas and example of the solar image with a sunspot with an area of 103 msh

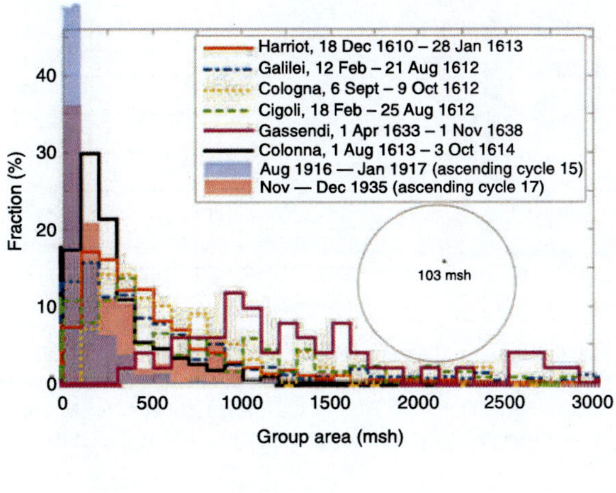

of the smallest portions of sunspot groups with an area of less than 100 msh, it is about 35%, while that of cycle 15 has one of the largest portions of small sunspot groups, about 70%. An example of a sunspot of an area of 103 msh is imprinted in Fig. 3.

Thereby, we can roughly estimate that the fraction of small sunspot groups should be from 35 to 70%. Reports by Pierre Gassendi (magenta stairs) lack sunspot groups of an area less than 300 msh. The wide abundance of large groups indicates that this observer apparently drew enlarged objects [17]. This, in turn, makes reconstructed sunspot group numbers and their areas poorly reliable. Observations by Sigismondo di Cologna are schematic and completely miss sunspot groups of an area less than 100 msh and significantly lack groups of an area 101–200 msh. Similarly, we can estimate the deficit of small groups in the observations of other astronomers. Depending on the observing method, instrument, style of drawing, etc., the uncertainty of sunspot group numbers reported by the considered here historical observers might make up from 20 to 85% [18].

A similar conclusion was made earlier by Ogurtsov [19]. Analysing reports of individual observers since the seventeenth century, he argued the uncertainty of the number of sunspot groups is about 30% and can be up to a factor of 2 and more in individual years. Also, the methodology of statistical processing of raw data significantly affects the average number of sunspot groups per year.

3 Butterfly Diagram

Figure 4 shows the current state of the art in the problem of reconstructing the latitudinal-time distribution of sunspots. In colors, data of various observers are depicted. Blue and red bars mark the times of minima and maxima of solar activity defined through proxy data [24–31]. Dashed ovals show the apparent butterfly wings.

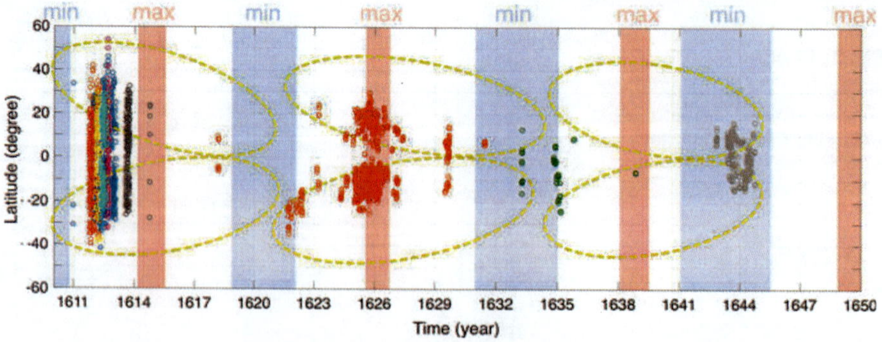

Fig. 4 Butterfly diagram of 1610–1650. Dark grey denotes observations by Colonna, magenta is Cologna, yellow is Cigoli [9], green show Gassendi's observations [17], lite green is Galilei [20], red is Scheiner [21], blue is Harriot [22], and light grey is Hevelius [23]. Blue and red bars correspond to the proposed solar minima and maxima. Ovals depict the apparent butterfly wings

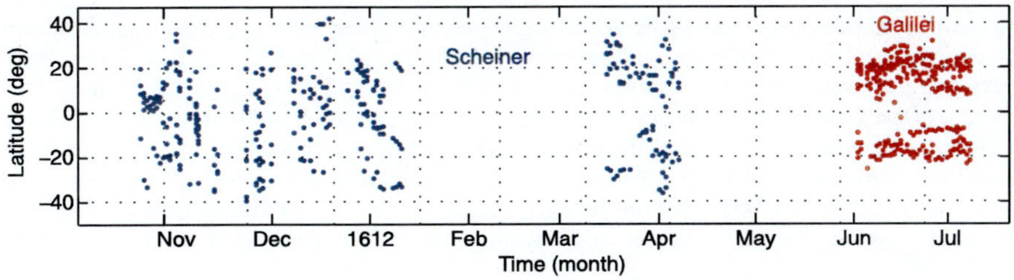

Fig. 5 Latitude-time diagram of sunspot observations by Scheiner [21] and Galilei [20]

Here, observations by Hevelius [23] on the eve of the Maunder minimum apparently belong to the declining phase of the solar cycle. Gassendi's observations [17] fall between two cycles when the activity of the southern hemisphere was stronger. Note that similar asymmetry was reported during the Maunder minimum. In the 1620s, observations by Scheiner [21] also exhibit the southern dominance. Soon after the invention of the telescope, the manuscripts of the 1610s abound in sunspot observations.

As an example, Fig. 5 compares early observations by Scheiner [21] and Galilei [20] in 1611–1612. While Scheiner schematically drew sunspots on small solar disks of 3–5 centimeters in diameter, Galilei, by means of the projection apparatus, provided large and detailed drawings [32] with the fine structure of umbra and penumbra. Uncertainty of sunspot mapping of Scheiner's drawings exceeds 5° of latitude, which leads to a populated equator. On contrary, Galilei's reports show the latitude-time distribution as we know it today. Note that the solar equator is much more populated by sunspots on all the schematic drawings of other observers that we have examined. Here, we would suggest that the paraglider shape of sunspot distribution is rather the result of the uncertainty of sunspot mapping than a real phenomenon.

4 Conclusions

In this work, we present state of the art of sunspot parameters reconstruction of the seventeenth century. Comparing sunspot group sorting on the historical drawings performed independently of the modern researchers, we conclude that discrepancy in sorting leads to uncertainty in 15–20% of the number of groups. Also, we compared the fraction distribution of sunspot group areas of historical and RGO datasets. If the proportion of small and large sunspot groups of the seventeenth and twentieth centuries were comparable, then the historical observations may lack from 20 to 85% of small groups. This finding agrees with the uncertainty of the number of sunspot groups of the seventeenth century evaluated by Ogurtsov [19]. Finally, we analyze the butterfly diagram of 1610–1650. We argue that on the schematic drawings, the solar equator is much more populated by sunspots assuming that this feature is the result of the uncertainty of sunspot mapping.

Acknowledgements We use data from the Royal Greenwich Observatory, United States Air Force, National Oceanic and Atmospheric Administration (RGO/USAF/NOAA: https://solarscience.msfc.nasa.gov/greenwch.shtml), the database by Hoyt and Schatten [1] provided by the National Geophysical Data Center (NOAA/NGDS: https://ngdc.noaa.gov/stp/SOLAR), the revised version of Greenwich Photoheliographic Results (GPR) sunspot catalogue provided by the Debrecen Heliophysical Observatory (DHO: https://fenyi.solarobs.unideb.hu/deb_obs_en.html), data from Historical Archive of SunSpot Observations by Victor M. S. Carrasco [23] (https://haso.unex.es). The reported study was funded by the Russian Science Foundation according to the research project 19-72-00053.

References

1. Hoyt, D.V., Schatten, K.H.: Group sunspot numbers: a new solar activity reconstruction. Solar Phys. **179**(1), 189–219 (1998)
2. Cliver, E.W., Ling, A.G.: The discontinuity circa 1885 in the group sunspot number. Solar Phys. **291**(9–10), 2763–2784 (2016)
3. Cliver, E.W., Clette, F., Svalgaard, L.: Recalibrating the sunspot number (SSN): the SSN workshops. Cent. Eur. Astrophys. Bull. **37**, 401–416 (2013)
4. Arlt, R., Vaquero, J.M.: Historical sunspot records. Living Rev. Sol. Phys. **17**(1), 1 (2020)
5. Neuhäuser, R., Geymeier, M., Arlt, R., Chapman, J.: Comparison of telescopic and naked-eye sunspots for the very small spots on February 15, 1900 and January 30, 1911. Astron. Nachr. **341**(4), 366–383 (2020)
6. Carrasco, V.M.S., Gallego, M.C., Villalba Álvarez, J., Vaquero, J.M.: A reanalysis of the number of sunspot groups recorded by Pierre Gassendi in the cycle before the Maunder minimum. Solar Phys. **296**(4), 59 (2021)
7. Hayakawa, H., Iju, T., Murata, K., Besser, B.P.: Daniel Mögling's sunspot observations in 1626–1629: a manuscript reference for the solar activity before the Maunder minimum. Astrophys. J. **909**(2), 194 (2021)
8. Hayakawa, H., Kuroyanagi, C., Carrasco, V.M.S., Uneme, S., Besser, B.P., Sôma, M., Imada, S.: Sunspot observations at the Eimmart observatory and in its neighborhood during the late Maunder minimum (1681–1718). Astrophys. J. **909**(2), 166 (2021)
9. Vokhmyanin, M., Arlt, R., Zolotova, N.: Sunspot positions and areas from observations by Cigoli, Galilei, Cologna, Scheiner, and Colonna in 1612–1614. Solar Phys. **296**(1), 4 (2021)
10. Miyahara, H., Tokanai, F., Moriya, T., Takeyama, M., Sakurai, H., Horiuchi, K., Hotta, H.: Gradual onset of the Maunder minimum revealed by high-precision carbon-14 analyses. Sci. Rep. **11**, 5482 (2021)
11. Pevtsov, A.A., Bertello, L., Nagovitsyn, Y.A., Tlatov, A.G., Pipin, V.V.: Long-term studies of photospheric magnetic fields on the Sun. J. Space Weather Space Clim. **11**, 4 (2021)
12. Ribes, J.C., Nesme-Ribes, E.: The solar sunspot cycle in the Maunder minimum AD1645 to AD1715. Astron. Astrophys. **276**, 549 (1993)
13. Arlt, R.: The butterfly diagram in the eighteenth century. Solar Phys. **255**(1), 143–153 (2009)
14. Hathaway, D.H.: The solar cycle. Living Rev. Sol. Phys. **12**(1), 4 (2015)
15. McIntosh, P.S.: The classification of sunspot groups. Solar Phys. **125**, 251–267 (1990)
16. Karachik, N.V., Pevtsov, A.A., Nagovitsyn, Y.A.: The effect of telescope aperture, scattered light and human vision on early measurements of sunspot and group numbers. Mon. Not. Roy. Astron. Soc. **488**(3), 3804–3809 (2019)
17. Vokhmyanin, M., Zolotova, N.: Sunspot positions and areas from observations by Pierre Gassendi. Solar Phys. **293**(11), 150 (2018)
18. Zolotova, N., Vokhmyanin, M.: On the Solar activity level in 1611–1613: sunspot goups and areas. Geomag. Aeron. (in press)

19. Ogurtsov, M.G.: Instrumental data on the sunspot formation in the 17th-18th centuries: correct information or approximations. Geomag. Aeron. **53**(5), 663–671 (2013)
20. Vokhmyanin, M., Zolotova, N.: Sunspot positions and areas from observations by Galileo Galilei. Solar Phys. **293**(2), 31 (2018)
21. Arlt, R., Pavai, V., Schmiel, C., Spada, F.: Sunspot positions, areas, and group tilt angles for 1611–1631 from observations by Christoph Scheiner. Astron. Astrophys. **595**, A104 (2016)
22. Vokhmyanin, M., Arlt, R., Zolotova, N.: Sunspot positions and areas from observations by Thomas Harriot. Solar Phys. **295**(3), 39 (2020)
23. Carrasco, V.M.S., Vaquero, J.M., Gallego, M.C., Muñoz-Jaramillo, A., de Toma, G., Galaviz, P., Arlt, R., Senthamizh Pavai, V., Sánchez-Bajo, F., Villalba Álvarez, J., Gómez, J.M.: Sunspot characteristics at the onset of the Maunder minimum based on the observations of Hevelius. Astrophys. J. **886**(1), 18 (2019)
24. Gleissberg, W.: A secular change in the shape of the spot-frequency curve. Observatory **65**, 244–244 (1944)
25. Gleissberg, W.: Secularly smoothed data on the minima and maxima of sunspot frequency. Solar Phys. **2**, 231–233 (1967)
26. Gleissberg, W., Damboldt, T., Schove, D.J.: Reflections of the Maunder minimum of sunspots. J. Br. Astron. Assoc. **89**, 440–449 (1979)
27. Schove, D.J.: The sunspot cycle, 649 B. C. T a. D. 2000. J. Geophys. Res. **60**, 127–146 (1955)
28. Schove, D.J.: Sunspot turning-points and aurorae since A. D. 1510. Solar Phys. **63**, 423–432 (1979)
29. Wittmann, A.: The sunspot cycle before the Maunder minimum. Astron. Astrophys. **66**, 93 (1978)
30. Vitinskij, Y.I., Kopetskij, M., Kuklin, G.V.: Statistics of the spot-forming activity of the sun. Glavnaya Redaktsiya Fiziko-Matematicheskoj Literatury, Nauka, Moskva (1986)
31. Lassen, K., Friis-Christensen, E.: Variability of the solar cycle length during the past five centuries and the apparent association with terrestrial climate. J. Atmos. Terr. Phys. **57**, 835–845 (1995)
32. Biagioli, M.: Galileo's instruments of credit: telescopes, images, secrecy. University of Chicago Press, Chicago, IL, USA (2006)